CHAPTER 4

10. $\dfrac{d}{dx}e^x = e^x, \quad \dfrac{d}{dx}e^{f(x)} = f'(x)e^{f(x)}$

11. $\dfrac{d}{dx}\ln x = \dfrac{1}{x}, \quad x > 0;$

$\quad \dfrac{d}{dx}\ln f(x) = f'(x) \cdot \dfrac{1}{f(x)}, \quad f(x) > 0$

12. $\dfrac{d}{dx}\ln |x| = \dfrac{1}{x}, \quad x < 0;$

$\quad \dfrac{d}{dx}\ln |f(x)| = f'(x) \cdot \dfrac{1}{f(x)}, \quad f(x) < 0$

13. If $\dfrac{dP}{dt} = kP$, then $P(t) = P_0 e^{kt}$.

14. If $\dfrac{dP}{dt} = -kP$, then $P(t) = P_0 e^{-kt}$.

15. $\dfrac{d}{dx}a^x = (\ln a)a^x$

16. $\dfrac{d}{dx}\log_a x = \dfrac{1}{\ln a} \cdot \dfrac{1}{x}, \quad x > 0;$

$\quad \dfrac{d}{dx}\log_a |x| = \dfrac{1}{\ln a} \cdot \dfrac{1}{x}, \quad x < 0$

17. $a^x = e^{x(\ln a)}$

Calculus

SIXTH EDITION

Calculus

SIXTH EDITION

MARVIN L. BITTINGER

Indiana University—Purdue University at Indianapolis

ADDISON-WESLEY PUBLISHING COMPANY

Reading, Massachusetts • Menlo Park, California • New York
Don Mills, Ontario • Wokingham, England • Amsterdam • Bonn • Sydney
Singapore • Tokyo • Madrid • San Juan • Milan • Paris

Football is like calculus.
Someone has to show you how to do it,
but once you've got er' figgered, why, she's easy.

Bert Jones, Former quarterback, *Baltimore Colts*
(Bert Jones majored in mathematics in college.)

Sponsoring Editor	Greg Tobin
Managing Editor	Karen Guardino
Production Supervisor	Jennifer Bagdigian
Senior Marketing Manager	Andrew Fisher
Text Design	Geri Davis, Quadrata, Inc.
Art, Design, Editorial, and Production Services	Geri Davis and Martha Morong, Quadrata, Inc.
Prepress Buying Manager	Sarah McCracken
Art Buyer	Joseph Vetere
Electronic Illustration	TechGraphics and Gail Hayes
Senior Manufacturing Manager	Roy Logan
Manufacturing Coordinator	Evelyn Beaton
Cover Design	Eileen Hoff
Composition	Beacon Graphics Corporation
Printer	R. R. Donnelley and Sons

Photo credits appear on page 582.

Library of Congress Cataloging-in-Publication Data

Bittinger, Marvin L.
    Calculus / Marvin L. Bittinger.—6th ed.
        p.   cm.
    Includes index.
    ISBN 0-201-59338-6
    1. Calculus.   I. Title.
QA303.B645   1995
515—dc20

95–12150
CIP

1 2 3 4 5 6 7 8 9 10—DOC—99 98 97 96 95 94

Preface

Appropriate for a one-term course, this text is an introduction to calculus as applied to business, economics, the life and physical sciences, the social sciences, and many general areas of interest to all students. A course in intermediate algebra is a prerequisite for the text, although Chapter 1 provides sufficient review to unify the diverse backgrounds of most students.

What's New in the Sixth Edition?

The style, format, and approach of the fifth edition have been retained in this new edition, but the text has been improved in the following ways.

TECHNOLOGY MATERIAL. Technology material has been integrated throughout the book. Graphing utilities have become more and more accessible in recent years. Since technology is effective in helping students become more successful in problem solving and allowing them time to explore, it is an important feature of this edition. This material, presented in an optional format, has been written to enhance student understanding as students learn the underlying algebraic concepts. The following items contain the technology material.

- *Technology Connection Features.* There are 96 features in the exposition of the text that illustrate the use of a grapher. Whenever appropriate, art that simulates graphs drawn using a graphing utility is included. Exercises for students to try with a grapher are also included in these features.

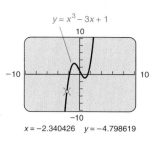

$y = x^3 - 3x + 1$

$x = -2.340426$ $y = -4.798619$

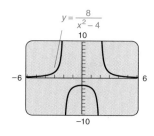

$y = \dfrac{8}{x^2 - 4}$

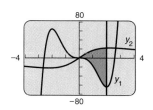

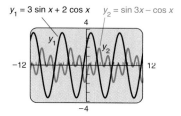

$y_1 = 3 \sin x + 2 \cos x$ $y_2 = \sin 3x - \cos x$

- *Technology Connection Exercises.* Most exercise sets contain technology-based exercises that can be found under the heading "Technology Connection." These exercises also appear in the Summary and Reviews and the Chapter Tests. The Printed Test Bank supplement includes technology-based exercises as well.

MODELING.   Enhanced use of mathematics modeling, developed in Section 1.6, is carried out throughout the book.

EXTENDED TECHNOLOGY APPLICATION FEATURES.   Each chapter ends with this feature, which highlights an application based on real-life data. The step-by-step analysis of problems presented in this feature can be assigned as a group project. The answers to the exercises in this feature are given in the Instructor's Manual/ Printed Test Bank.

WRITING EXERCISES.   Writing exercises enhance student understanding by asking students to explain mathematical concepts in their own words. These exercises, designated by the symbol ◈, appear in the synthesis section of most exercise sets. The answers to these exercises are given in the Instructor's Solutions Manual.

Content Features

USE OF COLOR.   The text uses full color in an extremely functional way, and not merely as window dressing. Color usage has been carried out in a methodical and precise manner so that it enhances the readability of the text for the student and the instructor. The following illustrates.

For example, when two curves are graphed using the same set of axes, one is usually red and the other blue with the red graph the curve of major importance. This is exemplified in the following graph from Chapter 1 (p. 52). Note that the equation labels are the same color as the curve. When the instructions say "Graph," the dots match the color of the curve.

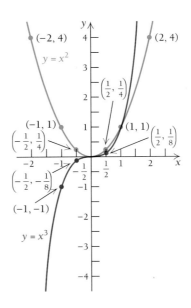

In the following figure from Chapter 2 (p. 115), we see how color is used to distinguish between secant and tangent lines. Throughout the text, blue is used for secant lines and red for tangent lines.

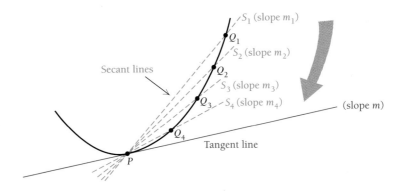

In the following graph and text development from Chapter 3 (pp. 189–190), the color red denotes substitution in equations and blue highlights maximum and minimum values (second coordinates). This specific use of color is carried out further, as

shown in the figure that follows. Note also that when dots are used for other emphasis than just mere plotting, they are black.

We then find second coordinates by substituting in the original function:

$$f(-3) = (-3)^3 + 3(-3)^2 - 9(-3) - 13 = 14;$$
$$f(1) = (1)^3 + 3(1)^2 - 9(1) - 13 = -18.$$

Thus there is a relative maximum at $(-3, 14)$ and a relative minimum at
$(1, -18)$. The relative extrema are shown in the graph below.

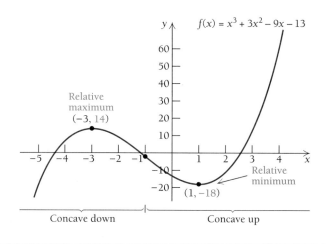

Beginning with integration in Chapter 5, the color amber is used to identify the
area, unless certain special concepts are considered in the initial learning process.
The following figure from Chapter 5 (p. 377) illustrates the vivid use of blue and red
for the curves and labels, and amber for the area.

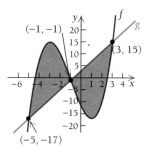

In Chapter 7 (p. 486), we use extra colors for clarity. The top of the surface is blue and the bottom of the surface is rose. This is carried out in all the examples in Section 7.4.

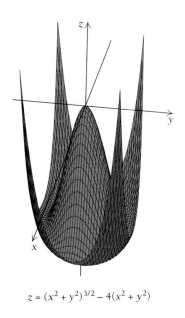

$$z = (x^2 + y^2)^{3/2} - 4(x^2 + y^2)$$

INTUITIVE APPROACH.   Although the word "intuitive" has many meanings and interpretations, its use here means "experience based." Throughout the text, when a particular concept is discussed, its presentation is designed so that the students' learning process is based on their earlier mathematical experience or a new experience presented by the author before the concept is formalized. This is illustrated by the following situations.

- The definition of the derivative in Chapter 2 is presented in the context of a discussion of average rates of change (see p. 104). This presentation is more accessible and realistic than the strictly geometric idea of slope.

- When maximum problems involving volume are introduced (see pp. 229–230), a function is derived that is to be maximized. Instead of forging ahead with the standard calculus solution, the student is first asked to stop, make a table of function values, graph the function, and then estimate the maximum value. This experience provides students with more insight into the problem. They recognize that not only do different dimensions yield different volumes, but also that the dimensions yielding the maximum volume may be conjectured or estimated as a result of the calculations.

- The understanding behind the definition of the number e in Chapter 4 is explained both graphically and through a discussion of continuously compounded interest (see pp. 271–272).

APPLICATIONS.   Relevant and factual applications drawn from a broad background of fields are integrated throughout the text as applied problems and exercises, and are also featured in separate application sections. These have been updated and expanded in this edition.

The applications in the exercise sets are grouped under headings that identify them as those of business and economics, life and physical sciences, social science, and general interest. This allows the instructor to gear the assigned exercises to a particular student and also allows the student to know whether a particular exercise applies to his or her major.

The following illustrates some examples of the applications.

- An early discussion in Chapter 1 of compound interest, total revenue, cost and profit, as well as supply and demand functions sets the stage for subsequent applications that permeate the book (see pp. 42, 141, and 233).

- All techniques of differentiation, including the Extended Power Rule, the Chain Rule, and material on higher derivatives are covered in Chapter 2. This is prior to the introduction of applications of differentiation in Chapter 3, which includes maximum–minimum applications (see pp. 227–244) and implicit differentiation and related rates (see pp. 251–258).

- When the exponential model is studied in Chapter 4, other applications, such as continuously compounded interest and the demand for natural resources, are also considered (see pp. 296–320). Growth and decay are covered in separate sections in Chapter 4 to allow room for the many worthwhile applications that relate to these concepts. The material on elasticity has been enhanced.

- Applications of integration are covered in a separate chapter and include a thorough coverage of continuous probability (see pp. 415–478).

- The Extended Technology Application feature includes interesting and topical situations such as the ecological effect of global warming (see pp. 78–80) and minimizing employees' travel time in a workplace (see pp. 535–536).

COVERAGE.   Many aspects of the treatment and coverage of topics in this text have been commended by reviewers. The following are a few examples.

- Trouble areas in algebra are integrated into coverage of traditional precalculus topics in Chapter 1.

- Relative maxima and minima (Sections 3.1–3.3) and absolute maxima and minima (Section 3.4) are covered in separate sections in Chapter 3, so that students obtain a gradual buildup of these topics as they consider graphing using calculus concepts (see pp. 172–227).

- The text also includes a section on differential equations (Section 6.7), coverage of probability (Sections 6.4 and 6.5), and a chapter on the trigonometric functions (Chapter 8).

Pedagogical Features

STUDENT-ORIENTED APPROACH.   As each new section begins, its objectives are stated in the margin. These can be spotted easily by the student, and they provide the answer to the typical question, "What material am I responsible for?"

VARIETY OF EXERCISES.   There are over 3500 exercises in this edition. The exercise sets are enhanced not only by the inclusion of many more business–economics applications, detailed art pieces, and extra graphs, but also by the following additional features.

- *Technology Connection Exercises.*   These exercises appear in the Technology Connection features (see pp. 133, 149, and 184) and at the ends of exercise sets (see pp. 145 and 370).

- *Synthesis Exercises.*   Synthesis exercises are included at the ends of most exercise sets and all Summary and Reviews and Chapter Tests. They require students to go beyond the immediate objectives of the section or chapter and are designed both to challenge students and to make then think about what they are learning (see pp. 243 and 385–386).

- *Thinking and Writing Exercises.*   These exercises appear in the synthesis section at the end of most exercise sets (see pp. 226, 278, and 295).

- *Applications.*   A section of applied problems is included in most exercise sets. The problems are grouped under headings that identify them as business and economics, life and physical sciences, social sciences, or general interest. Each problem is accompanied by a brief description of its subject matter (see pp. 241 and 305).

TESTS AND REVIEWS

- *Summary and Review.*   At the end of each chapter is a summary and review. These are designed to provide students with all the material they need for successful review. Answers are at the back of the book, together with section references so that students can easily find the correct material to restudy if they have difficulty with a particular exercise. In each summary and review there is a list of *Terms to Know*, accompanied by page references to help students key in on important concepts (see p. 473).

- *Tests.*   Each chapter ends with a chapter test that includes synthesis and technology questions. There is also a Cumulative Review at the end of the text that can also serve as a final examination. The answers, with section references to the chapter tests and the Cumulative Review, are at the back of the book. Six additional forms of each of the chapter tests and the final examination with answer keys appear, ready for classroom use, in the Instructor's Manual/Printed Test Bank.

Supplements for the Instructor

SUPPLEMENTARY CHAPTERS TO ACCOMPANY *CALCULUS AND ITS APPLICATIONS, SIXTH EDITION, BY MARVIN L. BITTINGER.*   These two chapters written by Marvin L. Bittinger and Michael A. Penna of Indiana University–Purdue University at Indianapolis are designed to present material on sequences and series and to expand the book's presentation of differential equations. Many applications along with optional technology material are included. This supplement is available to instructors and is for sale to students.

INSTRUCTOR'S MANUAL/PRINTED TEST BANK.   The Instructor's Manual/Printed Test Bank (IM/PTB) by Donna DeSpain contains six alternative test forms for each chapter test and six comprehensive final examinations with answers. The IM/PTB also has answers for the even-numbered exercises in the exercise sets. These can be easily copied and handed out to students. The answers to the odd-numbered exercises are at the back of the text.

INSTRUCTOR'S SOLUTIONS MANUAL.   This manual by Judith A. Penna contains worked-out solutions to *all* exercises in the exercise sets. A *Student's Solutions Manual* contains solutions to all odd-numbered exercises in the exercise sets (with the exception of the Thinking and Writing exercises). Either or both of these could be made available to the student.

SOFTWARE: COMPUTERIZED TESTING

Omni-Test[3].   Available in DOS-based and Macintosh formats, this easy-to-use software is developed exclusively for Addison-Wesley by ips Publishing, a leader in computerized testing and assessment. The system is:

- Algorithm driven—The program can automatically insert new numbers into the same equation, creating hundreds of variations of that equation.

- Keyed section by section to the text—Within the section, you may select questions that test individual objectives from that section.
- Flexible—The program allows you to enter your own questions by way of Omni Test³'s sophisticated editor, complete with mathematical notation.

Supplements for the Student

STUDENT'S SOLUTIONS MANUAL by Judith A. Penna. Complete worked-out solutions are provided in this manual for all odd-numbered exercises in the exercise sets (with the exception of the Thinking and Writing exercises). This supplement is available to instructors and is for sale to students.

Options for Integrating Technology

EXPLORING BUSINESS CALCULUS WITH A GRAPHING CALCULATOR by Charlene Beckmann. Containing laboratory explorations and problems to be completed with a graphing calculator, this supplement may be used with any graphing calculator. A Solutions Manual is also available.

THE CALCULUS EXPLORER. Consisting of 27 programs ranging from functions to vector fields, this software enables the instructor and student to use the computer as an "electronic chalkboard." *The Explorer* is highly interactive and allows for manipulation of variables and equations to provide graphical visualization of mathematical relationships that are not intuitively obvious. *The Explorer* provides user-friendly operation through an easy-to-use completely menu-driven system, extensive on-line documentation, superior graphics capability, and fast operation. An accompanying manual includes sections covering each program, including examples and exercises. Available for IBM PC/compatibles.

ANALYZER★. This program is a tool for exploring functions in calculus and many other disciplines. It can graph a function of a single variable and overlay graphs of other functions. It can differentiate, integrate, or iterate a function, and it can find roots, maxima and minima, and inflection points, as well as vertical asymptotes. In addition, *Analyzer★* can compose functions, graph polar and parametric equations, make families of curves, and make animated sequences with changing parameters. It exploits the unique flexibility of the Macintosh wherever possible, allowing input to be either numeric (from the keyboard) or graphic (with a mouse). *Analyzer★* runs on Macintosh II, Plus, or better.

Acknowledgments

Your author has taken many steps to ensure the accuracy of the manuscript. The graphic art pieces have been computer-generated. A number of devoted individuals comprised the team that was responsible for monitoring each step of the production process in such careful detail. In particular, I would like to thank Judith A. Beecher and Judith A. Penna for their helpful suggestions, proofreading, and checking of the art. Their efforts above and beyond the call of duty deserve infinite praise. I also wish to express my appreciation to my students for providing suggestions and criticisms so willingly during the preceding editions; to Mike Penna of IUPUI for his help with the computer graphics; and to Barbara Johnson and Michael Rosenborg for their precise checking and proofreading of the manuscript for mathematical accuracy. In addition, I would like to thank Scott Mortensen and Ken Hurley for their contributions to the technology material.

Finally, the following reviewers provided thoughtful and insightful comments that helped immeasurably in the revision of this text.

Yuanan Diao, *Kennesaw State College*

Mark Greenhalgh, *Fullerton College*

Paula Kemp, *Southwest Missouri State University*

J. Robert Knott, *University of Evansville*

Earl E. Kymala, *California State University, Sacramento*

Nancy Matthews, *University of Oklahoma*

Christine Merrin, *Utah Valley State College*

Charles Zimmerman, *Robert Morris College*

M. L. B.

Index of Applications

Index of Technology Connections

The following index lists 50 optional Technology Connections by topic. These features incorporate in-text comments, including exercises, that explain and encourage the use of technology to explore problem situations. Approximately 90% of the Technology Connection features are appropriate for graphing calculators; 10% lend themselves to the use of computer algebra systems. The Technology Connection features, along with the *Technology Connection Exercises* and *Extended Technology Applications,* described in the Preface, bolster the optional integration of technology into the course.

Contents

3 Applications of Differentiation   171

4 Exponential and Logarithmic Functions   265

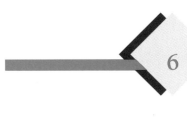

Calculus

SIXTH EDITION

1 Algebra Review, Functions, and Modeling

INTRODUCTION

Although this chapter is mainly a review of basic concepts of algebra used in calculus, it introduces many topics that we will consider several times throughout the text—functions and their graphs, supply and demand, total cost, revenue, and profit, and the concept of a mathematical model.

AN APPLICATION

While driving a car, you see a child suddenly cross the street unattended. Your brain registers the emergency and sends a signal to your foot to hit the brake. The car travels a distance D, in feet, during this time, where D is a function of the speed r, in miles per hour, that the car is traveling when you see the child. That reaction distance is a linear function given by

$$D(r) = \frac{11r + 5}{10}.$$

Find the reaction distance when the car is traveling at 50 mph.

THE MATHEMATICS

To solve the problem, we substitute 50 for r in the function

$$\underbrace{D(r) = \frac{11r + 5}{10}}.$$

    This is the formula for a *function*.

This problem appears as Exercise 60 in Exercise Set 1.4.

1.1 Exponents, Multiplying, and Factoring

OBJECTIVES

- Manipulate exponential expressions.
- Multiply and factor algebraic expressions.
- Solve applied problems involving compound interest.

Exponential Notation

Let us review the meaning of an expression

$$a^n,$$

where a is any real number and n is an integer; that is, n is a number in the set $\{\ldots, -3, -2, -1, 0, 1, 2, 3, \ldots\}$. The number a is called the **base** and n is called the **exponent.** When n is greater than 1, then

$$a^n = \underbrace{a \cdot a \cdot a \cdots a}_{n \text{ factors}}.$$

In other words, a^n is the product of n **factors,** each of which is a.

In later sections of the text, we will consider a^n when n is any real number.

EXAMPLE 1   Rename without exponents.

a) $4^3 = 4 \cdot 4 \cdot 4 = 64$

b) $(-2)^5 = (-2)(-2)(-2)(-2)(-2) = -32$

c) $(-2)^4 = (-2)(-2)(-2)(-2) = 16$

d) $-2^4 = -(2^4) = -(2)(2)(2)(2) = -16$

e) $(1.08)^2 = 1.08 \times 1.08 = 1.1664$

f) $\left(\dfrac{1}{2}\right)^3 = \dfrac{1}{2} \cdot \dfrac{1}{2} \cdot \dfrac{1}{2} = \dfrac{1}{8}$ ◆

We define an exponent of 1 as follows:

$$a^1 = a, \quad \text{for any real number } a.$$

In other words, any real number to the first power is that number itself.

We define an exponent of 0 as follows:

$$a^0 = 1, \quad \text{for any nonzero real number } a.$$

That is, any nonzero real number a to the zero power is 1.

EXAMPLE 2   Rename without exponents.

a) $(-2x)^0 = 1$ b) $(-2x)^1 = -2x$ c) $\left(\dfrac{1}{2}\right)^0 = 1$

d) $e^0 = 1$ e) $e^1 = e$ f) $\left(\dfrac{1}{2}\right)^1 = \dfrac{1}{2}$ ◆

The meaning of a negative integer as an exponent is as follows:

$$a^{-n} = \frac{1}{a^n}, \quad \text{for any nonzero real number } a.$$

That is, any nonzero real number a to the $-n$ power is the reciprocal of a^n.

EXAMPLE 3   Rename without negative exponents.

a) $2^{-5} = \dfrac{1}{2^5} = \dfrac{1}{2 \cdot 2 \cdot 2 \cdot 2 \cdot 2} = \dfrac{1}{32}$

b) $10^{-3} = \dfrac{1}{10^3} = \dfrac{1}{10 \cdot 10 \cdot 10} = \dfrac{1}{1000}$, or 0.001

c) $\left(\dfrac{1}{4}\right)^{-2} = \dfrac{1}{\left(\dfrac{1}{4}\right)^2} = \dfrac{1}{\dfrac{1}{4} \cdot \dfrac{1}{4}} = \dfrac{1}{\dfrac{1}{16}} = 1 \cdot \dfrac{16}{1} = 16$

d) $x^{-5} = \dfrac{1}{x^5}$

e) $e^{-k} = \dfrac{1}{e^k}$

f) $t^{-1} = \dfrac{1}{t^1} = \dfrac{1}{t}$ ◆

Properties of Exponents

Note the following:

$$b^5 \cdot b^{-3} = (b \cdot b \cdot b \cdot b \cdot b) \cdot \frac{1}{b \cdot b \cdot b} = \frac{b \cdot b \cdot b}{b \cdot b \cdot b} \cdot b \cdot b = 1 \cdot b \cdot b = b^2.$$

We could have obtained the same result by adding the exponents. This is true in general.

THEOREM 1

For any nonzero real number a and any integers n and m,

$$a^n \cdot a^m = a^{n+m}.$$

(To multiply when the bases are the same, add the exponents.)

EXAMPLE 4   Multiply.

a) $x^5 \cdot x^6 = x^{5+6} = x^{11}$

b) $x^{-5} \cdot x^6 = x^{-5+6} = x$

c) $2x^{-3} \cdot 5x^{-4} = 10x^{-3+(-4)} = 10x^{-7}$, or $\dfrac{10}{x^7}$

d) $r^2 \cdot r = r^{2+1} = r^3$ ◆

Note the following:

$$b^5 \div b^2 = \frac{b^5}{b^2} = \frac{b \cdot b \cdot b \cdot b \cdot b}{b \cdot b} = \frac{b \cdot b}{b \cdot b} \cdot b \cdot b \cdot b = 1 \cdot b \cdot b \cdot b = b^3.$$

We could have obtained the same result by subtracting the exponents. This is true in general.

THEOREM 2

For any nonzero real number a and any integers n and m,

$$\frac{a^n}{a^m} = a^{n-m}.$$

(To divide when the bases are the same, subtract the exponent in the denominator from the exponent in the numerator.)

EXAMPLE 5   Divide.

a) $\dfrac{a^3}{a^2} = a^{3-2} = a^1 = a$

b) $\dfrac{x^7}{x^7} = x^{7-7} = x^0 = 1$

c) $\dfrac{e^3}{e^{-4}} = e^{3-(-4)} = e^{3+4} = e^7$

d) $\dfrac{e^{-4}}{e^{-1}} = e^{-4-(-1)} = e^{-4+1} = e^{-3}$, or $\dfrac{1}{e^3}$ ◆

Note the following:

$$(b^2)^3 = b^2 \cdot b^2 \cdot b^2 = b^{2+2+2} = b^6.$$

We could have obtained the same result by multiplying the exponents. This is true in general.

> **THEOREM 3**
>
> For any nonzero real number a and any integers n and m,
>
> $$(a^n)^m = a^{nm}.$$
>
> (To raise a power to a power, multiply the exponents.)

EXAMPLE 6 Simplify.

a) $(x^{-2})^3 = x^{-2 \cdot 3} = x^{-6}$, or $\dfrac{1}{x^6}$

b) $(e^x)^2 = e^{2x}$

c) $(2x^4 y^{-5} z^3)^{-3} = 2^{-3} (x^4)^{-3} (y^{-5})^{-3} (z^3)^{-3}$

$$= \frac{1}{2^3} x^{-12} y^{15} z^{-9}, \text{ or } \frac{y^{15}}{8x^{12} z^9} \qquad \blacklozenge$$

Multiplication

The distributive laws are important in multiplying. These laws are as follows.

The Distributive Laws

For any numbers A, B, and C,

$$A(B + C) = AB + AC \quad \text{and} \quad A(B - C) = AB - AC.$$

EXAMPLE 7 Multiply.

a) $3(x - 5) = 3 \cdot x - 3 \cdot 5 = 3x - 15$

b) $P(1 + i) = P \cdot 1 + P \cdot i = P + Pi$

c) $(x - 5)(x + 3) = (x - 5)x + (x - 5)3$

$$= x \cdot x - 5x + 3x - 5 \cdot 3$$
$$= x^2 - 2x - 15$$

d) $(a + b)(a + b) = (a + b)a + (a + b)b$

$$= a \cdot a + ba + ab + b \cdot b$$
$$= a^2 + 2ab + b^2 \qquad \blacklozenge$$

The following formulas, which are obtained using the distributive laws, are also useful in multiplying.

$$(A + B)^2 = A^2 + 2AB + B^2, \qquad\qquad (1)$$
$$(A - B)^2 = A^2 - 2AB + B^2 \qquad\qquad (2)$$
$$(A - B)(A + B) = A^2 - B^2 \qquad\qquad (3)$$

EXAMPLE 8   Multiply.

a) $(x + h)^2 = x^2 + 2xh + h^2$

b) $(2x - t)^2 = (2x)^2 - 2(2x)t + t^2 = 4x^2 - 4xt + t^2$

c) $(3c + d)(3c - d) = (3c)^2 - d^2 = 9c^2 - d^2$ ◆

Factoring

Factoring is the reverse of multiplication. That is, to factor an expression, we find an equivalent expression that is a product. Always remember to look first for a common factor.

EXAMPLE 9   Factor.

a) $P + Pi = P \cdot 1 + P \cdot i = P(1 + i)$ We used a distributive law.

b) $2xh + h^2 = h(2x + h)$

c) $x^2 - 6xy + 9y^2 = (x - 3y)^2$

d) $x^2 - 5x - 14 = (x - 7)(x + 2)$ We looked for factors of -14 whose sum is -5.

e) $6x^2 + 7x - 5 = (2x - 1)(3x + 5)$ We first considered ways of factoring the first coefficient—for example, $(2x \quad)(3x \quad)$. Then we looked for factors of -5 such that when we multiply, we obtain the given expression.

f) $x^2 - 9t^2 = (x - 3t)(x + 3t)$ We used the formula $(A - B)(A + B) = A^2 - B^2$.   ◆

In later work, we will consider expressions like

$$(x + h)^2 - x^2.$$

To simplify this, first note that

$$(x + h)^2 = x^2 + 2xh + h^2.$$

Subtracting x^2 on both sides of this equation gives us

$$(x + h)^2 - x^2 = 2xh + h^2.$$

Factoring out an h on the right side, we get

$$\mathbf{(x + h)^2 - x^2 = h(2x + h).} \qquad\qquad (4)$$

Let us now use this result to compare two squares.

EXAMPLE 10   How close is $(3.1)^2$ to 3^2?

Solution   Substituting $x = 3$ and $h = 0.1$ in Eq. (4), we get

$$(3.1)^2 - 3^2 = 0.1(2 \cdot 3 + 0.1) = 0.1(6.1) = 0.61.$$

Thus, $(3.1)^2$ differs from 3^2 by 0.61.   ◆

Compound Interest

Suppose we invest P dollars at interest rate i, compounded annually. The amount A_1 in the account at the end of one year is given by

$$A_1 = P + Pi = P(1 + i) = Pr,$$

where, for convenience, we let

$$r = 1 + i.$$

Going into the second year, we have Pr dollars, so by the end of the second year, we would have the amount A_2 given by

$$A_2 = A_1 \cdot r = (Pr)r = Pr^2 = P(1 + i)^2.$$

Going into the third year, we have Pr^2 dollars, so by the end of the third year, we would have the amount A_3 given by

$$A_3 = A_2 \cdot r = (Pr^2)r = Pr^3 = P(1 + i)^3.$$

In general, we have the following.

> **THEOREM 4**
>
> If an amount P is invested at interest rate i, compounded annually, in t years it will grow to the amount A given by
>
> $$A = P(1 + i)^t.$$

EXAMPLE 11   *Business: Compound interest.* Suppose $1000 is invested at 8%, compounded annually. How much is in the account at the end of 2 years?

Solution   We substitute 1000 for P, 0.08 for i, and 2 for t into the equation $A = P(1 + i)^t$ and get

$$
\begin{aligned}
A &= 1000(1 + 0.08)^2 \\
&= 1000(1.08)^2 \\
&= 1000(1.1664) \\
&= \$1166.40.
\end{aligned}
$$

There is $1166.40 in the account after 2 years.   ◆

For interest that is compounded quarterly, we can find a formula like the one above.

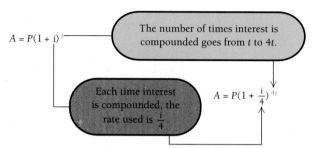

In general, the following theorem applies.

> **THEOREM 5**
>
> If a principal P is invested at interest rate i, compounded n times a year, in t years it will grow to an amount A given by
>
> $$A = P\left(1 + \frac{i}{n}\right)^{nt}.$$

EXAMPLE 12 *Business: Compound interest.* Suppose $1000 is invested at 8%, compounded quarterly. How much is in the account at the end of 2 years?

Solution We use the equation $A = P(1 + i/n)^{nt}$, substituting 1000 for P, 0.08 for i, 4 for n (compounding quarterly), and 2 for t. Then we get

$$A = 1000\left(1 + \frac{0.08}{4}\right)^{4 \times 2}$$

$$= 1000(1 + 0.02)^8$$

$$= 1000(1.02)^8$$

$$= 1000(1.171659381) \qquad \text{Using a calculator to approximate } (1.02)^8$$

$$= 1171.659381$$

$$\approx \$1171.66.$$

There is $1171.66 in the account after 2 years. ◆

A calculator with a $\boxed{y^x}$ or a $\boxed{\wedge}$ key and a ten-digit readout was used to find $(1.02)^8$ in Example 12. The number of places on a calculator may affect the accuracy of the answer. Thus you may occasionally find that your answers do not agree with those at the back of the book, which were found on a calculator with a ten-digit readout. In general, when using a calculator, do all your computations, and round only at the end, as in Example 12. Usually, your answer will agree to at least four digits. It might be wise to consult with your instructor on the accuracy required.

1.1 Exercise Set

Rename without exponents.

1. 5^3 2. 7^2 3. $(-7)^2$

4. $(-5)^3$ 5. $(1.01)^2$ 6. $(1.01)^3$

7. $\left(\dfrac{1}{2}\right)^4$ 8. $\left(\dfrac{1}{4}\right)^3$ 9. $(6x)^0$

10. $(6x)^1$ 11. t^1 12. t^0

13. $\left(\dfrac{1}{3}\right)^0$ 14. $\left(\dfrac{1}{3}\right)^1$

Rename without negative exponents.

15. 3^{-2} 16. 4^{-2}

17. $\left(\dfrac{1}{2}\right)^{-3}$ 18. $\left(\dfrac{1}{2}\right)^{-2}$

19. 10^{-1} 20. 10^{-4} 21. e^{-b}

22. t^{-k} 23. b^{-1} 24. h^{-1}

Multiply.

25. $x^2 \cdot x^3$ 26. $t^3 \cdot t^4$ 27. $x^{-7} \cdot x$

28. $x^5 \cdot x$ 29. $5x^2 \cdot 7x^3$ 30. $4t^3 \cdot 2t^4$

31. $x^{-4} \cdot x^7 \cdot x$ 32. $x^{-3} \cdot x \cdot x^3$ 33. $e^{-t} \cdot e^t$

34. $e^k \cdot e^{-k}$

Divide.

35. $\dfrac{x^5}{x^2}$ 36. $\dfrac{x^7}{x^3}$ 37. $\dfrac{x^2}{x^5}$

38. $\dfrac{x^3}{x^7}$ 39. $\dfrac{e^k}{e^k}$ 40. $\dfrac{t^k}{t^k}$

41. $\dfrac{e^t}{e^4}$ 42. $\dfrac{e^k}{e^3}$ 43. $\dfrac{t^6}{t^{-8}}$

44. $\dfrac{t^5}{t^{-7}}$ 45. $\dfrac{t^{-9}}{t^{-11}}$ 46. $\dfrac{t^{-11}}{t^{-7}}$

47. $\dfrac{ab(a^2b)^3}{ab^{-1}}$ 48. $\dfrac{x^2y^3(xy^3)^2}{x^{-3}y^2}$

Simplify.

49. $(t^{-2})^3$ 50. $(t^{-3})^4$ 51. $(e^x)^4$

52. $(e^x)^5$ 53. $(2x^2y^4)^3$ 54. $(2x^2y^4)^5$

55. $(3x^{-2}y^{-5}z^4)^{-4}$ 56. $(5x^3y^{-7}z^{-5})^{-3}$

57. $(-3x^{-8}y^7z^2)^2$ 58. $(-5x^4y^{-5}z^{-3})^4$

Multiply.

59. $5(x - 7)$ 60. $4(x - 3)$

61. $x(1 - t)$ 62. $x(1 + t)$

63. $(x - 5)(x - 2)$ 64. $(x - 4)(x - 3)$

65. $(a - b)(a^2 + ab + b^2)$

66. $(x^2 - xy + y^2)(x + y)$

67. $(2x + 5)(x - 1)$ 68. $(3x + 4)(x - 1)$

69. $(a - 2)(a + 2)$ 70. $(3x - 1)(3x + 1)$

71. $(5x + 2)(5x - 2)$ 72. $(t - 1)(t + 1)$

73. $(a - h)^2$ 74. $(a + h)^2$

75. $(5x + t)^2$ 76. $(7a - c)^2$

77. $5x(x^2 + 3)^2$

78. $-3x^2(x^2 - 4)(x^2 + 4)$

Use the following equation (Eq. 5) for Exercises 79–81.

$$
\begin{aligned}
(x + h)^3 &= (x + h)(x + h)^2 \\
&= (x + h)(x^2 + 2xh + h^2) \\
&= (x + h)x^2 + (x + h)2xh + (x + h)h^2 \\
&= x^3 + x^2h + 2x^2h + 2xh^2 + xh^2 + h^3 \\
&= x^3 + 3x^2h + 3xh^2 + h^3
\end{aligned}
\tag{5}
$$

79. $(a + b)^3$ 80. $(a - b)^3$

81. $(x - 5)^3$ 82. $(2x + 3)^3$

Factor.

83. $x - xt$ 84. $x + xh$

85. $x^2 + 6xy + 9y^2$ 86. $x^2 - 10xy + 25y^2$

87. $x^2 - 2x - 15$ 88. $x^2 + 8x + 15$

89. $x^2 - x - 20$ 90. $x^2 - 9x - 10$

91. $49x^2 - t^2$ 92. $9x^2 - b^2$

93. $36t^2 - 16m^2$ 94. $25y^2 - 9z^2$

95. $a^3b - 16ab^3$ 96. $2x^4 - 32$

97. $a^8 - b^8$ 98. $36y^2 + 12y - 35$

99. $10a^2x - 40b^2x$ 100. $x^3y - 25xy^3$

101. $2 - 32x^4$

102. $2xy^2 - 50x$

103. $9x^2 + 17x - 2$

104. $6x^2 - 23x + 20$

105. $x^3 + 8$
    (*Hint:* See Exercise 66.)

106. $a^3 - 27b^3$
    (*Hint:* See Exercise 65.)

107. $y^3 - 64t^3$

108. $m^3 + 1000p^3$

109. Use the equation

$$(x + h)^2 - x^2 = h(2x + h)$$

to answer the following.

a) How close is $(4.1)^2$ to 4^2?
b) How close is $(4.01)^2$ to 4^2?
c) How close is $(4.001)^2$ to 4^2?

110. From Eq. (5) above, it follows that

$$(x + h)^3 - x^3 = h(3x^2 + 3xh + h^2).$$

Use the equation to answer the following.

a) How close is $(4.1)^3$ to 4^3?
b) How close is $(4.01)^3$ to 4^3?
c) How close is $(4.001)^3$ to 4^3?

APPLICATIONS

◆ Business and Economics

111. *Compound interest.* Suppose $1000 is invested at 6%. How much is in the account at the end of 1 year, if interest is compounded:

a) annually?   b) semiannually?
c) quarterly?   d) daily?
e) hourly?

112. *Compound interest.* Suppose $1000 is invested at $8\frac{1}{2}\%$. How much is in the account at the end of 1 year, if interest is compounded:

a) annually?
b) semiannually?
c) quarterly?
d) daily?
e) hourly?

Determining monthly payments on a loan. If P dollars are borrowed for a home mortgage, the monthly payment M, made at the end of each month for n months, is given by

$$M = P\left[\frac{\dfrac{i}{12}\left(1 + \dfrac{i}{12}\right)^n}{\left(1 + \dfrac{i}{12}\right)^n - 1}\right],$$

where i = the annual interest rate and n = the total number of monthly payments.

113. The mortgage on a house is $43,000, the interest rate is $8\frac{3}{4}\%$, and the loan period is 25 years. What is the monthly payment?

114. The mortgage on a house is $100,000, the interest rate is $7\frac{1}{2}\%$, and the loan period is 30 years. What is the monthly payment?

SYNTHESIS

115. ◈ Explain why $a^3 - b^3 = (a - b)(a^2 + ab + b^2)$.

116. ◈ Explain why $(-1)^n = 1$ for any even integer n.

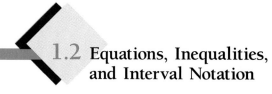

1.2 Equations, Inequalities, and Interval Notation

OBJECTIVES

• Solve equations, inequalities, and problems.
• Write interval notation.

Equations

Basic to the solution of many equations are two simple principles. We can add the same number on both sides of a true equation and still obtain a true equation. We can also multiply by any number on both sides of a true equation and obtain a true equation.

> **THE ADDITION PRINCIPLE**
>
> If an equation $a = b$ is true, then the equation $a + c = b + c$ is true for any number c.

> **THE MULTIPLICATION PRINCIPLE**
>
> If an equation $a = b$ is true, then the equation $ac = bc$ is true for any number c.

When solving an equation, we use these equation-solving principles and other properties of real numbers to get the variable alone on one side. Then it is easy to determine the solution.

EXAMPLE 1   Solve: $-\frac{5}{6}x + 10 = \frac{1}{2}x + 2$.

Solution   We first multiply on both sides by 6 to clear the fractions:

$$6\left(-\tfrac{5}{6}x + 10\right) = 6\left(\tfrac{1}{2}x + 2\right) \qquad \text{Using the Multiplication Principle}$$

$$6\left(-\tfrac{5}{6}x\right) + 6 \cdot 10 = 6\left(\tfrac{1}{2}x\right) + 6 \cdot 2 \qquad \text{Using the Distributive Law}$$

$$-5x + 60 = 3x + 12 \qquad \text{Simplifying}$$

$$60 = 8x + 12 \qquad \begin{array}{l}\text{Using the Addition Principle:} \\ \text{We add } 5x \text{ on both sides.}\end{array}$$

$$48 = 8x \qquad \text{Adding } -12 \text{ on both sides}$$

$$\tfrac{1}{8} \cdot 48 = \tfrac{1}{8} \cdot 8x \qquad \text{Multiplying by } \tfrac{1}{8} \text{ on both sides}$$

$$6 = x.$$

The variable is now alone, and we see that 6 is the solution. We can check by substituting 6 into the original equation.   ◆

To solve applied problems, we first translate to mathematical language, usually an equation. Then we solve the equation and check to see whether the solution to the equation is a solution to the problem.

EXAMPLE 2   *Life science: Weight gain.* After a 5% gain in weight, an animal weighs 693 lb. What was its original weight?

Solution   We first translate to an equation:

$$\underbrace{\text{(Original weight)}}_{w} + 5\% \underbrace{\text{(Original weight)}}_{w} = 693$$
$$w \quad + 5\% \quad w \quad = 693.$$

Now we solve the equation:

$$w + 5\%w = 693$$
$$1 \cdot w + 0.05w = 693$$
$$(1 + 0.05)w = 693$$
$$1.05w = 693$$
$$w = \frac{693}{1.05} = 660.$$

Check: $660 + 5\% \times 660 = 660 + 0.05 \times 660$
$$= 660 + 33 = 693.$$

The original weight of the animal was 660 lb.   ◆

The third principle for solving equations is the *Principle of Zero Products*.

THE PRINCIPLE OF ZERO PRODUCTS

For any numbers a and b, if $ab = 0$, then $a = 0$ or $b = 0$; and if $a = 0$ or $b = 0$, then $ab = 0$.

An equation being solved by this principle must have a 0 on one side and a product on the other. The solutions are then obtained by setting each factor equal to 0 and solving the resulting equations.

EXAMPLE 3   Solve: $3x(x - 2)(5x + 4) = 0$.

Solution

$$3x(x - 2)(5x + 4) = 0$$

$3x = 0$ or	$x - 2 = 0$ or	$5x + 4 = 0$

Using the Principle of Zero Products

$\frac{1}{3} \cdot 3x = \frac{1}{3} \cdot 0$ or	$x = 2$ or	$5x = -4$

Solving each separately

$x = 0$ or	$x = 2$ or	$x = -\frac{4}{5}$

The solutions are 0, 2, and $-\frac{4}{5}$.   ◆

Note that the Principle of Zero Products can be applied *only* when a product is 0. For example, although we may know that $ab = 8$, we *do not know* that $a = 8$ or $b = 8$.

EXAMPLE 4   Solve: $4x^3 = x$.

Solution

$$4x^3 = x$$

$$4x^3 - x = 0 \qquad \text{Adding } -x$$

$$x(4x^2 - 1) = 0$$

$$x(2x - 1)(2x + 1) = 0 \qquad \text{Factoring}$$

$$x = 0 \quad \text{or} \quad 2x - 1 = 0 \quad \text{or} \quad 2x + 1 = 0 \qquad \text{Using the Principle of Zero Products}$$

$$x = 0 \quad \text{or} \quad 2x = 1 \quad \text{or} \quad 2x = -1$$

$$x = 0 \quad \text{or} \quad x = \tfrac{1}{2} \quad \text{or} \quad x = -\tfrac{1}{2}$$

The solutions are 0, $\tfrac{1}{2}$, and $-\tfrac{1}{2}$.

Inequalities

The principles for solving inequalities are similar to those for solving equations. We can add the same number on both sides of an inequality. We can also multiply on both sides by the same nonzero number, but if that number is negative, we must reverse the inequality sign. The following are the inequality-solving principles.

THE INEQUALITY-SOLVING PRINCIPLES

If the inequality $a < b$ is true, then:

I1. $a + c < b + c$ is true, for any c.

I2. $a \cdot c < b \cdot c$, for any positive c.

I3. $a \cdot c > b \cdot c$, for any negative c.

Similar principles hold when $<$ is replaced by $\leq$ and $>$ is replaced by $\geq$.

EXAMPLE 5   Solve: $17 - 8x \geq 5x - 4$.

Solution

$$17 - 8x \geq 5x - 4$$

$$-8x \geq 5x - 21 \qquad \text{Adding } -17$$

$$-13x \geq -21 \qquad \text{Adding } -5x$$

$$-\tfrac{1}{13}(-13x) \leq -\tfrac{1}{13}(-21) \qquad \text{Multiplying by } -\tfrac{1}{13} \text{ and } \textit{reversing} \text{ the inequality sign}$$

$$x \leq \tfrac{21}{13}$$

Any number less than or equal to $\tfrac{21}{13}$ is a solution.

EXAMPLE 6   *Business: Total sales.* Raggs, Ltd., a clothing firm, determines that its total revenue, in dollars, from the sale of x suits is given by

$$200x + 50.$$

Determine the number of suits that the firm must sell so that its total revenue will be more than $70,050.

Solution   We translate to an inequality and solve:

$$200x + 50 > 70,050$$
$$200x > 70,000 \qquad \text{Adding } -50$$
$$x > 350. \qquad \text{Multiplying by } \tfrac{1}{200}$$

Thus the company's total revenue will exceed $70,050 when it sells more than 350 suits.   ◆

Interval Notation

The set of real numbers corresponds to the set of points on a line.

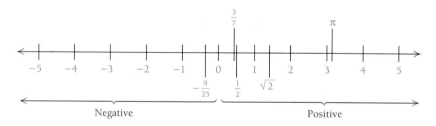

For real numbers a and b such that $a < b$ (a is to the left of b on a number line), we define the **open interval** (a, b) to be the set of numbers between, but not including, a and b. That is,

$(a, b) =$ the set of all numbers x such that $a < x < b$.

The inequality $a < x < b$ is true when both $a < x$ is true *and* $x < b$ is true.

The graph of (a, b) is shown in color above. The open circles and the parentheses indicate that a and b are not included. The numbers a and b are called **endpoints.**

The notation (a, b) is an example of what is called *interval notation.* We now define other kinds of *interval notation.*

The **closed interval** $[a, b]$ is the set of numbers between and including a and b. That is,

$[a, b] =$ the set of all numbers x such that $a \le x \le b$.

Intervals between houses are analogous to intervals on a number line.

The graph of $[a, b]$ is shown in color above. The solid circles and the brackets indicate that a and b are included.

There are two kinds of **half-open intervals** defined as follows:

$(a, b] = $ the set of all numbers x such that $a < x \leq b$.

Note that the open circle and the parenthesis in $(a, b]$ indicate that a is not included. The solid circle and the bracket indicate that b is included. Also,

$[a, b) = $ the set of all numbers x such that $a \leq x < b$.

Note that the solid circle and the bracket in $[a, b)$ indicate that a is included. The open circle and the parenthesis indicate that b is not included.

Some intervals are of unlimited extent in one or both directions. In such cases, we use the infinity symbol ∞. For example,

$[a, \infty) = $ the set of all numbers x such that $x \geq a$.

Note that ∞ is not a number and this is indicated with a parenthesis by the infinity symbol.

$(a, \infty) = $ the set of all numbers x such that $x > a$.

$(-\infty, b] = $ the set of all numbers x such that $x \leq b$.

$(-\infty, b) = $ the set of all numbers x such that $x < b$.

We can name the entire set of real numbers using $(-\infty, \infty)$.

$(-\infty, \infty)$

Any point in an interval that is not an endpoint is an **interior point.**

Note that all the points in an open interval are interior points.

![open interval diagram]

1.2 Exercise Set

Solve.

1. $-7x + 10 = 5x - 11$

2. $-8x + 9 = 4x - 70$

3. $5x - 17 - 2x = 6x - 1 - x$

4. $5x - 2 + 3x = 2x + 6 - 4x$

5. $x + 0.8x = 216$

6. $x + 0.5x = 210$

7. $x + 0.08x = 216$

8. $x + 0.05x = 210$

9. $2x(x + 3)(5x - 4) = 0$

10. $7x(x - 2)(2x + 3) = 0$

11. $x^2 + 1 = 2x + 1$ 12. $2t^2 = 9 + t^2$

13. $t^2 - 2t = t$ 14. $6x - x^2 = x$

15. $6x - x^2 = -x$ 16. $2x - x^2 = -x$

17. $9x^3 = x$ 18. $16x^3 = x$

19. $(x - 3)^2 = x^2 + 2x + 1$

20. $(x - 5)^2 = x^2 + x + 3$

21. $3 - x \leq 4x + 7$

22. $x + 6 \leq 5x - 6$

23. $5x - 5 + x > 2 - 6x - 8$

24. $3x - 3 + 3x > 1 - 7x - 9$

25. $-7x < 4$

26. $-5x \geq 6$

27. $5x + 2x \leq -21$

28. $9x + 3x \geq -24$

29. $2x - 7 < 5x - 9$

30. $10x - 3 \geq 13x - 8$

31. $8x - 9 < 3x - 11$

32. $11x - 2 \geq 15x - 7$

33. $8 < 3x + 2 < 14$

34. $2 < 5x - 8 \leq 12$

35. $3 \leq 4x - 3 \leq 19$

36. $9 \leq 5x + 3 < 19$

37. $-7 \leq 5x - 2 \leq 12$

38. $-11 \leq 2x - 1 < -5$

Write interval notation for each graph in Exercises 39–46.

39.

![number line graph for 39]

40.

![number line graph for 40]

41.

![number line graph for 41]

42.

![number line graph for 42]

43.

44.

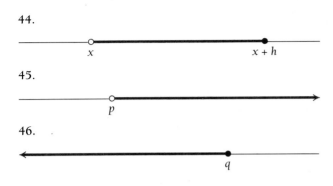

45.

46.

Write interval notation for each of the following.

47. The set of all numbers x such that $-3 \leq x \leq 3$

48. The set of all numbers x such that $-4 < x < 4$

49. The set of all numbers x such that $-14 \leq x < -11$

50. The set of all numbers x such that $6 < x \leq 20$

51. The set of all numbers x such that $x \leq -4$

52. The set of all numbers x such that $x > -5$

APPLICATIONS

◆ **Business and Economics**

53. *Investment increase.* An investment is made at $8\frac{1}{2}\%$, compounded annually. It grows to $705.25 at the end of 1 year. How much was invested originally?

54. *Investment increase.* An investment is made at 7%, compounded annually. It grows to $856 at the end of 1 year. How much was invested originally?

55. *Total revenue.* A firm determines that the total revenue, in dollars, from the sale of x units of a product is

$$3x + 1000.$$

Determine the number of units that must be sold so that its total revenue will be more than $22,000.

56. *Total revenue.* A firm determines that the total revenue, in dollars, from the sale of x units of a product is

$$5x + 1000.$$

Determine the number of units that must be sold so that its total revenue will be more than $22,000.

◆ **Life and Physical Sciences**

57. *Weight gain.* After a 6% gain in weight, an animal weighs 508.8 lb. What was its original weight?

58. *Weight gain.* After a 7% gain in weight, an animal weighs 363.8 lb. What was its original weight?

◆ **Social Sciences**

59. *Population increase.* After a 2% increase, the population of a city is 826,200. What was the former population?

60. *Population increase.* After a 3% increase, the population of a city is 741,600. What was the former population?

◆ **General Interest**

61. *Grade average.* To get a B in a course, a student's average must be greater than or equal to 80% (at least 80%) and less than 90%. On the first three tests, the student scores 78%, 90%, and 92%. Determine the scores on the fourth test that will guarantee a B.

62. *Grade average.* To get a C in a course, a student's average must be greater than or equal to 70% and less than 80%. On the first three tests, the student scores 65%, 83%, and 82%. Determine the scores on the fourth test that will guarantee a C.

SYNTHESIS

63. ◈ Explain the error or errors in the following:

$$-\frac{5}{6}x + 10 \leq \frac{1}{2}x + 2$$

$$5x - 60 \leq -3x - 12 \qquad (1)$$

$$5x \geq -3x + 48 \qquad (2)$$

$$8x \geq 48 \qquad (3)$$

$$x \geq 6. \qquad (4)$$

64. ◈ Explain the error or errors in the following:

$$x^2 - x = 6$$

$$x(x - 1) = 6 \qquad (1)$$

$$x = 6 \quad \text{or} \quad x - 1 = 6 \qquad (2)$$

$$x = 6 \quad \text{or} \qquad x = 7. \qquad (3)$$

The solutions are 6 and 7. $\qquad (4)$

1.3 Graphs and Functions

OBJECTIVES

* Given a function and several inputs, find the outputs.
* Graph an equation or a function.
* Determine whether a graph is that of a function.

A graph offers the opportunity to visualize relationships. We see graphs of all kinds in magazines and newspapers. Examples are shown below. The type of graph shown in the figure on the Dow Jones Industrial Average is the kind we will study most in calculus. It shows changes over a period of time. One topic we might be interested in is how the change of time affects the change in the Dow Jones average.

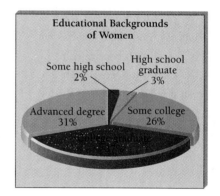

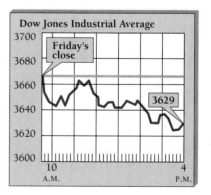

Graphs

Each point in the plane corresponds to an ordered pair of numbers. Note in the figure below that the pair (2, 5) is different from the pair (5, 2). This is why we call (2, 5) an **ordered pair.** The first member, 2, is called the **first coordinate** of the point, and the second member, 5, is called the **second coordinate.** Together these

are called the *coordinates of the point.*
The vertical line is often called the
y-axis, and the horizontal line is
often called the *x-axis.*

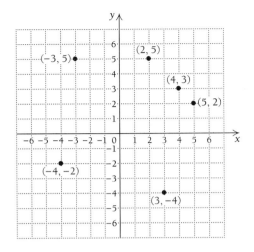

Graphs of Equations

A **solution** of an equation in two variables is an ordered pair of numbers that, when
substituted for the variables, gives a true sentence. If not directed otherwise, we usually take the variables in alphabetical *order.* For example, $(-1, 2)$ is a solution to the
equation $3x^2 + y = 5$, because when we substitute -1 for x and 2 for y, we get a
true sentence:

$$
\begin{array}{c|c}
3x^2 + y \;=\; 5 & \\
\hline
3(-1)^2 + 2 & 5 \\
3 + 2 & \\
5 & \text{TRUE}
\end{array}
$$

DEFINITION

The *graph* of an equation is a drawing that represents all the solutions of
the equation.

We obtain the graph of an equation by plotting enough ordered pairs (that are
solutions) to see a pattern. The graph could be a line, a curve (or curves), or some
other configuration.

EXAMPLE 1 Graph: $y = 2x + 1$.

Solution We first find some ordered pairs that are solutions and arrange them in a
table. To find an ordered pair, we can choose *any* number for x and then determine
y. For example, if we choose -2 for x, then $y = 2(-2) + 1 = -4 + 1 = -3$. We
substituted -2 for x in the equation $y = 2x + 1$. For balance, we make some negative choices for x, as well as some positive choices. If a number takes us off the graph

paper, we usually omit the pair from the graph.

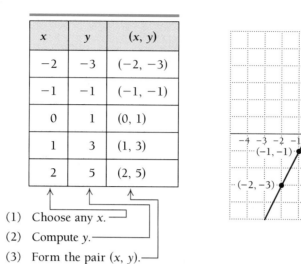

x	y	(x, y)
−2	−3	(−2, −3)
−1	−1	(−1, −1)
0	1	(0, 1)
1	3	(1, 3)
2	5	(2, 5)

(1) Choose any x. ⏋

(2) Compute y. ⏋

(3) Form the pair (x, y). ⏋

(4) Plot the points. ⏋

After we plot the points, we look for a pattern in the graph. If we had enough of the points, they would make a solid line. We can draw the line with a ruler and label it $y = 2x + 1$. ◆

EXAMPLE 2   Graph: $y = x^2 - 1$.

Solution

x	y	(x, y)
−2	3	(−2, 3)
−1	0	(−1, 0)
0	−1	(0, −1)
1	0	(1, 0)
2	3	(2, 3)

(1) Choose any x. ⏋

(2) Compute y. ⏋

(3) Form the pair (x, y). ⏋

(4) Plot the points. ⏋

This time the pattern of the points is a curve called a *parabola*. We fill in the pattern and obtain the graph. Note that we must plot enough points to see a pattern.   ◆

EXAMPLE 3   Graph: $x = y^2$.

Solution   In this case, x is expressed in terms of the variable y. Thus we first choose numbers for y and then compute x.

x	y	(x, y)
4	-2	$(4, -2)$
1	-1	$(1, -1)$
0	0	$(0, 0)$
1	1	$(1, 1)$
4	2	$(4, 2)$

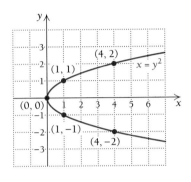

(1)   Choose any y.
(2)   Compute x.
(3)   Form the pair (x, y).
(4)   Plot the points.

We plot these points, keeping in mind that x is still the first coordinate and y the second. We look for a pattern and complete the graph.   ◆

   TECHNOLOGY CONNECTION

Graphers and Viewing Windows

We now begin to include some activities in text sections and exercise sets that utilize graphing calculators or computer graphing software. We will refer to all such calculators and software simply as **graphers.** Most activities are presented in a generic form. Be sure to check your user's manual or ask your instructor for the exact procedures.

One feature common to all graphers is the **viewing window.** This refers to the rectangular portion of the screen in which a graph appears. Windows are described by four numbers [L, R, B, T]

that represent the *Left* and *Right* endpoints of the x-axis and the *Bottom* and *Top* endpoints of the y-axis. A RANGE key is sometimes used to set these dimensions, but the method varies for each grapher. Axis-scaling notation like xScl = 1 and yScl = 30 means that there is 1 unit between "tick" marks on the x-axis and 30 units between tick marks on the y-axis. Too many tick marks can bold out or blur the unit markings in such a way that they cannot be distinguished.

The primary use for graphers is graphing

(continued)

TECHNOLOGY CONNECTION (*continued*)

equations. For example, let us graph the equation $y = x^3 - 3x + 1$. If we select the window $[-10, 10, -10, 10]$ and the axis scaling xScl = 2 and yScl = 2, we get the following graph.

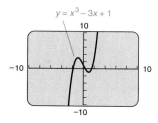

$y = x^3 - 3x + 1$

You may need to change viewing windows in order to "best" reveal the curvature of a graph. For example, each of the following is a graph of $f(x) = 3x^5 - 20x^3$, with a different viewing window. Which do you think is "best"?

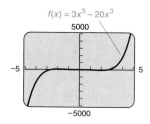

$f(x) = 3x^5 - 20x^3$

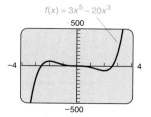

$f(x) = 3x^5 - 20x^3$

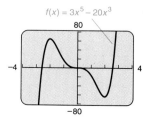

$f(x) = 3x^5 - 20x^3$

To graph an equation like $2x - 3y = 6$, some graphers require that the equation first be solved for y and put in the form "$y = \ldots$" Thus for $2x - 3y = 6$, we would graph $y = \frac{2}{3}x - 2$. For the equation $x = y^2$, we would first obtain $y = \pm\sqrt{x}$ and then graph the individual equations $y_1 = \sqrt{x}$ and $y_2 = -\sqrt{x}$. There are some graphers, however, that do not require solving for y.

Graph each of the following equations. First use the viewing window $[-10, 10, -10, 10]$. Then try another viewing box with your own dimensions. Keep trying until you find a window that, in your mind, best reveals the curvature of the graph.

$$y = 3 - 4x, \qquad\qquad 3x + 4y = 12,$$
$$y = x^2 - 4x + 3, \qquad\quad y^2 = 4 - x,$$
$$y = x^3 + 2x^2 - 4x - 13, \quad y^3 + 2x - 4 = 0$$

Functions

A **relation** is any set of ordered pairs. Thus the solutions of an equation in two variables form a relation.

A **function** is a special kind of relation. Such relations are of fundamental importance in calculus.

A FUNCTION AS AN INPUT–OUTPUT RELATION

DEFINITION

A *function* is a relation that assigns to each "input," or first coordinate, a unique "output," or second coordinate. The set of all input numbers is called the *domain*. The set of all output numbers is called the *range*.

EXAMPLE 4 Squaring numbers is a function. We can take any number x as an input. We square that number to find the output, x^2.

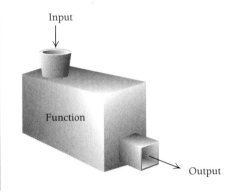

Input

Function

Output

Input	Output
-3	9
1.73	2.9929
k	k^2
$\sqrt{a}$	a
$1 + t$	$(1 + t)^2$, or $1 + 2t + t^2$

Think of a function as a machine, as shown here. Further, think of putting a member of the domain (an input) into the machine. The machine squares the input and gives you the output, a member of the range. The domain of this function is the set of all real numbers, because any real number can be squared. ◆

It is customary to use letters such as f and g to represent functions. Suppose that f is a function and x is a number in its domain. For the input x, we can name the output as

$$f(x), \quad \text{read} \quad \text{“}f \text{ of } x\text{,”} \quad \text{or} \quad \text{“the value of } f \text{ at } x\text{.”}$$

If f is the squaring function, then $f(3)$ is the output for the input 3. Thus, $f(3) = 3^2 = 9$.

EXAMPLE 5 The squaring function f is given by

$$f(x) = x^2.$$

Find $f(-3)$, $f(1)$, $f(k)$, $f(\sqrt{k})$, $f(1 + t)$, and $f(x + h)$.

Solution

$$f(-3) = (-3)^2 = 9,$$
$$f(1) = 1^2 = 1,$$
$$f(k) = k^2,$$
$$f(\sqrt{k}) = (\sqrt{k})^2 = k,$$
$$f(1 + t) = (1 + t)^2 = 1 + 2t + t^2,$$
$$f(x + h) = (x + h)^2 = x^2 + 2xh + h^2$$

To find $f(x + h)$, remember what the function does: It squares the input. Thus, $f(x + h) = (x + h)^2 = x^2 + 2xh + h^2$. This amounts to replacing x on both sides of $f(x) = x^2$ by $x + h$. ◆

EXAMPLE 6   A function f subtracts the square of an input from the input. A description of f is given by

$$f(x) = x - x^2.$$

Find $f(4)$ and $f(x + h)$.

Solution   We replace the x's on both sides by the inputs. Thus,

$$f(4) = 4 - 4^2 = 4 - 16 = -12;$$
$$f(x + h) = (x + h) - (x + h)^2$$
$$= x + h - (x^2 + 2xh + h^2)$$
$$= x + h - x^2 - 2xh - h^2.$$ ◆

Taking square roots is *not* a function, because an input can have more than one output. For example, the input 4 has two outputs, 2 and -2.

When a function is given by a formula, and nothing is said about the domain, its domain is understood to be the set of all numbers that can be substituted into the formula. For example, consider the reciprocal function

$$f(x) = \frac{1}{x}.$$

The only number that cannot be substituted into the formula is 0. We say that f is *not defined at* 0, *or* $f(0)$ *does not exist*. The domain consists of all nonzero real numbers.

EXAMPLE 7   Taking principal square roots (nonnegative roots) is a function. Let g be this function. Then g can be described as

$$g(x) = \sqrt{x}.$$

Recall from algebra that the symbol $\sqrt{a}$ represents the nonnegative square root of a for $a \geq 0$. There is only one such real-number root.

a) Find the domain of this function.

b) Find $g(0)$, $g(2)$, $g(a)$, $g(16)$, and $g(t + h)$.

Solution

a) The domain consists of numbers that can be substituted into the formula. We can take the principal square root of any nonnegative number. The principal square root of a negative number is not a real number. (Taking square roots of negative numbers would require us to consider complex numbers, which we will not cover in this text.) Thus the domain consists of all nonnegative numbers.

TECHNOLOGY CONNECTION

Functions, Function Values, and Input–Output Tables

Some graphers do not use the notation "$f(x) = \dots$" when defining a function. You may have to convert to the "$y = \dots$" notation.

With many graphers, you have the capability to create input–output tables. Usually the grapher requires a minimum value of x and a step-value. For example, you may enter a minimum value of $x = 2$ and step of 0.25. Next function values are created for $x = 2$, 2.25, 2.5, 2.75, 3, 3.25, and so on. An ERROR message is given when a function value is not defined. Consider the reciprocal function $f(x) = 1/(x - 3)$. Part of a table for this function is shown below. Note that we converted from $f(x) = 1/(x - 3)$ to $y = 1/(x - 3)$.

x	y
2	−1
2.25	−1.3333
2.5	−2
2.75	−4
3	ERROR
3.25	4
3.5	2
3.75	1.3333
4	1

Use your grapher to make a table for the function $f(x) = 1/(x^2 - 4)$ from $x = -3$ to $x = 3$ with a step-value of 0.5. Then create an input–output table for the function in Example 7.

b) $g(0) = \sqrt{0} = 0,$
         $g(2) = \sqrt{2},$
         $g(a) = \sqrt{a},$
         $g(16) = \sqrt{16} = 4,$
      $g(t + h) = \sqrt{t + h}$ ◆

A Function as a Mapping

We can also think of a function as a "mapping" of one set to another.

DEFINITION

A *function* is a mapping that associates with each number x in one set (called the *domain*) exactly one number y in another set (called the *range*).

Inputs
(Domain)

Outputs
(Range)

For example, the squaring function maps members of the set of real numbers to members of the set of nonnegative numbers, as shown below.

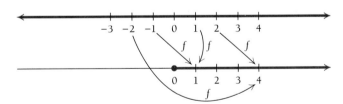

The statement

   $y = f(x)$

means that the number x is mapped to the number y by the function f. Functions are often implicit in certain equations. For example, consider

   $xy = 2.$

For any nonzero x, there is a unique number y that satisfies the equation. This yields a function that is given explicitly by

   $y = f(x) = \dfrac{2}{x}.$

The number of people at the beach at a given time is a function of the temperature, although a formula may not be readily available.

On the other hand, consider the equation

$$x = y^2.$$

A positive number x would be related to two values of y, namely $\sqrt{x}$ and $-\sqrt{x}$. Thus this equation is not an implicit description of a function that maps inputs x to outputs y.

Graphs of Functions

Consider again the squaring function. The input 3 is associated with the output 9. The input–output pair $(3, 9)$ is one point on the *graph* of this function.

DEFINITION

The *graph* of a function f is a drawing that represents all the input–output pairs $(x, f(x))$. In cases where the function is given by an equation, the graph of a function is the graph of the equation $y = f(x)$.

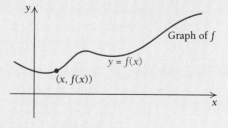

It is customary to locate input values (the domain) on the horizontal axis and output values (the range) on the vertical axis.

EXAMPLE 8   Graph: $f(x) = x^2 - 1$.

Solution

x	$f(x)$	$(x, f(x))$
-2	3	$(-2, 3)$
-1	0	$(-1, 0)$
0	-1	$(0, -1)$
1	0	$(1, 0)$
2	3	$(2, 3)$

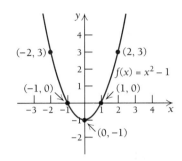

(1)   Choose any x.
(2)   Compute y.
(3)   Form the pair (x, y).
(4)   Plot the points.

We plot the input–output pairs from the table and, in this case, draw a curve to complete the graph.   ◆

The following graph illustrates how the idea of a mapping is connected with the graph of a function.

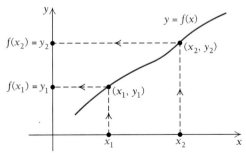

The Vertical-Line Test

Let us now determine how we can look at a graph and decide whether it is a graph of a function. We already know that

$$x = y^2$$

does not yield a function that maps a number x to a unique number y. In its graph, which follows, note that there is a point x_1 that has two outputs. This means that we have a vertical line that meets the graph in more than one place.

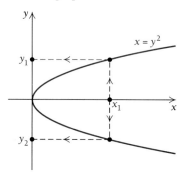

The Vertical-Line Test

A graph is the graph of a function provided that no vertical line meets the graph more than once.

EXAMPLE 9   Determine whether each of the following is a graph of a function.

a)

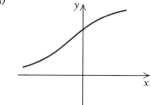

b)

c)

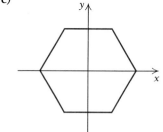

d)

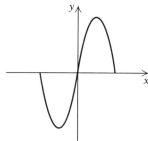

Solution

a) A function. No vertical line meets the graph more than once.

b) Not a function. A vertical line (in fact, many) meets the graph more than once.

c) Not a function.

d) A function.   ◆

Functions Defined Piecewise

Sometimes functions are defined piecewise. That is, there are different output formulas for different parts of the domain.

EXAMPLE 10   Graph the function defined as follows.

$$f(x) = \begin{cases} 4, & \text{for } x \leq 0 \\ & \text{(This means that for any input } x \text{ less than or equal to 0, the output is 4.)} \\ 4 - x^2, & \text{for } 0 < x \leq 2 \\ & \text{(This means that for any input } x \text{ greater than 0 and less than or equal to 2, the output is } 4 - x^2.) \\ 2x - 6, & \text{for } x > 2 \\ & \text{(This means that for any input } x \text{ greater than 2, the output is } 2x - 6.) \end{cases}$$

Solution   The graph of this function follows.

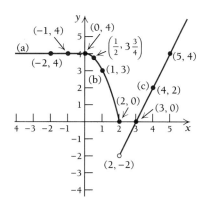

a) We graph $f(x) = 4$ for inputs less than or equal to 0 (that is, $x \leq 0$).

For $f(x) = 4$,

$f(-2) = 4$,

$f(-1) = 4$, and

$f(0) = 4$.

x	$f(x)$	$(x, f(x))$
-2	4	$(-2, 4)$
-1	4	$(-1, 4)$
0	4	$(0, 4)$

b) We graph $f(x) = 4 - x^2$ for inputs greater than 0 and less than or equal to 2 (that is, $0 < x \le 2$).

For $f(x) = 4 - x^2$,

$$f\left(\tfrac{1}{2}\right) = 4 - \left(\tfrac{1}{2}\right)^2 = 3\tfrac{3}{4},$$
$$f(1) = 4 - 1^2 = 3, \text{ and}$$
$$f(2) = 4 - 2^2 = 0.$$

x	$f(x)$	$(x, f(x))$
$\tfrac{1}{2}$	$3\tfrac{3}{4}$	$\left(\tfrac{1}{2}, 3\tfrac{3}{4}\right)$
1	3	(1, 3)
2	0	(2, 0)

The solid circle indicates that the point (2, 0) is part of the graph.

c) We graph $f(x) = 2x - 6$ for inputs greater than 2 (that is, $x > 2$).

For $f(x) = 2x - 6$,

$$f(3) = 2(3) - 6 = 0,$$
$$f(4) = 2(4) - 6 = 2, \text{ and}$$
$$f(5) = 2(5) - 6 = 4.$$

x	$f(x)$	$(x, f(x))$
3	0	(3, 0)
4	2	(4, 2)
5	4	(5, 4)

The open circle at $(2, -2)$ indicates that the point is not part of the graph. ◆

TECHNOLOGY CONNECTION

Graphs of Functions Defined Piecewise

To graph a function defined piecewise, consult your manual. Although some graphers do not have the capability, you might try the following way to enter the function formula for Example 10. It incorporates parenthetic descriptions of the intervals:

$$f(x) = 4 \; (x \le 0) \; + (4 - x^2) \; (x > 0)(x \le 2)$$
$$+ (2x - 6) \; (x > 2)$$

(*Note:* If your grapher provides this feature, then it probably also allows you to choose the mode in which the graph is displayed: CONNECTED, in which adjacent points are connected by segments; or DOT, in which only the points are plotted and they are not connected. For piecewise-defined functions, you should select the dot mode.)

Graph the following function.

$$f(x) = \begin{cases} x + 3, & \text{for } x \le -2, \\ 1, & \text{for } -2 < x \le 3, \\ x^2 - 10, & \text{for } x > 3 \end{cases}$$

Some Final Remarks

Almost all the functions in this text can be described by equations. Some functions, however, cannot. For example, there will be a function that assigns grades to students in this course, but that function will most likely not have a formula.

We sometimes use the terminology *y is a function of x.* This means that x is an input and y is an output. We often refer to x as the **independent variable** when it represents inputs and y as the **dependent variable** when it represents outputs. We may refer to "a function $y = x^2$," without naming it with a letter f. We may simply refer to x^2 (alone) as a function.

In calculus we will be studying how outputs of a function change when the inputs change.

1. A function f is given by

$$f(x) = 2x + 3.$$

This function takes a number x, multiplies it by 2, and adds 3.

a) Complete this table.

Input	Output
4.1	
4.01	
4.001	
4	

b) Find $f(5), f(-1), f(k), f(1 + t)$, and $f(x + h)$.

2. A function f is given by

$$f(x) = 3x - 1.$$

This function takes a number x, multiplies it by 3, and subtracts 1.

a) Complete this table.

Input	Output
5.1	
5.01	
5.001	
5	

b) Find $f(4), f(-2), f(k), f(1 + t)$, and $f(x + h)$.

3. A function g is given by

$$g(x) = x^2 - 3.$$

This function takes a number x, squares it, and subtracts 3. Find $g(-1), g(0), g(1), g(5), g(u),$ $g(a + h)$, and $g(1 - h)$.

4. A function g is given by

$$g(x) = x^2 + 4.$$

This function takes a number x, squares it, and adds 4. Find $g(-3), g(0), g(-1), g(7), g(v), g(a + h)$, and $g(1 - t)$.

5. A function f is given by

$$f(x) = (x - 3)^2.$$

This function takes a number x, subtracts 3 from it, and squares the result.

a) Find $f(4), f(-2), f(0), f(a), f(t + 1), f(t + 3)$, and $f(x + h)$.

b) Note that f could also be given by

$$f(x) = x^2 - 6x + 9.$$

Explain what this does to an input number x.

6. A function f is given by

$$f(x) = (x + 4)^2.$$

This function takes a number x, adds 4 to it, and squares the result.

a) Find $f(3), f(-6), f(0), f(k), f(t - 1), f(t - 4)$, and $f(x + h)$.

b) Note that f could also be given by

$$f(x) = x^2 + 8x + 16.$$

Explain what this does to an input number x.

Graph the function.

7. $f(x) = 2x + 3$

8. $f(x) = 3x - 1$

9. $g(x) = -4x$

10. $g(x) = -2x$

11. $f(x) = x^2 - 1$

12. $f(x) = x^2 + 4$

13. $g(x) = x^3$

14. $g(x) = \frac{1}{2}x^3$

Determine whether the graph is that of a function. (In Exercises 23–26, the vertical dashed lines are not part of the graph.)

15.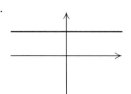

16.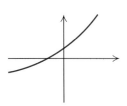

17.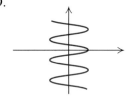

18.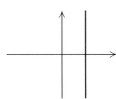

19.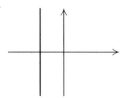

20.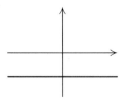

21.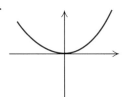

22.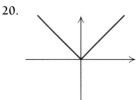

23.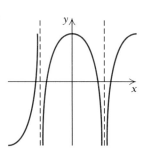

24.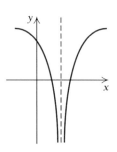

25.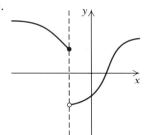

26.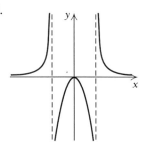

27. a) Graph $x = y^2 - 1$.
  b) Is this a function?

28. a) Graph $x = y^2 - 3$.
  b) Is this a function?

29. For $f(x) = x^2 - 3x$, find $f(x + h)$.

30. For $f(x) = x^2 + 4x$, find $f(x + h)$.

Graph.

31. $f(x) = \begin{cases} 1, & \text{for } x < 0, \\ -1, & \text{for } x \geq 0 \end{cases}$

32. $f(x) = \begin{cases} 2, & \text{for } x \text{ an integer}, \\ -2, & \text{for } x \text{ not an integer} \end{cases}$

33. $f(x) = \begin{cases} -3, & \text{for } x = -2, \\ x^2, & \text{for } x \neq -2 \end{cases}$

34. $f(x) = \begin{cases} -2x - 6, & \text{for } x \leq -2, \\ 2 - x^2, & \text{for } -2 < x < 2, \\ 2x - 6, & \text{for } x \geq 2 \end{cases}$

APPLICATIONS

◆   **Business and Economics**

35. *Total revenue.* Raggs, Ltd., a clothing firm, determines that its total revenue from the sale of x suits is

given by the function

$$R(x) = 200x + 50,$$

where $R(x)$ = the revenue, in dollars, from the sale of x suits. Find $R(10)$ and $R(100)$.

36. *Compound interest.* The amount of money in a savings account at 6%, compounded annually, depends on the initial investment x and is given by the function

$$A(x) = x + 6\%x,$$

where $A(x)$ = the amount in the account at the end of 1 year. Find $A(100)$ and $A(1000)$.

37. *Sales of musical recordings.* The following graph shows the unit sales, in millions, of three kinds of recording media: audiocassettes, compact discs, and longplaying records.

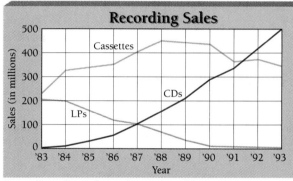

Source: Recording Industry Association of America

a) Is each graph that of a function?
b) Estimate the number of LP records sold in 1983 and in 1993.
c) In what year were the sales of LPs the same as that of CDs?
d) Estimate the sales of CDs in 1985 and in 1993.
e) In what year were the sales of cassettes the same as that of CDs?
f) In what year did the greatest sale of cassettes occur?
g) In what year did the greatest sale of CDs occur?

◆ **Life and Physical Sciences**

38. *Scaling stress factors.* In psychology a process called *scaling* is used to attach numerical ratings to a group of life experiences. In the following table, various events have been rated from 1 to 100 according to their stress levels.

Event	Scale of impact
Death of spouse	100
Divorce	73
Jail term	63
Marriage	50
Lost job	47
Pregnancy	40
Death of close friend	37
Loan over $10,000	31
Child leaving home	29
Change in schools	20
Loan less than $10,000	17
Christmas	12

a) Does the table constitute a function? Why or why not?
b) What are the inputs? What are the outputs?

SYNTHESIS

Solve for y in terms of x. Decide whether the resulting equation represents a function.

39. $2x + y - 16 = 4 - 3y + 2x$
40. $2y^2 + 3x = 4x + 5$
41. $(4y^{2/3})^3 = 64x$
42. $(3y^{3/2})^2 = 72x$

43. ◆ Suppose you were discussing the idea of a function to a friend. How would you explain it?
44. ◆ Explain the special aspects of the graph of a function.

  TECHNOLOGY CONNECTION

Graph the equation using the viewing window $[-10, 10, -10, 10]$, xScl = 1, and yScl = 1. Then try other viewing windows until you create a graph that "best" reveals the curvature to you. Graphing using calculus will be discussed in detail in Chapter 3.

45. $y = x + 200$

46. $9.6x + 4.2y = -100$

47. $x = y^2 + 1$

48. $f(x) = (x + 2)^2 - 3$

49. $f(x) = -(x - 3)^2 + 1$

50. $y^3 + (x - 5)^2 - 8x + 2 = 0$

51. $f(x) = (x - 3.4)^3 + 5.6$

52. $y = x^3 + 2x^2 - 4x - 13$

53. $y + x^3 - 6x^2 - 5 = 0$

1.4 Slope and Linear Functions

OBJECTIVES

- Graph equations of the type $x = a$ and $y = b$.
- Graph linear functions.
- Use equations and properties of lines and linear functions.

Horizontal and Vertical Lines

Let us consider graphs of equations $y = b$ and $x = a$.

EXAMPLE 1

a) Graph $y = 4$.

b) Decide whether the relation is a function.

Solution

a) The graph consists of all ordered pairs whose second coordinate is 4. To see how a pair such as $(-2, 4)$ could be a solution of $y = 4$, we can consider the equation above in the form

$$0x + y = 4.$$

Then $(-2, 4)$ is a solution because

$$0(-2) + 4 = 4$$

is true.

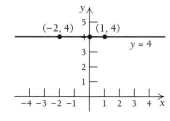

b) The vertical-line test holds. Thus this is a function. ◆

EXAMPLE 2

a) Graph $x = -3$.

b) Decide whether it is a function.

Solution

a) The graph consists of all ordered pairs whose first coordinate is -3. To see how a pair such as $(-3, 4)$ could be a solution of $x = -3$, we can consider the equation above in the form

$$x + 0y = -3.$$

Then $(-3, 4)$ is a solution because

$$(-3) + 0(4) = -3$$

is true.

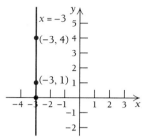

b) This is *not* a function. It fails the vertical-line test. The line itself meets the graph more than once—in fact, infinitely many times.   ◆

In general, we have the following.

> **THEOREM 6**
>
> The graph of $y = b$, a horizontal line, is the graph of a function.
> The graph of $x = a$, a vertical line, is not the graph of a function.

TECHNOLOGY CONNECTION

Squaring a Viewing Window

Consider the following $[-10, 10, -10, 10]$ viewing window. It is considered standard on most graphers.

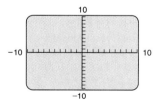

Note, however, that the distance between units is not visually the same on both axes. If we change to the window $[-6, 6, -4, 4]$, we get a graph for which it is.

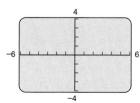

Creating such a window is called **squaring.** In this case, the length of the interval shown on the y-axis is two-thirds of the length of the interval on the x-axis. Using a squared window will be helpful in this section and in other parts of the book. You may have to consult your manual about how to square a window. On the TI-82, there is a ZSQUARE procedure for window squaring. On other graphers, you may simply have to use trial and error.

Use a square viewing window and your grapher to explore the effect of m when graphing $y = mx$. To do so, begin with the graph of $y = x$. Then, using the same set of axes, graph $y = 2x$, $y = 5x$, and $y = 10x$. What do you think the graph of $y = 128x$ will look like? Clear the screen. Now graph $y = \frac{1}{2}x$, $y = \frac{3}{25}x$, and $y = 0.0046x$. What do you think the graph of $y = 0.00000029x$ will look like?

Clear the screen and graph $y = -x$. Then, using the same set of axes, graph $y = -2x$, $y = -4x$, and $y = -10x$. What do you think the graph of $y = -200x$ will look like? Clear the screen. Now graph $y = -\frac{2}{3}x$, $y = -0.35x$, and $y = -0.042x$. What do you think the graph of $y = -0.000017x$ will look like?

The Equation $y = mx$

Consider the following table of numbers and look for a pattern.

x	1	-1	$-\frac{1}{2}$	2	-2	3	-7	5
y	3	-3	$-\frac{3}{2}$	6	-6	9	-21	15

Note that the ratio of the bottom number to the top one is 3. That is,

$$\frac{y}{x} = 3, \quad \text{or} \quad y = 3x.$$

Ordered pairs from the table can be used to graph the equation $y = 3x$ (see the figure at left). Note that this is a function.

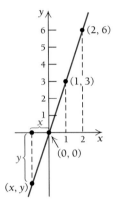

THEOREM 7

The graph of the function given by

$$y = mx \quad \text{or} \quad f(x) = mx$$

is the straight line through the origin $(0, 0)$ and the point $(1, m)$. The constant m is called the *slope* of the line.

Various graphs of $y = mx$ for positive m are shown below. Note that such graphs slant up from left to right. A line with large positive slope rises faster than a line with smaller positive slope.

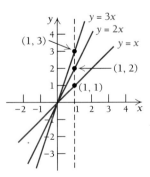

When $m = 0$, $y = 0x$, or $y = 0$. On the left at the top of the following page is a graph of $y = 0$. Note that this is both the x-axis and a horizontal line.

Lines of various slopes.

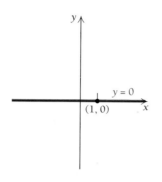
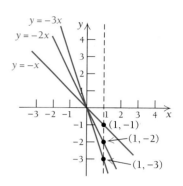

Graphs of $y = mx$ for negative m are shown on the right above. Note that such graphs slant down from left to right.

Direct Variation

There are many applications involving equations like $y = mx$, where m is some positive number. In such situations, we say that we have **direct variation,** and m (the slope) is called the **variation constant,** or **constant of proportionality.** Generally, only positive values of x and y are considered.

> **DEFINITION**
>
> The variable y *varies directly* as x if there is some positive constant m such that $y = mx$. We also say that y is *directly proportional* to x.

EXAMPLE 3 *Life science: Hair growth.* The number N of inches that human hair will grow is directly proportional to the time t, in months. Hair will grow 6 inches in 12 months.

a) Find an equation of variation.

b) How many months does it take for hair to grow 10 in.?

Solution

a) Since $N = mt$, then $6 = m(12)$ and $\frac{1}{2} = m$. Thus, $N = \frac{1}{2}t$.

b) To find how many months it takes for hair to grow 10 in., we solve

$$10 = \tfrac{1}{2}t$$

and get

$$20 = t.$$

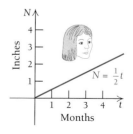

Thus it takes 20 months for hair to grow 10 in.

TECHNOLOGY CONNECTION

A grapher can be used to explore the effect of b when graphing $y = mx + b$. To do so, begin with the graph of $y = x$. Using the same set of axes, graph the lines $y = x + 3$ and $y = x - 4$. How do the lines differ from $y = x$? What do you think the line $y = x - 6$ will look like? Try graphing $y = -0.5x$, $y = -0.5x - 4$, and $y = -0.5x + 3$. Describe what happens to the graph of $y = -0.5x$ when a number b is added.

The Equation $y = mx + b$

Compare the graphs of the equations

$$y = 3x \quad \text{and} \quad y = 3x - 2$$

(see the following figure). Note that the graph of $y = 3x - 2$ is a shift 2 units down of the graph of $y = 3x$, and that $y = 3x - 2$ has y-intercept $(0, -2)$. Note also that the graph of $y = 3x - 2$ is a graph of a function.

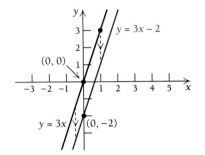

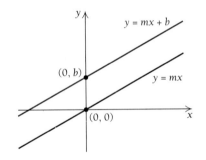

DEFINITION

A *linear function* is given by

$$y = mx + b \quad \text{or} \quad f(x) = mx + b$$

and has a graph that is the straight line parallel to $y = mx$ with y-intercept $(0, b)$. The constant m is called the *slope*. (See the figure at left.)

When $m = 0$, $y = 0x + b = b$, and we have what is known as a **constant function.** The graph of such a function is a horizontal line.

The Slope–Intercept Equation

Any nonvertical line l is uniquely determined by its slope m and its y-intercept $(0, b)$. In other words, the slope describes the "slant" of the line, and the y-intercept is the point at which the line crosses the y-axis. Thus we have the following definition.

DEFINITION

$y = mx + b$ is called the *slope–intercept equation* of a line.

EXAMPLE 4   Find the slope and the y-intercept of $2x - 4y - 7 = 0$.

Solution   We solve for y:

$$-4y = -2x + 7$$
$$y = \tfrac{1}{2}x - \tfrac{7}{4}$$

Slope: $\tfrac{1}{2}$ y-intercept: $\left(0, -\tfrac{7}{4}\right)$ ◆

The Point–Slope Equation

Suppose we know the slope of a line and some point on the line other than the y-intercept. We can still find an equation of the line.

EXAMPLE 5   Find an equation of the line with slope 3 containing the point $(-1, -5)$.

Solution   From the slope–intercept equation, we have

$$y = 3x + b,$$

so we must determine b. Since $(-1, -5)$ is on the line, it follows that

$$-5 = 3(-1) + b,$$

so $-2 = b$ and $y = 3x - 2.$ ◆

If a point (x_1, y_1) is on the line

$$y = mx + b, \tag{1}$$

it must follow that

$$y_1 = mx_1 + b. \tag{2}$$

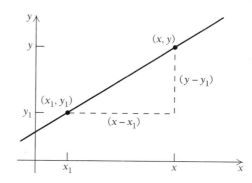

Subtracting Eq. (2) from Eq. (1) eliminates the b's, and we have

$$y - y_1 = (mx + b) - (mx_1 + b)$$
$$= mx + b - mx_1 - b$$
$$= mx - mx_1$$
$$= m(x - x_1).$$

> **DEFINITION**
>
> $y - y_1 = m(x - x_1)$ is called the *point–slope equation* of a line.

This definition allows us to write an equation of a line given its slope and the coordinates of *any* point on it.

EXAMPLE 6 Find an equation of the line with slope 3 containing the point $(-1, -5)$.

Solution Substituting in

$$y - y_1 = m(x - x_1),$$

we get

$$y - (-5) = 3[x - (-1)].$$

Simplifying and solving for y, we get the slope–intercept equation, as found in Example 5:

$$y + 5 = 3(x + 1)$$
$$y + 5 = 3x + 3$$
$$y = 3x + 3 - 5$$
$$y = 3x - 2.$$

Computing Slope

We now determine a method of computing the slope of a line when we know the coordinates of two of its points. Suppose that (x_1, y_1) and (x_2, y_2) are the coordinates of two different points, P_1 and P_2, respectively, on a line that is not vertical. Consider a right triangle with legs parallel to the axes, as shown in the following figure.

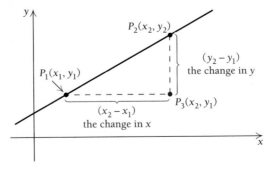

The point P_3 with coordinates (x_2, y_1) is the third vertex of the triangle. As we move from P_1 to P_2, y changes from y_1 to y_2. The change in y is $y_2 - y_1$. Similarly, the

Which lines have the same slope?

change in x is $x_2 - x_1$. The ratio of these changes is the slope. To see this, consider the point–slope equation,

$$y - y_1 = m(x - x_1).$$

Since (x_2, y_2) is on the line, it must follow that

$$y_2 - y_1 = m(x_2 - x_1).$$

Since the line is not vertical, the two x-coordinates must be different, so $x_2 - x_1$ is nonzero and we can divide by it to get the following theorem.

THEOREM 8

$$m = \frac{y_2 - y_1}{x_2 - x_1} = \frac{\text{change in } y}{\text{change in } x} = \text{slope of line containing points } (x_1, y_1) \text{ and } (x_2, y_2)$$

EXAMPLE 7   Find the slope of the line containing the points $(-2, 6)$ and $(-4, 9)$.

Solution   We have

$$m = \frac{y_2 - y_1}{x_2 - x_1} = \frac{6 - 9}{-2 - (-4)}$$

$$= \frac{-3}{2} = -\frac{3}{2}.$$

Note that it does not matter which point is taken first, so long as we subtract the coordinates in the same order. In this example, we can also find m as follows:

$$m = \frac{9 - 6}{-4 - (-2)} = \frac{3}{-2} = -\frac{3}{2}.$$

If a line is horizontal, the change in y for any two points is 0. Thus a horizontal line has slope 0. If a line is vertical, the change in x for any two points is 0. Thus the slope is *not defined* because we cannot divide by 0. A vertical line has no slope. Thus "0 slope" and "no slope" are two very distinct concepts.

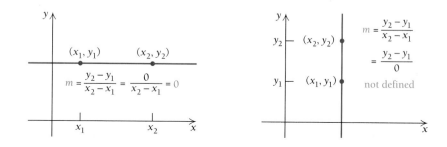

Applications of Linear Functions

Many applications are modeled by linear functions.

EXAMPLE 8 *Business: Total cost.* Raggs, Ltd., a clothing firm, has **fixed costs** of $10,000 per year. These costs, such as rent, maintenance, and so on, must be paid no matter how much the company produces. To produce x units of a certain kind of suit, it costs $20 per unit in addition to the fixed costs. That is, the **variable costs** for producing x of these units is $20x$ dollars. These are costs that are directly related to production, such as material, wages, fuel, and so on. Then the **total cost** $C(x)$ of producing x suits in a year is given by a function C:

$$C(x) = (\text{Variable costs}) + (\text{Fixed costs}) = 20x + 10{,}000.$$

a) Graph the variable-cost, the fixed-cost, and the total-cost functions.

b) What is the total cost of producing 100 suits? 400 suits?

c) How much more does it cost to produce 400 suits than 100 suits?

Solution

a) The variable-cost and fixed-cost functions appear to the left below. The total-cost function is shown to the right. From a practical standpoint, the domains of these functions are nonnegative integers 0, 1, 2, 3, and so on, since it does not make sense to make a negative number of suits or a fractional number of suits. It is common practice to draw the graphs as though the domains were the entire set of nonnegative real numbers.

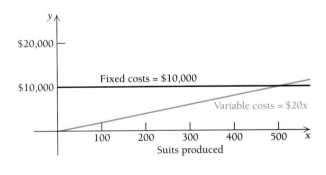

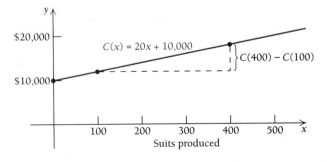

b) The total cost of producing 100 suits is

$$C(100) = 20 \cdot 100 + 10{,}000 = \$12{,}000.$$

The total cost of producing 400 suits is

$$C(400) = 20 \cdot 400 + 10{,}000$$
$$= \$18{,}000.$$

c) The extra cost of producing 400 suits rather than 100 suits is given by

$$C(400) - C(100) = \$18{,}000 - \$12{,}000$$
$$= \$6000.$$

EXAMPLE 9   *Business: Profit-and-loss analysis.* In reference to Example 8, Raggs, Ltd., determines that its total revenue from the sale of x suits is $80 per suit. That is, the total revenue $R(x)$ is given by the function

$$R(x) = 80x.$$

a) Graph $R(x)$ and $C(x)$ using the same set of axes.

b) The total profit $P(x)$ is given by a function P:

$$P(x) = (\text{Total revenue}) - (\text{Total costs}) = R(x) - C(x).$$

Determine $P(x)$ and draw its graph using the same set of axes.

c) The company will *break even* at that value of x for which $P(x) = 0$ (that is, no profit and no loss). This is where $R(x) = C(x)$. Find the **break-even value** of x.

Solution

a) The graphs of $R(x) = 80x$ and $C(x) = 20x + 10{,}000$ are shown below. When $C(x)$ is above $R(x)$, a loss will occur. This is shown by the color-shaded region. When $R(x)$ is above $C(x)$, a gain will occur. This is shown by the gray-shaded region.

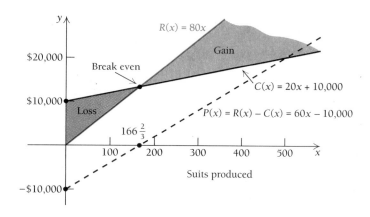

b) We see that

$$P(x) = R(x) - C(x) = 80x - (20x + 10,000)$$
$$= 60x - 10,000.$$

The graph of $P(x)$ is shown by the dashed line. The color dashed line shows a "negative" profit, or loss. The black dashed line shows a "positive" profit, or gain.

c) To find the break-even value, we solve $R(x) = C(x)$:

$$R(x) = C(x)$$
$$80x = 20x + 10,000$$
$$60x = 10,000$$
$$x = 166\tfrac{2}{3}.$$

How do we interpret the fractional answer, since it is not possible to produce $\tfrac{2}{3}$ of a suit? We simply round to 167. Estimates of break-even values are usually sufficient since companies want to operate well away from break-even values in order to maximize profit. ◆

1.4 Exercise Set

Graph.

1. $y = -4$

2. $y = -3.5$

3. $x = 4.5$

4. $x = 10$

Graph. Find the slope and the y-intercept.

5. $y = -3x$

6. $y = -0.5x$

7. $y = 0.5x$

8. $y = 3x$

9. $y = -2x + 3$

10. $y = -x + 4$

11. $y = -x - 2$

12. $y = -3x + 2$

Find the slope and the y-intercept.

13. $2x + y - 2 = 0$

14. $2x - y + 3 = 0$

15. $2x + 2y + 5 = 0$

16. $3x - 3y + 6 = 0$

Find an equation of the line:

17. with $m = -5$, containing $(1, -5)$.

18. with $m = 7$, containing $(1, 7)$.

19. with $m = -2$, containing $(2, 3)$.

20. with $m = -3$, containing $(5, -2)$.

21. with y-intercept $(0, -6)$ and slope $\tfrac{1}{2}$.

22. with y-intercept $(0, 7)$ and slope $\tfrac{4}{3}$.

23. with slope 0, containing $(2, 3)$.

24. with slope 0, containing $(4, 8)$.

Find the slope of the line containing the given pair of points, if it exists.

25. $(-4, -2)$ and $(-2, 1)$

26. $(-2, 1)$ and $(6, 3)$

27. $(2, -4)$ and $(4, -3)$

28. $(-5, 8)$ and $(5, -3)$

29. $(3, -7)$ and $(3, -9)$

30. $(-4, 2)$ and $(-4, 10)$

31. $(2, 3)$ and $(-1, 3)$

32. $\left(-6, \tfrac{1}{2}\right)$ and $\left(-7, \tfrac{1}{2}\right)$

33. $(x, 3x)$ and $(x + h, 3(x + h))$

34. $(x, 4x)$ and $(x + h, 4(x + h))$

35. $(x, 2x + 3)$ and $(x + h, 2(x + h) + 3)$

36. $(x, 3x - 1)$ and $(x + h, 3(x + h) - 1)$

37.–48. Find an equation of the line containing each pair of points in Exercises 25–36.

APPLICATIONS

◆   **Business and Economics**

49. **Investment.** A person makes an investment of P dollars at 8%. After 1 year, it grows to an amount A.

a) Show that A is directly proportional to P.
b) Find A when $P = \$100$.
c) Find P when $A = \$259.20$.

50. **Profit-and-loss analysis.** A ski manufacturer is planning a new line of skis. For the first year, the fixed costs for setting up the new production line are $22,500. The variable costs for producing each pair of skis are estimated at $40. The sales department projects that 3000 pairs can be sold during the first year at a price of $85 per pair.

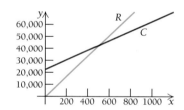

a) Formulate a function $C(x)$ for the total cost of producing x pairs of skis.
b) Formulate a function $R(x)$ for the total revenue from the sale of x pairs of skis.
c) Formulate a function $P(x)$ for the total profit from the production and sale of x pairs of skis.
d) What profit or loss will the company realize if the expected sales of 3000 pairs occurs?
e) How many pairs must the company sell in order to break even?

51. **Profit-and-loss analysis.** Boxowitz, Inc., a computer firm, is planning to sell a new graphing calculator. For the first year, the fixed costs for setting up the new production line are $100,000. The variable costs for producing each calculator are estimated at $20. The sales department projects that 150,000 calculators can be sold during the first year at a price of $45 each.

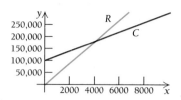

a) Formulate a function $C(x)$ for the total cost of producing x calculators.
b) Formulate a function $R(x)$ for the total revenue from the sale of x calculators.
c) Formulate a function $P(x)$ for the total profit from the production and sale of x calculators.
d) What profit or loss will the firm realize if the expected sales of 150,000 calculators occurs?
e) How many calculators must the firm sell in order to break even?

52. **Straight-line depreciation.** A company buys an office machine for $5200 on January 1 of a given year. The machine is expected to last for 8 years, at the end of which time its *trade-in value*, or *salvage value*, will be $1100. If the company figures the decline in value to be the same each year, then the *book value*, or *salvage value*, after t years, $0 \le t \le 8$, is given by the linear function

$$V(t) = C - t\left(\frac{C - S}{N}\right),$$

where $C =$ the original cost of the item ($5200), $N =$ the number of years of expected life (8), and $S =$ the salvage value ($1100).

a) Find the linear function for the straight-line depreciation of the office machine.
b) Find the salvage value after 0 years, 1 year, 2 years, 3 years, 4 years, 7 years, and 8 years.

53. **Profit-and-loss analysis.** A college student decides to mow lawns in the summer. The initial cost of the lawnmower is $250. Gasoline and maintenance costs are $1 per lawn.

a) Formulate a function $C(x)$ for the total cost of mowing x lawns.
b) The student determines that the total-profit function for the lawnmowing business is given by $P(x) = 9x - 250$. Find a function for the total revenue from mowing x lawns. How much does the student charge per lawn?
c) How many lawns must the student mow before making a profit?

54. *Sales commissions.* A person applying for a sales position is offered alternative salary plans:

Plan A: A base salary of $600 per month plus a commission of 4% of the gross sales for the month.

Plan B: A base salary of $700 per month plus a commission of 6% of the gross sales for the month in excess of $10,000.

a) For each plan, formulate a function that expresses monthly earnings as a function of gross sales x.

b) For what values of gross sales is plan B preferable?

Muscle weight is directly proportional to body weight.

◆ Life and Physical Sciences

55. *Energy conservation.* The R-factor of home insulation is directly proportional to its thickness T.

a) Find an equation of variation if $R = 12.51$ when $T = 3$ in.

b) What is the R-factor for insulation that is 6 in. thick?

56. *Nerve impulse speed.* Impulses in nerve fibers travel at a speed of 293 ft/sec. The distance D traveled in t sec is given by $D = 293t$. How long would it take an impulse to travel from the brain to the toes of a person who is 6 ft tall?

57. *Brain weight.* The weight B of a human's brain is directly proportional to his or her body weight W.

a) It is known that a person who weighs 200 lb has a brain that weighs 5 lb. Find an equation of variation expressing B as a function of W.

b) Express the variation constant as a percent and interpret the resulting equation.

c) What is the weight of the brain of a person who weighs 120 lb?

58. *Muscle weight.* The weight M of the muscles in a human is directly proportional to his or her body weight W.

a) It is known that a person who weighs 200 lb has 80 lb of muscles. Find an equation of variation expressing M as a function of W.

b) Express the variation constant as a percent and interpret the resulting equation.

c) What is the muscle weight of a person who weighs 120 lb?

59. *Stopping distance on glare ice.* The stopping distance (at some fixed speed) of regular tires on glare ice is given by a linear function of the air temperature F,

$$D(F) = 2F + 115,$$

where $D(F) =$ the stopping distance, in feet, when the air temperature is F, in degrees Fahrenheit.

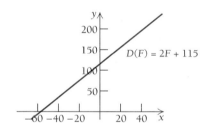

a) Find $D(0°)$, $D(-20°)$, $D(10°)$, and $D(32°)$.

b) Explain why the domain should be restricted to the interval $[-57.5°, 32°]$.

60. *Reaction time.* While driving a car, you see a child suddenly cross the street unattended. Your brain registers the emergency and sends a signal to your foot to hit the brake. The car travels a distance D, in feet, during this time, where D is a function of the speed r, in miles per hour, that the car is traveling when you see the child. That reaction distance is a linear function given by

$$D(r) = \frac{11r + 5}{10}.$$

a) Find $D(5)$, $D(10)$, $D(20)$, $D(50)$, and $D(65)$.

b) Graph $D(r)$.

c) What is the domain of the function? Explain.

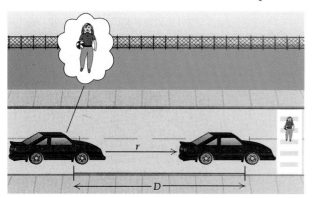

61. **Estimating heights.** An anthropologist can use certain linear functions to estimate the height of a male or female, given the length of certain bones. The *humerus* is the bone from the elbow to the shoulder. Let x = the length of the humerus, in centimeters. Then the height, in centimeters, of a male with a humerus of length x is given by

$$M(x) = 2.89x + 70.64.$$

The height, in centimeters, of a female with a humerus of length x is given by

$$F(x) = 2.75x + 71.48.$$

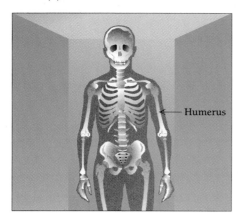

A 26-cm humerus was uncovered in a ruins.

a) If we assume it was from a male, how tall was he?
b) If we assume it was from a female, how tall was she?

◆ **Social Sciences**

62. **Urban population.** The population of a town is P. After a growth of 2%, its new population is N.

a) Assuming that N is directly proportional to P, find an equation of variation.
b) Find N when $P = 200,000$.
c) Find P when $N = 367,200$.

63. **Median age of women at first marriage.** In general, our society is marrying at a later age. The median age of women at first marriage can be approximated by the linear function

$$A(t) = 0.08t + 19.7,$$

where $A(t)$ = the median age of women at first marriage the tth year after 1950. Thus, $A(0)$ is the median age of women at first marriage in the year 1950, $A(50)$ is the median age in 2000, and so on.

a) Find $A(0)$, $A(1)$, $A(10)$, $A(30)$, and $A(50)$.
b) What will be the median age of women at first marriage in 1998?
c) Graph $A(t)$.

SYNTHESIS

64. ◈ Explain and compare the situations in which you would use the slope–intercept equation rather than the point–slope equation.

65. ◈ Discuss and relate the concepts of fixed cost, total cost, total revenue, and total profit.

 TECHNOLOGY CONNECTION

66. *Bread consumption.* The number N of one-pound loaves of bread consumed per person t years after 1995 is given by the function

$$N(t) = 0.6514t + 53.1599.$$

a) Find the consumption of bread per person in 1998 and in 2000.
b) Use your grapher to sketch a graph of the function.

67. Graph some of the total-revenue, total-cost, and total-profit functions in this exercise set using the same set of axes. Identify regions of profit and loss.

1.5 Other Types of Functions

OBJECTIVES

• Graph functions and solve application problems.
• Manipulate radical expressions and rational exponents.
• Determine the domain of a rational function and graph certain rational functions.
• Find the equilibrium point given a supply function and a demand function.

Quadratic Functions

> **DEFINITION**
>
> A *quadratic function* f is given by
> $$f(x) = ax^2 + bx + c, \quad \text{where } a \neq 0.$$

We have already considered some quadratic functions—for example, $f(x) = x^2$ and $g(x) = x^2 - 1$. We graph quadratic functions using the following information.

The graph of a quadratic function $f(x) = ax^2 + bx + c$ is called a *parabola*.

a) It is always a cup-shaped curve, like those in Examples 1 and 2.

b) It has a turning point, or *vertex*, at a point whose first coordinate is given by

$$x = -\frac{b}{2a}.$$

c) It has the vertical line $x = -b/2a$ as a line of symmetry (not part of the graph).

d) It opens up if $a > 0$ or opens down if $a < 0$.

EXAMPLE 1   Graph: $y = x^2 - 2x - 3$.

Solution   Let us first find the vertex, or turning point. The x-coordinate of the vertex is

$$x = -\frac{b}{2a}$$

$$= -\frac{-2}{2(1)} = 1.$$

Substituting 1 for x in the equation, we find the second coordinate of the vertex:

$$y = x^2 - 2x - 3$$
$$= 1^2 - 2(1) - 3$$
$$= 1 - 2 - 3$$
$$= -4.$$

The vertex is $(1, -4)$. The vertical line $x = 1$ is the line of symmetry of the graph. We choose some x-values on each side of the vertex, compute y-values, plot the points, and graph the parabola.

$$y = x^2 - 2x - 3$$

x	y	
1	-4	← Vertex
0	-3	
2	-3	
3	0	
4	5	
-1	0	
-2	5	

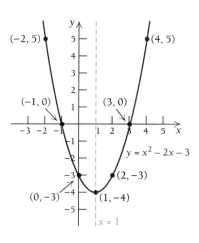

EXAMPLE 2   Graph: $y = -2x^2 + 10x - 7$.

Solution   Let us first find the vertex, or turning point. The x-coordinate of the vertex is

$$x = -\frac{b}{2a}$$

$$= -\frac{10}{2(-2)} = \frac{5}{2}.$$

Substituting $\frac{5}{2}$ for x in the equation, we find the second coordinate of the vertex:

$$y = -2x^2 + 10x - 7$$
$$= -2\left(\tfrac{5}{2}\right)^2 + 10\left(\tfrac{5}{2}\right) - 7$$
$$= -2\left(\tfrac{25}{4}\right) + 25 - 7 = \tfrac{11}{2}.$$

The vertex is $\left(\frac{5}{2}, \frac{11}{2}\right)$, and the line of symmetry is $x = \frac{5}{2}$. We choose some x-values on each side of the vertex, compute y-values, plot the points, and graph the parabola.

Use the procedure of Examples 1 and 2 to graph each of the following functions:

$y = x^2 - 6x + 4,$

$f(x) = -2x^2 + 4x + 1.$

Then use your grapher to check your work. Create an input–output table for each function, if your grapher has the capability.

$y = -2x^2 + 10x - 7$

x	y
$\frac{5}{2}$	$\frac{11}{2}$	← Vertex
0	-7
1	1
2	5
3	5
4	1
5	-7

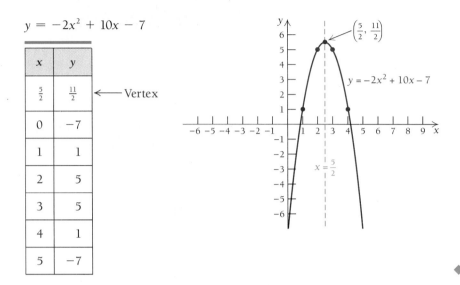

First coordinates of points at which a quadratic function intersects the x-axis (x-intercepts), if they exist, can be found by solving the quadratic equation $ax^2 + bx + c = 0$. If real-number solutions exist, they can be found using the *quadratic formula*.

THEOREM 9

The Quadratic Formula

The solutions of any quadratic equation $ax^2 + bx + c = 0$, $a \neq 0$, are given by

$$x = \frac{-b \pm \sqrt{b^2 - 4ac}}{2a}.$$

When solving a quadratic equation, first try to factor and then use the Principle of Zero Products, as we did in Section 1.2. When factoring is not possible or seems difficult, try the quadratic formula. It will always give the solutions. When $b^2 - 4ac < 0$, there are no real-number solutions, but there are solutions in an expanded number system called the *complex numbers*. In this text, we will be considering only real-number solutions.

EXAMPLE 3 Solve: $3x^2 - 4x = 2$.

Solution We first find standard form $ax^2 + bx + c = 0$, and then determine a, b, and c:

$$3x^2 - 4x - 2 = 0,$$

$$a = 3, \quad b = -4, \quad c = -2.$$

We then use the quadratic formula:

$$x = \frac{-b \pm \sqrt{b^2 - 4ac}}{2a}$$

$$= \frac{-(-4) \pm \sqrt{(-4)^2 - 4(3)(-2)}}{2 \cdot 3}$$

$$= \frac{4 \pm \sqrt{16 + 24}}{6} = \frac{4 \pm \sqrt{40}}{6}$$

$$= \frac{4 \pm \sqrt{4 \cdot 10}}{6} = \frac{4 \pm 2\sqrt{10}}{6}$$

$$= \frac{2(2 \pm \sqrt{10})}{2 \cdot 3} = \frac{2 \pm \sqrt{10}}{3}.$$

The solutions are $(2 + \sqrt{10})/3$ and $(2 - \sqrt{10})/3$.   ◆

Polynomial Functions

Linear and quadratic functions are part of a general class of *polynomial functions*.

> **DEFINITION**
>
> A *polynomial function* f is given by
>
> $$f(x) = a_n x^n + a_{n-1} x^{n-1} + \cdots + a_2 x^2 + a_1 x^1 + a_0,$$
>
> where n is a nonnegative integer and $a_n, a_{n-1}, \ldots, a_1, a_0$ are real numbers, called the *coefficients* of the polynomial.

The following are examples of polynomial functions:

$$f(x) = -5, \qquad \text{(A constant function)}$$
$$f(x) = 4x + 3, \qquad \text{(A linear function)}$$
$$f(x) = -x^2 + 2x + 3, \qquad \text{(A quadratic function)}$$
$$f(x) = 2x^3 - 4x^2 + x + 1. \qquad \text{(A cubic function)}$$

In general, graphing polynomial functions other than linear and quadratic functions is difficult unless we use a grapher. We use calculus to sketch such graphs in Chapter 3. Some **power functions,** such as

$$y = ax^n,$$

are relatively easy to graph.

EXAMPLE 4   Using the same set of axes, graph $y = x^2$ and $y = x^3$.

Solution   We set up a table of values, plot the points, and then draw the graphs.

x	x^2	x^3
-2	4	-8
-1	1	-1
$-\frac{1}{2}$	$\frac{1}{4}$	$-\frac{1}{8}$
0	0	0
$\frac{1}{2}$	$\frac{1}{4}$	$\frac{1}{8}$
1	1	1
2	4	8

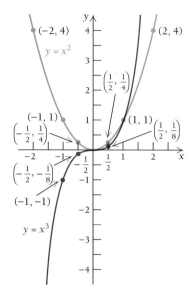

TECHNOLOGY CONNECTION

The Trace and Zoom Features

Trace. Once a graph has been created, we can investigate some of its points by using a trace feature that most graphers offer. (Some have the feature but it is called by a different name. Check your manual.) Consider the graph of $y = x^3 - 3x + 1$ on the viewing window $[-10, 10, -10, 10]$ with xScl = 2 and yScl = 2.

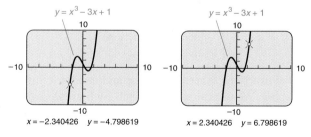

When the TRACE key is pressed, a cursor (often blinking) appears somewhere on the graph, while the x- and y- coordinates are shown elsewhere on the

screen (in this case, below the graph.) These coordinates will change as the cursor moves along the graph.

Graph the equation $y = x^3 - 3x + 1$ on your grapher as described above. Find some other points on this graph using the trace feature. Then graph $f(x) = -x^3 + 6x^2 + 5$ on the viewing window $[-3, 10, -10, 50]$ with xScl = 1 and yScl = 10. Find some points on this graph using the trace feature.

Zoom. Suppose we want to solve the equation $x^3 - 3x + 1 = 0$. We can find an estimate for the solution using a grapher. Indeed, some graphers have programs that find such an estimate with the use of a few keystrokes. For others, we can use a graphical approach. A solution of such an equation occurs at a point where the graph crosses the x-axis. This point is called an x-intercept; it has the general form $(a, 0)$ and occurs at an x-value a for which $y = 0$. If a

(continued)

TECHNOLOGY CONNECTION *(continued)*

function f is being considered, then the number a is called a **zero** of the function. A zero is an input a for which $f(a) = 0$. We can use a grapher to approximate values of the intercepts. This procedure involves the use of a ZOOM feature, which can actually be used to find approximate coordinates of any point on a curve. A zoom feature allows a graph to be magnified; that is, the viewing window shows a smaller section of the graph.

Let us examine the graph of $y = x^3 - 3x + 1$ considered above. It appears to have a zero near -2. Now let us zoom in on the graph near that point.

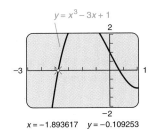

$x = -1.893617 \quad y = -0.109253$

We change the viewing window to $[-3, 1, -2, 2]$. Moving the tracing cursor along the graph, we see that a zero seems to occur somewhere between -2

and -1.5. We can magnify the area around the point where the zero seems to occur by using the zoom feature. Then we can use the trace feature along the magnified graph to get a better estimate of the x-intercept. By repeated use of the zoom and trace features, we can get the y-value closer and closer to 0. We can obtain a viewing window like the following.

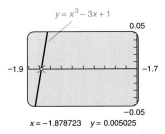

$x = -1.878723 \quad y = 0.005025$

We now find that an estimate for the zero is about -1.878.

Use your grapher to estimate the other x-intercepts, or zeros, of $y = x^3 - 3x + 1$. Then estimate all the x-intercepts, or zeros, of $f(x) = -x^3 + 6x^2 + 5$. Also, consider Example 3. Estimate the x-intercepts, or zeros, of $y = 3x^2 - 4x - 2$. How do your answers relate to the results of Example 3?

Rational Functions

> **DEFINITION**
>
> Functions given by the quotient, or ratio, of two polynomials are called *rational functions*.

The following are examples of rational functions:

$$f(x) = \frac{x^2 - 9}{x - 3},$$

$$g(x) = \frac{x^2 - 16}{x + 4},$$

$$h(x) = \frac{x - 3}{x^2 - x - 2}.$$

The domain of a rational function is restricted to those input values that do not re-sult in division by zero. Thus for f above, the domain consists of all real numbers except 3. To determine the domain of h, we set the denominator equal to 0 and solve:

$$x^2 - x - 2 = 0$$
$$(x + 1)(x - 2) = 0$$
$$x = -1 \quad \text{or} \quad x = 2.$$

Therefore, -1 and 2 are not in the domain. The domain of h consists of all real numbers except -1 and 2.

The graphing of most rational functions is rather complicated and is best dealt with using the tools of calculus that we will develop in Chapters 2 and 3. At this point, we will consider graphs that are fairly easy and leave the more complicated graphs for Chapter 3.

EXAMPLE 5   Graph: $f(x) = \dfrac{x^2 - 9}{x - 3}$.

Solution   This particular function can be simplified before we graph it. We do so by factoring the numerator and removing a factor of 1 as follows:

$$f(x) = \frac{x^2 - 9}{x - 3}$$
$$= \frac{(x - 3)(x + 3)}{x - 3}$$
$$= \frac{x - 3}{x - 3} \cdot \frac{x + 3}{1}$$
$$= x + 3, \quad x \neq 3.$$

This simplification assumes that x is not 3. The number 3 is not in the domain because it would result in division by zero. Thus we can express the function as follows:

$$y = f(x) = x + 3, \quad x \neq 3.$$

To find function values, we substitute any value for x other than 3. We make calcula-tions as in the following table and draw the graph. The open circle at the point $(3, 6)$ indicates that it is not part of the graph.

x	1	0	2	-3	4	-1	-2
y	4	3	5	0	7	2	1

TECHNOLOGY CONNECTION

Use your grapher and different viewing windows to graph

$$f(x) = \frac{x^2 - 9}{x - 3}.$$

Describe what happens on your grapher. If available, try both dot and connected modes. Use the zoom and trace features to find the coordinates of the point not in the domain. Does the "hole" in the graph ever become evident? What happens on an input–output table?

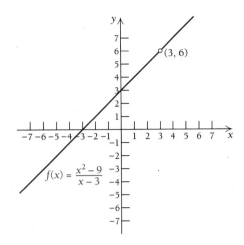

One important class of rational functions is given by $y = k/x$.

EXAMPLE 6   Graph: $y = 1/x$.

Solution   We make a table of values, plot the points, and then draw the graph.

x	y
-3	$-\frac{1}{3}$
-2	$-\frac{1}{2}$
-1	-1
$-\frac{1}{2}$	-2
$-\frac{1}{4}$	-4
$\frac{1}{4}$	4
$\frac{1}{2}$	2
1	1
2	$\frac{1}{2}$
3	$\frac{1}{3}$

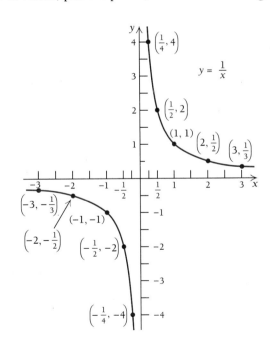

In Example 6, note that 0 is not in the domain of the function because it would yield a denominator of zero. The function is decreasing over the intervals $(-\infty, 0)$ and $(0, \infty)$. It is an example of **inverse variation.**

DEFINITION

y varies inversely as x if there is some positive number k such that $y = k/x$. We also say that y is *inversely proportional* to x.

EXAMPLE 7 *Business: Stocks and gold.* Certain economists theorize that stock prices are inversely proportional to the price of gold. That is, when the price of gold goes up, the prices of stock go down; and when the price of gold goes down, the prices of stock go up. Let us assume that the Dow Jones Industrial Average, D, an index of the overall price of stock, is inversely proportional to the price of gold, G, in dollars per ounce. One day the Dow Jones was 3639 and the price of gold was $364 per ounce. What will the Dow Jones be if the price of gold drops to $300?

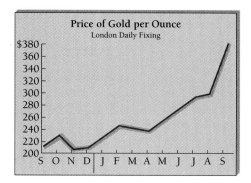

Solution

a) We know that $D = k/G$, so $3639 = k/364$ and $k = 1{,}324{,}596$. Thus,

$$D = \frac{1{,}324{,}596}{G}.$$

b) We substitute 300 for G and compute D:

$$D = \frac{1{,}324{,}596}{300} \approx 4415.$$

Warning! Do not put too much "stock" in the equation of this example. It is meant to give us an idea of economic relationships. An equation to predict the stock market accurately has not been found!   ◆

Absolute-Value Functions

The following is an example of an absolute-value function and its graph. The absolute value of a number is its distance from 0 on a number line. We denote the absolute value of a number x as $|x|$.

EXAMPLE 8   Graph: $f(x) = |x|$.

Solution   We make a table of values, plot the points, and then draw the graph.

x	-3	-2	-1	0	1	2	3
$f(x)$	3	2	1	0	1	2	3

We can think of this function as being defined piecewise by considering the definition of absolute value:

$$f(x) = |x| = \begin{cases} x, & \text{if } x \geq 0, \\ -x, & \text{if } x < 0. \end{cases}$$

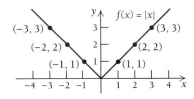

Square-Root Functions

The following is an example of a square-root function and its graph.

EXAMPLE 9   Graph: $f(x) = -\sqrt{x}$.

Solution   The domain of this function is just the nonnegative numbers—the interval $[0, \infty)$. You can find approximate values of square roots on your calculator.
    We set up a table of values, plot the points, and then draw the graph.

x	0	1	2	3	4	5	10
$f(x)$, or $-\sqrt{x}$	0	-1	-1.4	-1.7	-2	-2.2	-3.2

TECHNOLOGY CONNECTION

Use your grapher to graph both $g(x) = \sqrt{x}$ and $f(x) = \sqrt{2x + 3}$. You might have to enter $\sqrt{x}$ as $x \wedge (1/2)$. Consult your manual. Then use the procedure described in Example 9 to find the domain of each function. Check the result on your grapher. How might you use a grapher to find the domain of a function?

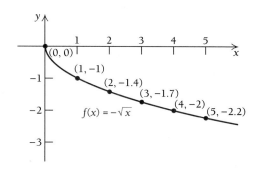

Power Functions with Rational Exponents

We are motivated to define rational exponents so that the laws we discussed in Section 1.1 still hold. For example, if the laws of exponents are to hold, we would have

$$a^{1/2} \cdot a^{1/2} = a^{1/2+1/2} = a^1 = a.$$

Thus we are led to define $a^{1/2}$ as $\sqrt{a}$. Similarly, we are led to define $a^{1/3}$ as the cube root of a, $\sqrt[3]{a}$. In general,

$$a^{1/n} = \sqrt[n]{a}, \quad \text{provided } \sqrt[n]{a} \text{ is defined.}$$

Again, if the laws of exponents are to hold, we would have

$$\sqrt[n]{a^m} = (a^m)^{1/n} = (a^{1/n})^m = a^{m/n}.$$

An expression $a^{-m/n}$ is defined by

$$a^{-m/n} = \frac{1}{a^{m/n}} = \frac{1}{\sqrt[n]{a^m}}.$$

EXAMPLE 10   Convert to rational exponents.

a) $\sqrt[3]{x^2} = x^{2/3}$

b) $\sqrt[4]{y} = y^{1/4}$

c) $\dfrac{1}{\sqrt[3]{b^5}} = \dfrac{1}{b^{5/3}} = b^{-5/3}$

d) $\dfrac{1}{\sqrt{x}} = \dfrac{1}{x^{1/2}} = x^{-1/2}$

e) $\sqrt{x^8} = x^{8/2}$, or x^4

EXAMPLE 11   Convert to radical notation.

a) $x^{1/3} = \sqrt[3]{x}$

b) $t^{6/7} = \sqrt[7]{t^6}$

c) $x^{-2/3} = \dfrac{1}{x^{2/3}} = \dfrac{1}{\sqrt[3]{x^2}}$

d) $e^{-1/4} = \dfrac{1}{e^{1/4}} = \dfrac{1}{\sqrt[4]{e}}$

EXAMPLE 12   Simplify.

a) $8^{5/3} = (8^{1/3})^5 = (\sqrt[3]{8})^5 = 2^5 = 32$

b) $81^{3/4} = (81^{1/4})^3 = (\sqrt[4]{81})^3 = 3^3 = 27$ ◆

TECHNOLOGY CONNECTION

Convert $x^{2/3}$ to radical notation. Then graph $f(x) = x^{2/3}$. You might have to enter $x^{2/3}$ as $(x \wedge 2) \wedge (1/3)$ or $(\text{abs}(x)) \wedge (2/3)$. Consult your manual. Some graphers will not graph a function like this for negative inputs, even though such numbers are in the domain of the function. What happens with your grapher?

Earlier when we graphed $f(x) = \sqrt{x}$, we were also graphing $f(x) = x^{1/2}$, or $f(x) = x^{0.5}$. The power functions

$$f(x) = ax^k, \quad k \text{ fractional},$$

do arise in application. For example, the *home range* of an animal is defined as the region to which the animal confines its movements. It has been hypothesized in statistical studies* that the area H of that region can be approximated using the body weight W of an animal by the function

$$H = W^{1.41}.$$

We can approximate function values using a power key $\boxed{y^x}$ on a calculator.

W	0	10	20	30	40	50
H	0	26	68	121	182	249

We see that

$$H = W^{1.41} = W^{141/100} = \sqrt[100]{W^{141}}.$$

The graph is shown below. Note that this is an increasing function. As body weight increases, the area over which the animal moves increases.

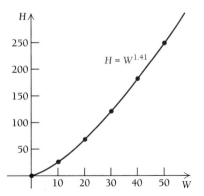

Caribou in their territorial area.

---

*See J. M. Emlen, *Ecology: An Evolutionary Approach*, p. 200 (Reading, MA: Addison-Wesley, 1973).

Supply and Demand Functions

Supply and demand in economics are modeled by increasing and decreasing functions. Although specific scientific formulas for these concepts are not generally known, the notions of increasing and decreasing help us understand the ideas.

DEMAND FUNCTIONS

Look at the following table and graph. They show the relationship between the price p per bag of sugar and the quantity x of 5-lb bags that the consumer will buy at that price.

Demand schedule

Price (p) per 5-lb bag	Quantity (x) of 5-lb bags, in millions
$5	4
4	5
3	7
2	.10
1	15

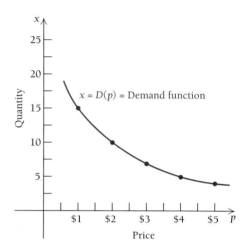

Note that as the price per bag increases, the quantity demanded by the consumer decreases; and as the price per bag decreases, the quantity demanded by the consumer increases. Thus it is natural to think of x as a function of p,

$$x = D(p),$$

and to graph the function with the price p on the horizontal axis and the quantity x on the vertical axis. Thus, for a **demand function** D, $D(p)$ is the quantity x of units demanded by the consumer when the price is p.

In some situations, it may be appropriate to consider the demand function as $D(x) = p$, where price is a function of the quantity demanded.

SUPPLY FUNCTIONS

Look at the following table and graph. They show the relationship between the price p per bag of sugar and the quantity x of 5-lb bags that the seller is willing to supply, or sell, at that price. Note that as the price per bag increases, the more the seller is willing to supply; and as the price per bag decreases, the less the seller is willing to supply. Thus it is natural to think of x as a function of p,

$$x = S(p),$$

Supply schedule

Price (p) per 5-lb bag	Quantity (x) of 5-lb bags, in millions
$1	0
2	10
3	16
4	20
5	22

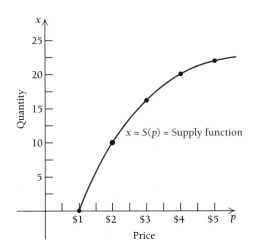

and to graph the function with the price p on the horizontal axis and the quantity x on the vertical axis. Thus, for a **supply function** S, $S(p)$ is the quantity x of items that the seller is willing to supply, or sell, at price p.

In some situations, it may be appropriate to consider the supply function as $S(x) = p$, where the price is a function of the the quantity supplied.

Let us now look at these curves together. As price increases, supply increases and demand decreases; and as price decreases, supply decreases and demand increases. The point of intersection of the two curves (p_E, x_E) is called the **equilibrium point.** The equilibrium price p_E (in this case, $2 per bag) is the point at which the amount x_E (in this case, 10 million bags) that the seller willingly supplies is the same as the amount that the consumer willingly demands. The situation is analogous to a buyer and seller haggling over the sale of an item. The equilibrium point, or selling price, is what they finally agree on.

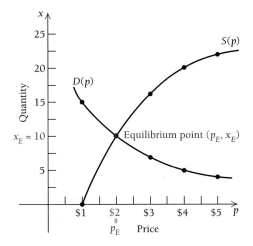

TECHNOLOGY CONNECTION

In Example 13, we have a *system of equations*. Some graphers and software packages have programs that solve systems of equations. A grapher can also be used for a graphical solution to a system, especially if it has a zoom feature that allows a small portion of the graph to be magnified. After graphing both equations, use the zoom feature to magnify the intersection (shrink the window's dimensions if your grapher lacks the zoom feature). By doing this repeatedly and using the trace feature, you can determine the coordinates of the point of intersection to the desired accuracy.

Use your grapher to find the equilibrium point for the demand and supply functions

$$D(p) = 1123.6 - 61.4p \quad \text{and}$$
$$S(p) = 201.8 + 4.6p.$$

EXAMPLE 13 *Economics: Equilibrium point.* Find the equilibrium point for the demand and supply functions

$$D(p) = 300 - 49p \quad \text{and} \quad S(p) = 100 + p.$$

Solution   To find the equilibrium point, we set $D(p) = S(p)$ and solve:

$$300 - 49p = 100 + p$$
$$-50p = -200$$
$$p = \frac{-200}{-50} = 4.$$

Thus, $p_E = \$4$. To find x_E, we substitute p_E into either $D(p)$ or $S(p)$. We choose $S(p)$:

$$x_E = S(p_E) = S(4) = 100 + 4 = 104.$$

Thus the equilibrium quantity is 104 units and the equilibrium point is ($\$4$, 104).   ◆

1.5 Exercise Set

Using the same set of axes, graph the pair of equations.

1. $y = \frac{1}{2}x^2$ and $y = -\frac{1}{2}x^2$

2. $y = \frac{1}{4}x^2$ and $y = -\frac{1}{4}x^2$

3. $y = x^2$ and $y = (x - 1)^2$

4. $y = x^2$ and $y = (x - 3)^2$

5. $y = x^2$ and $y = (x + 1)^2$

6. $y = x^2$ and $y = (x + 3)^2$

7. $y = |x|$ and $y = |x + 3|$

8. $y = |x|$ and $y = |x + 1|$

9. $y = x^3$ and $y = x^3 + 1$

10. $y = x^3$ and $y = x^3 - 1$

11. $y = \sqrt{x}$ and $y = \sqrt{x + 1}$

12. $y = \sqrt{x}$ and $y = \sqrt{x - 2}$

Graph.

13. $y = x^2 - 4x + 3$

14. $y = x^2 - 6x + 5$

15. $y = -x^2 + 2x - 1$

16. $y = -x^2 - x + 6$

17. $y = \dfrac{2}{x}$

18. $y = \dfrac{3}{x}$

19. $y = \dfrac{-2}{x}$

20. $y = \dfrac{-3}{x}$

21. $y = \dfrac{1}{x^2}$

22. $y = \dfrac{1}{x - 1}$

23. $y = \sqrt[3]{x}$

24. $y = \dfrac{1}{|x|}$

25. $f(x) = \dfrac{x^2 - 9}{x + 3}$

26. $g(x) = \dfrac{x^2 - 4}{x - 2}$

27. $f(x) = \dfrac{x^2 - 1}{x - 1}$

28. $g(x) = \dfrac{x^2 - 25}{x + 5}$

Solve.

29. $x^2 - 2x = 2$ **30.** $x^2 - 2x + 1 = 5$

31. $x^2 + 6x = 1$ **32.** $x^2 + 4x = 3$

33. $4x^2 = 4x + 1$ **34.** $-4x^2 = 4x - 1$

35. $3y^2 + 8y + 2 = 0$ **36.** $2p^2 - 5p = 1$

Convert to rational exponents.

37. $\sqrt{x^3}$ **38.** $\sqrt{x^5}$

39. $\sqrt[5]{a^3}$ **40.** $\sqrt[4]{b^2}$

41. $\sqrt[7]{t}$ **42.** $\sqrt[8]{c}$

43. $\dfrac{1}{\sqrt[3]{t^4}}$ **44.** $\dfrac{1}{\sqrt[5]{b^6}}$

45. $\dfrac{1}{\sqrt{t}}$ **46.** $\dfrac{1}{\sqrt{m}}$

47. $\dfrac{1}{\sqrt{x^2 + 7}}$ **48.** $\sqrt{x^3 + 4}$

Convert to radical notation.

49. $x^{1/5}$ **50.** $t^{1/7}$

51. $y^{2/3}$ **52.** $t^{2/5}$

53. $t^{-2/5}$ **54.** $y^{-2/3}$

55. $b^{-1/3}$ **56.** $b^{-1/5}$

57. $e^{-17/6}$ **58.** $m^{-19/6}$

59. $(x^2 - 3)^{-1/2}$ **60.** $(y^2 + 7)^{-1/4}$

Simplify.

61. $9^{3/2}$ **62.** $16^{5/2}$ **63.** $64^{2/3}$

64. $8^{2/3}$ **65.** $16^{3/4}$ **66.** $25^{5/2}$

Determine the domain of the function.

67. $f(x) = \dfrac{x^2 - 25}{x - 5}$ **68.** $f(x) = \dfrac{x^2 - 4}{x + 2}$

69. $f(x) = \dfrac{x^3}{x^2 - 5x + 6}$ **70.** $f(x) = \dfrac{x^4 + 7}{x^2 + 6x + 5}$

71. $f(x) = \sqrt{5x + 4}$ **72.** $f(x) = \sqrt{2x - 6}$

APPLICATIONS

◆ **Business and Economics**

Find the equilibrium point for the given demand and supply functions.

73. $D(p) = \dfrac{8 - p}{2}$, $S(p) = p - 2$

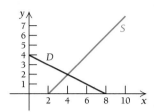

74. $D(p) = 800 - 43p$, $S(p) = 210 + 16p$

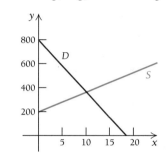

75. $D(p) = 1000 - 10p$, $S(p) = 250 + 5p$

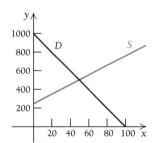

76. $D(p) = 8800 - 30p$, $S(p) = 7000 + 15p$

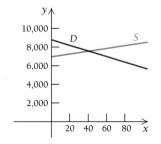

77. $D(p) = \dfrac{5}{p}$, $S(p) = \dfrac{p}{5}$

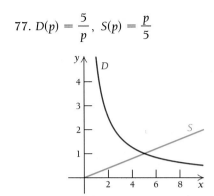

78. $D(p) = \dfrac{4}{p}$, $S(p) = \dfrac{p}{4}$

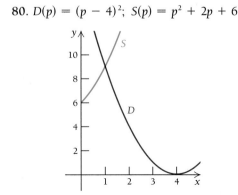

79. $D(p) = (p - 3)^2$, $S(p) = p^2 + 2p + 1$

80. $D(p) = (p - 4)^2$; $S(p) = p^2 + 2p + 6$

81. $D(p) = 5 - p$, $S(p) = \sqrt{p + 7}$

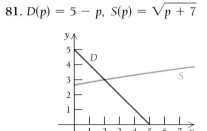

82. $D(p) = 7 - p$, $S(p) = 2\sqrt{p + 1}$

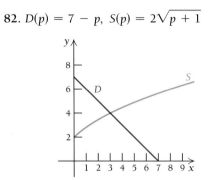

83. *Stock prices and prime rate.* It is theorized that the price per share of a stock is inversely proportional to the prime (interest) rate. Recently, the price per share S of Silicon Graphics, Inc., was $23\frac{7}{8}$ and the prime rate was $7\frac{1}{4}\%$. The prime rate R rose to $8\frac{1}{2}\%$. What would the price per share be if the assumption of inverse proportionality is correct?

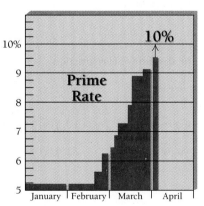

84. *Demand.* The quantity sold, x, of a certain kind of fax machine is inversely proportional to the price, p. It was found that 240,000 fax machines will be sold if the price is \$320 each. How many will be sold if the price is \$480 each?

◆   Life and Physical Sciences

85. *Fruit stacking.* The number of oranges in a pile of the type shown below is approximated by the function

$$f(x) = \frac{1}{6}x^3 + \frac{1}{2}x^2 + \frac{1}{3}x,$$

where $f(x)$ = the number of oranges and x = the number of layers. Each layer is an equilateral triangle. Find the number of oranges when the number of layers is 7, 10, and 12.

86. *Territorial area.* The *territorial area* of an animal is defined to be its defended region, or exclusive region. For example, a lion has a certain region over which it is ruler. It has been hypothesized in statistical studies (see the footnote on p. 59) that the area T of that region can be approximated using the body weight W of the animal by the power function

$$T = W^{1.31}.$$

Complete the table of approximate function values and graph the function.

W	0	10	20	30	40	50	100	150
T	0	20						

SYNTHESIS

87. *Life science: Pollution control.* Pollution control has become a very important concern in all countries. If controls are not put in force, it has been predicted that the function

$$P = 1000t^{5/4} + 14,000$$

will describe the average pollution, in particles of pollution per cubic centimeter, in most cities at time t, in years, where $t = 0$ corresponds to 1970 and $t = 28$ corresponds to 1998.

a) Predict the pollution in 1998; in 2000; in 2010.
b) Graph the function over the interval $[0, 40]$.

88. ◈   At most, how many y-intercepts can a function have? Explain.

89. ◈   Explain the difference between a rational function and a polynomial function. Is every polynomial function a rational function?

   TECHNOLOGY CONNECTION

Use a graphical procedure to approximate the zeros of the function to three decimal places. If your grapher also has an equation-solving feature, use it to check your approximations.

90. $f(x) = x^3 - x$
   (Also, use algebra to find the zeros of the function.)
91. $f(x) = x^3 - 2x^2 - 2$
92. $f(x) = 2x^3 - x^2 - 14x - 10$
93. $f(x) = \frac{1}{2}(|x - 4| + |x - 7|) - 4$
94. $f(x) = x^4 + 4x^3 - 36x^2 - 160x + 300$
95. $f(x) = \sqrt{7 - x^2}$
96. $f(x) = |x + 1| + |x - 2| - 5$
97. $f(x) = |x + 1| + |x - 2|$
98. $f(x) = |x + 1| + |x - 2| - 3$
99. $f(x) = x^8 + 8x^7 - 28x^6 - 56x^5 + 70x^4 + 56x^3 - 28x^2 - 8x + 1$
100. Find the equilibrium point for the demand and supply functions

$$D(p) = 83 - p \quad \text{and}$$
$$S(p) = 24\sqrt{p} + 1.9.$$

Use your grapher to solve the system of equations. Solve each equation for y if appropriate.

101. $y = -4.56x + 12.95,$
$\quad\quad y = 7.88x - 6.77$

102. $y = 10x - 30,$
$\quad\quad y = x^4 - x^3 - 24x^2 + 4x + 80$

103. $y = x^6 + 4x^5 - 54x^4 - 160x^3 + 641x^2$
$\quad\quad\quad + 828x - 1260,$
$\quad\quad y = -450x + 800$

1.6 Mathematical Modeling

OBJECTIVE

• Use curve fitting to find a model for a set of data; then use the model to make predictions.

What Is a Mathematical Model?

When the essential parts of a problem are described in mathematical language, we say that we have a **mathematical model.** For example, the arithmetic of the natural numbers constitutes a mathematical model for situations in which counting is the essential ingredient. Situations in which calculus can be brought to bear often require the use of equations and functions, and typically there is concern with the way a change in one variable affects a change in another.

Mathematical models are abstracted from real-world situations (see the figure below). Procedures within the mathematical model then give results that allow one to predict what will happen in that real-world situation. To the extent that these predictions are inaccurate or the results of experimentation do not conform to the model, the model is in need of modification.

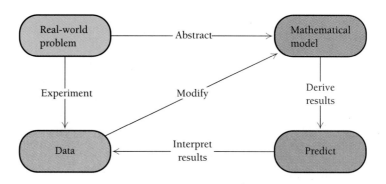

Mathematical modeling is an ongoing, possibly everchanging, process. This is often the case. For example, finding a mathematical model that will enable accurate prediction of population growth is not a simple problem. Surely any population model one might devise will need to be altered as further relevant information is acquired.

Although models can reveal worthwhile information, one must always be cautious using them. An interesting case in point is a study* showing that world records in *any* running race can be modeled by a linear function. In particular, for the mile run,

$$R = -0.00582x + 15.3476,$$

where R = the world record, in minutes, and x = the year. Roger Bannister shocked the world in 1954 by breaking the 4-minute mile. Had people been aware of this model, they would not have been shocked, for when we substitute 1954 for x, we get

$$R = -0.00582(1954) + 15.3476$$
$$= 3.97532$$
$$\approx 3{:}58.5.$$

The actual record was 3:59.4. Although this model will continue for 40 to 50 years to be worthwhile in predicting the world record in the mile run, we see that we cannot get meaningful answers to some questions. For example, we could use the model to find when the 1-minute mile will be broken. We set $R = 1$ and solve for x:

$$1 = -0.00582x + 15.3476$$
$$2465 = x.$$

Most track athletes would assure us that the 1-minute mile is beyond human capability. In fact, at the time of this writing, experienced runners think the record will never reach 3:40.0, the current world record being 3:44.39. Going to an even further extreme, we see that the model predicts that the 0-minute mile will be run in 2637. In conclusion, one must be careful in the use of any model. (You will develop this model in Exercise Set 7.5.)

Curve Fitting

We will develop many kinds of mathematical models in this text. In general, the idea is to find a function that fits observations (data), theoretical reasoning, and common sense as well as possible. We call this **curve fitting.** We now consider a somewhat oversimplified procedure to get the idea of curve fitting. In Section 7.5, we will see how calculus can be used to develop a curve-fitting procedure called *linear regression*.

---

*Ryder, H. W., H. J. Carr, and P. Herget, "Future Performance in Footracing," *Scientific American* **234** (June 1976): 109–119.

The four types of function shown below fit many situations.

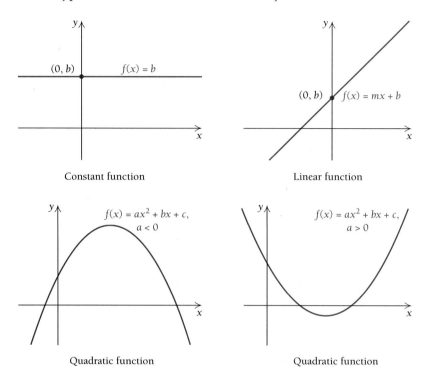

How can we decide what function might fit the data for a particular model? One simple way is to graph the data and look over the result. Simply stated, we "eyeball" the situation. The following graphs show the relationship between the death rate of men and the average number of hours per day that the men slept. Compare these graphs with those in the figure above. Which type of function seems to correspond to the graphs in this example? It can help to draw line segments between the data points.

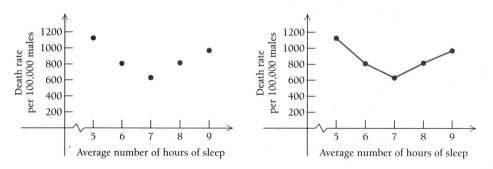

Note that the rate drops and then rises. This suggests that the data might fit a quadratic function.

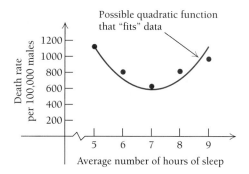

Now let us look at a graph that shows the life expectancy of women for various years.

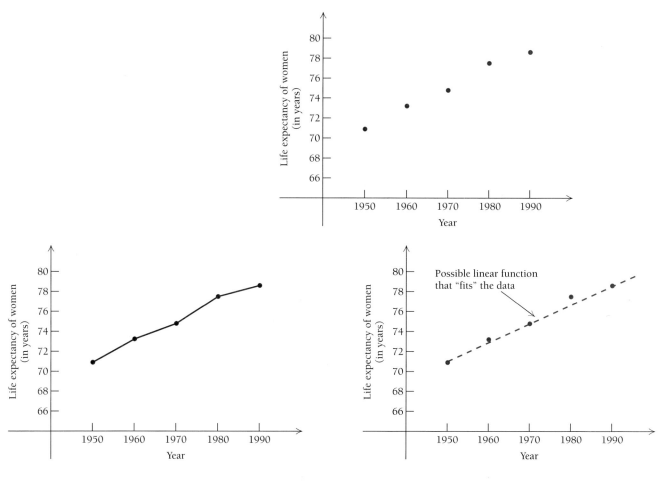

Note that the data points seem to be rising in a manner that might fit a linear function, although a "perfect" straight line cannot be drawn through the data points.

The following is a procedure that sometimes works for finding mathematical models.

Curve Fitting

Given a set of data:

1. Graph the data.
2. Look at the data and the graph to determine whether a known function seems to fit.
3. Find a function that fits the data by using data points to derive the constants.

EXAMPLE 1   *Life science: Life expectancy of women.* The following table lists data that show the life expectancy of women for various years. We graphed the data above and noted that it seemed to fit a linear function. Let us now find a linear function, or model, and make some predictions.

a) Find a linear function that fits the data.

b) Use the model to predict the life expectancy of women in the year 2000.

c) Data are missing from the table regarding the year 1985. What was the life expectancy of women in 1985?

In the table, we let $t = $ the number of years since 1950. This means that for 1950, $t = 0$, and for 1990, $t = 40$.

Year (t)	Life expectancy of women, in years
1950 ($t = 0$)	70.9
1960 ($t = 10$)	73.2
1970 ($t = 20$)	74.8
1980 ($t = 30$)	77.5
1990 ($t = 40$)	78.6

Solution

a) We consider the linear model

$$E = mt + b, \tag{1}$$

where $E = $ the life expectancy of women t years after 1950. To derive the constants (or parameters) m and b, we choose two data points. Although this choice is subjective, you should not pick a point that deviates greatly from the general pattern. We choose the points (10, 73.2) and (30, 77.5). Since the points are to be solutions of Eq. (1), it follows that

$$73.2 = m \cdot 10 + b, \quad \text{or} \quad 73.2 = 10m + b, \tag{2}$$

$$77.5 = m \cdot 30 + b, \quad \text{or} \quad 77.5 = 30m + b. \tag{3}$$

This is a system of equations. We can now subtract each side of Eq. (2) from Eq. (3) to eliminate b:

$$4.3 = 20m.$$

Then we have

$$m = \frac{4.3}{20} = 0.215.$$

Substituting 0.215 for m in Eq. (3), we can solve for b:

$$77.5 = 30(0.215) + b$$
$$77.5 = 6.45 + b$$
$$71.05 = b.$$

Substituting these values of m and b into Eq. (1), we get the function (model) given by

$$E = 0.215t + 71.05. \tag{4}$$

b) The life expectancy of women in 2000 is found by letting $t = 50$ in Eq. (4):

$$E = 0.215(50) + 71.05$$
$$= 81.8.$$

c) The life expectancy of women in 1985 is found by letting $t = 35$ in Eq. (4):

$$E = 0.215(35) + 71.05$$
$$= 78.575. \qquad \blacklozenge$$

EXAMPLE 2 *Life science: Hours of sleep and death rate.* The following table relates death rate y, in deaths per 100,000 males, to the average number x of hours of sleep. We noted earlier that the data seem to fit a quadratic function. In the procedure that follows, we need only three data points to derive that quadratic model. We choose (5, 1121), (7, 626), and (9, 967). For the given set of data:

a) Find a quadratic model.

b) Use the model to find the death rate for males who sleep 2 hr; 8 hr; 10 hr.

Average number of hours of sleep (x)	Death rate per 100,000 males (y)
5	1121
6	805
7	626
8	813
9	967

Solution

a) We consider the quadratic model

$$y = ax^2 + bx + c. \tag{5}$$

To derive the constants (or parameters) a, b, and c, we use the three data points (5, 1121), (7, 626), and (9, 967). Since these points are to be solutions to Eq. (5), it follows that

$$1121 = a \cdot 5^2 + b \cdot 5 + c, \quad \text{or} \quad 1121 = 25a + 5b + c,$$
$$626 = a \cdot 7^2 + b \cdot 7 + c, \quad \text{or} \quad 626 = 49a + 7b + c,$$
$$967 = a \cdot 9^2 + b \cdot 9 + c, \quad \text{or} \quad 967 = 81a + 9b + c.$$

We solve this system of three equations in three variables using procedures of algebra and get

$$a = 104.5, \quad b = -1501.5, \quad \text{and} \quad c = 6016.$$

Substituting these values of a, b, and c into Eq. (5), we get the function (model) given by

$$y = 104.5x^2 - 1501.5x + 6016.$$

b) The death rate for 2 hr is given by

$$y = 104.5(2)^2 - 1501.5(2) + 6016 = 3431.$$

The death rate for 8 hr is given by

$$y = 104.5(8)^2 - 1501.5(8) + 6016 = 692.$$

The death rate for 10 hr is given by

$$y = 104.5(10)^2 - 1501.5(10) + 6016 = 1451. \qquad \blacklozenge$$

 TECHNOLOGY CONNECTION

In Examples 1 and 2, we are not able to take into account *all* the data in finding the models. One procedure that does use all the data is called **linear regression.** Some graphers can do linear regression. We develop that procedure carefully in Section 7.5. For now, you can use it to fit data to certain functions. With an appropriate grapher, we may also have the advantage of curve-fitting polynomials of third degree (cubic) and fourth degree (quartic) to data. Use a grapher that performs linear regression. Consult your manual.

Consider all the data to find a linear function that fits the data and complete the other parts of Example 1. Compare your results.

Consider all the data to find a quadratic function that fits the data and complete the other parts of Example 2. Compare your results.

Consider all the data to find a cubic polynomial function $y = ax^3 + bx^2 + cx + d$ that fits the data and complete the other parts of Example 2. Compare your results.

1.6 Exercise Set

Use the model for the world record in the mile run (see page 67).

1. What will the world record be in 1998?
2. When will the world record be 3:42.0?
   (*Hint:* 3:42.0 = 3.70.)

For each of Exercises 3–8, parts (a) and (b) are:

a) Graph the data given.
b) Look at the data and determine whether one of the four models discussed in this section seems to fit.

APPLICATIONS

◆   **Business and Economics**

3. *Fast-food spending.* People are eating fast food more and more often. The following data relate the amount spent *S*, in billions of dollars, to recent years, *t*.

Year (t)	Fast-food spending (S), in billions
1990 ($t = 0$)	84
1991 ($t = 1$)	85
1992 ($t = 2$)	90
1993 ($t = 3$)	95

c) Use the data points (1, 85) and (3, 95) to find a linear function that fits the data.
d) Use the linear function to predict fast-food spending in 1998; in 2004.

4. *Sales of country music.* In recent years, sales of

Year (t)	Percentage (P) of country music sales
1989 ($t = 0$)	6.8%
1990 ($t = 1$)	7.9%
1991 ($t = 2$)	11.7%
1992 ($t = 3$)	15.4%
1993 ($t = 4$)	17.5%

The data in Exercises 3 and 4 are factual.

country music have increased according to the data in the preceding table. It relates the percentage *P* of all music sales to time *t*, in years, where $t = 0$ corresponds to 1989.

c) Use the data points (2, 11.7) and (4, 17.5) to find a linear function that fits the data.
d) Use the linear function to predict the percentage of country music sales in 1999; in 2005.

5. *Total sales.*

Year (t)	Total sales (S), in dollars
1	$100,310
2	100,290
3	100,305
4	100,280

c) Use the data points (1, $100,310) and (2, $100,290) to find a linear function that fits the data.
d) This data set approximates a constant function. What procedure, apart from that of part (c), could you use to find the constant?

6. *Total sales.*

Year (t)	Sales (S), in dollars
1	$10,000
2	21,000
3	27,000
4	37,000

c) Use the data points (1, $10,000) and (4, $37,000) to find a linear function that fits the data.
d) Use the model to predict the sales of the company in the fifth year.

◆   General Interest

The problems in Exercises 7 and 8 are based on factual data.

7. **VCR counter readings.** The instruction booklet for a video cassette recorder (VCR) includes a table relating the counter readings and the time, in hours, that the tape has run.

Time (t) that the tape has run	Counter readings (N)
0	000
1	300
2	500
3	675
4	800

c) Use the data points (0, 0), (2, 500), and (4, 800) to find a quadratic equation that fits the data.
d) The counter readings do not give times for programs on the half hour. Use the function to find the counter reading after the tape has run for $\frac{1}{2}$ hr; $1\frac{1}{2}$ hr; $2\frac{1}{2}$ hr.

8. **Accidents at certain speeds.**

Travel speed (x), in miles per hour	Number (N) of vehicles involved in an accident at nighttime (for every 100 million miles of travel)
20	10,000
30	2,000
40	400
50	250
60	250
70	350
80	1,500

c) Use the data points (20, 10,000), (50, 250), and (80, 1500) to find a quadratic function that fits the data.
d) Use the model to find the number of vehicles involved in an accident at 30 mph.

TECHNOLOGY CONNECTION

Use a grapher that performs linear regression.

9. **Fast-food spending.** Use the data from Exercise 3.
   a) Find a cubic polynomial function $y = ax^3 + bx^2 + cx + d$ that fits the data.
   b) Use the cubic polynomial function to predict fast-food spending in 2004. Compare your answer to that of Exercise 3.

10. **Prices of personal computers.** The price P of a personal computer has varied greatly in recent years. The data in this table relate price P to time t, where $t = 0$ corresponds to 1981.

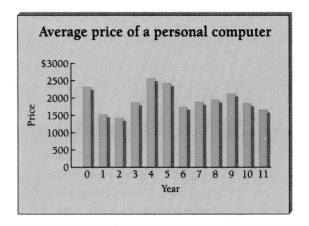

Average price of a personal computer

Year (t)	Average price (P) of a personal computer
1981 ($t = 0$)	$2290
1982 ($t = 1$)	1500
1983 ($t = 2$)	1400
1984 ($t = 3$)	1850
1985 ($t = 4$)	2540
1986 ($t = 5$)	2400
1987 ($t = 6$)	1720
1988 ($t = 7$)	1860
1989 ($t = 8$)	1930
1990 ($t = 9$)	2100
1991 ($t = 10$)	1820
1992 ($t = 11$)	1640

a) Find a cubic polynomial function
   $y = ax^3 + bx^2 + cx + d$ that fits the data.
b) Graph the cubic polynomial function.
c) Use the cubic polynomial function to predict
   the price of a personal computer in 1996.
d) Find a quartic polynomial function
   $y = ax^4 + bx^3 + cx^2 + dx + e$ that fits the
   data.

e) Graph the quartic polynomial function.
f) Use the quartic polynomial function to predict
   the price of a personal computer in 1996.
g) ◆ Discuss the merits of which function to use
   to predict the price of a personal computer in
   1996.

1 Chapter Summary and Review

TERMS TO KNOW

Base, p. 2
Exponent, p. 2
Factor, p. 2
Distributive law, p. 5
Factoring, p. 6
Compound interest, p. 7
Addition Principle, p. 11
Multiplication Principle, p. 11
Principle of Zero Products, p. 12
Inequality, p. 13
Open interval, p. 14
Closed interval, p. 14
Half-open interval, p. 15
Endpoint, p. 14
Interior point, p. 16
Ordered pair, p. 18
First coordinate, p. 18
Second coordinate, p. 18
Solution, p. 19
Graph, pp. 19, 26

Grapher, p. 21
Viewing window, p. 21
Relation, p. 22
Function, pp. 22, 25
Input, p. 22
Output, p. 22
Domain, p. 22
Range, p. 22
Vertical-line test, p. 28
Functions defined piecewise, p. 29
Independent variable, p. 30
Dependent variable, p. 30
Slope, pp. 36, 38, 41
Direct variation, p. 37
Variation constant, p. 37
Linear function, p. 38
Constant function, p. 38
Slope–intercept equation, p. 38
Point–slope equation, p. 40
Fixed costs, p. 42

Variable costs, p. 42
Total cost, p. 42
Break-even value, p. 43
Quadratic function, p. 48
Parabola, p. 48
Vertex, p. 48
Quadratic formula, p. 50
Polynomial function, p. 51
Power function, p. 51
Trace, p. 52
Zoom, p. 52
Rational function, p. 53
Inverse variation, p. 56
Absolute-value function, p. 57
Square-root function, p. 57
Demand function, p. 60
Supply function, p. 61
Equilibrium point, p. 61
Mathematical model, p. 66
Curve fitting, p. 67

REVIEW EXERCISES

These review exercises are for test preparation. They can
also be used as a lengthened practice test. Answers are at
the back of the book. The answers also contain bracketed
section references, which tell you where to restudy if your
answer is incorrect.

1. Rename without an exponent: $(2.01)^2$.

2. Divide: $\dfrac{y^{-2}}{y^{-5}}$.

3. Multiply: $y^{-2} \cdot y^{-5}$.

4. Simplify: $(3t^{-2}m^4)^5$.

Multiply.

5. $(3x - t)^2$

6. $(3x - t)(3x + t)$

7. $(3x - t)(4x + 5t)$

Factor.

8. $x^2 - 2x - 48$

9. $25x^2 - 16t^2$

10. $a^2 - 8ab + 16b^2$

11. $21x^2 - 19x + 4$

12. *Business: Compound interest.* Suppose $1100 is invested at 10%, compounded semiannually. How much is in the account at the end of 1 year?

13. *Business: Compound interest.* Suppose $4000 is invested at 12%, compounded annually. How much is in the account at the end of 2 years?

Solve.

14. $3 - 6x \le 5x - 8$

15. $3 - 6x = 5x - 8$

16. $16x^2 - 25 = 0$

17. $x(x - 3)(2x + 5) = 0$

18. Write interval notation for the set of all numbers x such that $-6 < x \le 1$.

19. Write interval notation for the set of all positive numbers.

A function f is given by $f(x) = 2x^2 - x + 3$. Find each of the following.

20. $f(-2)$

21. $f(1 + h)$

22. $f(0)$

A function f is given by $f(x) = (1 - x)^2$. Find each of the following.

23. $f(-5)$

24. $f(2 - h)$

25. $f(4)$

Graph.

26. $y = |x + 1|$

27. $f(x) = (x - 2)^2$

28. $f(x) = \dfrac{x^2 - 16}{x + 4}$

Determine whether each of the following is a graph of a function.

29.

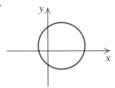

30.

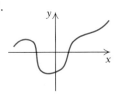

31.

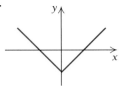

Graph.

32. $x = -2$

33. $y = 4 - 2x$

34. Write an equation of the line containing the points $(4, -2)$ and $(-7, 5)$.

35. Write an equation of the line with slope 8, containing the point $\left(\frac{1}{2}, 11\right)$.

36. For the linear equation $y = 3 - \frac{1}{6}x$, find the slope and the y-intercept.

37. *Life science: Muscle weight and body weight.* The weight M of muscles in a human is directly proportional to its body weight W. It is known that a person who weighs 150 lb has 60 lb of muscles. What is the muscle weight of a person who weighs 180 lb?

38. *Business: Profit-and-loss analysis.* A furniture manufacturer has fixed costs of $80,000 for producing a certain type of classroom chair for universities. The variable cost for producing x chairs is $16x$ dollars. The chairs will be sold at a price of $28 each.

a) Formulate a function $R(x)$ for the total revenue from the sale of x chairs.

b) Formulate a function $C(x)$ for the total cost of producing x chairs.

c) Formulate a function $P(x)$ for the total profit from the production and sale of x chairs.

d) What profit or loss will the company realize from expected sales of 8000 chairs?

e) How many chairs must the company sell in order to break even?

39. Convert to radical notation: $y^{1/6}$.

40. Convert to rational exponents: $\sqrt[20]{x^3}$.

41. Simplify: $25^{3/2}$.

Find the domain of the function.

42. $f(x) = \sqrt{18 - 6x}$

43. $g(x) = \dfrac{1}{x^2 - 1}$

44. **Economics: Equilibrium point.** Find the equilibrium point if $D(p) = p^2 - 11p + 51, 0 \le p \le 5$, and $S(p) = p^2 + 18$.

45. Derive a linear function that fits the data points $(6, -1)$ and $(0, 10)$.

46. Derive a quadratic function that fits the data points $(1, 4), (3, 8),$ and $(-1, 8)$.

47. Graph:

$$f(x) = \begin{cases} x - 5, & \text{for } x > 2, \\ x + 3, & \text{for } x \le 2. \end{cases}$$

SYNTHESIS

48. Simplify: $(64^{5/3})^{-1/2}$.

TECHNOLOGY CONNECTION

Use your grapher to graph each of the following.

49. $f(x) = x^3 - 4x$

50. $f(x) = \sqrt[3]{|9 - x^2|} - 1$

Find the zeros of the function.

51. $f(x) = x^3 - 4x$

52. $f(x) = \sqrt[3]{|9 - x^2|} - 1$

53. Solve this system of equations:

$$2.34x + 5.71y = -12.45,$$
$$-9.08x + 14.45y = 9.45.$$

1 Chapter Test

1. Rename without a negative exponent: e^{-k}.

2. Divide: $\dfrac{e^{-5}}{e^8}$.

3. Multiply: $(x + h)^2$.

4. Factor: $25x^2 - t^2$.

5. **Business: Compound interest.** A person makes an investment at 13%, compounded annually. It has grown to $1039.60 at the end of 1 year. How much was originally invested?

6. Solve: $-3x < 12$.

7. A function is given by $f(x) = x^2 - 4$. Find (a) $f(-3)$ and (b) $f(x + h)$.

8. What are the slope and the y-intercept of $y = -3x + 2$?

9. Find an equation of the line with slope $\frac{1}{4}$, containing the point $(8, -5)$.

10. Find the slope of the line containing the points $(-2, 3)$ and $(-4, -9)$.

11. **Life science: Body fluids.** The weight F of fluids in a human is directly proportional to body weight W. It is

known that a person who weighs 180 lb has 120 lb of fluids. Find an equation of variation expressing F as a function of W.

12. **Business: Profit-and-loss analysis.** A music company has fixed costs of $10,000 for producing a CD master. Thereafter, the variable costs are $0.50 per CD for duplicating from the CD master. Revenue from each CD is expected to be $7.50.

 a) Formulate a function $C(x)$ for the total cost of producing x CDs.
 b) Formulate a function $R(x)$ for the total revenue from the sale of x CDs.
 c) Formulate a function $P(x)$ for the total profit from the production and sale of x CDs.
 d) How many CDs must the company sell in order to break even?

13. **Economics: Equilibrium point.** Find the equilibrium point for the demand and supply functions

$$D(p) = (p - 7)^2, 0 \le p \le 7,$$

and

$$S(p) = p^2 + p + 4.$$

14. Graph: $y = 4/x$.

15. Convert to rational exponents: $1/\sqrt{t}$.

16. Convert to radical notation: $t^{-3/5}$.

17. Graph: $f(x) = \dfrac{x^2 - 1}{x + 1}$.

Determine the domain of the function.

18. $f(x) = \dfrac{x^2 + 20}{(x - 2)(x + 7)}$

19. $f(x) = \sqrt{5x + 10}$

20. Find a linear function that fits the data points (1, 3) and (2, 7).

21. Find a quadratic function that fits the data points (1, 5), (2, 9), and (3, 4).

22. Write interval notation for this graph.

23. Graph:

$$f(x) = \begin{cases} x^2 + 2, & \text{for } x \geq 0, \\ x^2 - 2, & \text{for } x < 0. \end{cases}$$

SYNTHESIS

24. *Economics: Demand.* The demand function for a product is given by

$$x = D(p) = 800 - p^3, \quad \$0 \leq p \leq \$9.28.$$

a) Find the number of units sold when the price per unit is $6.50.

b) Find the price per unit when 720 units are sold.

TECHNOLOGY CONNECTION

Use your grapher to graph each of the following.

25. $f(x) = x^3 - 9x^2 + 27x + 50$

26. $f(x) = \sqrt[3]{|4 - x^2|} + 1$

Find the zeros of the function.

27. $f(x) = x^3 - 9x^2 + 27x + 50$

28. $f(x) = \sqrt[3]{|4 - x^2|} + 1$

29. Find the equilibrium point for the demand and supply functions

$$D(p) = 1000 - 60p \quad \text{and} \quad S(p) = 200 + 4p.$$

EXTENDED TECHNOLOGY APPLICATION

The Ecological Effect of Global Warming

Extended Technology Applications occur at the end of each chapter. They are designed to consider certain applications in greater depth; make use of curve fitting, regression, and modeling using a grapher; and allow for possible group or collaborative learning.

Ecologists continue to be concerned about global warming, that is, the trend of average global temperatures to rise over recent years. One possible effect of such a warming is the melting of the icecaps. Concern is so great that in 1992, over 150 nations signed a treaty designed to reduce the emission of gases believed to compound the effects of global warming.

One cause of global warming may be the so-called *greenhouse effect.* Smog and pollution in the atmosphere create a cover over the surface of the earth.

Greenhouse Effect

The heat of the sun hits the surface and bounces off, but tends not to leave the surface because of the cover of pollution. Since 1959, the carbon dioxide content of the atmosphere has increased 13%. The primary sources of this increase are the exhaust of automobiles and deforestation.

Note the data in the table to the right regarding average global temperature and the concentration of atmospheric carbon dioxide. Let's examine the data, make a graph, find some models, and make predictions.

To find a mathematical model of global warming, let's first graph the temperature data and look for a pattern.

Year	Average global temperature (in degrees Fahrenheit, F)	Carbon dioxide (in parts per million)
0. 1975	58.91°	331.0
1. 1976	58.62°	332.0
2. 1977	59.29°	333.7
3. 1978	59.16°	335.3
4. 1979	59.25°	336.7
5. 1980	59.50°	338.5
6. 1981	59.70°	339.8
7. 1982	59.13°	341.0
8. 1983	59.52°	342.6
9. 1984	59.20°	344.3
10. 1985	59.20°	345.7
11. 1986	59.29°	347.0
12. 1987	59.58°	348.8
13. 1988	59.63°	351.4
14. 1989	59.45°	352.8
15. 1990	59.85°	354.0
16. 1991	59.74°	355.4
17. 1992	59.23°	356.2
18. 1993	59.36°	357.0

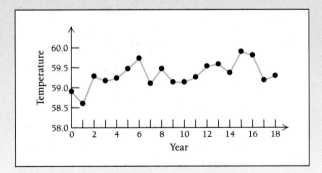

If we look at the graph above and try to determine a model, we see erratic behavior, but there is an overall trend toward an increase. Indeed, if we draw a line through the data as shown to the right, we see that the line does seem to have a positive slope and to increase.

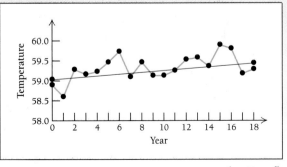

(continued)

EXERCISES

1. a) Use a grapher that performs linear regression. Consider all the data to find a linear function that fits the data.
   b) Graph the linear function.
   c) Use the linear function to predict the average global temperature in 1994, 1996, 1999, 2000, and 2050.
   d) Use the linear function to predict when the average global temperature will reach 60°, 70°, and 80°.

2. a) Consider all the data to find a cubic polynomial function $y = ax^3 + bx^2 + cx + d$ that fits the data.

   b) Graph the cubic polynomial function.
   c) Use the cubic function to predict the average global temperature in 1994, 1996, 1999, 2000, and 2050.
   d) Use the cubic function to predict when the average global temperature will reach 60°, 70°, and 80°.

3. a) Consider all the data to find a quartic polynomial function

   $$y = ax^4 + bx^3 + cx^2 + dx + e$$

   that fits the data.
   b) Graph the quartic polynomial function.
   c) Use the quartic function to predict the average global temperature in 1994, 1996, 1999, 2000, and 2050.
   d) Use the quartic function to predict when the average global temperature will reach 60°, 70°, and 80°.

4. Ecologists are forewarning that average global temperature will reach between 60.8° and 66.2° by the year 2050. How well does each of your regression models confirm that prediction?

5. Discuss the merits of which type of function to use to predict global warming.

6. Consider the data on concentration of atmospheric carbon dioxide. Fit a linear, a cubic, and a quartic polynomial to the data and make three predictions of the carbon dioxide concentration in the year 2020. Discuss the merits of each prediction.

2 Differentiation

INTRODUCTION

With this chapter, we begin our study of calculus. The first concepts we consider are those of limits and continuity. Then we apply those concepts to establishing the first of the two building blocks of calculus: differentiation.

Differentiation is a process that takes a formula for a function and derives a formula for another function, called a *derivative,* that allows us to find the slope of the tangent line to a curve at a point. We also find that a derivative can represent an instantaneous rate of change. Throughout the chapter, we will learn various techniques for finding derivatives.

AN APPLICATION

The temperature T of a person during an illness is given by

$$T(t) = \frac{4t}{t^2 + 1} + 98.6,$$

where T = the temperature, in degrees Fahrenheit, at time t, in hours.

Find the rate of change of the temperature with respect to time.

THE MATHEMATICS

The rate of change is given by

$$\underbrace{T'(t) = \frac{4 - 4t^2}{(t^2 + 1)^2}}_{\uparrow}.$$

This function is a *derivative.*

This problem appears as Exercise 49 in Exercise Set 2.7.

2.1 Limits and Continuity

OBJECTIVES

• Find limits of functions, if they exist.
• Determine continuity of a function from its graph.
• Determine continuity at a point.

In this section we give an intuitive (meaning "based on prior and present experience") treatment of two important concepts: *limits* and *continuity*.

Limits

Suppose that a defensive football team is on its own 5-yd line and is given a penalty. Such a penalty is half the distance to the goal. Suppose it keeps getting such a penalty. Then the offense moves from the 5-yd line to the $2\frac{1}{2}$-yd line to the $1\frac{1}{4}$-yd line, and so on. Note that it can never score under this premise, but its distance to the goal continues to get closer and closer to 0. We say that the *limit* is 0.

One important aspect of the study of calculus is the analysis of how function values, or outputs, change when the inputs change. Basic to this study is the notion of limit. Suppose that the inputs get closer and closer to some number. If the corresponding outputs get closer and closer to a number, then that number is called a *limit*.

Consider the function f given by

$$f(x) = 2x + 3.$$

Suppose we select input numbers x closer and closer to the number 4, and look at the output numbers $2x + 3$. Study the input–output table and the graph that follow.

In the table and the graph, we see that as input numbers approach 4, but do not equal 4, from the left, output numbers approach 11.

x	$f(x)$
2	7
3.6	10.2
3.9	10.8
3.99	10.98
3.999	10.998
$\cdots$	$\cdots$
4.001	11.002
4.01	11.02
4.1	11.2
4.8	12.6
5	13

These inputs approach 4 from the left.

These inputs approach 4 from the right.

These outputs approach 11.

These outputs approach 11.

$4 \leftarrow \cdots \qquad \cdots \rightarrow 11$

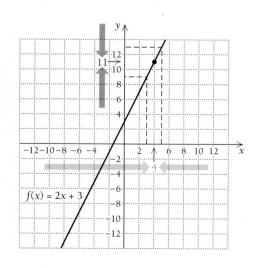

$f(x) = 2x + 3$

As input numbers approach 4, but do not equal 4, from the right, output numbers approach 11. Thus we say:

As *x approaches* 4,   2*x* + 3 *approaches* 11.

An arrow, →, is often used for the word "approaches." Thus the statement above can be written:

As *x* → 4,   2*x* + 3 → 11.

The number 11 is said to be the *limit* of 2*x* + 3 as *x* approaches 4, but does not equal 4. We can abbreviate this statement as follows:

$$\lim_{x \to 4} (2x + 3) = 11.$$

This is read, "The limit, as *x* approaches 4, of 2*x* + 3 is 11."

DEFINITION

A function *f* has the *limit L* as *x* approaches *a*, written

$$\lim_{x \to a} f(x) = L,$$

if *all* values *f*(*x*) for *f* are close to *L* for values of *x* that are arbitrarily close, but not equal, to *a*.

 TECHNOLOGY CONNECTION

Consider $f(x) = 3x - 1$ and complete this table. Do not substitute 6 into the formula.

x	f(x)
5	
5.8	
5.9	
5.99	
5.999	
6.001	
6.01	
6.1	
6.4	
7	

6 ←   → ?

Use your grapher to graph $f(x) = 3x - 1$. Now using the trace feature, move the cursor from left to right so that the *x*-coordinate approaches 6 from the left. What number does the *y*-coordinate approach? Now move the cursor from right to left so that the *x*-coordinate approaches 6 from the right. Use the table and the graph to find $\lim_{x \to 6} f(x)$. Use a similar procedure to find

$$\lim_{x \to 2} f(x) \quad \text{and} \quad \lim_{x \to -1} f(x).$$

Consider $f(x) = x^3 - 2x - 2$. Find $f(1.9)$, $f(1.99)$, $f(1.999)$, $f(2.1)$, $f(2.01)$, and $f(2.001)$. Next, graph *f* and use the graph and the trace feature to find

$$\lim_{x \to 2} f(x) \quad \text{and} \quad \lim_{x \to -1} f(x).$$

TECHNOLOGY CONNECTION

Consider $f(x) = 1/(x - 3)$ and complete this table.

Inputs, x	Outputs, $f(x)$
1	
2	
2.9	
2.99	
2.999	
3.001	
3.01	
3.1	
3.8	
4	

Use your grapher to graph $f(x) = 1/(x - 3)$. Now use the table, the graph, and the trace feature to find

$$\lim_{x \to 3} f(x), \quad \lim_{x \to 1} f(x), \quad \text{and} \quad \lim_{x \to 4} f(x).$$

Consider $f(x) = (x^2 - 1)/(x^2 + x - 6)$. Find $f(1.9), f(1.99), f(1.999), f(2.1), f(2.01)$, and $f(2.001)$. Next, graph f. Then use the table, the graph, and the trace feature to find

$$\lim_{x \to 2} f(x) \quad \text{and} \quad \lim_{x \to -4} f(x).$$

EXAMPLE 1   Consider the function

$$F(x) = \frac{1}{x}.$$

Find each of the following limits, if they exist.

a) $\lim\limits_{x \to 3} F(x)$ **b)** $\lim\limits_{x \to 0} F(x)$

Solution

a) From the graph below, we see that as inputs x approach 3 from either the left or the right, outputs $1/x$ approach $\frac{1}{3}$. We can also check this using an input–output table. Thus we have

$$\lim_{x \to 3} F(x) = \frac{1}{3}.$$

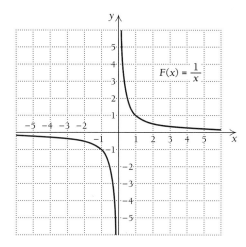

b) From either the graph or an input–output table, we see that as inputs x approach 0 from the left, outputs become more and more negative without bound. Similarly, as inputs x approach 0 from the right, outputs get larger and larger without bound. We say that

$$\lim_{x \to 0} \frac{1}{x} \quad \textit{does not exist.} \qquad \blacklozenge$$

The following is very important to keep in mind when determining whether a limit exists.

In order for a limit to exist, both of the limits from the left and the right must exist and be the same.

We use the notation

$$\lim_{x \to a^+} f(x) \quad \text{to indicate the limit from the right}$$

and $$\lim_{x \to a^-} f(x) \quad \text{to indicate the limit from the left.}$$

Then in order for a limit to exist, both of the limits above must exist and be the same.

EXAMPLE 2   Consider the function H defined as follows:

$$H(x) = \begin{cases} 2x + 2, & \text{for } x < 1, \\ 2x - 2, & \text{for } x \geq 1. \end{cases}$$

Graph the function and find each of the following limits, if they exist.

a) $\lim_{x \to 1} H(x)$ b) $\lim_{x \to -3} H(x)$

Solution   The graph is shown at right.

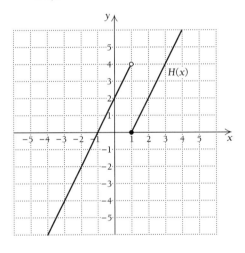

a) As inputs x approach 1 from the left, outputs $H(x)$ approach 4. Thus the limit from the left is 4. That is,

$$\lim_{x \to 1^-} H(x) = 4.$$

But as inputs x approach 1 from the right, outputs $H(x)$ approach 0. Thus the limit from the right is 0. That is,

$$\lim_{x \to 1^+} H(x) = 0.$$

Since the limit from the left, 4, is not the same as the limit from the right, 0, we say that

$$\lim_{x \to 1} H(x) \quad \text{does not exist.}$$

b) As inputs x approach -3 from the left, outputs $H(x)$ approach -4, so the limit from the left is -4. That is,

$$\lim_{x \to -3^-} H(x) = -4.$$

As inputs x approach -3 from the right, outputs $H(x)$ approach -4, so the limit from the right is -4. That is,

$$\lim_{x \to -3^+} H(x) = -4.$$

Since the limits from the left and from the right exist and are the same, we have

$$\lim_{x \to -3} H(x) = -4. \qquad \blacklozenge$$

The following is also important in finding limits.

The limit at a number a *does not depend* on the function value at a even if that function value, $f(a)$, exists. That is, whether or not a limit exists at a has nothing to do with the function value $f(a)$.

EXAMPLE 3   Consider the function G defined as follows:

$$G(x) = \begin{cases} 5, & \text{for } x = 1, \\ x + 1, & \text{for } x \neq 1. \end{cases}$$

Graph the function and find each of the following limits, if they exist.

a) $\lim_{x \to 1} G(x)$ **b)** $\lim_{x \to 3} G(x)$

Solution   The graph follows.

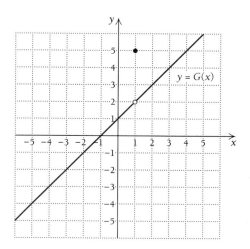

a) As inputs x approach 1 from the left, outputs $G(x)$ approach 2, so the limit from the left is 2. As inputs x approach 1 from the right, outputs $G(x)$ approach 2, so the limit from the right is 2. Since the limit from the left, 2, is the same as the limit from the right, 2, we have

$$\lim_{x \to 1} G(x) = 2.$$

Note that the limit, 2, is not the same as the function value at 1, which is $G(1) = 5$.

b) We have

$$\lim_{x \to 3} G(x) = 4.$$

We also know that

$$G(3) = 4.$$

In this case, the function value and the limit are the same.   ◆

Continuity

When the limit of a function is the same as its function value, it satisfies a condition called **continuity at a point.** We now consider the concept of continuity.

The following are the graphs of functions that are *continuous* over the whole real line $(-\infty, \infty)$.

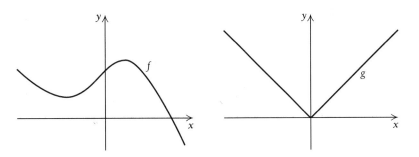

Note that there are no "jumps" or holes in the graphs. For now we will use a some-what intuitive definition of continuity, which we will refine. We say that a function is **continuous over, or on, some interval** of the real line if its graph can be traced without lifting the pencil from the paper. If a function has one point at which it fails to be continuous, then we say that it is *not continuous* over the whole real line. Graphs of functions that are *not* continuous (F, G, and H) over the whole real line follow. For functions G and H, the open circle indicates that the point is not part of the graph.

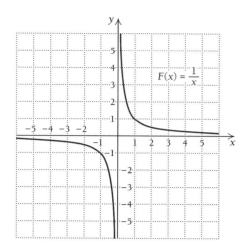

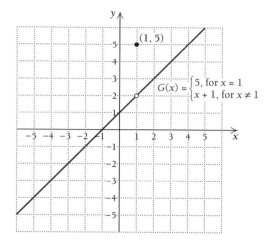

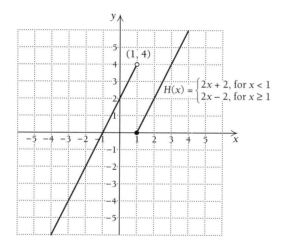

A continuous curve.

In each case, the graph *cannot* be traced without lifting the pencil from the paper. However, each case represents a different situation. Let us discuss why each case fails to be continuous over the whole real line.

The function F fails to be continuous over the whole real line $(-\infty, \infty)$. Since F is not defined at $x = 0$, the point $x = 0$ is not part of the domain, so $F(0)$ does not exist and there is no point $(0, F(0))$ on the graph. Thus there is no point to trace at $x = 0$. However, F is continuous on the intervals $(-\infty, 0)$ and $(0, \infty)$.

The function G is not continuous over the whole real line since it is not continuous at $x = 1$. Let us trace the graph of G starting to the left of $x = 1$. As x approaches 1, $G(x)$ seems to approach 2. However, at $x = 1$, $G(x)$ *jumps* up to 5, whereas to the right of $x = 1$, $G(x)$ *drops* back to some value close to 2. Thus G is discontinuous at $x = 1$.

*The function H is not continuous over the whole real line since it is not continuous at x = 1. Let us trace the graph of H starting to the left of x = 1. As x approaches 1, H(x) approaches 4. However, at x = 1, H(x) *drops* down to 0; and just to the right of x = 1, H(x) is close to 0. Since the limit from the left is 4 and the limit from the right is 0, we see that lim $_{x \to 1}$ f(x) does not exist. Thus H(x) is discontinuous at x = 1.*

Limits and Continuity

After working through Example 3, we might ask, "When can we substitute to find a limit?" The answer lies in the following definition, which also provides a more formal definition of continuity.

DEFINITION

A function f is *continuous at x = a* if:

a) f(a) exists,   (The output f(a) exists.)

b) lim $_{x \to a}$ f(x) exists, and   (The limit as x → a exists.)

c) lim $_{x \to a}$ f(x) = f(a).   (The limit is the same as the output.)

A function is *continuous over an interval I* if it is continuous at each point in I.

Note that we will also use the notation "lim $_{x \to a}$ f(x)" for a limit.

EXAMPLE 4   Determine whether the function given by

$$f(x) = 2x + 3$$

is continuous at x = 4.

Solution   This function is continuous at 4 because:

a) f(4) exists,   (f(4) = 11)

b) lim $_{x \to 4}$ f(x) exists,   [lim $_{x \to 4}$ f(x) = 11 (as shown earlier)],

c) lim $_{x \to 4}$ f(x) = 11 = f(4).

In fact, f(x) = 2x + 3 is continuous at any point on the real line.   ◆

EXAMPLE 5   Determine whether the function in Example 1 is continuous at x = 0.

Solution   The function is *not* continuous at x = 0 because f(0) does not exist.   ◆

EXAMPLE 6   Determine whether the function in Example 2 is continuous at x = 1.

Solution   The function is not continuous at $x = 1$ because $\lim_{x \to 1} H(x)$ does not exist.   ◆

EXAMPLE 7   Determine whether the function in Example 3 is continuous at $x = 1$.

Solution   The function is not continuous at $x = 1$ because $G(1) = 5$, but $\lim_{x \to 1} G(x) = 2$.   ◆

CONTINUITY PRINCIPLES

The following continuity principles, which we will not prove, allow us to determine whether a function is continuous.

C1. Any constant function is continuous (such a function never varies).

C2. For any positive integer n and any continuous function f, $[f(x)]^n$ and $\sqrt[n]{f(x)}$ are continuous. When n is even, the inputs of f in $\sqrt[n]{f(x)}$ are restricted to inputs x for which $f(x) \geq 0$.

C3. If $f(x)$ and $g(x)$ are continuous, then so are $f(x) + g(x)$, $f(x) - g(x)$, and $f(x) \cdot g(x)$.

C4. If $f(x)$ and $g(x)$ are continuous, so is $g(x)/f(x)$, so long as the inputs x do not yield outputs $f(x) = 0$.

EXAMPLE 8   Provide an argument to show that

$$f(x) = x^2 - 3x + 2$$

is continuous.

Solution   First we note that x is a continuous function. We determine this by the definition of continuity. We know that x^2 is continuous by C2. The constant function 3 is continuous by C1 and the function x is continuous, so the product $3x$ is continuous by C3. Thus, $x^2 - 3x$ is continuous by C3, and since the constant 2 is continuous, we can apply C3 again to show that $x^2 - 3x + 2$ is continuous.   ◆

In similar fashion, we can show that any polynomial, such as

$$f(x) = x^4 - 5x^3 + x^2 - 7,$$

is continuous.

A rational function is a quotient of two polynomials

$$r(x) = \frac{g(x)}{f(x)}.$$

Thus by C4, a rational function is continuous so long as the inputs x are not such that $f(x) = 0$. Thus a function like

$$r(x) = \frac{x^4 - 5x^3 + x^2 - 7}{x - 3}$$

is continuous for all values of x except $x = 3$.

2.1 Exercise Set

Determine whether each of the following is continuous.

1.

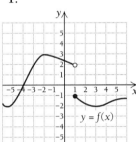

2.

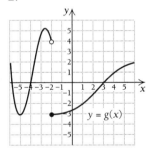

3.

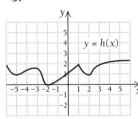

4.

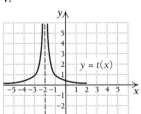

Use the graphs and functions in Exercises 1–4 to answer each of the following.

5. a) Find $\lim\limits_{x \to 1^+} f(x)$, $\lim\limits_{x \to 1^-} f(x)$, and $\lim\limits_{x \to 1} f(x)$.

 b) Find $f(1)$.
 c) Is f continuous at $x = 1$?
 d) Find $\lim\limits_{x \to -2} f(x)$.
 e) Find $f(-2)$.
 f) Is f continuous at $x = -2$?

6. a) Find $\lim\limits_{x \to 1^+} g(x)$, $\lim\limits_{x \to 1^-} g(x)$, and $\lim\limits_{x \to 1} g(x)$.

 b) Find $g(1)$.
 c) Is g continuous at $x = 1$?
 d) Find $\lim\limits_{x \to -2} g(x)$.
 e) Find $g(-2)$.
 f) Is g continuous at $x = -2$?

7. a) Find $\lim\limits_{x \to 1} h(x)$.

 b) Find $h(1)$.
 c) Is h continuous at $x = 1$?
 d) Find $\lim\limits_{x \to -2} h(x)$.
 e) Find $h(-2)$.
 f) Is h continuous at $x = -2$?

8. a) Find $\lim\limits_{x \to 1} t(x)$.

 b) Find $t(1)$.
 c) Is t continuous at $x = 1$?
 d) Find $\lim\limits_{x \to -2} t(x)$.
 e) Find $t(-2)$.
 f) Is t continuous at $x = -2$?

In Exercises 9–12, use the graphs to find the limits and answer the related questions.

9. Consider $f(x) = \begin{cases} 4 - x, & \text{for } x \ne 1, \\ 2, & \text{for } x = 1. \end{cases}$

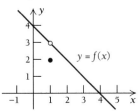

 a) Find $\lim\limits_{x \to 1^+} f(x)$.

b) Find $\lim_{x \to 1^-} f(x)$.

c) Find $\lim_{x \to 1} f(x)$.

d) Find $f(1)$.

e) Is f continuous at $x = 1$?

f) Is f continuous at $x = 2$?

10. Consider $f(x) = \begin{cases} 4 - x^2, & \text{for } x \neq -2, \\ 3, & \text{for } x = -2. \end{cases}$

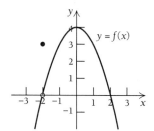

a) Find $\lim_{x \to -2^+} f(x)$.

b) Find $\lim_{x \to -2^-} f(x)$.

c) Find $\lim_{x \to -2} f(x)$.

d) Find $f(-2)$.

e) Is f continuous at $x = -2$?

f) Is f continuous at $x = 1$?

11. Refer to the graph of f below to determine whether each statement is true or false.

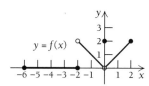

a) $\lim_{x \to -2^+} f(x) = 1$

b) $\lim_{x \to -2^-} f(x) = 0$

c) $\lim_{x \to -2^-} f(x) = \lim_{x \to -2^+} f(x)$

d) $\lim_{x \to -2} f(x)$ exists.

e) $\lim_{x \to -2} f(x) = 2$

f) $\lim_{x \to 0} f(x) = 0$

g) $f(0) = 2$

h) f is continuous at $x = -2$.

i) f is continuous at $x = 0$.

j) f is continuous at $x = -1$.

12. Refer to the graph below to determine whether each statement is true or false.

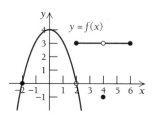

a) $\lim_{x \to 2^-} f(x) = 3$

b) $\lim_{x \to 2^+} f(x) = 0$

c) $\lim_{x \to 2^-} f(x) = \lim_{x \to 2^+} f(x)$

d) $\lim_{x \to 2} f(x)$ exists.

e) $\lim_{x \to 4} f(x)$ exists.

f) $\lim_{x \to 4} f(x) = f(4)$

g) f is continuous at $x = 4$.

h) f is continuous at $x = 0$.

i) $\lim_{x \to 3} f(x) = \lim_{x \to 5} f(x)$

j) f is continuous at $x = 2$.

APPLICATIONS

◆ **Business and Economics**

The postage function. Postal rates are 32¢ for the first ounce and 23¢ for each additional ounce or fraction thereof. Formally speaking, if x is the weight of a letter in ounces, then $p(x)$ is the cost of mailing the letter, where

$$p(x) = 32¢, \quad \text{if } 0 < x \leq 1,$$
$$p(x) = 55¢, \quad \text{if } 1 < x \leq 2,$$
$$p(x) = 78¢, \quad \text{if } 2 < x \leq 3,$$

and so on, up to 11 ounces (at which point postal cost also depends on distance). The graph of p is shown here.

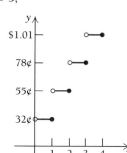

13. Is p continuous at 1? at $1\frac{1}{2}$? at 2? at 2.53?

14. Is p continuous at 3? at $3\frac{1}{4}$? at 4? at 3.98?

Using the graph of the postage function, find each of the following limits, if it exists.

15. $\lim\limits_{x\to1^-} p(x)$, $\lim\limits_{x\to1^+} p(x)$, $\lim\limits_{x\to1} p(x)$

16. $\lim\limits_{x\to2^-} p(x)$, $\lim\limits_{x\to2^+} p(x)$, $\lim\limits_{x\to2} p(x)$

17. $\lim\limits_{x\to2.3} p(x)$

18. $\lim\limits_{x\to1/2} p(x)$

Total cost. The total cost $C(x)$ of producing x units of a product is shown in the graph below. Note that the curve increases gradually, but jumps suddenly at a certain value. This is probably the point at which new machines must be purchased and extra employees must be employed.

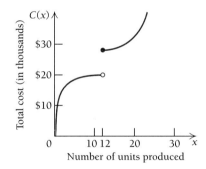

Using the graph, find each of the following limits, if it exists.

19. $\lim\limits_{x\to12^-} C(x)$, $\lim\limits_{x\to12^+} C(x)$, $\lim\limits_{x\to12} C(x)$

20. $\lim\limits_{x\to20^-} C(x)$, $\lim\limits_{x\to20^+} C(x)$, $\lim\limits_{x\to20} C(x)$

21. Is C continuous at 12? at 20?

22. Is C continuous at 6? at 18?

◆ **Social Sciences**

A learning curve. In psychology one often takes a certain amount of time t to learn a task. Suppose that the goal is to do a task perfectly and that you are practicing the ability to master it. After a certain time period, what is known to psychologists as an "I've got it!" experience occurs, and you are able to perform the task perfectly.

Where do you think this happens on the learning curve below?

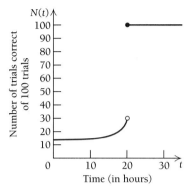

Using the graph, find each of the following limits, if it exists.

23. $\lim\limits_{t\to20^+} N(t)$, $\lim\limits_{t\to20^-} N(t)$, $\lim\limits_{t\to20} N(t)$

24. $\lim\limits_{t\to30^-} N(t)$, $\lim\limits_{t\to30^+} N(t)$, $\lim\limits_{t\to30} N(t)$

25. Is N continuous at 20? at 30?

26. Is N continuous at 10? at 26?

SYNTHESIS

27. ◈ Discuss three ways in which a function may not be continuous at a point a. Draw graphs to illustrate your discussion.

TECHNOLOGY CONNECTION

Use a grapher that can graph functions defined piecewise.

28. Graph the function f given by

$$f(x) = \begin{cases} -3, & \text{for } x = -2, \\ x^2, & \text{for } x \neq -2. \end{cases}$$

Use the graph and the trace feature to find each of the following limits, if it exists.

a) $\lim\limits_{x\to-2^+} f(x)$

b) $\lim\limits_{x\to-2^-} f(x)$

c) $\lim\limits_{x\to-2} f(x)$

d) $\lim_{x \to 2^+} f(x)$

e) $\lim_{x \to 2^-} f(x)$

f) Does $\lim_{x \to -2} f(x) = f(-2)$?

g) Does $\lim_{x \to 2} f(x) = f(2)$?

29. Graph the function f given by

$$f(x) = \begin{cases} 2 - x^2, & \text{for } x \geq 0, \\ x^2 - 2, & \text{for } x < 0. \end{cases}$$

Use the graph and the trace feature to find each of the following limits, if it exists.

a) $\lim_{x \to 0} f(x)$

b) $\lim_{x \to -2} f(x)$

30. ◆ Graph the function f given by

$$f(x) = \begin{cases} -3, & \text{for } x = -2, \\ x^2, & \text{for } x \neq -2. \end{cases}$$

Using the trace feature, determine whether this function is continuous at $x = -2$. Explain.

31. ◆ Graph the function f given by

$$f(x) = \begin{cases} 2 - x^2, & \text{for } x \geq 0, \\ x^2 - 2, & \text{for } x < 0. \end{cases}$$

Using the trace feature, determine whether this function is continuous at $x = 0$. Explain.

2.2 More on Limits

OBJECTIVE

• Find limits using input–output tables, algebra, or graphs.

Using Limit Principles

If a function is continuous at a, we can substitute to find the limit.

EXAMPLE 1 Find $\lim_{x \to 2} (x^4 - 5x^3 + x^2 - 7)$.

Solution It follows from the Continuity Principles that $x^4 - 5x^3 + x^2 - 7$ is continuous. Thus we can find the limit by substitution:

$$\lim_{x \to 2} (x^4 - 5x^3 + x^2 - 7) = 2^4 - 5 \cdot 2^3 + 2^2 - 7$$
$$= 16 - 40 + 4 - 7$$
$$= -27. \qquad \blacklozenge$$

EXAMPLE 2 Find $\lim_{x \to 0} \sqrt{x^2 - 3x + 2}$.

Solution Using the Continuity Principles, we have shown that $x^2 - 3x + 2$ is continuous for all values of x. When we restrict x to values for which $x^2 - 3x + 2$ is nonnegative, it follows from Principle C2 that $\sqrt{x^2 - 3x + 2}$ is continuous. Since $x^2 - 3x + 2$ is nonnegative when $x = 0$, we can substitute to find the limit:

$$\lim_{x \to 0} \sqrt{x^2 - 3x + 2} = \sqrt{0^2 - 3 \cdot 0 + 2}$$
$$= \sqrt{2}. \qquad \blacklozenge$$

There are Limit Principles that correspond to the Continuity Principles. We can use them to find limits when we are uncertain of the continuity of a function at a given point.

LIMIT PRINCIPLES

If $\lim_{x \to a} f(x) = L$ and $\lim_{x \to a} g(x) = M$, then we have the following.

L1. $\lim_{x \to a} c = c$.

(The limit of a constant is the constant.)

L2. $\lim_{x \to a} [f(x)]^n = \left[\lim_{x \to a} f(x) \right]^n = L^n$,

$\lim_{x \to a} \sqrt[n]{f(x)} = \sqrt[n]{\lim_{x \to a} f(x)} = \sqrt[n]{L}$, for any positive integer.

When n is even, the inputs x in $\sqrt[n]{f(x)}$ must be restricted to those inputs x for which $f(x) \geq 0$.

(The limit of a power is the power of the limit, and the limit of a root is the root of the limit.)

L3. $\lim_{x \to a} [f(x) \pm g(x)] = \lim_{x \to a} f(x) \pm \lim_{x \to a} g(x) = L \pm M$.

(The limit of a sum or a difference is the sum or the difference of the limits.)

$\lim_{x \to a} [f(x) \cdot g(x)] = \left[\lim_{x \to a} f(x) \right] \cdot \left[\lim_{x \to a} g(x) \right] = L \cdot M$.

(The limit of a product is the product of the limits.)

L4. $\lim_{x \to a} \dfrac{g(x)}{f(x)} = \dfrac{\lim_{x \to a} g(x)}{\lim_{x \to a} f(x)} = \dfrac{M}{L}$, provided $L \neq 0$.

(The limit of a quotient is the quotient of the limits.)

L5. $\lim_{x \to a} cf(x) = c \cdot \lim_{x \to a} f(x) = c \cdot L$.

(The limit of a constant times a function is the constant times the limit.)

Principle L5 can actually be proved using Principles L1 and L3, but we state it for emphasis.

EXAMPLE 3   Find

$$\lim_{x \to -3} \frac{x^2 - 9}{x + 3}.$$

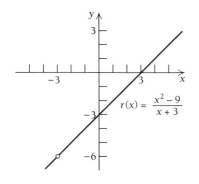

$$r(x) = \frac{x^2 - 9}{x + 3}$$

Solution   Consider the graph of the rational function $r(x) = (x^2 - 9)/(x + 3)$, shown at left. We can see that the limit is -6. But let us also examine the limit using the Limit Principles.

The function $(x^2 - 9)/(x + 3)$ is not continuous at $x = -3$. We use some algebraic simplification and then some Limit Principles:

$$\lim_{x \to -3} \frac{x^2 - 9}{x + 3} = \lim_{x \to -3} \frac{(x + 3)(x - 3)}{x + 3}$$

$$= \lim_{x \to -3} (x - 3) \qquad \text{Simplifying, assuming } x \neq -3$$

$$= \lim_{x \to -3} x - \lim_{x \to -3} 3 \qquad \text{By L3}$$

$$= -3 - 3 = -6. \qquad \blacklozenge$$

TECHNOLOGY
CONNECTION

Find each limit, if it exists. Use an input–output table, the graph, and the trace feature.

$$\lim_{x \to -2} (x^4 - 5x^3 + x^2 - 7),$$

$$\lim_{x \to 1} \sqrt{x^2 + 3x + 4},$$

$$\lim_{x \to 3} \frac{x - 3}{x^2 - 9}$$

In Section 2.3, we will encounter expressions with two variables, x and h. Our interest is in those limits where x is fixed as a constant and h approaches zero.

EXAMPLE 4   Find $\lim_{h \to 0} (3x^2 + 3xh + h^2)$.

Solution   We treat x as a constant since we are interested only in the way in which the expression varies when $h \to 0$. We use the Limit Principles to find that

$$\lim_{h \to 0} (3x^2 + 3xh + h^2) = 3x^2 + 3x(0) + 0^2 = 3x^2.$$

The reader can check any limit about which there is uncertainty by using an input–output table. The following is a table for this limit.

h	$3x^2 + 3xh + h^2$	
1	$3x^2 + 3x \cdot 1 + 1^2,$	or $3x^2 + 3x + 1$
0.8	$3x^2 + 3x(0.8) + (0.8)^2,$	or $3x^2 + 2.4x + 0.64$
0.5	$3x^2 + 3x(0.5) + (0.5)^2,$	or $3x^2 + 1.5x + 0.25$
0.1	$3x^2 + 3x(0.1) + (0.1)^2,$	or $3x^2 + 0.3x + 0.01$
0.01	$3x^2 + 3x(0.01) + (0.01)^2,$	or $3x^2 + 0.03x + 0.0001$
0.001	$3x^2 + 3x(0.001) + (0.001)^2,$	or $3x^2 + 0.003x + 0.000001$

From the pattern in the table, it appears that

$$\lim_{h \to 0} (3x^2 + 3xh + h^2) = 3x^2. \qquad \blacklozenge$$

Limits and Infinity

In Example 1 of Section 2.1, we discussed that

$$\lim_{x \to 0} \frac{1}{x} \quad \text{does not exist.}$$

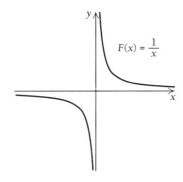

$$F(x) = \frac{1}{x}$$

Looking at the graph, we note that as x approaches 0 from the right, the outputs increase without bound. These numbers do not approach any real number, though it might be said that the limit from the right is infinity (∞). That is,

$$\lim_{x \to 0^+} \frac{1}{x} = \infty.$$

As x approaches 0 from the left, the outputs become more and more negative without bound. These numbers do not approach any real number, though it might be said that the limit from the left is negative infinity ($-\infty$). That is,

$$\lim_{x \to 0^-} \frac{1}{x} = -\infty.$$

Keep in mind that ∞ and $-\infty$ are not real numbers. We associate ∞ with numbers increasing without bound in a positive direction and $-\infty$ with numbers decreasing without bound in a negative direction.

Limits Involving Infinity

Occasionally we need to determine limits when the inputs get larger and larger without bound, that is, when they approach infinity. In such cases, we are finding *limits at infinity*. Such a limit is expressed as

$$\lim_{x \to \infty} f(x).$$

EXAMPLE 5   Find

$$\lim_{x \to \infty} \frac{3x - 1}{x}.$$

Solution   One way to find such a limit is to use an input–output table, as follows.

Inputs, x	1	10	50	100	2000
Outputs, $\dfrac{3x-1}{x}$	2.0	2.9	2.98	2.99	2.9995

TECHNOLOGY
CONNECTION

Graph $f(x) = (2x + 5)/x$.
Then find $f(4)$, $f(20)$, $f(80)$,
$f(200)$, $f(1000)$, and
$f(10,000)$. Next, use the
trace feature and move the
cursor along the graph from
left to right, considering
positive values of x. What
happens to the y-coordinate
as x gets larger and larger?
Find

$$\lim_{x\to\infty} \frac{2x+5}{x}.$$

As the inputs get larger and larger without bound, the outputs get closer and closer to 3. Thus,

$$\lim_{x\to\infty} \frac{3x-1}{x} = 3.$$

Another way to find this limit is to use some algebra and the fact that as $x \to \infty$, $b/ax^n \to 0$, for any positive integer n.

We multiply by 1, using $(1/x) \div (1/x)$. This amounts to dividing both the numerator and the denominator by x:

$$\lim_{x\to\infty} \frac{3x-1}{x} = \lim_{x\to\infty} \frac{3x-1}{x} \cdot \frac{(1/x)}{(1/x)} = \lim_{x\to\infty} \frac{(3x-1)\dfrac{1}{x}}{x\cdot\dfrac{1}{x}}$$

$$= \lim_{x\to\infty} \frac{3x\cdot\dfrac{1}{x} - 1\cdot\dfrac{1}{x}}{1} = \lim_{x\to\infty}\left(3 - \frac{1}{x}\right) = 3 - 0 = 3. \quad \blacklozenge$$

EXAMPLE 6   Find

$$\lim_{x\to\infty} \frac{3x^2 - 7x + 2}{7x^2 + 5x + 1}.$$

Solution   The function involved is a quotient of two polynomials. The *degree* of a polynomial is its highest power. Note that the degree of each polynomial above is 2.

The highest power of x in the denominator is x^2. We divide both the numerator and the denominator by x^2:

$$\lim_{x\to\infty} \frac{3x^2 - 7x + 2}{7x^2 + 5x + 1} = \lim_{x\to\infty} \frac{3 - \dfrac{7}{x} + \dfrac{2}{x^2}}{7 + \dfrac{5}{x} + \dfrac{1}{x^2}} = \frac{3 - 0 + 0}{7 + 0 + 0} = \frac{3}{7}. \quad \blacklozenge$$

EXAMPLE 7   Find

$$\lim_{x\to\infty} \frac{5x^2 + 7x + 9}{3x^3 + 2x - 4}.$$

Solution   The highest power of x in the denominator is x^3. We divide the numerator and the denominator by x^3:

$$\lim_{x \to \infty} \frac{5x^2 + 7x + 9}{3x^3 + 2x - 4} = \lim_{x \to \infty} \frac{\dfrac{5}{x} + \dfrac{7}{x^2} + \dfrac{9}{x^3}}{3 + \dfrac{2}{x^2} - \dfrac{4}{x^3}} = \frac{0 + 0 + 0}{3 + 0 - 0} = \frac{0}{3} = 0.$$ ◆

EXAMPLE 8   Find

$$\lim_{x \to \infty} \frac{7x^5 - 8x - 6}{3x^2 + 5x + 2}.$$

Solution   The highest power of x in the denominator is x^2. We divide the numerator and the denominator by x^2:

$$\lim_{x \to \infty} \frac{7x^5 - 8x - 6}{3x^2 + 5x + 2} = \lim_{x \to \infty} \frac{7x^3 - \dfrac{8}{x} - \dfrac{6}{x^2}}{3 + \dfrac{5}{x} + \dfrac{2}{x^2}} = \frac{\lim_{x \to \infty} 7x^3 - 0 - 0}{3 + 0 + 0} = \infty.$$

The numerator, in this case, increases without bound positively while the denominator approaches 3. This can be checked with an input–output table. Thus the limit is ∞. ◆

TECHNOLOGY CONNECTION

Limits Involving Rational Functions

Graphers, for all their advantages, also have their disadvantages. One of them can occur with certain kinds of rational functions. In connected mode, graphers connect plotted points with line segments. In dot mode, they simply plot dots representing coordinates of points. The connected mode can lead to an incorrect graph. For example, consider the following graphs of $f(x) = 8/(x^2 - 4)$.

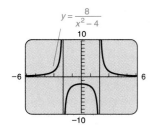

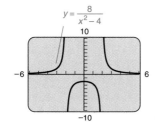

The graph on the left is not the graph of a function, because the vertical lines $x = -2$ and $x = 2$ have been included. The graph on the right is correct.

This disadvantage should not deter you from using a grapher for such functions. But you may need to apply your math knowledge before you accept a graph at face value. If you have a choice when graphing rational functions, use the dot mode.

Use input–output tables, the graph, and the trace feature to find each of the following limits. Then use the analytic procedure of Examples 6–8.

$$\lim_{x \to \infty} \frac{2x^2 + x - 7}{3x^2 - 4x + 1}, \qquad \lim_{x \to \infty} \frac{5x + 4}{2x^3 - 3},$$

$$\lim_{x \to \infty} \frac{5x^2 - 2}{4x + 5}$$

2.2 Exercise Set

Find the limit. Use any method: algebra, graphs, or input–output tables.

1. $\lim\limits_{x \to 1} (x^2 - 3)$

2. $\lim\limits_{x \to 1} (x^2 + 4)$

3. $\lim\limits_{x \to 0} \dfrac{3}{x}$

4. $\lim\limits_{x \to 0} \dfrac{-4}{x}$

5. $\lim\limits_{x \to 3} (2x + 5)$

6. $\lim\limits_{x \to 4} (5 - 3x)$

7. $\lim\limits_{x \to -5} \dfrac{x^2 - 25}{x + 5}$

8. $\lim\limits_{x \to -4} \dfrac{x^2 - 16}{x + 4}$

9. $\lim\limits_{x \to -2} \dfrac{5}{x}$

10. $\lim\limits_{x \to -5} \dfrac{-2}{x}$

11. $\lim\limits_{x \to 2} \dfrac{x^2 + x - 6}{x - 2}$

12. $\lim\limits_{x \to -4} \dfrac{x^2 - x - 20}{x + 4}$

13. $\lim\limits_{x \to 5} \sqrt[3]{x^2 - 17}$

14. $\lim\limits_{x \to 2} \sqrt{x^2 + 5}$

15. $\lim\limits_{x \to 1} (x^4 - x^3 + x^2 + x + 1)$

16. $\lim\limits_{x \to 2} (2x^5 - 3x^4 + x^3 - 2x^2 + x + 1)$

17. $\lim\limits_{x \to 2} \dfrac{1}{x - 2}$

18. $\lim\limits_{x \to 1} \dfrac{1}{(x - 1)^2}$

19. $\lim\limits_{x \to 2} \dfrac{3x^2 - 4x + 2}{7x^2 - 5x + 3}$

20. $\lim\limits_{x \to -1} \dfrac{4x^2 + 5x - 7}{3x^2 - 2x + 1}$

21. $\lim\limits_{x \to 2} \dfrac{x^2 + x - 6}{x^2 - 4}$

22. $\lim\limits_{x \to 4} \dfrac{x^2 - 16}{x^2 - x - 12}$

23. $\lim\limits_{x \to 0} \dfrac{1}{x^2}$

24. $\lim\limits_{x \to 0} \dfrac{-4}{x^2}$

25. $\lim\limits_{h \to 0} (6x^2 + 6xh + 2h^2)$

26. $\lim\limits_{h \to 0} (10x + 5h)$

27. $\lim\limits_{h \to 0} \dfrac{-2x - h}{x^2(x + h)^2}$

28. $\lim\limits_{h \to 0} \dfrac{-5}{x(x + h)}$

29. $\lim\limits_{x \to \infty} \dfrac{2x - 4}{5x}$

30. $\lim\limits_{x \to \infty} \dfrac{3x + 1}{4x}$

31. $\lim\limits_{x \to \infty} \left(5 - \dfrac{2}{x}\right)$

32. $\lim\limits_{x \to \infty} \left(7 + \dfrac{3}{x}\right)$

33. $\lim\limits_{x \to \infty} \dfrac{2x - 5}{4x + 3}$

34. $\lim\limits_{x \to \infty} \dfrac{6x + 1}{5x - 2}$

35. $\lim\limits_{x \to \infty} \dfrac{2x^2 - 5}{3x^2 - x + 7}$

36. $\lim\limits_{x \to \infty} \dfrac{4 - 3x - 12x^2}{1 + 5x + 3x^2}$

37. $\lim\limits_{x \to \infty} \dfrac{4 - 3x}{5 - 2x^2}$

38. $\lim\limits_{x \to \infty} \dfrac{6x^2 - x}{4x^4 - 3x^3}$

39. $\lim\limits_{x \to \infty} \dfrac{8x^4 - 3x^2}{5x^2 + 6x}$

40. $\lim\limits_{x \to \infty} \dfrac{6x^4 - x^3}{4x^2 - 3x^3}$

41. $\lim\limits_{x \to \infty} \dfrac{6x^4 - 5x^2 + 7}{8x^6 + 4x^3 - 8x}$

42. $\lim\limits_{x \to \infty} \dfrac{4x^2 - 8x + 11}{7x^3 + 2x - 3}$

43. $\lim\limits_{x \to \infty} \dfrac{11x^5 + 4x^3 - 6x + 2}{6x^3 + 5x^2 + 3x - 1}$

44. $\lim\limits_{x \to \infty} \dfrac{7x^9 - 6x^3 + 2x^2 - 10}{2x^6 + 4x^2 - x + 23}$

45. Consider

$$f(x) = \begin{cases} 1, & \text{for } x \neq 2, \\ -1, & \text{for } x = 2. \end{cases}$$

Find each of the following.

a) $\lim\limits_{x \to 0} f(x)$

b) $\lim\limits_{x \to 2^-} f(x)$

c) $\lim\limits_{x \to 2^+} f(x)$

d) $\lim\limits_{x \to 2} f(x)$

e) Is f continuous at 0? at 2?

46. Consider

$$g(x) = \begin{cases} -4, & \text{for } x = 3, \\ 2x + 5, & \text{for } x \neq 3. \end{cases}$$

Find each of the following.

a) $\lim\limits_{x \to 3^-} g(x)$

b) $\lim\limits_{x \to 3^+} g(x)$

c) $\lim\limits_{x \to 3} g(x)$

d) $\lim\limits_{x \to 2} g(x)$

e) Is g continuous at 3? at 2?

APPLICATIONS

◆   **Business and Economics**

47. *Depreciation.* A new conveyor system costs $10,000. In any year, it depreciates 8% of its value at the beginning of that year.
    a) What is the annual depreciation in each of the first 5 years?
    b) What is the total depreciation at the end of 10 years?
    c) What is the limit of the sum of the annual depreciation costs?

48. *Depreciation.* A new car costs $26,000. In any year, it depreciates 30% of its value at the beginning of that year.
    a) What is the annual depreciation in each of the first 5 years?
    b) What is the total depreciation at the end of 10 years?
    c) What is the limit of the sum of the annual depreciation costs?

49. *Average cost.* The average cost per unit in dollars for a company to produce x units of a product is given by the function

$$A(x) = \frac{13x + 100}{x}.$$

Find $\lim_{x \to \infty} A(x)$. Explain the meaning of this limit.

◆   **Life and Physical Sciences**

50. *Medical dosage.* The function

$$N(t) = \frac{0.8t + 1000}{5t + 4}$$

gives the body concentration $N(t)$, in parts per million, of a certain dosage of medication after time t, in hours. Find $\lim_{t \to \infty} N(t)$. Explain the meaning of this limit.

◆   **General Interest**

51. *Baseball: Earned-run average.* A pitcher's *earned-run average* (the average number of runs given up every 9 innings, or 1 game) is given by

$$E = 9 \cdot \frac{n}{i},$$

where n = the number of earned runs allowed and

i = the number of innings pitched. Suppose we fix the number of earned runs allowed at 4 and let i vary. We get a function given by

$$E(i) = 9 \cdot \frac{4}{i}.$$

a) Complete the following table, rounding to two decimal places.

Innings pitched (i)	Earned-run average (E)
9	
8	
7	
6	
5	
4	
3	
2	
1	
$\frac{2}{3}$ (2 outs)	
$\frac{1}{3}$ (1 out)	

b) Find $\lim_{i \to 0^+} E(i)$.
c) On the basis of parts (a) and (b), determine a pitcher's earned run average if 4 runs were allowed and there were 0 outs.

_____

Find the limit, if it exists.

52. $\lim\limits_{x\to-\infty} \dfrac{-3x^2 + 5}{2 - x}$

53. $\lim\limits_{x\to0} \dfrac{|x|}{x}$

54. $\lim\limits_{x\to-2} \dfrac{x^3 + 8}{x^2 - 4}$

55. $\lim\limits_{x\to\infty} \dfrac{-6x^3 + 7x}{2x^2 - 3x - 10}$

56. $\lim\limits_{x\to-\infty} \dfrac{-6x^3 + 7x}{2x^2 - 3x - 10}$

57. $\lim\limits_{x\to1} \dfrac{x^3 - 1}{x^2 - 1}$

58. $\lim\limits_{x\to-\infty} \dfrac{7x^5 + x - 9}{6x + x^3}$

59. $\lim\limits_{x\to-\infty} \dfrac{2x^4 + x}{x + 1}$

   TECHNOLOGY CONNECTION

Further Use of Input–Output Tables to Find Limits

Graphers that create input–output tables can be especially useful to find certain limits. Consider

$$\lim\limits_{x\to0} \dfrac{\sqrt{1 + x} - 1}{x}.$$

Part of a table for this function starting at $x = -1$ for Step $= 0.5$ is shown below on the left:

x	y
−1	1
−0.5	0.585786
0	ERROR
0.5	0.449490
1	0.414214
1.5	0.387426
2	0.366025

x	y
−0.03	0.503807
−0.02	0.502525
−0.01	0.501256
0	ERROR
0.01	0.498756
0.02	0.497525
0.03	0.496305

By using smaller and smaller step values and beginning closer to the limit that we are approaching, we can refine the table and obtain a better estimate of the limit. Part of a table for Step $= 0.01$ is shown above on the right.

   We might assert now that the limit is 0.5, which it is. We can verify this further by graphing

in dot mode, using zoom and trace with smaller viewing windows. We see that

$$\lim\limits_{x\to0} \dfrac{\sqrt{1 + x} - 1}{x} = 0.5.$$

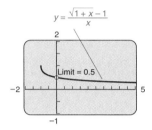

Using a grapher that creates input–output tables, find each of the following limits, if it exists. Start with Step $= 0.1$. Then move to 0.01, 0.001, and 0.0001. When you think you know the limit, graph, and then use the trace feature to further verify your assertion.

60. $\lim\limits_{x\to1} \dfrac{\sqrt{x} - 1}{x - 1}$

61. $\lim\limits_{a\to-2} \dfrac{a^2 - 4}{\sqrt{a^2 + 5} - 3}$

62. $\lim\limits_{x\to0} \dfrac{\sqrt{4 + x} - \sqrt{4 - x}}{x}$

63. $\lim\limits_{x\to0} \dfrac{\sqrt{3 - x} - \sqrt{3}}{x}$

64. $\lim\limits_{x\to0} \dfrac{\sqrt{7 + 2x} - \sqrt{7}}{x}$

65. $\lim\limits_{x\to1} \dfrac{x - \sqrt[4]{x}}{x - 1}$

66. $\lim\limits_{x\to0} \dfrac{7 - \sqrt{49 - x^2}}{x}$

67. $\lim\limits_{x\to4} \dfrac{2 - \sqrt{x}}{4 - x}$

68. ◈ Graph the function

$$f(x) = \dfrac{\sqrt{x^2 + 3x + 2}}{x - 3}.$$

Use several different viewing windows.

a) Estimate $\lim_{x \to \infty} f(x)$ and $\lim_{x \to -\infty} f(x)$ using the graph and input–output tables as needed to refine your estimates. Explain.

b) Describe the outputs of the function on the interval $(-2, -1)$. Explain.

c) What appears to be the domain of the function? Explain.

d) Find $\lim_{x \to -2^-} f(x)$ and $\lim_{x \to -1^+} f(x)$. Explain.

2.3 Average Rates of Change

OBJECTIVES

• Compute an average rate of change.
• Find a simplified difference quotient.

A car travels 110 miles in 2 hours. Its *average speed* is 110 mi/2 hr, or 55 mi/hr. This is the *average rate of change* of distance with respect to time. The concept of average rate of change applies to many situations, as we will see in the following examples.

The graph below shows the total production of suits by Raggs, Ltd., during one morning of work. Industrial psychologists have found curves like this typical of the production of factory workers.

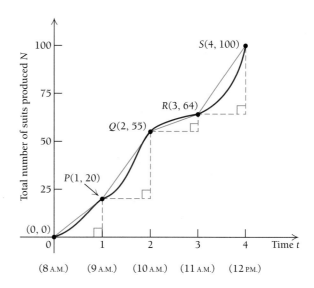

EXAMPLE 1 *Business: Production.* What was the number of suits produced from 9 A.M. to 10 A.M.?

Solution   At 9 A.M., 20 suits had been produced. At 10 A.M., 55 suits had been produced. In the hour from 9 A.M. to 10 A.M., the number of suits produced was

$$55 \text{ suits} - 20 \text{ suits}, \quad \text{or} \quad 35 \text{ suits}.$$

Note that 35 is the slope of the line from P to Q.   ◆

EXAMPLE 2   *Business: Average rate of change.* What was the average number of suits produced per hour from 9 A.M. to 11 A.M.?

Solution   We have

$$\frac{64 \text{ suits} - 20 \text{ suits}}{11 \text{ A.M.} - 9 \text{ A.M.}} = \frac{44 \text{ suits}}{2 \text{ hr}} = 22 \frac{\text{suits}}{\text{hr}} \text{ (suits per hour).}$$

Note that 22 is the slope of the line from P to R. The line from P to R is not shown in the graph. ◆

Let us consider a function $y = f(x)$ and two inputs x_1 and x_2. The *change in input*, or the *change in x*, is

$$x_2 - x_1.$$

The *change in output*, or the *change in y*, is

$$y_2 - y_1,$$

where $y_1 = f(x_1)$ and $y_2 = f(x_2)$.

DEFINITION

The *average rate of change of y with respect to x*, as x changes from x_1 to x_2, is the ratio of the change in output to the change in input:

$$\frac{y_2 - y_1}{x_2 - x_1}, \quad \text{where } x_2 \neq x_1.$$

If we look at a graph of the function, we see that

$$\frac{y_2 - y_1}{x_2 - x_1} = \frac{f(x_2) - f(x_1)}{x_2 - x_1}$$

and that this is the slope of the line from $P(x_1, y_1)$ to $Q(x_2, y_2)$. The line $\overleftrightarrow{PQ}$ is called a **secant line.**

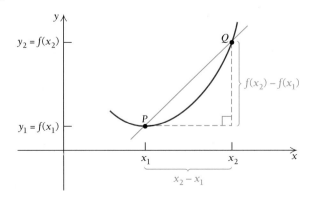

EXAMPLE 3   For $y = f(x) = x^2$, find the average rate of change as:

a) x changes from 1 to 3;

b) x changes from 1 to 2;

c) x changes from 2 to 3.

Solution   The graph below is not necessary to the computations, but gives us a look at two of the secant lines whose slopes are being computed.

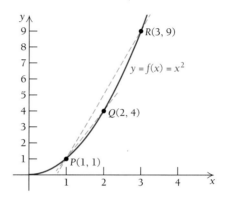

a) When $x_1 = 1$,

$$y_1 = f(x_1) = f(1) = 1^2 = 1;$$

and when $x_2 = 3$,

$$y_2 = f(x_2) = f(3) = 3^2 = 9.$$

The average rate of change is

$$\frac{y_2 - y_1}{x_2 - x_1} = \frac{f(x_2) - f(x_1)}{x_2 - x_1}$$

$$= \frac{9 - 1}{3 - 1}$$

$$= \frac{8}{2} = 4.$$

b) When $x_1 = 1$,

$$y_1 = f(x_1) = f(1) = 1^2 = 1;$$

and when $x_2 = 2$,

$$y_2 = f(x_2) = f(2) = 2^2 = 4.$$

The average rate of change is

$$\frac{4 - 1}{2 - 1} = \frac{3}{1} = 3.$$

c) When $x_1 = 2$,

$$y_1 = f(x_1) = f(2)$$
$$= 2^2 = 4;$$

and when $x_2 = 3$,

$$y_2 = f(x_2) = f(3)$$
$$= 3^2 = 9.$$

The average rate of change is

$$\frac{9 - 4}{3 - 2} = \frac{5}{1} = 5.$$ ◆

For a linear function, the average rates of change are the same for any choice of x_1 and x_2; that is, they are equal to the slope m of the line. As we saw in Example 3, a function that is not linear has average rates of change that vary with the choice of x_1 and x_2.

Difference Quotients

Let us now simplify our notation a bit by eliminating the subscripts. Instead of x_1, we will write simply x.

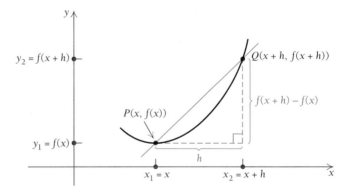

To get from x_1, or x, to x_2, we move a distance h. Thus, $x_2 = x + h$. Then the average rate of change, also called a **difference quotient,** is given by

$$\frac{y_2 - y_1}{x_2 - x_1} = \frac{f(x_2) - f(x_1)}{x_2 - x_1}$$
$$= \frac{f(x + h) - f(x)}{(x + h) - x}$$
$$= \frac{f(x + h) - f(x)}{h}.$$

> **DEFINITION**
>
> The average rate of change of f with respect to x is also called the *difference quotient*. It is given by
> $$\frac{f(x + h) - f(x)}{h}, \quad \text{where } h \neq 0.$$
> The difference quotient is equal to the slope of the line from a point $P(x, f(x))$ to a point $Q(x + h, f(x + h))$.

EXAMPLE 4   For $f(x) = x^2$, find the difference quotient when:

a) $x = 5$ and $h = 3$;

b) $x = 5$ and $h = 0.1$.

Solution

a) We substitute $x = 5$ and $h = 3$ into the formula:

$$\frac{f(x + h) - f(x)}{h} = \frac{f(5 + 3) - f(5)}{3} = \frac{f(8) - f(5)}{3}.$$

Now $f(8) = 8^2 = 64$ and $f(5) = 5^2 = 25$, and we have

$$\frac{f(8) - f(5)}{3} = \frac{64 - 25}{3} = \frac{39}{3} = 13.$$

The difference quotient is 13. It is also the slope of the line from $(5, 25)$ to $(8, 64)$.

b) We substitute $x = 5$ and $h = 0.1$ into the formula:

$$\frac{f(x + h) - f(x)}{h} = \frac{f(5 + 0.1) - f(5)}{0.1} = \frac{f(5.1) - f(5)}{0.1}.$$

Now $f(5.1) = (5.1)^2 = 26.01$ and $f(5) = 25$, and we have

$$\frac{f(5.1) - f(5)}{0.1} = \frac{26.01 - 25}{0.1} = \frac{1.01}{0.1} = 10.1. \qquad \blacklozenge$$

For the function in Example 4, let us find a general form of the difference quotient. This will allow more efficient computations.

EXAMPLE 5   For $f(x) = x^2$, find a **simplified** form of the **difference quotient.** Then find the value of the difference quotient when $x = 5$ and $h = 0.1$.

Solution   We have

$$f(x) = x^2,$$

so

$$f(x + h) = (x + h)^2 = x^2 + 2xh + h^2.$$

Then

$$f(x + h) - f(x) = (x^2 + 2xh + h^2) - x^2 = 2xh + h^2.$$

Thus,

$$\frac{f(x + h) - f(x)}{h} = \frac{2xh + h^2}{h} = \frac{h(2x + h)}{h} = 2x + h, \quad h \neq 0.$$

It is important to note that a difference quotient is defined only when $h \neq 0$. The simplification above is valid only for nonzero values of h.

When $x = 5$ and $h = 0.1$,

$$\frac{f(x + h) - f(x)}{h} = 2x + h = 2 \cdot 5 + 0.1 = 10 + 0.1 = 10.1. \qquad \blacklozenge$$

EXAMPLE 6   For $f(x) = x^3$, find a simplified form of the difference quotient.

Solution   Now $f(x) = x^3$, so

$$f(x + h) = (x + h)^3$$
$$= x^3 + 3x^2h + 3xh^2 + h^3.$$

(This was shown in Exercise Set 1.1.) Then

$$f(x + h) - f(x) = (x^3 + 3x^2h + 3xh^2 + h^3) - x^3$$
$$= 3x^2h + 3xh^2 + h^3.$$

Thus,

$$\frac{f(x + h) - f(x)}{h} = \frac{3x^2h + 3xh^2 + h^3}{h}$$
$$= \frac{h(3x^2 + 3xh + h^2)}{h}$$
$$= 3x^2 + 3xh + h^2, \quad h \neq 0.$$

Again, this is true *only* for $h \neq 0$. $\blacklozenge$

EXAMPLE 7   For $f(x) = 3/x$, find a simplified form of the difference quotient.

Solution   Now

$$f(x) = \frac{3}{x},$$

so

$$f(x + h) = \frac{3}{x + h}.$$

Then

$$f(x + h) - f(x) = \frac{3}{x + h} - \frac{3}{x}$$

$$= \frac{3}{x + h} \cdot \frac{x}{x} - \frac{3}{x} \cdot \frac{x + h}{x + h} \quad \text{Here we are multiplying by 1 to get a common denominator.}$$

$$= \frac{3x - 3(x + h)}{x(x + h)}$$

$$= \frac{3x - 3x - 3h}{x(x + h)}$$

$$= \frac{-3h}{x(x + h)}.$$

Thus,

$$\frac{f(x + h) - f(x)}{h} = \frac{\dfrac{-3h}{x(x + h)}}{h}$$

$$= \frac{-3h}{x(x + h)} \cdot \frac{1}{h} = \frac{-3}{x(x + h)}, \quad h \neq 0.$$

This is true *only* for $h \neq 0$. ◆

2.3 Exercise Set

For the functions in each of Exercises 1–12, (a) find a simplified difference quotient and (b) complete the following table.

x	h	$\dfrac{f(x + h) - f(x)}{h}$
4	2	
4	1	
4	0.1	
4	0.01	

1. $f(x) = 7x^2$
2. $f(x) = 5x^2$
3. $f(x) = -7x^2$
4. $f(x) = -5x^2$
5. $f(x) = 7x^3$
6. $f(x) = 5x^3$
7. $f(x) = \dfrac{5}{x}$
8. $f(x) = \dfrac{4}{x}$
9. $f(x) = -2x + 5$
10. $f(x) = 2x + 3$
11. $f(x) = x^2 - x$
12. $f(x) = x^2 + x$

APPLICATIONS

◆ **Business and Economics**

13. *Utility.* Utility is a type of function that occurs in economics. When a consumer receives x units of a

certain product, a certain amount of pleasure, or utility, U, is derived from them. Below is a typical graph of a utility function.

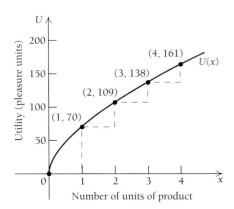

a) Find the average rate of change of U as x changes from 0 to 1; from 1 to 2; from 2 to 3; from 3 to 4.
b) Why do you think the average rates of change are decreasing?

14. *Advertising results.* The following graph shows a typical response to advertising. After an amount a is spent on advertising, the company sells $N(a)$ units of a product.

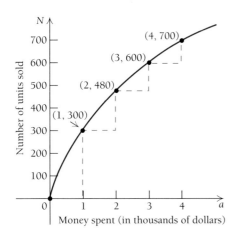

a) Find the average rate of change of N as a changes from 0 to 1; from 1 to 2; from 2 to 3; from 3 to 4.
b) Why do you think the average rates of change are decreasing?

15. *Total revenue.* A firm determines that the total revenue from the sale of x units of a certain product is given by

$$R(x) = -0.01x^2 + 1000x,$$

where $R(x)$ is in dollars.
a) Find $R(301)$.
b) Find $R(300)$.
c) Find $R(301) - R(300)$.
d) Find $\dfrac{R(301) - R(300)}{301 - 300}$.

16. *Total cost.* A firm determines that the total cost C of producing x units of a certain product is given by

$$C(x) = -0.05x^2 + 50x,$$

where $C(x)$ is in dollars.
a) Find $C(301)$.
b) Find $C(300)$.
c) Find $C(301) - C(300)$.
d) Find $\dfrac{C(301) - C(300)}{301 - 300}$.

17. *Total revenue of Pepsico.* Pepsico is a company that owns many fast-food chains, including Taco Bell and Pizza Hut. It sells other food items through subsidiaries such as Frito-Lay and Pepsi-Cola. It has experienced great sales growth during the past few years, as shown in the following chart.

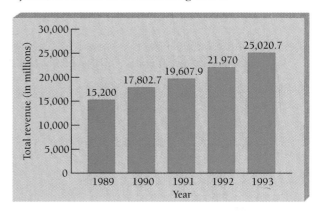

a) In 1990, the total revenue was $17,802.7 million. In 1993, it was $25,020.7 million. Find the average rate of change from 1990 to 1993.
b) In 1986, the total revenue was $930 million. Find the average rate of change of total revenue from 1986 to 1993.

◆   **Life and Physical Sciences**

18. *Temperature during an illness.* The temperature T, in degrees Fahrenheit, of a patient during an illness is shown in the graph below, where t = the time in days.

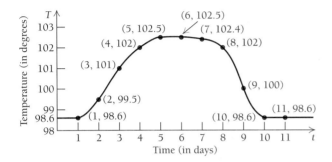

a) Find the average rate of change of T as t changes from 1 to 10. Using this rate of change, would you know that the person was sick?
b) Find the rate of change of T with respect to t, as t changes from 1 to 2; from 2 to 3; from 3 to 4; from 4 to 5; from 5 to 6; from 6 to 7; from 7 to 8; from 8 to 9; from 9 to 10; from 10 to 11.
c) When do you think the temperature began to rise?
d) When do you think the temperature reached its peak?
e) When do you think the temperature began to subside?
f) When was the temperature back to normal?

19. *Memory.* The total number of words $M(t)$ that a person can memorize in time t, in minutes, is shown in the following graph.

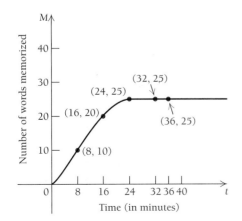

a) Find the average rate of change of M as t changes from 0 to 8; from 8 to 16; from 16 to 24; from 24 to 32; from 32 to 36.
b) Why do the average rates of change become 0 after 24 min?

20. *Average velocity.* A car is at a distance s, in miles, from its starting point in t hours, given by

$$s(t) = 10t^2.$$

a) Find $s(2)$ and $s(5)$.
b) Find $s(5) - s(2)$. What does this represent?
c) Find the average rate of change of distance with respect to time as t changes from $t_1 = 2$ to $t_2 = 5$. This is known as *average velocity*, or *speed*.

21. *Average velocity.* An object is dropped from a certain height. It is known that it will fall a distance s, in feet, in t seconds, given by

$$s(t) = 16t^2.$$

a) How far will the object fall in 3 sec?
b) How far will the object fall in 5 sec?
c) What is the average rate of change of distance with respect to time during the period from 3 to 5 sec? This is also *average velocity*, or *speed*.

22. *Gas mileage.* At the beginning of a trip, the odometer on a car reads 30,680 and the car has a full tank of gas. At the end of the trip, the odometer reads 30,970. It takes 20 gal of gas to fill the tank again.

a) What is the average rate of consumption (that is, the rate of change of the number of miles with respect to the number of gallons)?
b) What is the average rate of change of the number of gallons with respect to the number of miles?

◆   **Social Sciences**

23. *Population growth.* The two curves shown in the figure at the top of the following page describe the number of people in each of two countries at time t, in years.

a) Find the average rate of change of each population (the number of people in the population) with respect to time t as t changes from 0 to 4. This is often called an *average growth rate*.
b) If the calculation in part (a) were the only one made, would we detect the fact that the populations were growing differently?

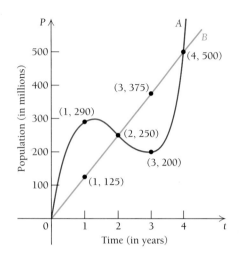

c) Find the average rates of change of each population as t changes from 0 to 1; from 1 to 2; from 2 to 3; from 3 to 4.

d) For which population does the statement "the population grew by 125 million each year" convey the least information about what really took place?

24. Marriage rate. It is known that in 1982, there were 2,495,000 marriages. In 1993, there were 2,362,000 marriages. Find the average rate of change of marriages with respect to time.

25. Divorce rate. It is known that in 1982, there were 1,180,000 divorces. In 1993, there were 1,215,000 divorces. Find the average rate of change of divorces with respect to time.

SYNTHESIS

Find the simplified difference quotient.

26. $f(x) = mx + b$

27. $f(x) = ax^2 + bx + c$

28. $f(x) = ax^3 + bx^2$

29. $f(x) = \sqrt{x}$

$\left(\text{Hint: Multiply by 1 using } \dfrac{\sqrt{x + h} + \sqrt{x}}{\sqrt{x + h} + \sqrt{x}}.\right)$

30. $f(x) = x^4$

31. $f(x) = \dfrac{1}{x^2}$

32. $f(x) = \dfrac{1}{1 - x}$

33. $f(x) = \dfrac{x}{1 + x}$

34. $f(x) = \sqrt{3 - 2x}$

35. ◈ **Teenage smoking.** In 1990, the percentage of teenagers who smoked daily was 19.2%. In 1992, it was 17%, and in 1993, it was 19.0%.

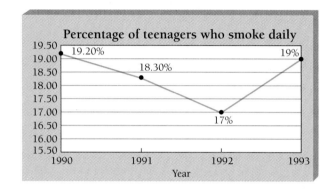

a) Find the average rate of change from 1990 to 1992. Was this change positive or negative? Relate this to slope.

b) Find the average rate of change from 1992 to 1993. Was this change positive or negative? Relate this to slope.

c) What, in your opinion, are the reasons for the changes?

36. ◈ **Meat consumption.** The following graph shows the changes in consumption of beef and chicken over several years (future years are predictions).

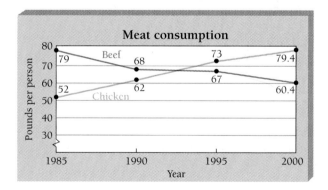

a) In what year was the consumption of each type of meat about the same?

b) What was the average rate of change in beef consumption from 1985 to 2000? Was this positive or negative?

c) What was the average rate of change in chicken consumption from 1985 to 2000? Was this positive or negative?

d) What, in your opinion, are the reasons for the changes?

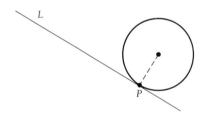

2.4 Differentiation Using Limits

Tangent Lines

A line tangent to a circle is a line that touches the circle exactly once.

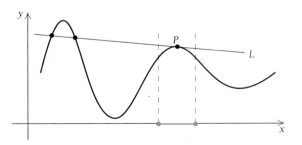

The path of the loose ski follows a tangent line to the ski chute at the point where it breaks loose.

This definition becomes unworkable with other curves. For example, consider the curve shown in Fig. 1. Line *L* touches the curve at point *P* but meets the curve at other places as well. It is considered a tangent line, but "touching at one point" cannot be its definition.

Note in Fig. 1 that over a small interval containing *P*, line *L* does touch the curve exactly once. This is still not a suitable definition of a *tangent line* because it allows a line like *M* in Fig. 2 to be a tangent line, which we will not accept.

FIGURE 1

TECHNOLOGY
CONNECTION

Graph $f(x) = 3x^5 - 20x^3$
with the viewing window
$[-3, 3, -80, 80]$ and
$xScl = 1$, $yScl = 10$. Then
also graph the lines
$y_1 = -7x - 10$,
$y_2 = -44x + 28$, and
$y_3 = -45x + 28$. Which
appear to be tangent lines to
the graph of f at the point
$(1, -17)$? Why might you
reject a line? Use smaller
viewing windows to try to
refine your guess.

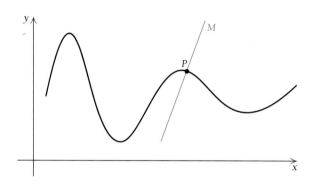

FIGURE 2

Later we will give a definition of a tangent line, but for now we will rely on intuition. In Fig. 3, lines L_1 and L_2 are not tangent lines. All the others are.

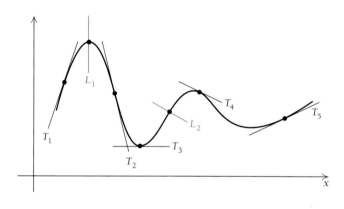

FIGURE 3

Why Do We Study Tangent Lines?

The reason for our study of tangent lines will become evident in Chapter 3. For now, look at the following graph of a total-profit function. Note that the largest (or maximum) value of the function occurs at the point where the graph has a horizontal tangent; that is, where the tangent line has slope 0.

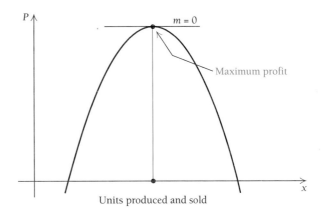

Differentiation Using Limits

We will define *tangent line* in such a way that it makes sense for *any* curve. To do this, we use the notion of limit.

We obtain the line tangent to the curve at point P by considering secant lines through P and neighboring points Q_1, Q_2, and so on. As the points Q approach P, the secant lines approach the tangent line. Each secant line has a slope. The slopes of the secant lines approach the slope of the tangent line. In fact, we *define* the **tangent line** as the line that contains the point P and has slope m, where m is the limit of the slopes of the secant lines as the points Q approach P.

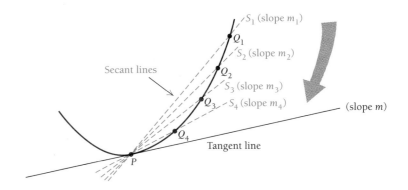

How might we calculate the limit m? Suppose P has coordinates $(x, f(x))$ in Fig. 4 on the following page. Then the first coordinate of Q is x plus some number h, or $x + h$. The coordinates of Q are $(x + h, f(x + h))$.

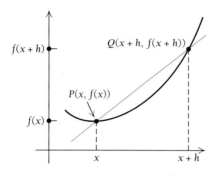

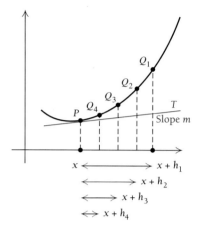

FIGURE 4   FIGURE 5

From Section 2.3, we know that the slope of the secant line $\overleftrightarrow{PQ}$ is given by the difference quotient

$$\frac{f(x + h) - f(x)}{h}.$$

Now, as we see in Fig. 5, as the points Q approach P, $x + h$ approaches x. That is, h approaches 0. Thus we have the following.

The slope of the tangent line $= m = \lim\limits_{h \to 0} \dfrac{f(x + h) - f(x)}{h}$.

The formal definition of the *derivative of a function f* can now be given. We will designate the derivative at x as $f'(x)$, rather than $m(x)$.

DEFINITION

For a function $y = f(x)$, its *derivative* at x is defined as

$$f'(x) = \lim_{h \to 0} \frac{f(x + h) - f(x)}{h},$$

provided the limit exists. If $f'(x)$ exists, then we say that f is *differentiable* at x.

"Nothing in this world is so powerful as an idea whose time has come."
                    Victor Hugo

This is the basic definition of *differential calculus.*

Let us now calculate some formulas for derivatives. That is, given a formula for a function f, we will be trying to find a formula for f'.

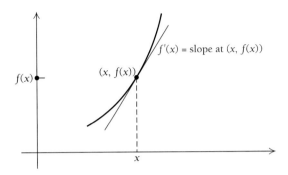

There are three steps in calculating a derivative.

1. Write down the difference quotient $[f(x + h) - f(x)]/h$.
2. Simplify the difference quotient.
3. Find the limit as h approaches 0.

A formula for the derivative of a linear function

$$f(x) = mx + b$$

is $$f'(x) = m.$$

Let us verify this using the definition.

EXAMPLE 1   For $f(x) = mx + b$, find $f'(x)$.

Solution   We follow the steps in the box above. Thus,

1. $$\frac{f(x + h) - f(x)}{h} = \frac{[m(x + h) + b] - (mx + b)}{h}$$

2. $$\frac{f(x + h) - f(x)}{h} = \frac{mx + mh + b - mx - b}{h}$$

$$= \frac{mh}{h} = m, \quad h \neq 0$$

3. $$\lim_{h \to 0} \frac{f(x + h) - f(x)}{h} = \lim_{h \to 0} m = m,$$

since m represents a constant.

Thus if $f(x) = mx + b$, then $f'(x) = m$. ◆

Consider the graph of $y = x^2$ that follows. Tangent lines are drawn at various points on the graph. Let $m(x) =$ the slope at the point $(x, f(x))$. Estimate the slope of each line and complete the table. Can you guess a formula for $m(x)$?

Lines	x	$m(x)$
L_1	-1	
L_2	$-\frac{1}{2}$	
L_3	0	
L_4	$\frac{1}{2}$	
L_5	1	

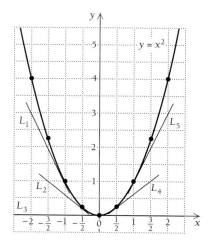

Now let us see if we can find a formula for the derivative of

$$f(x) = x^2.$$

We first find the slope of the tangent line at $x = 4$. That would be $f'(4)$. Then we find the general formula for $f'(x)$.

EXAMPLE 2   For $f(x) = x^2$, find $f'(4)$.

Solution   We have

1. $\dfrac{f(4 + h) - f(4)}{h} = \dfrac{(4 + h)^2 - 4^2}{h}$

2. $\dfrac{f(4 + h) - f(4)}{h} = \dfrac{16 + 8h + h^2 - 16}{h}$

$$= \dfrac{8h + h^2}{h}$$

$$= \dfrac{h(8 + h)}{h}$$

$$= 8 + h, \quad h \neq 0$$

3. $\displaystyle\lim_{h \to 0} \dfrac{f(4 + h) - f(4)}{h} = \lim_{h \to 0} (8 + h) = 8.$

Thus, $f'(4) = 8$.   ◆

EXAMPLE 3   For $f(x) = x^2$, find (the general formula) $f'(x)$.

Solution

1. We have

$$\frac{f(x + h) - f(x)}{h} = \frac{(x + h)^2 - x^2}{h}.$$

2. In Example 5 of Section 2.3, we showed how this difference quotient can be simplified to

$$\frac{f(x + h) - f(x)}{h} = 2x + h.$$

3. We want to find

$$\lim_{h \to 0} \frac{f(x + h) - f(x)}{h} = \lim_{h \to 0} (2x + h).$$

As $h \to 0$, we see that $2x + h \to 2x$. Thus,

$$\lim_{h \to 0} (2x + h) = 2x,$$

and we have

$$f'(x) = 2x,$$

which tells us, for example, that at $x = -3$, the curve has a tangent line whose slope is

$$f'(-3) = 2(-3), \quad \text{or} \quad -6.$$

We can say, simply, "The curve has slope -6 at the point $(-3, 9)$."

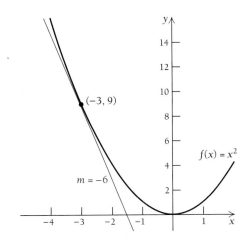

EXAMPLE 4   For $f(x) = x^3$, find $f'(x)$. Then find $f'(-1)$ and $f'(1.5)$.

Solution

1. We have

$$\frac{f(x + h) - f(x)}{h} = \frac{(x + h)^3 - x^3}{h}.$$

2. In Example 6 of Section 2.3, we showed how this difference quotient can be simplified to

$$\frac{f(x + h) - f(x)}{h} = 3x^2 + 3xh + h^2.$$

3. We then have

$$\lim_{h \to 0} \frac{f(x + h) - f(x)}{h} = \lim_{h \to 0} (3x^2 + 3xh + h^2) = 3x^2.$$

(An input–output table for this is shown in Example 4 of Section 2.2.) Thus, for $f(x) = x^3$, we have $f'(x) = 3x^2$. Then

$$f'(-1) = 3(-1)^2 = 3 \quad \text{and} \quad f'(1.5) = 3(1.5)^2 = 6.75.$$

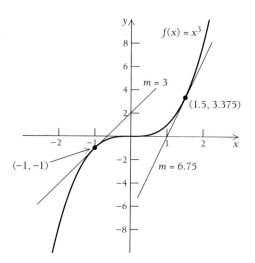

EXAMPLE 5   For $f(x) = 3/x$:

a) Find $f'(x)$.

b) Find $f'(1)$ and $f'(2)$.

c) Find an equation of the tangent line to the curve at the point $(1, 3)$.

d) Find an equation of the tangent line to the curve at the point $\left(2, \frac{3}{2}\right)$.

Solution

a) 1. We have

$$\frac{f(x + h) - f(x)}{h} = \frac{[3/(x + h)] - (3/x)}{h}.$$

2. In Example 7 of Section 2.3, we showed that this difference quotient can be simplified to

$$\frac{f(x + h) - f(x)}{h} = \frac{-3}{x(x + h)}.$$

3. We want to find

$$\lim_{h \to 0} \frac{f(x + h) - f(x)}{h} = \lim_{h \to 0} \frac{-3}{x(x + h)}.$$

As $h \to 0$, $x + h \to x$, so we have

$$f'(x) = \lim_{h \to 0} \frac{-3}{x(x + h)} = \frac{-3}{x^2}.$$

b) Then

$$f'(1) = \frac{-3}{1^2} = -3$$

and

$$f'(2) = \frac{-3}{2^2} = -\frac{3}{4}.$$

c) We know that the point $(1, 3)$ is on the graph of the function because $f(1) = 3$. To find an equation of the tangent line to the curve at the point $(1, 3)$, we use the fact that the slope at $x = 1$ is -3, as we found in the preceding work. Now we have

Point: $(1, 3)$,

Slope: -3.

We substitute into the point–slope equation (see Section 1.4):

$$y - y_1 = m(x - x_1)$$
$$y - 3 = -3(x - 1)$$
$$y = -3x + 3 + 3$$
$$y = -3x + 6.$$

The equation of the tangent line to the curve at $x = 1$ is $y = -3x + 6$ (see the figure at the top of the following page).

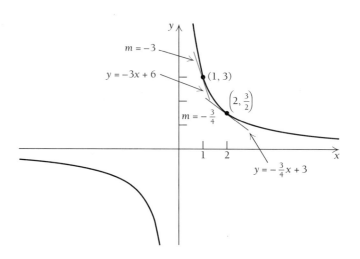

d) To find an equation of the tangent line to the curve at $x = 2$, we use the fact that the slope at $x = 2$ is $-\frac{3}{4}$, as we found in part (b). Now we have

> Point: $\left(2, \frac{3}{2}\right)$,
>
> Slope: $-\frac{3}{4}$.

We substitute into the point–slope equation (see Section 1.4):

$$y - y_1 = m(x - x_1)$$
$$y - \tfrac{3}{2} = -\tfrac{3}{4}(x - 2)$$
$$y = -\tfrac{3}{4}x + \tfrac{3}{2} + \tfrac{3}{2}$$
$$y = -\tfrac{3}{4}x + 3.$$

The equation of the tangent line to the curve at $x = 2$ is $y = -\frac{3}{4}x + 3$.

Because $f(0)$ does not exist, we cannot evaluate the difference quotient

$$\frac{f(0 + h) - f(0)}{h}.$$

Thus, $f'(0)$ does not exist. We say that "f is not differentiable at 0."   ◆

When a function is not defined at a point, it is not differentiable at that point. Also, if a function is discontinuous at a point, it is not differentiable at that point.

It can happen that a function f is defined and continuous at a point but that its derivative f' is not. The function f given by

$$f(x) = |x|$$

is an example. Note that

$$f(0) = |0| = 0,$$

so the function is defined at 0.

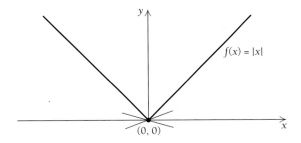

Suppose we try to draw a tangent line at (0, 0). A function like this with a corner (not smooth) would seem to have many tangent lines at (0, 0), and thus many slopes. The derivative at such a point would not be unique. Let us try to calculate the derivative at 0.

Since

$$f'(x) = \lim_{h \to 0} \frac{|x + h| - |x|}{h},$$

at $x = 0$, we have

$$f'(0) = \lim_{h \to 0} \frac{|0 + h| - |0|}{h} = \lim_{h \to 0} \frac{|h|}{h}.$$

h	$\dfrac{|h|}{h}$
2	$\dfrac{|2|}{2}$, or $\dfrac{2}{2}$, or 1
1	1
0.1	1
0.01	1
0.001	1

h	$\dfrac{|h|}{h}$
-2	$\dfrac{|-2|}{-2}$, or $\dfrac{2}{-2}$, or -1
-1	-1
-0.1	-1
-0.01	-1
-0.001	-1

Look at the input–output tables. Note that as h approaches 0 from the right, $|h|/h$ approaches 1, but as h approaches 0 from the left, $|h|/h$ approaches -1. Thus,

$$\lim_{h \to 0} \frac{|h|}{h} \quad \text{does not exist,}$$

so $f'(0)$ does not exist.

If a function has a "sharp point" or "corner," it will not have a derivative at that point.

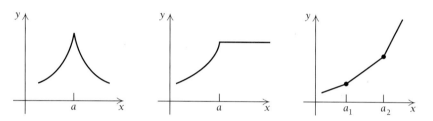

A function may also fail to be differentiable at a point by having a vertical tangent at that point. For example, the function shown below has a vertical tangent at point a. Recall that since the slope of a vertical line is undefined, there is no derivative at such a point.

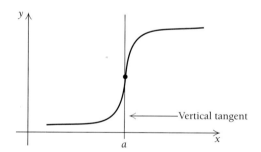

The function $f(x) = |x|$ illustrates the fact that although a function may be continuous at each point in an interval I, it may not be differentiable at each point in I. That is, continuity does not imply differentiability. On the other hand, if we know that a function is differentiable at each point in an interval I, then it is continuous over I. That is, if $f'(a)$ exists, then f is continuous at a. The function $f(x) = x^2$ is an example of a function that is differentiable over the interval $(-\infty, \infty)$ and is thus continuous everywhere. Also, if a function is discontinuous at some point a, then it is not differentiable at a. Thus when we know that a function is differentiable over an interval, it is *smooth* in the sense that there are no "sharp points," "corners," or "breaks" in the graph.

2.4 Exercise Set

In Exercises 1–14:

a) Graph the function.

b) Draw tangent lines to the graph at points whose x-coordinates are −2, 0, and 1.

c) Find $f'(x)$.

d) Find $f'(-2)$, $f'(0)$, and $f'(1)$. How do these slopes compare with those of the lines you drew in part (b)?

1. $f(x) = 5x^2$

2. $f(x) = 7x^2$

3. $f(x) = -5x^2$

4. $f(x) = -7x^2$

5. $f(x) = x^3$

6. $f(x) = -x^3$

7. $f(x) = 2x + 3$

8. $f(x) = -2x + 5$

9. $f(x) = -4x$

10. $f(x) = \dfrac{1}{2}x$

11. $f(x) = x^2 + x$

12. $f(x) = x^2 - x$

13. $f(x) = \dfrac{1}{x}$

14. $f(x) = \dfrac{5}{x}$

15. Find $f'(x)$ for $f(x) = mx$.

16. Find $f'(x)$ for $f(x) = ax^2 + bx + c$.

17. Find an equation of the tangent line to the graph of $f(x) = x^2$ at the point $(3, 9)$, at $(-1, 1)$, and at $(10, 100)$.

18. Find an equation of the tangent line to the graph of $f(x) = x^3$ at the point $(-2, -8)$, at $(0, 0)$, and at $(4, 64)$.

19. Find an equation of the tangent line to the graph of $f(x) = 5/x$ at the point $(1, 5)$, at $(-1, -5)$, and at $(100, 0.05)$.

20. Find an equation of the tangent line to the graph of $f(x) = 2/x$ at the point $(-1, -2)$, at $(2, 1)$, and at $\left(10, \frac{1}{5}\right)$.

21. Find an equation of the tangent line to the graph of $f(x) = 4 - x^2$ at the point $(-1, 3)$, at $(0, 4)$, and at $(5, -21)$.

22. Find an equation of the tangent line to the graph of $f(x) = x^2 - 2x$ at the point $(-2, 8)$, at $(1, -1)$, and at $(4, 8)$.

List the points in the graph at which each function is not differentiable.

23.

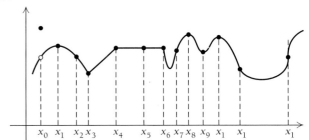

24.

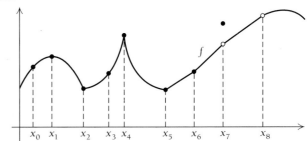

APPLICATIONS

◆ **Business and Economics**

25. *The postage function.* Consider the postage function defined in Exercise Set 2.1. At what values is the function not differentiable?

SYNTHESIS

Find $f'(x)$.

26. $f(x) = x^4$

27. $f(x) = \dfrac{1}{x^2}$

28. $f(x) = \dfrac{1}{1 - x}$

29. $f(x) = \dfrac{x}{1 + x}$

30. $f(x) = \sqrt{x}$

$\left(\text{Multiply by 1, using } \dfrac{\sqrt{x + h} + \sqrt{x}}{\sqrt{x + h} + \sqrt{x}}. \right)$

31. Consider the function f given by

$$f(x) = \dfrac{x^2 - 9}{x + 3}.$$

For what values is this function not differentiable?

32. ◈ Which of the following appear to be tangent lines? Try to explain why or why not.

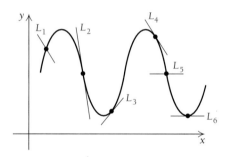

33. ◈ In the figure at the top of the next column, use a blue colored pencil and draw each secant line from point P to the points Q. Then use a red colored pencil and draw a tangent line to the curve at P. Describe what happens.

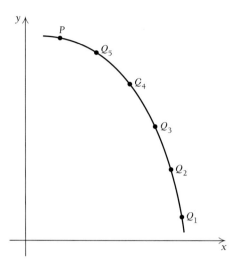

TECHNOLOGY CONNECTION

34. *Business: Growth of an investment.* A company determines that the value of an investment is V, in millions of dollars, after time t, in years, where V is given by

$$V(t) = 5t^3 - 30t^2 + 45t + 5\sqrt{t}.$$

Note: Computers and calculators usually graph functions using only the variables y and x, so you may need to change the variables when entering this function.

a) Graph V over the interval $[0, 5]$.

b) Find the equation of the secant line passing through the points $(1, V(1))$ and $(5, V(5))$. Then sketch this secant line using the same axes as in step (a).

c) Find the average rate of change of the investment between year 1 and year 5.

d) Repeat steps (b) and (c) for the pairs of points $(1, V(1))$ and $(4, V(4))$; $(1, V(1))$ and $(3, V(3))$; $(1, V(1))$ and $(1.5, V(1.5))$.

e) What appears to be the slope of the tangent line to the graph at the point $(1, V(1))$?

2.5 Differentiation Techniques: The Power and Sum–Difference Rules

Leibniz's Notation

When y is a function of x, we will also designate the derivative, $f'(x)$, as*

$$\frac{dy}{dx},$$

which is read "the derivative of y with respect to x." This notation was invented by the German mathematician Leibniz. It does *not* mean dy divided by dx! (That is, we cannot interpret dy/dx as a quotient until meanings are given to dy and dx, which we will not do here.) For example, if $y = x^2$, then

$$\frac{dy}{dx} = 2x.$$

We can also write

$$\frac{d}{dx} f(x)$$

to denote the derivative of f with respect to x. For example,

$$\frac{d}{dx} x^2 = 2x.$$

The value of dy/dx when $x = 5$ can be denoted by

$$\left. \frac{dy}{dx} \right|_{x=5}.$$

Thus for $dy/dx = 2x$,

$$\left. \frac{dy}{dx} \right|_{x=5} = 2 \cdot 5, \quad \text{or} \quad 10.$$

In general, for $y = f(x)$,

$$\left. \frac{dy}{dx} \right|_{x=a} = f'(a).$$

---

*The notation $D_x y$ is also used.

The Power Rule

In the remainder of this section, we will develop rules and techniques for efficient differentiation.

Look for a pattern in the following table, which contains functions and derivatives that we have found in previous work.

Function	Derivative
x^2	$2x^1$
x^3	$3x^2$
x^4	$4x^3$
$\dfrac{1}{x} = x^{-1}$	$-1 \cdot x^{-2} = \dfrac{-1}{x^2}$
$\dfrac{1}{x^2} = x^{-2}$	$-2 \cdot x^{-3} = \dfrac{-2}{x^3}$

Perhaps you have discovered the following theorem.

Historical Note: The German mathematician and philosopher Gottfried Wilhelm von Leibniz (1646–1716) and the English mathematician, philosopher, and physicist Sir Isaac Newton (1642–1727) are both credited with the invention of the calculus, though each made the invention independently of the other. Newton used the dot notation $\dot{y}$ for dy/dt, where y is a function of time, and this notation is still used, though it is not as common as Leibniz's notation.

THEOREM 1

The Power Rule

For any real number k,

$$\frac{d}{dx}x^k = k \cdot x^{k-1}.$$

We proved this theorem for the cases $k = 2$, 3, and -1 in Examples 3 and 4 and Exercise 13, respectively, of Section 2.4. We will not prove the other cases in this text. Note that this rule holds no matter what the exponent. That is, to differentiate x^k, we write the exponent k as the coefficient, followed by x with an exponent 1 less than k.

① Write the exponent as the coefficient.

$$\boxed{\begin{array}{c} ① \\ x^k \\ k \cdot x^{k-1} \quad ② \end{array}}$$

② Subtract 1 from the exponent.

EXAMPLE 1 $\dfrac{d}{dx}x^5 = 5x^4$ ◆

EXAMPLE 2
$$\frac{d}{dx}x = 1 \cdot x^{1-1}$$
$$= 1 \cdot x^0 = 1$$

◆

EXAMPLE 3
$$\frac{d}{dx}x^{-4} = -4 \cdot x^{-4-1}$$
$$= -4x^{-5}, \quad \text{or } -4 \cdot \frac{1}{x^5}, \quad \text{or } -\frac{4}{x^5}$$

◆

The Power Rule also allows us to differentiate $\sqrt{x}$. To do so, it helps to first convert to an expression with a rational exponent.

EXAMPLE 4
$$\frac{d}{dx}\sqrt{x} = \frac{d}{dx}x^{1/2} = \frac{1}{2} \cdot x^{(1/2)-1}$$
$$= \frac{1}{2}x^{-1/2}, \quad \text{or } \frac{1}{2} \cdot \frac{1}{x^{1/2}}, \quad \text{or } \frac{1}{2} \cdot \frac{1}{\sqrt{x}}, \quad \text{or } \frac{1}{2\sqrt{x}}$$

◆

EXAMPLE 5
$$\frac{d}{dx}x^{-2/3} = -\frac{2}{3}x^{(-2/3)-1}$$
$$= -\frac{2}{3}x^{-5/3}, \quad \text{or } -\frac{2}{3}\frac{1}{x^{5/3}}, \quad \text{or } -\frac{2}{3\sqrt[3]{x^5}}$$

◆

The Derivative of a Constant Function

Look at the graph of the constant function $F(x) = c$ shown below. What is the slope at each point on the graph?

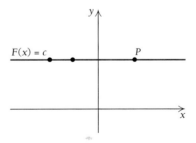

We now have the following.

THEOREM 2

The derivative of a constant function is 0. That is, $\frac{d}{dx}c = 0$.

Proof: Let F be the function given by $F(x) = c$. Then

$$\frac{F(x + h) - F(x)}{h} = \frac{c - c}{h} = \frac{0}{h} = 0.$$

The difference quotient is always 0. Thus, as h approaches 0, the limit of the difference quotient approaches 0, so $F'(x) = 0$.

The Derivative of a Constant Times a Function

Now let us consider differentiating functions such as

$$f(x) = 5x^2 \quad \text{and} \quad g(x) = -7x^4.$$

Note that we already know how to differentiate x^2 and x^4. Let us again look for a pattern in the results of Exercise Set 2.3.

Function	Derivative
$5x^2$	$10x$
$4x^{-1}$	$-4x^{-2}$
$-7x^2$	$-14x$
$5x^3$	$15x^2$

Perhaps you have discovered the following.

THEOREM 3

The derivative of a constant times a function is the constant times the derivative of the function. Using derivative notation, we can write this as

$$\frac{d}{dx}[c \cdot f(x)] = c \cdot f'(x).$$

Proof: Let F be the function given by $F(x) = cf(x)$. Then

$$\frac{F(x + h) - F(x)}{h} = \frac{cf(x + h) - cf(x)}{h}$$

$$= c\left[\frac{f(x + h) - f(x)}{h}\right].$$

As h approaches 0, the limit of the preceding expression is the same as c times $f'(x)$. Thus, $F'(x) = cf'(x)$.

Combining this rule with the Power Rule allows us to find many derivatives.

EXAMPLE 6 $\dfrac{d}{dx} 5x^4 = 5\dfrac{d}{dx}x^4 = 5 \cdot 4 \cdot x^{4-1} = 20x^3$ ◆

EXAMPLE 7 $\dfrac{d}{dx}(-9x) = -9\dfrac{d}{dx}x = -9 \cdot 1 = -9$ ◆

With practice you will be able to differentiate many such functions in one step.

EXAMPLE 8 $\dfrac{d}{dx}\dfrac{-4}{x^2} = \dfrac{d}{dx}(-4x^{-2}) = -4 \cdot \dfrac{d}{dx}x^{-2}$

$$= -4(-2)x^{-2-1}$$

$$= 8x^{-3}, \quad \text{or} \quad \dfrac{8}{x^3}$$ ◆

EXAMPLE 9 $\dfrac{d}{dx}(-x^{0.7}) = -1 \cdot \dfrac{d}{dx}x^{0.7}$

$$= -1 \cdot 0.7 \cdot x^{0.7-1}$$

$$= -0.7x^{-0.3}$$ ◆

The Derivative of a Sum or a Difference

In Exercise 11 of Exercise Set 2.4, you found that the derivative of

$$f(x) = x^2 + x$$

is

$$f'(x) = 2x + 1.$$

Note that the derivative of x^2 is $2x$, the derivative of x is 1, and the sum of these derivatives is $f'(x)$. This illustrates the following.

THEOREM 4

The Sum–Difference Rule

Sum: The derivative of a sum is the sum of the derivatives:

   If $F(x) = f(x) + g(x)$,   then $F'(x) = f'(x) + g'(x)$.

Difference: The derivative of a difference is the difference of the derivatives:

   If $F(x) = f(x) - g(x)$,   then $F'(x) = f'(x) - g'(x)$.

Proof: For the Sum Rule, we have

$$\frac{F(x + h) - F(x)}{h} = \frac{[f(x + h) + g(x + h)] - [f(x) + g(x)]}{h}$$

$$= \frac{f(x + h) - f(x)}{h} + \frac{g(x + h) - g(x)}{h}.$$

As h approaches 0, the two terms on the right approach $f'(x)$ and $g'(x)$, respectively, so their sum approaches $f'(x) + g'(x)$. Thus, $F'(x) = f'(x) + g'(x)$.

The proof of the Difference Rule is similar.

Any function that is a sum or a difference of several terms can be differentiated term by term.

EXAMPLE 10 $\dfrac{d}{dx}(3x + 7) = \dfrac{d}{dx}(3x) + \dfrac{d}{dx}(7)$

$$= 3\frac{d}{dx}(x) + \frac{d}{dx}(7)$$

$$= 3 \cdot 1 + 0$$

$$= 3 \qquad\qquad\qquad \blacklozenge$$

EXAMPLE 11 $\dfrac{d}{dx}(5x^3 - 3x^2) = \dfrac{d}{dx}(5x^3) - \dfrac{d}{dx}(3x^2)$

$$= 5\frac{d}{dx}x^3 - 3\frac{d}{dx}x^2$$

$$= 5 \cdot 3x^2 - 3 \cdot 2x$$

$$= 15x^2 - 6x \qquad\qquad\qquad \blacklozenge$$

EXAMPLE 12 $\dfrac{d}{dx}\left(24x - \sqrt{x} + \dfrac{2}{x}\right) = \dfrac{d}{dx}(24x) - \dfrac{d}{dx}(\sqrt{x}) + \dfrac{d}{dx}\left(\dfrac{2}{x}\right)$

$$= 24 \cdot \frac{d}{dx}x - \frac{d}{dx}x^{1/2} + 2 \cdot \frac{d}{dx}x^{-1}$$

$$= 24 \cdot 1 - \frac{1}{2}x^{(1/2)-1} + 2(-1)x^{-1-1}$$

$$= 24 - \frac{1}{2}x^{-1/2} - 2x^{-2}$$

$$= 24 - \frac{1}{2\sqrt{x}} - \frac{2}{x^2} \qquad\qquad\qquad \blacklozenge$$

A *word of caution*! The derivative of

$$f(x) + c,$$

a function plus a constant, is just the derivative of the function,

$$f'(x).$$

The derivative of

$$c \cdot f(x),$$

a function times a constant, is the constant times the derivative

$$c \cdot f'(x).$$

That is, for a product the constant is retained, but for a sum it is not.

TECHNOLOGY CONNECTION

Many graphers can draw tangent lines automatically. For example, here is the graph of $f(x) = -\frac{1}{3}x^3 + 6x^2 - 11x - 50$. It shows a tangent at $x = 12$.

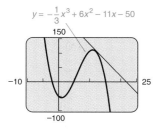

$$y = -\frac{1}{3}x^3 + 6x^2 - 11x - 50$$

Consult your manual. Graph each of the following functions and draw tangent lines at various points. Then estimate those x-coordinates at which tangent lines are horizontal. In Examples 13 and 14, you will learn to use calculus techniques to obtain exact values.

$$f(x) = -x^3 + 6x^2,$$
$$f(x) = \frac{1}{3}x^3 - 2x^2 + 4x,$$
$$f(x) = x\sqrt{4 - x^2}$$

Slopes of Tangent Lines

It is important to be able to determine points at which the tangent line to a curve has a certain slope, that is, points at which the derivative attains a certain value.

EXAMPLE 13 Find the points on the graph of $y = -x^3 + 6x^2$ at which the tangent line is horizontal.

Solution A horizontal tangent has slope 0. Thus we seek the values of x for which $dy/dx = 0$. That is, we want to find x such that

$$-3x^2 + 12x = 0.$$

We factor and solve:

$$-3x(x - 4) = 0$$
$$-3x = 0 \quad \text{or} \quad x - 4 = 0$$
$$x = 0 \quad \text{or} \qquad x = 4.$$

We are to find the points *on the graph,* so we must determine the second coordinates from the original equation, $y = -x^3 + 6x^2$.

For $x = 0$, $y = -0^3 + 6 \cdot 0^2 = 0$.

For $x = 4$, $y = -4^3 + 6 \cdot 4^2 = -64 + 96 = 32$.

Thus the points we are seeking are (0, 0) and (4, 32). This is shown on the graph.

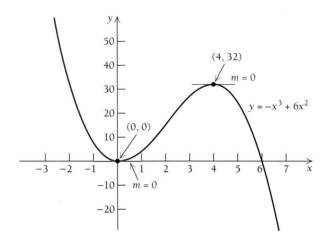

EXAMPLE 14   Find the points on the graph of $y = -x^3 + 6x^2$ at which the tangent has slope 6.

Solution   We want to find values of x for which $dy/dx = 6$. That is, we want to find x such that

$$-3x^2 + 12x = 6.$$

To solve, we add -6 on both sides and get

$$-3x^2 + 12x - 6 = 0.$$

We can simplify this equation by multiplying by $-\frac{1}{3}$ since each term has a common factor of -3. This gives us

$$x^2 - 4x + 2 = 0.$$

This is a quadratic equation, not readily factorable, so we use the quadratic formula, where $a = 1$, $b = -4$, and $c = 2$:

$$x = \frac{-b \pm \sqrt{b^2 - 4ac}}{2a}$$

$$= \frac{-(-4) \pm \sqrt{(-4)^2 - 4 \cdot 1 \cdot 2}}{2 \cdot 1}$$

$$= \frac{4 \pm \sqrt{8}}{2} = \frac{2 \cdot 2 \pm 2\sqrt{2}}{2 \cdot 1}$$

$$= \frac{2}{2} \cdot \frac{2 \pm \sqrt{2}}{1} = 2 \pm \sqrt{2}.$$

The solutions are $2 + \sqrt{2}$ and $2 - \sqrt{2}$.

We determine the second coordinates from the original equation. For $x = 2 + \sqrt{2}$,

$$
\begin{aligned}
y &= -(2 + \sqrt{2})^3 + 6(2 + \sqrt{2})^2 \\
&= -[(2 + \sqrt{2})^2(2 + \sqrt{2})] + 6(4 + 4\sqrt{2} + 2) \\
&= -[(6 + 4\sqrt{2})(2 + \sqrt{2})] + 6(6 + 4\sqrt{2}) \\
&= -[12 + 6\sqrt{2} + 8\sqrt{2} + 8] + 36 + 24\sqrt{2} \\
&= -[20 + 14\sqrt{2}] + 36 + 24\sqrt{2} \\
&= -20 - 14\sqrt{2} + 36 + 24\sqrt{2} \\
&= 16 + 10\sqrt{2}.
\end{aligned}
$$

Similarly, for $x = 2 - \sqrt{2}$,

$$
y = 16 - 10\sqrt{2}.
$$

Thus the points we are seeking are $(2 + \sqrt{2}, 16 + 10\sqrt{2})$ and $(2 - \sqrt{2}, 16 - 10\sqrt{2})$. This is shown on the graph.

TECHNOLOGY CONNECTION

Graph $y = \frac{1}{3}x^3 - 2x^2 + 4x$ and draw tangent lines at various points. Estimate points at which the tangent line is horizontal. Then use the calculus (analytic techniques) of Examples 13 and 14 to find the exact results.

Use a square viewing window. Estimate the points on the graph of $y = \frac{1}{3}x^3 - 2x^2 + 4x$ at which the tangent line has slope 3. Then use the calculus (analytic techniques) of Examples 13 and 14 to find the exact result.

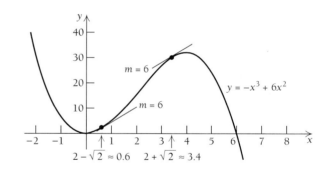

2.5 Exercise Set

Find $\dfrac{dy}{dx}$.

1. $y = x^7$

2. $y = x^8$

3. $y = 15$

4. $y = 78$

5. $y = 4x^{150}$

6. $y = 7x^{200}$

7. $y = x^3 + 3x^2$

8. $y = x^4 - 7x$

9. $y = 8\sqrt{x}$

10. $y = 4\sqrt{x}$

11. $y = x^{0.07}$

12. $y = x^{0.78}$

13. $y = \dfrac{1}{2}x^{4/5}$

14. $y = -4.8x^{1/3}$

15. $y = x^{-3}$

16. $y = x^{-4}$

17. $y = 3x^2 - 8x + 7$

18. $y = 4x^2 - 7x + 5$

19. $y = \sqrt[4]{x} - \dfrac{1}{x}$

20. $y = \sqrt[5]{x} - \dfrac{2}{x}$

Find $f'(x)$.

21. $f(x) = 0.64x^{2.5}$

22. $f(x) = 0.32x^{12.5}$

23. $f(x) = \dfrac{5}{x} - x$

24. $f(x) = \dfrac{4}{x} - x$

25. $f(x) = 4x - 7$

26. $f(x) = 7x + 11$

27. $f(x) = 4x + 9$

28. $f(x) = 7x - 14$

29. $f(x) = \dfrac{x^4}{4}$

30. $f(x) = \dfrac{x^3}{3}$

31. $f(x) = -0.01x^2 - 0.5x + 70$

32. $f(x) = -0.01x^2 + 0.4x + 50$

33. $f(x) = 3x^{-2/3} + x^{3/4} + x^{6/5} + \dfrac{8}{x^3}$

34. $f(x) = x^{-3/4} - 3x^{2/3} + x^{5/4} + \dfrac{2}{x^4}$

35. $f(x) = \dfrac{2}{x} - \dfrac{x}{2}$

36. $f(x) = \dfrac{x}{5} + \dfrac{5}{x}$

37. $f(x) = \dfrac{16}{x} - \dfrac{8}{x^3} + \dfrac{1}{x^4}$

38. $f(x) = \dfrac{20}{x^5} + \dfrac{1}{x^3} - \dfrac{2}{x}$

39. $f(x) = \sqrt{x} + \sqrt[3]{x} - \sqrt[4]{x} + \sqrt[5]{x}$

40. $f(x) = \dfrac{x^5 - 3x^4 + 2x^3 - 5x^2 - 8x + 4}{x^2}$

41. Find an equation of the tangent line to the graph of $f(x) = x^3 - 2x + 1$ at the point $(2, 5)$, at $(-1, 2)$, and at $(0, 1)$.

42. Find an equation of the tangent line to the graph of $f(x) = x^2 - \sqrt{x}$ at the point $(1, 0)$, at $(4, 14)$, and at $(9, 78)$.

For each function, find the points on the graph at which the tangent line is horizontal.

43. $y = x^2$

44. $y = -x^2$

45. $y = -x^3$

46. $y = x^3$

47. $y = 3x^2 - 5x + 4$

48. $y = 5x^2 - 3x + 8$

49. $y = -0.01x^2 - 0.5x + 70$

50. $y = -0.01x^2 + 0.4x + 50$

51. $y = 2x + 4$

52. $y = -2x + 5$

53. $y = 4$

54. $y = -3$

55. $y = -x^3 + x^2 + 5x - 1$

56. $y = -\frac{1}{3}x^3 + 6x^2 - 11x - 50$

57. $y = \frac{1}{3}x^3 - 3x + 2$

58. $y = x^3 - 6x + 1$

For each function, find the points on the graph at which the tangent line has slope 1.

59. $y = 20x - x^2$

60. $y = 6x - x^2$

61. $y = -0.025x^2 + 4x$

62. $y = -0.01x^2 + 2x$

63. $y = \frac{1}{3}x^3 + 2x^2 + 2x$

64. $y = \frac{1}{3}x^3 - x^2 - 4x + 1$

SYNTHESIS

65. Find the points on the graph of

$$y = x^4 - \tfrac{4}{3}x^2 - 4$$

at which the tangent line is horizontal.

66. Find the points on the graph of

$$y = 2x^6 - x^4 - 2$$

at which the tangent line is horizontal.

Find dy/dx. Each of the following can be differentiated using the rules developed in this section, but some algebra may be required beforehand.

67. $y = x(x - 1)$

68. $y = (x - 1)(x + 1)$

69. $y = (x - 2)(x + 3)$

70. $y = \dfrac{5x^2 - 8x + 3}{8}$

71. $y = \dfrac{x^5 + x}{x^2}$

72. $y = (5x)^2$

73. $y = (-4x)^3$

74. $y = \sqrt{7x}$

75. $y = \sqrt[3]{8x}$

76. $y = (x - 3)^2$

77. $y = (x + 1)^3$

78. $y = (x - 2)^3(x + 1)$

79. Prove Theorem 4 (Difference Rule).

80. ◈ Write a paragraph comparing the Power Rule, the Sum–Difference Rule, and the rules for differentiating a constant or a constant times a function.

81. ◈ Write a short biographical paper on the lives of Leibniz and/or Newton. Emphasize the contributions they made to many areas of science and society.

TECHNOLOGY CONNECTION

Graph each of the following. Draw tangent lines at various points. Estimate those values at which tangent lines are horizontal.

82. $f(x) = x^4 - 3x^2 + 1$

83. $f(x) = 1.6x^3 - 2.3x - 3.7$

84. $f(x) = 10.2x^4 - 6.9x^3$

85. $f(x) = \dfrac{5x^2 + 8x - 3}{3x^2 + 2}$

Some graphers can graph both a function f and its derivative f' using the same viewing window. For each of the following functions, graph f and f'.

86. $f(x) = 20x^3 - 3x^5$ **87.** $f(x) = x^4 - 3x^2 + 1$

88. $f(x) = x^3 - 2x - 2$ **89.** $f(x) = x^4 - x^3$

90. $f(x) = \dfrac{4x}{x^2 + 1}$

91. $f(x) = \dfrac{5x^2 + 8x - 3}{3x^2 + 2}$

2.6 Instantaneous Rates of Change

OBJECTIVES

* Given a formula for distance, find velocity and acceleration.
* Find instantaneous rates of change.

A car travels 108 miles in 2 hours. Its *average speed* (or *average velocity*) is 108 mi/2 hr, or 54 mi/hr. This is the *average rate of change* of distance with respect to time. At various times during the trip, however, the speedometer did not read 54. Thus we say that 54 is the *average*. A snapshot of the speedometer taken at any instant would indicate *instantaneous* speed, or **instantaneous rate of change.**

Some of the newer automobiles describe fuel economy by giving average miles per gallon and instantaneous miles per gallon.

Average rates of change are given by difference quotients. If distance s is a function of time t and if h is the duration of the trip, then average velocity is given by

$$\text{Average velocity} = \frac{s(t + h) - s(t)}{h}.$$

Instantaneous rates of change are found by letting h approach 0. Thus,

$$\text{Instantaneous velocity} = \lim_{h \to 0} \frac{s(t + h) - s(t)}{h} = s'(t).$$

EXAMPLE 1 *Physical science: Velocity.* An object travels in such a way that distance s (in miles) from the starting point is a function of time t (in hours) as follows:

$$s(t) = 10t^2.$$

a) Find the average velocity between the times $t = 2$ and $t = 5$.

b) Find the (instantaneous) velocity when $t = 4$.

Solution

a) From $t = 2$ to $t = 5$, $h = 3$, so

$$\begin{aligned}
\frac{s(t + h) - s(t)}{h} &= \frac{s(2 + 3) - s(2)}{3} \\
&= \frac{s(5) - s(2)}{3} \\
&= \frac{10 \cdot 5^2 - 10 \cdot 2^2}{3} \\
&= \frac{250 - 40}{3} \\
&= \frac{210}{3} = 70 \frac{\text{mi}}{\text{hr}}.
\end{aligned}$$

An instantaneous velocity.

b) The instantaneous velocity is given by $s'(t) = 20t$. Thus,

$$s'(4) = 20 \cdot 4 = 80 \frac{\text{mi}}{\text{hr}}.$$

We generally use the letter v for velocity. Thus we have the following.

> **DEFINITION**
>
> $$\text{Velocity} = v(t) = \lim_{h \to 0} \frac{s(t + h) - s(t)}{h} = s'(t)$$

The rate of change of velocity is called **acceleration.** We generally use the letter a for acceleration. Thus the following definition applies.

> **DEFINITION**
>
> $$\text{Acceleration} = a(t) = v'(t)$$

EXAMPLE 2 *Physical science: Distance, velocity, and acceleration.* For $s(t) = 10t^2$, find $v(t)$ and $a(t)$, where s is in miles and t is in hours. Then find the distance, velocity, and acceleration when $t = 4$ hr.

Solution We have

$$v(t) = s'(t) = 20t,$$
$$a(t) = v'(t) = 20.$$

Then

$$s(4) = 10(4)^2 = 160 \text{ miles},$$
$$v(4) = 20(4) = 80 \text{ mi/hr}, \quad \text{and}$$
$$a(4) = 20 \text{ mi/hr}^2.$$

If this distance function applies to a vehicle, then at time $t = 4$ hr, the distance is 160 miles, the velocity, or instantaneous speed, is 80 mi/hr, and the acceleration is 20 miles per hour per hour, which we abbreviate as 20 mi/hr^2.   ◆

In general, derivatives give instantaneous rates of change.

DEFINITION

If y is a function of x, then the (instantaneous) *rate of change of y with respect to x* is given by the derivative

$$\frac{dy}{dx} = f'(x) = \lim_{h \to 0} \frac{f(x + h) - f(x)}{h}.$$

EXAMPLE 3 *Life science: Volume of a cancer tumor.* The spherical volume V of a cancer tumor is given by

$$V(r) = \tfrac{4}{3}\pi r^3,$$

where $r = $ the radius of the tumor, in centimeters.

a) Find the rate of change of the volume with respect to the radius.

b) Find the rate of change of the volume at $r = 1.2$ cm.

Solution

a) $\dfrac{dV}{dr} = V'(r) = 3 \cdot \dfrac{4}{3} \cdot \pi r^2 = 4\pi r^2$

(This turns out to be the surface area.)

b) $V'(1.2) = 4\pi(1.2)^2 = 5.76\pi \approx 18\dfrac{\text{cm}^3}{\text{cm}} = 18 \text{ cm}^2$ ◆

EXAMPLE 4   *Life science: Population growth.* The initial population in a bacteria colony is 10,000. After t hours, the colony has grown to a number $P(t)$ given by

$$P(t) = 10,000(1 + 0.86t + t^2).$$

a) Find the rate of change of the population P with respect to time t. This is also known as the **growth rate.**

b) Find the number of bacteria present after 5 hr. Also, find the growth rate when $t = 5$.

Solution

a) Note that $P(t) = 10,000 + 8600t + 10,000t^2$. Then

$$P'(t) = 8600 + 20,000t.$$

b) The number of bacteria present when $t = 5$ is given by

$$P(5) = 10,000 + 8600 \cdot 5 + 10,000 \cdot 5^2 = 303,000.$$

The growth rate when $t = 5$ is given by

$$P'(5) = 8600 + 20,000 \cdot 5 = 108,600 \; \frac{\text{bacteria}}{\text{hr}}.$$

Thus at $t = 5$, there are 303,000 bacteria present, and the colony is growing at the rate of 108,600 bacteria per hour.   ◆

Rates of Change in Business and Economics

In the study of business and economics, we are frequently interested in how such quantities as cost, revenue, and profit change with an increase in product quantity. In particular, we are interested in what is called **marginal*** cost or profit (or whatever). This term is used to signify the *rate of change with respect to quantity*. Thus, if

$$C(x) = \text{the } total\ cost \text{ of producing } x \text{ units of a product}$$
$$\text{(usually considered in some time period),}$$

then $C'(x) = \text{the } \textbf{marginal cost}$

$$= \text{the rate of change of the total cost with respect to}$$
$$\text{the number of units, } x, \text{ produced.}$$

Let us think about these interpretations. The total cost of producing 5 units of a product is $C(5)$. The rate of change $C'(5)$ is the cost per unit at that stage in the production process. That this cost per unit does not include fixed costs is seen in this

---

*The term "marginal" comes from the Marginalist School of Economic Thought, which originated in Austria for the purpose of applying mathematics and statistics to the study of economics.

example:

$$C(x) = \underbrace{(x^2 + 4x)}_{\text{Variable costs}} \quad + \quad \underbrace{\$10,000.}_{\text{Fixed costs (constant)}}$$

Because the derivative of a constant is 0,

$$C'(x) = 2x + 4.$$

This verifies an economic principle stating that the fixed costs of a company have no effect on marginal cost.

Following are some other marginal functions. Recall that

$$R(x) = \text{the } total \ revenue \text{ from the sale of } x \text{ units.}$$

Then

$$R'(x) = \text{the } \textbf{marginal revenue}$$
$$= \text{the rate of change of the total revenue with respect to the number of units, } x, \text{ sold.}$$

Also,

$$P(x) = \text{the } total \ profit \text{ from the production and sale of } x \text{ units of a product}$$
$$= R(x) - C(x).$$

Then

$$P'(x) = \text{the } \textbf{marginal profit}$$
$$= \text{the rate of change of the total profit with respect to the number of units, } x, \text{ produced and sold}$$
$$= R'(x) - C'(x).$$

 TECHNOLOGY CONNECTION

Business: Marginal Revenue, Cost, and Profit

Using the same viewing window, graph these total-revenue and total-cost functions:

$$R(x) = 50x - 0.5x^2 \quad \text{and}$$
$$C(x) = 10x + 3.$$

Then find $P(x)$ and graph it using the same viewing window. Find $R'(x)$, $C'(x)$, and $P'(x)$ and graph them using a new viewing window. Then find $R(40)$, $C(40)$, $P(40)$, $R'(40)$, $C'(40)$, and $P'(40)$. Is the marginal revenue constant?

EXAMPLE 5   *Business: Marginal revenue, cost, and profit.* Given

$$R(x) = 50x, \quad \text{and}$$
$$C(x) = 2x^3 - 12x^2 + 40x + 10,$$

find each of the following.

a) Total profit $P(x)$

b) Total revenue $R(2)$, cost $C(2)$, and profit $P(2)$ from the production and sale of 2 units of the product

c) Marginal revenue $R'(x)$, cost $C'(x)$, and profit $P'(x)$

d) Marginal revenue $R'(2)$, cost $C'(2)$, and profit $P'(2)$ at that point in the process where 2 units have been produced and sold

Solution

a) The total profit $P(x) = R(x) - C(x) = 50x - (2x^3 - 12x^2 + 40x + 10)$
$$= -2x^3 + 12x^2 + 10x - 10$$

b) $R(2) = 50 \cdot 2 = \$100$ (the total revenue from the sale of the first 2 units);

$C(2) = 2 \cdot 2^3 - 12 \cdot 2^2 + 40 \cdot 2 + 10 = \$58$ (the total cost of producing the first 2 units);

$P(2) = R(2) - C(2) = \$100 - \$58 = \$42$ (the total profit from the production and sale of the first 2 units)

c) The marginal revenue $R'(x) = 50$;

The marginal cost $C'(x) = 6x^2 - 24x + 40$;

The marginal profit $P'(x) = R'(x) - C'(x) = 50 - (6x^2 - 24x + 40)$
$$= -6x^2 + 24x + 10$$

d) $R'(2) = \$50$ per unit;
$C'(2) = 6 \cdot 2^2 - 24 \cdot 2 + 40 = \$16$ per unit;
$P'(2) = \$50 - \$16 = \$34$ per unit

Note that the marginal revenue in this example is constant. No matter how much is produced and sold, the revenue per unit stays the same. This may not always be the case. Also note that $C'(2)$, or $16 per unit, is not the average cost per unit, which is given by

$$\frac{\text{Total cost of producing 2 units}}{2 \text{ units}} = \frac{\$58}{2} = \$29 \text{ per unit.} \qquad \blacklozenge$$

In general, we have the following.

DEFINITION

$$A(x) = \text{the } \textit{average cost} \text{ of producing } x \text{ units} = \frac{C(x)}{x}$$

Let us look at a typical marginal-cost function C' and its associated total-cost function C.

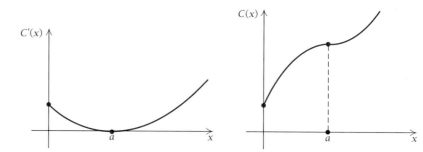

Marginal cost normally decreases as more units are produced until it reaches some minimum value at a, and then it increases. (This is probably due to factors such as paying overtime or buying more machinery.) Since $C'(x)$ represents the slope of $C(x)$ and is positive and decreasing up to a, the graph of $C(x)$ turns down as x goes from 0 to a. Then past a, it turns up.

2.6 Exercise Set

1. Given

$$s(t) = t^3 + t,$$

where s is in feet and t is in seconds, find each of the following.

a) $v(t)$
b) $a(t)$
c) The velocity and acceleration when $t = 4$ sec

2. Given

$$s(t) = 3t + 10,$$

where s is in miles and t is in hours, find each of the following.

a) $v(t)$
b) $a(t)$
c) The velocity and acceleration when $t = 2$ hr. When the distance function is given by a linear function, we have *uniform motion*.

APPLICATIONS

◆   **Business and Economics**

3. *Marginal revenue, cost, and profit.* Given

$$R(x) = 5x \quad \text{and}$$
$$C(x) = 0.001x^2 + 1.2x + 60,$$

find each of the following.

a) $P(x)$
b) $R(100)$, $C(100)$, and $P(100)$
c) $R'(x)$, $C'(x)$, and $P'(x)$
d) $R'(100)$, $C'(100)$, and $P'(100)$

4. *Marginal revenue, cost, and profit.* Given

$$R(x) = 50x - 0.5x^2 \quad \text{and}$$
$$C(x) = 4x + 10,$$

find each of the following.

a) $P(x)$
b) $R(20)$, $C(20)$, and $P(20)$
c) $R'(x)$, $C'(x)$, and $P'(x)$
d) $R'(20)$, $C'(20)$, and $P'(20)$

5. *Advertising.* A firm estimates that it will sell N units of a product after spending a dollars on advertising, where

$$N(a) = -a^2 + 300a + 6$$

and a is in thousands of dollars.

a) What is the rate of change of the number of units sold with respect to the amount spent on advertising?

b) How many units will be sold after spending
$10,000 on advertising?

c) What is the rate of change at $a = 10$?

6. **Sales.** A company determines that monthly sales S, in
thousands of dollars, after t months of marketing a
product is given by

$$S(t) = 2t^3 - 40t^2 + 220t + 160.$$

a) Find the monthly sales after 1 month; 4 months;
6 months; 9 months; 20 months.

b) Find the rate of change $S'(t)$.

c) Find the rate of change at $t = 1$; $t = 4$; $t = 6$;
$t = 9$; $t = 20$.

7. **Marginal productivity.** An employee's monthly
productivity, in numbers of units produced, M, is
found to be a function of the number of years of
service, t. For a certain product, the productivity
function is given by

$$M(t) = -2t^2 + 100t + 180.$$

a) Find the productivity of an employee after
5 years, 10 years, 25 years, and 45 years of
service.

b) Find the marginal productivity.

c) Find the marginal productivity at $t = 5$; $t = 10$;
$t = 25$; $t = 45$.

8. **Supply.** A supply function for a certain product is
given by

$$S(p) = 0.08p^3 + 2p^2 + 10p + 11.$$

a) Find the rate of change of supply with respect to
price, dS/dp.

b) How many units will the seller allow to be sold
when the price is $3 per unit?

c) What is the rate of change at $p = 3$?

9. **Demand.** A demand function for a certain product is
given by

$$D(p) = 100 - \sqrt{p}.$$

a) Find the rate of change of quantity with respect to
price, dD/dp.

b) How many units will the consumer want to buy
when the price is $25 per unit?

c) What is the rate of change at $p = 25$?

◆   **Life and Physical Sciences**

10. **Stopping distance on glare ice.** The stopping distance
on glare ice (at some fixed speed) of regular tires is
given by a linear function of the air temperature F,

$$D(F) = 2F + 115,$$

where $D(F) = $ the stopping distance, in feet, when the
air temperature is F, in degrees Fahrenheit. Find the
rate of change of the stopping distance D with respect
to the air temperature F.

11. **Healing wound.** The circumference C, in centimeters,
of a healing wound is given by

$$C(r) = 2\pi r,$$

where $r = $ the radius, in centimeters. Find the rate of
change of the circumference with respect to the
radius.

12. **Healing wound.** The circular area A, in square
centimeters, of a healing wound is given by

$$A(r) = \pi r^2,$$

where $r = $ the radius, in centimeters. Find the rate of
change of the area with respect to the radius.

13. **Temperature during an illness.** The temperature T of a
person during an illness is given by

$$T(t) = -0.1t^2 + 1.2t + 98.6,$$

where $T = $ the temperature, in degrees Fahrenheit, at
time t, in days.

a) Find the rate of change of the temperature with respect to time.

b) Find the temperature at $t = 1.5$ days.

c) Find the rate of change at $t = 1.5$ days.

14. **Blood pressure.** For a certain dosage of x cubic centimeters (cc) of a drug, the resulting blood pressure B
is approximated by

$$B(x) = 0.05x^2 - 0.3x^3.$$

Find the rate of change of the blood pressure with respect to the dosage. Such a rate of change is often
called the *sensitivity*.

15. **Territorial area.** The territorial area T of an animal
is defined as its defended, or exclusive, region. The
area T of that region can be approximated using the

animal's body weight W by

$$T = W^{1.31}$$

(see Section 1.5). Find dT/dW.

These bears may be determining territorial area.

16. **Home range.** The home range H of an animal is de-fined as the region to which the animal confines its movements. The area of that region can be approxi-mated using the animal's body weight W by

$$H = W^{1.41}$$

(see Section 1.5). Find dH/dW.

17. **Sensitivity.** The reaction R of the body to a dose Q of medication is often represented by the general function

$$R(Q) = Q^2\left(\frac{k}{2} - \frac{Q}{3}\right),$$

where k is a constant and R is in millimeters, if the reaction is a change in blood pressure, or in degrees of temperature, if the reaction is a change in tempera-ture. The rate of change dR/dQ is defined to be the *sensitivity.* Find a formula for the sensitivity.

◆   **Social Sciences**

18. **Population growth rate.** The population of a city grows from an initial size of 100,000 to an amount P given by

$$P(t) = 100{,}000 + 2000t^2,$$

where t is in years.

a) Find the growth rate.
b) Find the number of people in the city after 10 years (at $t = 10$ yr).
c) Find the growth rate at $t = 10$.

19. **Median age of women at first marriage.** The median age of women at first marriage can be approximated by the linear function

$$A(t) = 0.08t + 19.7,$$

where $A(t) =$ the median age of women at first mar-riage the tth year after 1950. Find the rate of change of the median age A with respect to time t.

◆   **General Interest**

20. **View to the horizon.** The view V, or distance, in miles that one can see to the horizon from a height h, in feet, is given by

$$V = 1.22\sqrt{h}.$$

a) Find the rate of change of V with respect to h.
b) How far can one see to the horizon from an air-plane window from a height of 40,000 ft?
c) Find the rate of change at $h = 40{,}000$.

SYNTHESIS

21. ◈   Explain and compare average rate of change and instantaneous rate of change.

22. ◈   Discuss as many interpretations as you can of the derivative of a function at a point x.

TECHNOLOGY CONNECTION

Graph each total-revenue and total-cost function. Then graph the marginal-revenue and marginal-cost functions.

23. $R(x) = 50x - 0.5x^2,\ C(x) = 4x + 10$

24. $R(x) = 5x,\ C(x) = 0.001x^2 + 1.2x + 60$

25. $R(x) = 9x,\ C(x) = x^3 - 6x^2 + 15x$

26. $R(x) = 50x - 0.5x^2,\ C(x) = 10x + 3$

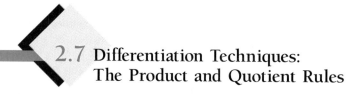

2.7 Differentiation Techniques: The Product and Quotient Rules

The Product Rule

The derivative of a sum is the sum of the derivatives, but the derivative of a product is *not* the product of the derivatives. To see this, consider x^2 and x^5. The product is x^7, and the derivative of this product is $7x^6$. The individual derivatives are $2x$ and $5x^4$, and the product of these derivatives is $10x^5$, which is not $7x^6$.

The following is the rule for finding the derivative of a product.

THEOREM 5

The Product Rule

Suppose $F(x) = f(x) \cdot s(x)$, where $f(x)$ is the "first" function and $s(x)$ is the "second" function. Then

$$F'(x) = f(x) \cdot s'(x) + f'(x) \cdot s(x).$$

The derivative of a product is the first function times the derivative of the second function, plus the derivative of the first function times the second function.

Let us check this for $x^2 \cdot x^5$. There are five steps.

$$
\begin{array}{c}
x^2 \bullet x^5 \\
\hline
\text{① ②   ③④} \\
x^2 \cdot 5x^4 + 2x \cdot x^5 \\
= 5x^6 + 2x^6 \quad \text{⑤} \\
= 7x^6
\end{array}
$$

1. Write down the first factor.
2. Multiply it by the derivative of the second factor.
3. Write down the derivative of the first factor.
4. Multiply it by the second factor.
5. Add the result of steps (1) and (2) to the result of steps (3) and (4).

EXAMPLE 1   Find $\dfrac{d}{dx}[(x^4 - 2x^3 - 7)(3x^2 - 5x)]$.

Solution   We have

$$\frac{d}{dx}(x^4 - 2x^3 - 7)(3x^2 - 5x) = (x^4 - 2x^3 - 7)(6x - 5)$$
$$+ (4x^3 - 6x^2)(3x^2 - 5x).$$

Note that we could have multiplied the polynomials and then differentiated, avoiding the use of the Product Rule, but this would have been more work.   ◆

EXAMPLE 2   Given $F(x) = (x^2 + 4x - 11)(7x^3 - \sqrt{x})$, find $F'(x)$.

Solution   We rewrite this as

$$F(x) = (x^2 + 4x - 11)(7x^3 - x^{1/2}).$$

Then using the Product Rule, we have

$$F'(x) = (x^2 + 4x - 11)\left(21x^2 - \tfrac{1}{2}x^{-1/2}\right) + (2x + 4)(7x^3 - x^{1/2}).$$ ◆

The Quotient Rule

The derivative of a quotient is *not* the quotient of the derivatives. To see why, consider x^5 and x^2. The quotient x^5/x^2 is x^3, and the derivative of this quotient is $3x^2$. The individual derivatives are $5x^4$ and $2x$, and the quotient of these derivatives $5x^4/2x$ is $(5/2)x^3$, which is not $3x^2$.

The rule for differentiating quotients is as follows.

THEOREM 6

The Quotient Rule

If

$$Q(x) = \frac{N(x)}{D(x)},$$

then

$$Q'(x) = \frac{D(x) \cdot N'(x) - D'(x) \cdot N(x)}{[D(x)]^2}.$$

The derivative of a quotient is the denominator times the derivative of the numerator, minus the derivative of the denominator times the numerator, all divided by the square of the denominator.

(If we think of the function in the numerator as the first function and the function in the denominator as the second function, then we can reword the Quotient Rule as "the second function times the derivative of the first function minus the derivative of the second function times the first function, all divided by the square of the second function.")

The Quotient Rule is illustrated below.

$$\frac{D(x) \cdot N'(x) \quad - \quad D'(x) \cdot N(x)}{[D(x)]^2}$$

1. Write down the denominator.

2. Multiply the denominator by the derivative of the numerator.

3. Write a minus sign.

4. Write down the derivative of the denominator.

5. Multiply it by the numerator.

6. Divide by the square of the denominator.

EXAMPLE 3   For $Q(x) = x^5/x^3$, find $Q'(x)$.

Solution

$$Q'(x) = \frac{x^3 \cdot 5x^4 - 3x^2 \cdot x^5}{(x^3)^2}$$

$$= \frac{5x^7 - 3x^7}{x^6}$$

$$= \frac{2x^7}{x^6} = 2x$$

EXAMPLE 4   Differentiate $f(x) = \dfrac{1 + x^2}{x^3}$.

Solution

$$f'(x) = \frac{x^3 \cdot 2x - 3x^2(1 + x^2)}{(x^3)^2}$$

$$= \frac{2x^4 - 3x^2 - 3x^4}{x^6}$$

$$= \frac{-x^4 - 3x^2}{x^6}$$

$$= \frac{-x^2 \cdot x^2 - 3x^2}{x^6}$$

$$= \frac{x^2(-x^2 - 3)}{x^2 \cdot x^4}$$

$$= \frac{-x^2 - 3}{x^4}$$

EXAMPLE 5 Differentiate $f(x) = \dfrac{x^2 - 3x}{x - 1}$.

Solution

$$f'(x) = \frac{(x - 1)(2x - 3) - 1(x^2 - 3x)}{(x - 1)^2}$$

$$= \frac{2x^2 - 5x + 3 - x^2 + 3x}{(x - 1)^2}$$

$$= \frac{x^2 - 2x + 3}{(x - 1)^2}$$

It is not necessary to multiply out $(x - 1)^2$ to get $x^2 - 2x + 1$.   ◆

$$-x^2(3x+2) = -3x^3 -2x^2$$
$$-9x^2 -4x$$
$$-x(9x+4)$$

TECHNOLOGY CONNECTION

Checking Derivatives Graphically

How can we use a grapher to check calculus, or analytic, ways of finding derivatives, as in Example 5? One way is to enter the function f. Then graph the derivative f'. This assumes that your grapher graphs f' automatically. Now graph your analytic result and compare the graphs. If they agree, then we have a "partial" check of the result. We consider such a check to be partial because we usually cannot draw complete graphs, so it is possible that there might be some location "way out" at which the two graphs do not agree.

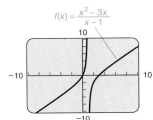

$f(x) = \dfrac{x^2 - 3x}{x - 1}$

$f'(x) = \dfrac{x^2 - 2x + 3}{(x - 1)^2}$

Suppose we had incorrectly calculated the derivative to be $y_2 = (x^2 - 2x - 8)/(x - 1)^2$. We see in the following that the graphs do not agree:

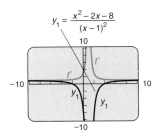

$y_1 = \dfrac{x^2 - 2x - 8}{(x - 1)^2}$

Using a grapher that automatically graphs derivatives, graph f and f' for

$$f(x) = \frac{x^2 - 4x}{x + 2}.$$

Then decide graphically which of the following seems to be the correct derivative:

$$y_1 = \frac{-x^2 - 4x - 8}{(x + 2)^2},$$

$$y_2 = \frac{x^2 - 4x + 8}{(x^2 - 4x)^2},$$

$$y_3 = \frac{x^2 + 4x - 8}{(x + 2)^2}.$$

Check the results of Example 4 graphically.

Business: Demand and Revenue

A company determines that the demand function for a certain product is given by $D(p) = 200 - p$. Find an expression for the total revenue $R(p)$. Then find the marginal revenue $R'(p)$. Graph all three functions.

An Application

We discussed earlier that it is typical for a total-revenue function to vary depending on the number of units x sold. Let us see what can determine this. Recall the consumer's demand function, or price function, $x = D(p)$, discussed in Section 1.5. It is the quantity x of units of a product that a consumer will purchase when the price is p dollars per unit. This is typically a decreasing function.

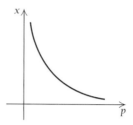

The total revenue when the price is p dollars per unit is then

$$R(p) = (\text{Number of units sold}) \cdot (\text{Price charged per unit}),$$

or $$R(p) = x \cdot p = D(p) \cdot p = p\,D(p).$$

A typical graph of a revenue function follows.

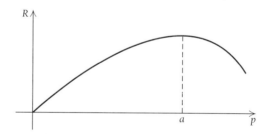

As the price increases, $D(p)$ decreases. Because we have a product $p \cdot D(p)$, the revenue typically rises for a while as p increases, but tapers off as $D(p)$ gets smaller and smaller.

Using the Product Rule, we can obtain an expression for the marginal revenue $R'(p)$ in terms of p and $D'(p)$. If

$$R(p) = p\,D(p),$$

then $$R'(p) = p \cdot D'(p) + 1 \cdot D(p) = p\,D'(p) + D(p).$$

You need not memorize this. You can merely repeat the Product Rule where necessary.

2.7 Exercise Set

Differentiate.

1. $y = x^3 \cdot x^8$, two ways

2. $y = x^4 \cdot x^9$, two ways

3. $y = \dfrac{-1}{x}$, two ways

4. $y = \dfrac{1}{x}$, two ways

5. $y = \dfrac{x^8}{x^5}$, two ways

6. $y = \dfrac{x^9}{x^5}$, two ways

7. $y = (8x^5 - 3x^2 + 20)(8x^4 - 3\sqrt{x})$

8. $f(x) = (7x^6 + 4x^3 - 50)(9x^{10} - 7\sqrt{x})$

9. $f(x) = x(300 - x)$, two ways

10. $f(x) = x(400 - x)$, two ways

11. $f(x) = (4\sqrt{x} - 6)(x^3 - 2x + 4)$

12. $f(x) = (\sqrt[3]{x} - 5x^2 + 4)(4x^2 + 11x - 5)$

13. $f(x) = (x + 3)^2$ [*Hint:* $(x + 3)^2 = (x + 3)(x + 3)$.]

14. $f(x) = (5x - 4)^2$

15. $f(x) = (x^3 - 4x)^2$

16. $f(x) = (3x^2 - 4x + 5)^2$

17. $f(x) = 5x^{-3}(x^4 - 5x^3 + 10x - 2)$

18. $f(x) = 6x^{-4}(6x^3 + 10x^2 - 8x + 3)$

19. $f(x) = \left(x + \dfrac{2}{x}\right)(x^2 - 3)$

20. $f(x) = (4x^3 - x^2)\left(x - \dfrac{5}{x}\right)$

21. $f(x) = \dfrac{x}{300 - x}$

22. $f(x) = \dfrac{x}{400 - x}$

23. $f(x) = \dfrac{3x - 1}{2x + 5}$

24. $f(x) = \dfrac{2x + 3}{x - 5}$

25. $y = \dfrac{x^2 + 1}{x^3 - 1}$

26. $y = \dfrac{x^3 - 1}{x^2 + 1}$

27. $y = \dfrac{x}{1 - x}$

28. $y = \dfrac{x}{3 - x}$

29. $y = \dfrac{x - 1}{x + 1}$

30. $y = \dfrac{x + 2}{x - 2}$

31. $f(x) = \dfrac{1}{x - 3}$

32. $f(x) = \dfrac{1}{x + 2}$

33. $f(x) = \dfrac{3x^2 + 2x}{x^2 + 1}$

34. $f(x) = \dfrac{3x^2 - 5x}{x^2 - 1}$

35. $f(x) = \dfrac{3x^2 - 5x}{x^8}$, two ways

36. $f(x) = \dfrac{3x^2 + 2x}{x^5}$, two ways

37. $g(x) = \dfrac{4x + 3}{\sqrt{x}}$

38. $g(x) = \dfrac{6x^2 - 3x}{3\sqrt{x}}$

39. Find an equation of the tangent line to the graph of $y = 8/(x^2 + 4)$ at the point $(0, 2)$ and at $(-2, 1)$.

40. Find an equation of the tangent line to the graph of $y = 4x/(1 + x^2)$ at the point $(0, 0)$ and at $(-1, -2)$.

APPLICATIONS

◆ **Business and Economics**

41. *Marginal demand.* The demand function for a certain product is given by

$$D(p) = \dfrac{2p + 300}{10p + 11}.$$

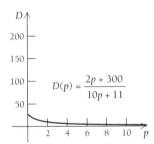

a) Find the marginal demand $D'(p)$.
b) Find $D'(4)$.

42. *Marginal revenue.* A company sells fix-it-yourself books. It finds that its total revenue from the sale of

x books, where x is in thousands, is given by

$$R(x) = \frac{120{,}000\sqrt{x}}{4 + x^{3/2}}.$$

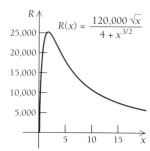

a) Find the marginal revenue $R'(x)$.
b) Find $R'(4)$.

Marginal revenue. In each of Exercises 43–46, a demand function $x = D(p)$ is given. Find (a) the total revenue $R(p)$ and (b) the marginal revenue $R'(p)$.

43. $D(p) = 400 - p$

44. $D(p) = 500 - p$

45. $D(p) = \dfrac{4000}{p} + 3$

46. $D(p) = \dfrac{3000}{p} + 5$

47. *Marginal average cost.* In Section 2.6, we defined the average cost of producing x units of a product in terms of the total cost $C(x)$ by

$$A(x) = \frac{C(x)}{x}.$$

Use the Quotient Rule to find a general expression for marginal average cost $A'(x)$.

48. *Marginal demand.* In this section, we determined that

$$R(p) = p\,D(p).$$

Then

$$D(p) = \frac{R(p)}{p} = \text{the number of units sold}$$
when the price is p dollars per unit
$$= \text{the demand function.}$$

Use the Quotient Rule to find a general expression for marginal demand $D'(p)$.

◆ **Life and Physical Sciences**

49. *Temperature during an illness.* The temperature T of a person during an illness is given by

$$T(t) = \frac{4t}{t^2 + 1} + 98.6,$$

where $T =$ the temperature, in degrees Fahrenheit, at time t, in hours.

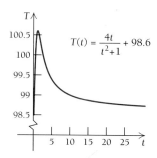

a) Find the rate of change of the temperature with respect to time.
b) Find the temperature at $t = 2$ hr.
c) Find the rate of change at $t = 2$ hr.

◆ **Social Sciences**

50. *Population growth.* The population P, in thousands, of a small city is given by

$$P(t) = \frac{500t}{2t^2 + 9},$$

where $t =$ the time, in months.

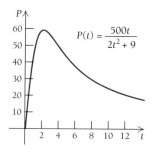

a) Find the growth rate.
b) Find the population after 12 months.
c) Find the growth rate at $t = 12$ months.

SYNTHESIS

Differentiate.

51. $f(x) = \dfrac{x^3}{\sqrt{x} - 5}$

52. $g(t) = \dfrac{1 + \sqrt{t}}{t^5 + 3}$

53. $f(v) = \dfrac{3}{1 + v + v^2}$

54. $g(z) = \dfrac{1 + z + z^2}{1 - z + z^2}$

55. $p(t) = \dfrac{t}{1 - t + t^2 - t^3}$

56. $f(x) = \dfrac{\dfrac{2}{3x} - 1}{\dfrac{3}{x^2} + 5}$

57. $h(x) = \dfrac{x^3 + 5x^2 - 2}{\sqrt{x}}$

58. $y(t) = 5t(t - 1)(2t + 3)$

59. $f(x) = x(3x^3 + 6x - 2)(3x^4 + 7)$

60. $g(x) = (x^3 - 8) \cdot \dfrac{x^2 + 1}{x^2 - 1}$

61. $f(t) = (t^5 + 3) \cdot \dfrac{t^3 - 1}{t^3 + 1}$

62. $f(x) = \dfrac{(x^2 + 3x)(x^5 - 7x^2 - 3)}{x^4 - 3x^3 - 5}$

63. $f(x) = \dfrac{(2x^2 + 3)(4x^3 - 7x + 2)}{x^7 - 2x^6 + 9}$

64. $s(t) = \dfrac{5t^8 - 2t^3}{(t^5 - 3)(t^4 + 7)}$

65. ◆ Try to develop a rule for finding the derivative of the product of three functions. Describe the rule in words.

TECHNOLOGY CONNECTION

For each function, graph f and f'. Then estimate points at which the tangent line is horizontal.

66. $f(x) = x^2(x - 2)(x + 2)$

67. $f(x) = \left(x + \dfrac{2}{x}\right)(x^2 - 3)$

68. $f(x) = \dfrac{x^3 - 1}{x^2 + 1}$

69. $f(x) = \dfrac{0.3x}{0.04 + x^2}$

70. $f(x) = \dfrac{0.01x^2}{x^4 + 0.0256}$

71. $f(x) = \dfrac{4x}{x^2 + 1}$

72. Decide graphically which of the following seems to be the correct derivative of the function of Exercise 71.

$$y_1 = \dfrac{2}{x},$$

$$y_2 = \dfrac{4 - 4x}{x^2 + 1},$$

$$y_3 = \dfrac{4 - 4x^2}{(x^2 + 1)^2},$$

$$y_4 = \dfrac{4x^2 - 4}{(x^2 + 1)^2}$$

2.8 The Chain Rule

- Find the composition of functions.
- Differentiate using the Extended Power Rule and the Chain Rule.

The Extended Power Rule

How can we differentiate more complicated functions such as

$$y = (1 + x^2)^3,$$
$$y = (1 + x^2)^{89},$$

or

$$y = (1 + x^2)^{1/3}?$$

For $(1 + x^2)^3$, we can expand and then differentiate. Although this could be done for $(1 + x^2)^{89}$, it would certainly be time-consuming, and such an expansion would not work for $(1 + x^2)^{1/3}$. Not knowing a rule, we might conjecture that the derivative of the function $y = (1 + x^2)^3$ is

$$3(1 + x^2)^2. \tag{1}$$

To check this, we expand $(1 + x^2)^3$ and then differentiate. From Exercise Set 1.1, we know that $(a + h)^3 = a^3 + 3a^2h + 3ah^2 + h^3$, so

$$(1 + x^2)^3 = 1^3 + 3 \cdot 1^2 \cdot (x^2)^1 + 3 \cdot 1 \cdot (x^2)^2 + (x^2)^3$$
$$= 1 + 3x^2 + 3x^4 + x^6.$$

(We could also have done this by finding $(1 + x^2)^2$ and then multiplying again by $1 + x^2$.) It follows that

$$\frac{dy}{dx} = 6x + 12x^3 + 6x^5$$
$$= (1 + 2x^2 + x^4)6x$$
$$= 3(1 + x^2)^2 \cdot 2x. \tag{2}$$

Comparing this with Eq. (1), we see that the Power Rule is not sufficient for such a differentiation. Note that the factor $2x$ in the actual derivative, Eq. (2), is the derivative of the "inside" function, $1 + x^2$. This is consistent with the following new rule.

> **THEOREM 7**
>
> **The Extended Power Rule**
>
> Suppose $g(x)$ is a function of x. Then for any real number k,
>
> $$\frac{d}{dx}[g(x)]^k = k[g(x)]^{k-1} \cdot \frac{d}{dx}g(x).$$

Let us differentiate $(1 + x^3)^5$. There are three steps to carry out.

$(1 + x^3)^5$ **1.** Mentally block out the "inside" function, $1 + x^3$.

$5(1 + x^3)^4$ **2.** Differentiate the "outside" function, $(1 + x^3)^5$.

$5(1 + x^3)^4 \cdot 3x^2$ **3.** Multiply by the derivative of the "inside" function.

$= 15x^2(1 + x^3)^4$

Step (3) is most commonly overlooked. Try not to forget it!

TECHNOLOGY
CONNECTION

Use a grapher that automatically graphs derivatives. For Example 1, graph f and f'. Check the results of Example 1 graphically.

EXAMPLE 1   Differentiate: $f(x) = (1 + x^3)^{1/2}$.

Solution

$$\frac{d}{dx}(1 + x^3)^{1/2} = \frac{1}{2}(1 + x^3)^{1/2-1} \cdot 3x^2$$

$$= \frac{1}{2}(1 + x^3)^{-1/2} \cdot 3x^2$$

$$= \frac{3x^2}{2\sqrt{1 + x^3}}$$

EXAMPLE 2   Differentiate: $y = (1 - x^2)^3 - (1 - x^2)^2$.

Solution   Here we combine the Difference Rule and the Extended Power Rule:

$$\frac{dy}{dx} = 3(1 - x^2)^2(-2x) - 2(1 - x^2)(-2x). \quad \text{We differentiate each term using the Extended Power Rule.}$$

Thus,

$$\frac{dy}{dx} = -6x(1 - x^2)^2 + 4x(1 - x^2)$$

$$= x(1 - x^2)[-6(1 - x^2) + 4] \quad \text{Here we factor out } x(1 - x^2).$$

$$= x(1 - x^2)[-6 + 6x^2 + 4]$$

$$= x(1 - x^2)(6x^2 - 2)$$

$$= 2x(1 - x^2)(3x^2 - 1).$$

EXAMPLE 3   Differentiate: $f(x) = (x - 5)^4(7 - x)^{10}$.

Solution   Here we combine the Product Rule and the Extended Power Rule:

$$f'(x) = (x - 5)^4 \cdot 10(7 - x)^9(-1) + 4(x - 5)^3(1)(7 - x)^{10}$$

$$= -10(x - 5)^4(7 - x)^9 + 4(x - 5)^3(7 - x)^{10}$$

$$= 2(x - 5)^3(7 - x)^9[-5(x - 5) + 2(7 - x)] \quad \text{We factor out } 2(x - 5)^3(7 - x)^9.$$

$$= 2(x - 5)^3(7 - x)^9[-5x + 25 + 14 - 2x]$$

$$= 2(x - 5)^3(7 - x)^9(39 - 7x).$$

EXAMPLE 4 Differentiate: $f(x) = \sqrt[4]{\dfrac{x+3}{x-1}}$.

Solution We must use the Quotient Rule to differentiate the inside function, $(x+3)/(x-1)$:

$$\frac{d}{dx}\sqrt[4]{\frac{x+3}{x-1}} = \frac{d}{dx}\left(\frac{x+3}{x-1}\right)^{1/4} = \frac{1}{4}\left(\frac{x+3}{x-1}\right)^{1/4-1}\left[\frac{(x-1)1-1(x+3)}{(x-1)^2}\right]$$

$$= \frac{1}{4}\left(\frac{x+3}{x-1}\right)^{-3/4}\left[\frac{x-1-x-3}{(x-1)^2}\right]$$

$$= \frac{1}{4}\left(\frac{x+3}{x-1}\right)^{-3/4}\cdot\frac{-4}{(x-1)^2}$$

$$= \left(\frac{x+3}{x-1}\right)^{-3/4}\cdot\frac{-1}{(x-1)^2}.$$

Composition of Functions and the Chain Rule

The Extended Power Rule is a special case of a more general rule called the *Chain Rule*. Before discussing it, we define the *composition* of functions.

There is a function g that gives a correspondence between women's shoe sizes in the United States and those in Italy. The function is given by $g(x) = 2(x + 12)$, where x is a shoe size in the United States and $g(x)$ is a shoe size in Italy. For example, a shoe size of 4 in the United States corresponds to a shoe size of $g(4) = 2(4 + 12)$, or 32 in Italy.

g		f	
$g(x) = 2(x + 12)$		$f(x) = \frac{1}{2}x - 14$	

U.S.	Italy	Britain
4	32	2
5	34	3
6	36	4
7	38	5
8	40	6

h

$h(x) = ?$

There is also a function f that gives a correspondence between women's shoe sizes in Italy and those in Britain. The function is given by $f(x) = \frac{1}{2}x - 14$, where x

is a shoe size in Italy and $f(x)$ is the corresponding shoe size in Britain. For example, a shoe size of 32 in Italy corresponds to a shoe size of $f(32) = \frac{1}{2}(32) - 14$, or 2 in Britain.

It seems reasonable to assume that a shoe size of 4 in the United States corresponds to a shoe size of 2 in Britain and that there is a function h that describes this correspondence. Can we find a formula for h? Looking at the tables above, we might guess that such a formula is $h(x) = x - 2$, and that is indeed correct. For more complicated formulas, however, we need to do some algebra.

A shoe size x in the United States corresponds to a shoe size $g(x)$ in Italy, where

$$g(x) = 2(x + 12).$$

Now $2(x + 12)$ is a shoe size in Italy. If we replace x in $f(x)$ by $2(x + 12)$, we can find the corresponding shoe size in Britain:

$$\begin{aligned}
f(g(x)) &= \tfrac{1}{2}[2(x + 12)] - 14 \\
&= \tfrac{1}{2}[2x + 24] - 14 \\
&= x + 12 - 14 \\
&= x - 2.
\end{aligned}$$

This gives a formula for h: $h(x) = x - 2$. Thus a shoe size of 4 in the United States corresponds to a shoe size of $h(4) = 4 - 2$, or 2 in Britain. The function h is called the **composition** of f and g and is denoted $f \circ g$.

DEFINITION

The *composed* function $f \circ g$, the *composition* of f and g, is defined as

$$f \circ g(x) = f(g(x)).$$

We can visualize the composition of functions as shown below.

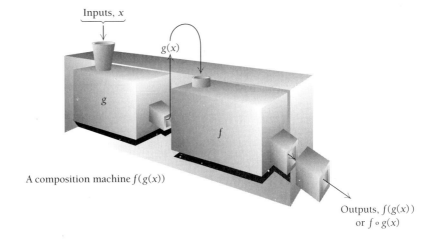

Inputs, x

$g(x)$

g

f

A composition machine $f(g(x))$

Outputs, $f(g(x))$
or $f \circ g(x)$

To find $f \circ g(x)$, we substitute $g(x)$ for x in $f(x)$.

EXAMPLE 5   Given $f(x) = x^3$ and $g(x) = 1 + x^2$, find $f \circ g(x)$ and $g \circ f(x)$.

Solution   Consider each function separately:

$$f(x) = x^3 \qquad \text{This function cubes each input.}$$

and $\qquad g(x) = 1 + x^2. \qquad \text{This function adds 1 to the square of each input.}$

a) $f \circ g$ first does what g does (adds 1 to the square) and then does what f does (cubes). We find $f(g(x))$ by substituting $g(x)$ for x:

$$\begin{aligned} f \circ g(x) = f(g(x)) = f(1 + x^2) & \qquad \text{Substituting } 1 + x^2 \text{ for } x \\ & = (1 + x^2)^3 \\ & = 1 + 3x^2 + 3x^4 + x^6. \end{aligned}$$

b) $g \circ f$ first does what f does (cubes) and then does what g does (adds 1 to the square). We find $g(f(x))$ by substituting $f(x)$ for x:

$$\begin{aligned} g \circ f(x) = g(f(x)) = g(x^3) & \qquad \text{Substituting } x^3 \text{ for } x \\ & = 1 + (x^3)^2 \\ & = 1 + x^6. \end{aligned}$$ ◆

EXAMPLE 6   Given $f(x) = \sqrt{x}$ and $g(x) = x - 1$, find $f \circ g(x)$ and $g \circ f(x)$.

Solution

$$\begin{aligned} f \circ g(x) = f(g(x)) = f(x - 1) = \sqrt{x - 1}, \\ g \circ f(x) = g(f(x)) = g(\sqrt{x}) = \sqrt{x} - 1 \end{aligned}$$ ◆

How do we differentiate the composition of functions? The following theorem tells us.

THEOREM 8

The Chain Rule

The derivative of the composition $f \circ g$ is given by

$$\frac{d}{dx}[f \circ g(x)] = \frac{d}{dx}[f(g(x))] = f'(g(x)) \cdot \frac{d}{dx} g(x).$$

Note that the Extended Power Rule is a special case of the Chain Rule. Consider the function $f(x) = x^k$. Then for any other function $g(x)$, $f \circ g(x) = [g(x)]^k$ and the derivative of the composition is

$$\frac{d}{dx}[g(x)]^k = k[g(x)]^{k-1} \cdot \frac{d}{dx} g(x).$$

The Chain Rule often appears in another form. Suppose $y = f(u)$ and $u = g(x)$. Then

$$\frac{dy}{dx} = \frac{dy}{du} \cdot \frac{du}{dx}.$$

EXAMPLE 7   Given $y = 2 + \sqrt{u}$ and $u = x^3 + 1$, find dy/du, du/dx, and dy/dx.

Solution   First we find dy/du and du/dx:

$$\frac{dy}{du} = \frac{1}{2}u^{-1/2} \quad \text{and} \quad \frac{du}{dx} = 3x^2.$$

Then

$$\frac{dy}{dx} = \frac{dy}{du} \cdot \frac{du}{dx}$$

$$= \frac{1}{2\sqrt{u}} \cdot 3x^2$$

$$= \frac{3x^2}{2\sqrt{x^3 + 1}}. \qquad \text{Substituting } x^3 + 1 \text{ for } u \qquad \blacklozenge$$

It is important to be able to recognize how a function can be expressed as a composition.

EXAMPLE 8   Find $f(x)$ and $g(x)$ such that $h(x) = f \circ g(x)$:

$$h(x) = (4x^3 - 7)^6.$$

Solution   This is $4x^3 - 7$ to the 6th power. Two functions that can be used for the composition are $f(x) = x^6$ and $g(x) = 4x^3 - 7$. We can check by forming the composition:

$$h(x) = f \circ g(x) = f(g(x)) = f(4x^3 - 7) = (4x^3 - 7)^6.$$

This is the most "obvious" answer to the question. There can be other less obvious answers—for example, $f(x) = x^3$ and $g(x) = (4x^3 - 7)^2$. $\qquad \blacklozenge$

2.8 Exercise Set

Differentiate.

1. $y = (1 - x)^{55}$

2. $y = (1 - x)^{100}$

3. $y = \sqrt{1 + 8x}$

4. $y = \sqrt{1 - x}$

5. $y = \sqrt{3x^2 - 4}$

6. $y = \sqrt{4x^2 + 1}$

7. $y = (3x^2 - 6)^{-40}$

8. $y = (4x^2 + 1)^{-50}$

9. $y = x\sqrt{2x + 3}$

10. $y = x\sqrt{4x - 7}$

11. $y = x^2\sqrt{x - 1}$

12. $y = x^3\sqrt{x + 1}$

13. $y = \dfrac{1}{(3x + 8)^2}$

14. $y = \dfrac{1}{(4x + 5)^2}$

15. $f(x) = (1 + x^3)^3 - (1 + x^3)^4$

16. $f(x) = (1 + x^3)^5 - (1 + x^3)^4$

17. $f(x) = x^2 + (200 - x)^2$

18. $f(x) = x^2 + (100 - x)^2$

19. $f(x) = (x + 6)^{10}(x - 5)^4$

20. $f(x) = (x - 4)^8(x + 3)^9$

21. $f(x) = (x - 4)^8(3 - x)^4$

22. $f(x) = (x + 6)^{10}(5 - x)^9$

23. $f(x) = -4x(2x - 3)^3$

24. $f(x) = -5x(3x + 5)^6$

25. $f(x) = \sqrt{\dfrac{1 - x}{1 + x}}$

26. $f(x) = \sqrt{\dfrac{3 + x}{2 - x}}$

27. $f(x) = \left(\dfrac{3x - 1}{5x + 2}\right)^4$

28. $f(x) = \left(\dfrac{x}{x^2 + 1}\right)^3$

29. $f(x) = \sqrt[3]{x^4 + 3x^2}$

30. $f(x) = \sqrt{x^2 + 5x}$

31. $f(x) = (2x^3 - 3x^2 + 4x + 1)^{100}$

32. $f(x) = (7x^4 + 6x^3 - x)^{204}$

33. $g(x) = \left(\dfrac{2x + 3}{5x - 1}\right)^{-4}$

34. $h(x) = \left(\dfrac{1 - 3x}{2 - 7x}\right)^{-5}$

35. $f(x) = \sqrt{\dfrac{x^2 + 1}{x^2 - 1}}$

36. $f(x) = \sqrt[3]{\dfrac{4 - x^3}{1 - x^2}}$

37. $f(x) = \dfrac{(2x + 3)^4}{(3x - 2)^5}$

38. $f(x) = \dfrac{(5x - 4)^7}{(6x + 1)^3}$

39. $f(x) = 12(2x + 1)^{2/3}(3x - 4)^{5/4}$

40. $y = 6\sqrt[3]{x^2 + x}(x^4 - 6x)^3$

Find $\dfrac{dy}{du}, \dfrac{du}{dx}$, and $\dfrac{dy}{dx}$.

41. $y = \sqrt{u}$ and $u = x^2 - 1$

42. $y = \dfrac{15}{u^3}$ and $u = 2x + 1$

43. $y = u^{50}$ and $u = 4x^3 - 2x^2$

44. $y = \dfrac{u + 1}{u - 1}$ and $u = 1 + \sqrt{x}$

45. $y = u(u + 1)$ and $u = x^3 - 2x$

46. $y = (u + 1)(u - 1)$ and $u = x^3 + 1$

47. Find an equation for the tangent line to the graph of $y = \sqrt{x^2 + 3x}$ at the point $(1, 2)$.

48. Find an equation for the tangent line to the graph of $y = (x^3 - 4x)^{10}$ at the point $(2, 0)$.

49. Consider

$$f(x) = \dfrac{x^2}{(1 + x)^5}.$$

a) Find $f'(x)$ using the Quotient Rule and the Extended Power Rule.

b) Note that $f(x) = x^2(1 + x)^{-5}$. Find $f'(x)$ using the Product Rule and the Extended Power Rule.

c) Compare your answers to parts (a) and (b).

50. Consider

$$g(x) = (x^3 + 5x)^2.$$

a) Find $g'(x)$ using the Extended Power Rule.

b) Note that $g(x) = x^6 + 10x^4 + 25x^2$. Find $g'(x)$.

c) Compare your answers to parts (a) and (b).

Find $f \circ g(x)$ and $g \circ f(x)$.

51. $f(x) = 3x^2 + 2$, $g(x) = 2x - 1$

52. $f(x) = 4x + 3$, $g(x) = 2x^2 - 5$

53. $f(x) = 4x^2 - 1$, $g(x) = \dfrac{2}{x}$

54. $f(x) = \dfrac{3}{x}$, $g(x) = 2x^2 + 3$

55. $f(x) = x^2 + 1$, $g(x) = x^2 - 1$

56. $f(x) = \dfrac{1}{x^2}$, $g(x) = x + 2$

Find $f(x)$ and $g(x)$ such that $h(x) = f \circ g(x)$. Answers may vary.

57. $h(x) = (3x^2 - 7)^5$

58. $h(x) = \dfrac{1}{\sqrt{7x + 2}}$

59. $h(x) = \dfrac{x^3 + 1}{x^3 - 1}$

60. $h(x) = (\sqrt{x} + 5)^4$

Do Exercises 61–64 in two ways. First, use the Chain Rule to find the answer. Next, check your answer by finding $f(g(x))$, taking the derivative, and substituting.

61. $f(u) = u^3$, $g(x) = u = 2x^4 + 1$
Find $(f \circ g)'(-1)$.

62. $f(u) = \dfrac{u + 1}{u - 1}, \quad g(x) = u = \sqrt{x}$

Find $(f \circ g)'(4)$.

63. $f(u) = \sqrt[3]{u}, \quad g(x) = u = 1 - 3x^2$

Find $(f \circ g)'(2)$.

64. $f(u) = 2u^5, \quad g(x) = u = \dfrac{3 - x}{4 + x}$

Find $(f \circ g)'(-10)$.

APPLICATIONS

◆   **Business and Economics**

65. *Marginal cost.* A total-cost function is given by

$$C(x) = 1000\sqrt{x^3 + 2}.$$

Find the marginal costs $C'(x)$ and $C'(10)$.

66. *Marginal revenue.* A total-revenue function is given by

$$R(x) = 2000\sqrt{x^2 + 3}.$$

Find the marginal revenues $R'(x)$ and $R'(20)$.

67. *Marginal profit.* Use the total-cost and total-revenue functions in Exercises 65 and 66 to find the marginal profit.

68. *Utility.* Utility is a type of function or concept considered in economics. When a consumer receives x units of a product, a certain amount of pleasure, or utility, U, is derived. Suppose for a new kind of video game device that the following is the utility function related to the number of cartridges obtained:

$$U(x) = 80\sqrt{\dfrac{2x + 1}{3x + 4}}.$$

Find the marginal utility.

69. *Compound interest.* If $1000 is invested at interest rate i, compounded annually, it will grow in 3 years to an amount A given by

$$A = \$1000(1 + i)^3$$

(see Section 1.1). Find the rate of change, dA/di.

70. *Compound interest.* If $1000 is invested at interest rate i, compounded quarterly, it will grow in 5 years to an amount A given by

$$A = \$1000\left(1 + \dfrac{i}{4}\right)^{20}.$$

Find the rate of change, dA/di.

71. *Marginal demand.* Suppose that the demand function for a product is given by

$$D(p) = \dfrac{80{,}000}{p}$$

and that price p is a function of time given by $p = 1.6t + 9$, where t is in days.

a) Find the demand as a function of time t.
b) Find the marginal demand as a function of time.
c) Find the rate of change of the quantity demanded when $t = 100$ days.

72. *Marginal profit.* A company is selling microcomputers. It determines that its total profit is given by

$$P(x) = 0.08x^2 + 80x + 260,$$

where x = the number of units produced and sold. Suppose that x is a function of time, in months, where $x = 5t + 1$.

a) Find the total profit as a function of time t.
b) Find the marginal profit as a function of time.
c) Find the rate of change of total profit when $t = 48$ months.

SYNTHESIS

Differentiate.

73. $y = \sqrt[3]{x^3 - 6x + 1}$

74. $s = \sqrt[4]{t^4 + 3t^2 + 8}$

75. $y = \dfrac{x}{\sqrt{x - 1}}$

76. $y = \dfrac{(x + 1)^2}{(x^2 + 1)^3}$

77. $u = \dfrac{(1 + 2v)^4}{v^4}$

78. $y = x\sqrt{1 + x^2}$

79. $y = \dfrac{\sqrt{1 - x^2}}{1 - x}$

80. $w = \dfrac{u}{\sqrt{1 + u^2}}$

81. $y = \left(\dfrac{x^2 - x - 1}{x^2 + 1}\right)^3$

82. $y = \sqrt{1 + \sqrt{x}}$

83. $s = \dfrac{\sqrt{t} - 1}{\sqrt{t} + 1}$

84. $y = x^{2/3} \cdot \sqrt[3]{1 + x^2}$

85. ◈ The following is the beginning of an alternative proof of the Quotient Rule for finding the derivative of a quotient that uses the Product Rule and the Power Rule. Complete the proof, giving reasons for each step.

Proof. Let

$$Q(x) = \frac{N(x)}{D(x)}.$$

Then

$$Q(x) = N(x) \cdot [D(x)]^{-1}.$$

Therefore, ...

86. ◈ Describe composition of functions in as many ways as possible.

TECHNOLOGY CONNECTION

For each function, graph f and f' on the given interval. Then estimate points at which the tangent line is horizontal.

87. $f(x) = 1.68x\sqrt{9.2 - x^2}$; $[-3, 3]$

88. $f(x) = \sqrt{6x^3 - 3x^2 - 48x + 45}$; $[-3, 3]$

Find the derivative of each of the following functions analytically. Then, using a grapher that automatically graphs derivatives, graph f and f' and check the results graphically.

89. $f(x) = x\sqrt{4 - x^2}$ 90. $f(x) = \dfrac{4x}{\sqrt{x - 10}}$

2.9 Higher-Order Derivatives

OBJECTIVE

• Find derivatives of higher order.

Consider the function given by

$$y = f(x) = x^5 - 3x^4 + x.$$

Its derivative f' is given by

$$y' = f'(x) = 5x^4 - 12x^3 + 1.$$

The derivative function f' can also be differentiated. We can think of its derivative as the rate of change of the slope of the tangent lines of f. We use the notation f'' for the derivative $(f')'$. We call f'' the *second derivative* of f. It is given by

$$y'' = f''(x) = 20x^3 - 36x^2.$$

Continuing in this manner, we have

$$f'''(x) = 60x^2 - 72x, \qquad \text{The third derivative of } f$$
$$f''''(x) = 120x - 72, \qquad \text{The fourth derivative of } f$$
$$f'''''(x) = 120. \qquad \text{The fifth derivative of } f$$

When notation like $f'''(x)$ gets lengthy, we abbreviate it using a symbol in parentheses. Thus, $f^{(n)}(x)$ is the nth derivative. For the function above,

$$f^{(4)}(x) = 120x - 72,$$
$$f^{(5)}(x) = 120,$$
$$f^{(6)}(x) = 0, \quad \text{and}$$
$$f^{(n)}(x) = 0, \quad \text{for any integer } n \geq 6.$$

Leibniz's notation for the second derivative of a function given by $y = f(x)$ is

$$\frac{d^2y}{dx^2}, \quad \text{or} \quad \frac{d}{dx}\left(\frac{dy}{dx}\right),$$

TECHNOLOGY CONNECTION

Some graphers will sketch the graph of a function f and any successive derivative. Consider

$$f(x) = x^6 - 5x^5 + 5x^4 + 20x.$$

Graph f and each of its first six derivatives. Use $[-2, 4, -50, 140]$ as a viewing window, with xScl = 1 and yScl = 20. Describe what happens.

read "the second derivative of y with respect to x." The 2's in this notation *are not* exponents. If $y = x^5 - 3x^4 + x$, then

$$\frac{d^2y}{dx^2} = 20x^3 - 36x^2.$$

Leibniz's notation for the third derivative is d^3y/dx^3; for the fourth derivative, d^4y/dx^4; and so on:

$$\frac{d^3y}{dx^3} = 60x^2 - 72x, \qquad \frac{d^4y}{dx^4} = 120x - 72,$$

$$\frac{d^5y}{dx^5} = 120.$$

EXAMPLE 1   For $y = 1/x$, find d^2y/dx^2.

Solution   We have $y = x^{-1}$, so

$$\frac{dy}{dx} = -1 \cdot x^{-1-1} = -x^{-2}, \quad \text{or} \quad -\frac{1}{x^2}.$$

Then

$$\frac{d^2y}{dx^2} = (-2)(-1)x^{-2-1} = 2x^{-3}, \quad \text{or} \quad \frac{2}{x^3}. \qquad \blacklozenge$$

EXAMPLE 2   For $y = (x^2 + 10x)^{20}$, find y' and y''.

Solution   To find y', we use the Extended Power Rule:

$$\begin{aligned}
y' &= 20(x^2 + 10x)^{19}(2x + 10) \\
&= 20(x^2 + 10x)^{19} \cdot 2(x + 5) \\
&= 40(x^2 + 10x)^{19}(x + 5).
\end{aligned}$$

To find y'', we use the Product Rule and the Extended Power Rule:

$$\begin{aligned}
y'' &= 40(x^2 + 10x)^{19}(1) + 19 \cdot 40(x^2 + 10x)^{18}(2x + 10)(x + 5) \\
&= 40(x^2 + 10x)^{19} + 760(x^2 + 10x)^{18} \cdot 2(x + 5)(x + 5) \\
&= 40(x^2 + 10x)^{19} + 1520(x^2 + 10x)^{18}(x + 5)^2 \\
&= 40(x^2 + 10x)^{18}[(x^2 + 10x) + 38(x + 5)^2] \\
&= 40(x^2 + 10x)^{18}[x^2 + 10x + 38(x^2 + 10x + 25)] \\
&= 40(x^2 + 10x)^{18}[39x^2 + 390x + 950]. \qquad \blacklozenge
\end{aligned}$$

Acceleration can be regarded as a second derivative. As an object moves, its distance from a fixed point after time t is some function of the time, say, $s(t)$. Then

$$v(t) = s'(t) = \text{the velocity at time } t$$

and $a(t) = v'(t) = s''(t) = \text{the acceleration at time } t.$

Whenever a quantity is a function of time, the first derivative gives the rate of change with respect to time and the second derivative gives the acceleration. For example, if $y = P(t)$ gives the number of people in a population at time t, then $P'(t)$ represents how fast the size of the population is changing and $P''(t)$ gives the acceleration in the size of the population.

2.9 Exercise Set

Find d^2y/dx^2.

1. $y = 3x + 5$

2. $y = -4x + 7$

3. $y = -\dfrac{1}{x}$

4. $y = -\dfrac{3}{x}$

5. $y = x^{1/4}$

6. $y = \sqrt{x}$

7. $y = x^4 + \dfrac{4}{x}$

8. $y = x^3 - \dfrac{3}{x}$

9. $y = x^{-3}$

10. $y = x^{-4}$

11. $y = x^n$

12. $y = x^{-n}$

13. $y = x^4 - x^2$

14. $y = x^4 + x^3$

15. $y = \sqrt{x-1}$

16. $y = \sqrt{x+1}$

17. $y = ax^2 + bx + c$

18. $y = \sqrt{a-bx}$

19. $y = (x^2 - 8x)^{43}$

20. $y = (x^3 + 15x)^{20}$

21. $y = (x^4 - 4x^2)^{50}$

22. $y = (x^2 - 3x + 1)^{12}$

23. $y = x^{2/3} + 4x$

24. $y = x^{5/4} - x^{4/5}$

25. $y = (x - 8)^{3/4}$

26. $y = \dfrac{x^5}{40} - \dfrac{x^3}{36}$

27. $y = \dfrac{1}{x^2} + \dfrac{2}{x^3}$

28. $y = \dfrac{4}{x^5} - \dfrac{7}{3x^4}$

29. For $y = x^4$, find d^4y/dx^4.

30. For $y = x^5$, find d^4y/dx^4.

31. For $y = x^6 - x^3 + 2x$, find d^5y/dx^5.

32. For $y = x^7 - 8x^2 + 2$, find d^6y/dx^6.

33. For $y = (x^2 - 5)^{10}$, find d^2y/dx^2.

34. For $y = x^k$, find d^5y/dx^5.

35. If s is a distance given by $s(t) = t^3 + t^2 + 2t$, find the acceleration.

36. If s is a distance given by $s(t) = t^4 + t^2 + 3t$, find the acceleration.

APPLICATIONS

◆ Life and Physical Sciences

37. *Population growth.* A population grows from an initial size of 100,000 to an amount $P(t)$, given by

$$P(t) = 100,000(1 + 0.6t + t^2).$$

What is the acceleration in the size of the population?

38. *Population growth.* A population grows from an initial size of 100,000 to an amount $P(t)$, given by

$$P(t) = 100,000(1 + 0.4t + t^2).$$

What is the acceleration in the size of the population?

Synthesis

Find y', y'', and y'''.

39. $y = x^{-1} + x^{-2}$

40. $y = \dfrac{1}{1 - x}$

41. $y = x\sqrt{1 + x^2}$

42. $y = 3x^5 + 8\sqrt{x}$

43. $y = \dfrac{3x - 1}{2x + 3}$

44. $y = \dfrac{1}{\sqrt{x - 1}}$

45. $y = \dfrac{x}{\sqrt{x - 1}}$

46. $y = \dfrac{\sqrt{x} - 1}{\sqrt{x} + 1}$

Find $f''(x)$.

47. $f(x) = \dfrac{x}{x - 1}$

48. $f(x) = \dfrac{1}{1 + x^2}$

Find the first through the fifth derivatives. Be sure to simplify at each stage before continuing.

49. $f(x) = \dfrac{x - 1}{x + 2}$

50. $f(x) = \dfrac{x + 3}{x - 2}$

TECHNOLOGY CONNECTION

For each function, graph f, f', and f'' on the given interval. Analyze and compare the behavior of each function.

51. $f(x) = 0.1x^4 - x^2 + 0.4$; $[-5, 5]$

52. $f(x) = -x^3 + 3x$; $[-3, 3]$

53. $f(x) = x^4 + x^3 - 4x^2 - 2x + 4$; $[-3, 2]$

54. $f(x) = x^3 - 3x^2 + 2$; $[-2, 4]$

55. ◆ Consider the function f given by

$$f(x) = \frac{5x^2 + 8x - 3}{3x^2 + 2}.$$

Graph f and its first four derivatives. Describe what happens. How does this situation differ from graphing a polynomial and its successive derivatives?

2 Chapter Summary and Review

Terms to Know

REVIEW EXERCISES

These exercises are for test preparation. They can also be used as a lengthened practice test. Answers are at the back of the book. The answers also contain bracketed section references, which tell you where to restudy if your answer is incorrect.

Find the limit, if it exists.

1. $\lim\limits_{x \to -2} \dfrac{8}{x}$

2. $\lim\limits_{x \to 1} (4x^3 - x^2 + 7x)$

3. $\lim\limits_{x \to -7} \dfrac{x^2 + 4x - 21}{x + 7}$

4. $\lim\limits_{x \to \infty} \dfrac{5x + 6}{x}$

5. $\lim\limits_{x \to \infty} \dfrac{6x^2 + 5x - 7}{3x^4 - 2x + 4}$

Determine whether the graph is continuous.

6.

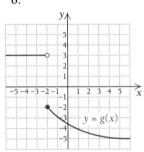

7.

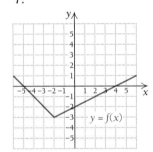

For the function in Exercise 6, answer the following.

8. Find $\lim\limits_{x \to 1} g(x)$.

9. Find $g(1)$.

10. Is g continuous at 1?

11. Find $\lim\limits_{x \to -2} g(x)$.

12. Find $g(-2)$.

13. Is g continuous at -2?

14. For $f(x) = x^3 - 2x$, find the average rate of change as x changes from -2 to 1.

15. Find a simplified difference quotient for $g(x) = -3x + 5$.

16. Find a simplified difference quotient for $f(x) = 2x^2 - 3$.

17. Find an equation of the tangent line to the graph of $y = x^2 + 3x$ at the point $(-1, -2)$.

18. Find the points on the graph of $y = -x^2 + 8x - 11$ at which the tangent line is horizontal.

19. Find the points on the graph of $y = 5x^2 - 49x + 12$ at which the tangent line has slope 1.

Find dy/dx.

20. $y = 4x^5$

21. $y = 3\sqrt[3]{x}$

22. $y = \dfrac{-8}{x^8}$

23. $y = 15x^{2/5}$

24. $y = 0.1x^7 - 3x^4 - x^3 + 6$

Differentiate.

25. $f(x) = \dfrac{1}{6}x^6 + 8x^4 - 5x$

26. $y = \dfrac{x^3 + x}{x}$

27. $y = \dfrac{x^2 + 8}{8 - x}$

28. $g(x) = (5 - x)^2(2x - 1)^5$

29. $f(x) = (x^5 - 2)^7$

30. $f(x) = x^2(4x + 3)^{3/4}$

31. For $y = x^3 - \dfrac{2}{x}$, find $\dfrac{d^5y}{dx^5}$.

32. For $y = x^7 + 3x^2$, find $\dfrac{d^4y}{dx^4}$.

33. Given $s(t) = t + t^4$, find each of the following.
 a) $v(t)$
 b) $a(t)$
 c) The velocity and the acceleration when $t = 2$ sec

34. *Business: Marginal revenue, cost, and profit.* Given $R(x) = 40x$ and $C(x) = 8x^2 - 7x - 10$, find each of the following.
 a) $P(x)$
 b) $R(20)$, $C(20)$, and $P(20)$
 c) $R'(x)$, $C'(x)$, and $P'(x)$
 d) $R'(20)$, $C'(20)$, and $P'(20)$

35. *Life science: Growth rate.* The population of a city grows from an initial size of 10,000 to an amount P, given by $P = 10,000 + 50t^2$, where t is in years.
 a) Find the growth rate.

b) Find the number of people in the city after 20 years (at $t = 20$ yr).

c) Find the growth rate at $t = 20$.

36. Find $f \circ g(x)$ and $g \circ f(x)$, given that $f(x) = x^2 + 5$ and $g(x) = 1 - 2x$.

SYNTHESIS

37. Find $\lim\limits_{x \to \infty} \dfrac{2 - 5x^5}{4 + 3x^6}$.

38. Differentiate $y = \dfrac{x\sqrt{1 + 3x}}{1 + x^3}$.

TECHNOLOGY CONNECTION

Use a grapher that creates input–output tables. Find each of the following limits. Start with Step $= 0.1$ and then go to 0.01, 0.001, and 0.0001. When you think you know the limit, graph, and use the trace feature to further verify your assertion.

39. $\lim\limits_{x \to 11} \dfrac{\sqrt{x - 2} - 3}{x - 11}$ **40.** $\lim\limits_{x \to 1} \dfrac{2 - \sqrt{x + 3}}{x - 1}$

41. Graph f and f' on the given interval. Then estimate points at which the tangent line to f is horizontal.

$$f(x) = 3.8x^5 - 18.6x^3; \quad [-3, 3]$$

2 Chapter Test

Determine whether the function is continuous.

1.

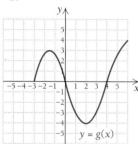

2.

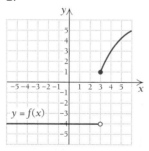

For the function in Question 2, answer the following.

3. Find $\lim\limits_{x \to 3} f(x)$.   **4.** Find $f(3)$.

5. Is f continuous at 3?   **6.** Find $\lim\limits_{x \to 4} f(x)$.

7. Find $f(4)$.   **8.** Is f continuous at 4?

Find the limit, if it exists.

9. $\lim\limits_{x \to 1} (3x^4 - 2x^2 + 5)$ **10.** $\lim\limits_{x \to 1} \dfrac{x - 1}{x^2 - 1}$

11. $\lim\limits_{x \to 0} \dfrac{7}{x}$ **12.** $\lim\limits_{x \to \infty} \dfrac{4x - 3}{x}$

13. $\lim\limits_{x \to \infty} \dfrac{2x^5 - 4x + 1}{6x^3 + 5}$

14. Find a simplified difference quotient for

$$f(x) = 3x^2 + 1.$$

15. Find an equation of the tangent line to the graph of $y = x + (4/x)$ at the point $(4, 5)$.

16. Find the points on the graph of $y = x^3 - 3x^2$ at which the tangent line is horizontal.

Find dy/dx.

17. $y = x^{84}$ **18.** $y = 10\sqrt{x}$

19. $y = \dfrac{-10}{x}$ **20.** $y = x^{5/4}$

21. $y = -0.5x^2 + 0.61x + 90$

Differentiate.

22. $y = \dfrac{1}{3}x^3 - x^2 + 2x + 4$

23. $y = \dfrac{2x - 5}{x^4}$

24. $f(x) = \dfrac{x}{5 - x}$

25. $f(x) = (x + 3)^4 (7 - x)^5$

26. $y = (x^5 - 4x^3 + x)^{-5}$

27. $f(x) = x\sqrt{x^2 + 5}$

28. For $y = x^4 - 3x^2$, find $\dfrac{d^3y}{dx^3}$.

29. *Business: Marginal revenue, cost, and profit.* Given $R(x) = 50x$ and $C(x) = 0.001x^2 + 1.2x + 60$, find each of the following.

   a) $P(x)$
   b) $R(10)$, $C(10)$, and $P(10)$
   c) $R'(x)$, $C'(x)$, and $P'(x)$
   d) $R'(10)$, $C'(10)$, and $P'(10)$

30. *Social sciences: Memory.* In a certain memory experiment, a person is able to memorize M words after t minutes, where $M = -0.001t^3 + 0.1t^2$.

   a) Find the rate of change of the number of words memorized with respect to time.
   b) How many words are memorized during the first 10 min (at $t = 10$)?
   c) What is the memory rate at $t = 10$ min?

31. Find $f \circ g(x)$ and $g \circ f(x)$, given that $f(x) = x + x^2$ and $g(x) = x^3$.

SYNTHESIS

32. Find $\displaystyle\lim_{x \to 2} \dfrac{x^3 - 8}{x - 2}$.

33. Differentiate $y = (1 - 3x)^{2/3}(1 + 3x)^{1/3}$.

TECHNOLOGY CONNECTION

34. Use a grapher that creates input–output tables. Find the following limit. Start with Step $= 0.1$ and then go to 0.01, 0.001, and 0.0001. When you think you know the limit, graph, and use the trace feature to further verify your assertion.

$$\lim_{x \to 0} \dfrac{\sqrt{5x + 25} - 5}{x}$$

35. Graph f and f' on the given interval. Then estimate points at which the tangent line to f is horizontal.

$$f(x) = 5x^3 - 30x^2 + 45x + 5\sqrt{x}; \ [0, 5]$$

EXTENDED TECHNOLOGY APPLICATION

Path of a Baseball: The Tale of the Tape

Have you ever been to a baseball game and seen a home run ball strike some obstacle after it has cleared the fence? Suppose the ball hits the obstacle at a location that is 60 ft above the ground at a distance of 400 ft from home plate. The scoreboard would display a message reading, "According to the tale of the tape, the ball would have traveled 442 ft." Did you ever wonder how the calculation is made? The answer is related to the curve formed by the path of a baseball.

Whatever the path of a well-hit baseball is, it is *not* the graph of a parabola,

$$f(x) = ax^2 + bx + c.$$

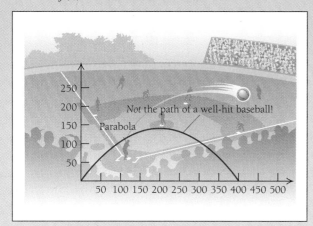

A well-hit baseball follows a path close to the following. This curve must be skewed to the direction in which the ball is hit.

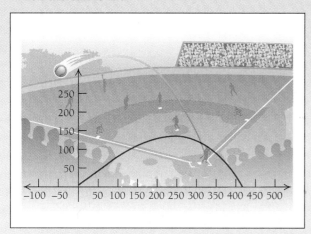

One reason that the ball does not follow the graph of a parabola is that a well-hit ball has backspin. This fact, combined with the frictional effect of the stitches on the ball with the air, tends to make the path of the ball "skewed" toward the direction in which it is hit. In effect, the path follows that of a skewed parabola.

Let's see if we can model the path of a baseball. Consider the following data. Assume that (0, 4.5) is the point at home plate at which the ball is hit,

roughly 4.5 ft off the ground. Also, assume that the ball has hit an obstacle 60 ft above the ground and 400 ft from home plate.

Horizontal distance, x (in feet)	0	50	100	200	285	300	360	400
Vertical distance, y (in feet)	4.5	43	82	130	142	134	100	60

EXERCISES

You will need a grapher that performs curve fitting and finds zeros or roots of functions.

1. Plot the points and connect them with line segments. This can be done on some graphers using a STAT PLOTS command. Does it seem to fit the preceding graph?

2. **a)** Use all the data from the table and your grapher to find a cubic polynomial function $y = ax^3 + bx^2 + cx + d$ that fits the data.
   b) Graph the function over the interval $[0, 500]$.
   c) Analyze the validity of using the curve to fit the data in the table.
   d) Predict the horizontal distance from home plate at which the ball would have hit the ground had it not hit the obstacle.
   e) Find the rate of change of the vertical height with respect to the horizontal distance.
   f) Find the points at which the graph has a horizontal tangent.

3. **a)** Use all the data to find a quartic polynomial function $y = ax^4 + bx^3 + cx^2 + dx + e$ that fits the data.
   b) Graph the function over the interval $[0, 500]$.
   c) Analyze the validity of using the curve to fit the data in the table.
   d) Predict the horizontal distance from home plate at which the ball would have hit the ground had it not hit the obstacle.
   e) Find the rate of change of the vertical height with respect to the horizontal distance.
   f) Find the points at which the graph has a horizontal tangent.

4. a) Although your grapher probably cannot fit such a function to the data, assume that the equation

$$y = 0.0015x\sqrt{202,500 - x^2}$$

has been found using some type of curve-fitting technique. Graph the function over the interval [0, 500].

  b) Predict the horizontal distance from home plate at which the ball would have hit the ground had it not hit the obstacle.

  c) Find the rate of change of the vertical height with respect to the horizontal distance.

  d) Find the points at which the graph has a horizontal tangent.

5. Compare the answers in Exercises 2(d), 3(d), and 4(b). Discuss the relative merits of using the quartic model in Exercise 3 with the model in Exercise 4 to make the prediction.

Tale of the tape. Actually, scoreboard operators in the major leagues use different models to predict the distance that a home run will travel. The models are linear and are related to the trajectory of the ball, that is, how high the ball is hit. See the following graph.

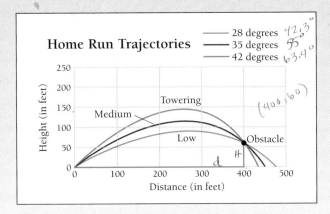

Suppose a ball hits an obstacle d feet horizontally from home plate at a vertical height of H feet. Then the estimated horizontal distance that the ball would have traveled is given by

$$D = kH + d,$$

where D = the estimated distance, H = the vertical height of the ball when obstructed, and d = the horizontal distance of the ball when obstructed. Then if we consider the trajectory of the ball, the formula becomes as follows:

Low trajectory: $D = 1.1H + d$,

Medium trajectory: $D = 0.7H + d$,

Towering trajectory: $D = 0.5H + d$.

6. For a ball striking an obstacle at $d = 400$ ft and $H = 60$ ft, estimate how far the ball would have traveled if it were low trajectory, medium trajectory, and towering trajectory.

7. In 1953, Hall-of-Famer Mickey Mantle hit a towering home run in old Griffith Stadium in Washington, D.C., that hit a sign 60 ft tall and 460 ft from home plate. The press asserted at the time that the ball would have traveled 565 ft. Is this estimate valid?

*Many thanks to Robert K. Adair, professor of physics at Yale University, for many of the ideas presented in this section.

3 Applications of Differentiation

INTRODUCTION

In this chapter, we learn many applications of differentiation. We learn to find maximum and minimum values of functions, and that skill allows us to solve many kinds of problems in which we need to find the largest and/or smallest value of a function. We also apply our differentiation skills to graphing functions.

AN APPLICATION

An employee's monthly production M, in number of units produced, is found to be a function of the number t of years of service. For a certain product, a productivity function is given by

$$M(t) = -2t^2 + 100t + 180, \quad 0 \le t \le 40.$$

Find the maximum productivity and the year in which it is achieved.

THE MATHEMATICS

If we find the first derivative, set it equal to 0, and solve, that value of x will yield the years in which the maximum productivity is achieved:

$$M'(t) = -4t + 100 = 0.$$

This problem appears as Exercise 77 in Exercise Set 3.4.

3.1 Using First Derivatives to Find Maximum and Minimum Values and Sketch Graphs

OBJECTIVES

• Find relative extrema of a continuous function using the First-Derivative Test.
• Sketch graphs of continuous functions.

The graph below illustrates a typical life cycle of a retail product. Note that the number of items sold varies with respect to time. Sales begin at a small level and increase to a point of maximum sales after which they decline and taper off to a low level, probably due to the effect of new competitive products. Then the company rejuvenates the product by making product improvements. Think about products such as black and white TVs, color TVs, records, compact discs, and CD-Rom computers. Where might each be in a typical life cycle and is that curve appropriate?

The graph of a typical life cycle illustrates many of the ideas that we will consider in this chapter.

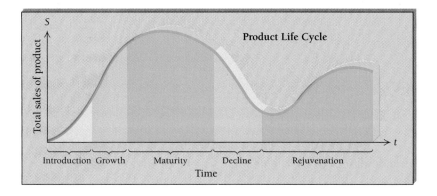

Finding the largest and smallest values of a function—that is, the maximum and minimum values—has extensive application. The first and second derivatives of a function are tools of calculus that give us information about the shape of a graph that may be helpful in finding maximum and minimum values of functions and in graphing functions. Throughout this section, we will assume that the functions f are continuous, but this does not necessarily imply that f' and f'' are continuous.

Increasing and Decreasing Functions

If the graph of a function rises from left to right on an interval I, it is said to be **increasing** on I. If the graph drops from left to right, it is said to be **decreasing** on I. We can describe this mathematically as follows.

Graph

$$f(x) = -\tfrac{1}{3}x^3 + 6x^2 - 11x - 50$$

and

$$f'(x) = -x^2 + 12x - 11$$

using a viewing window of $[-10, 25, -100, 150]$. Then use the trace feature, moving from left to right along each graph. As you move the cursor from left to right, note that the x-coordinate always increases. If a function is increasing on an interval, the y-coordinate will be increasing. If a function is decreasing on an interval, the y-coordinate will be decreasing.

Over what intervals is the function increasing?

Over what intervals is the function decreasing?

Over what intervals is the derivative positive?

Over what intervals is the derivative negative?

What can you conjecture?

DEFINITION

A function f is *increasing* on I, if for every a and b in I,

  if $a < b$,   then $f(a) < f(b)$.

(If the input a is less than the input b, then the output for a is less than the output for b.)

A function f is *decreasing* on I, if for every a and b in I,

  if $a < b$,   then $f(a) > f(b)$.

(If the input a is less than the input b, then the output for a is greater than the output for b.)

Note that the directions of the inequalities stay the same for an increasing function, but they differ for a decreasing function.

The following figures show examples of the preceding definition.

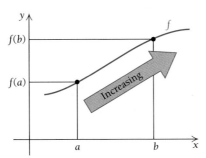

f is an increasing function.
If $a < b$, then $f(a) < f(b)$.
$f'(x) > 0$ for all x.

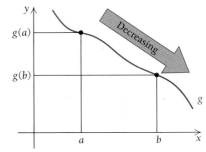

g is a decreasing function.
If $a < b$, then $g(a) > g(b)$.
$g'(x) < 0$ for all x.

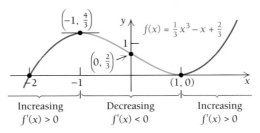

f is increasing on the intervals $(-\infty, -1)$ and $(1, \infty)$ and decreasing on the interval $(-1, 1)$.

In Chapter 1, we saw how the slope of a linear function determines whether that function is increasing or decreasing (or neither). For a general function, the derivative yields similar information.

The following theorem shows how we can use derivatives to determine whether a function is increasing or decreasing.

THEOREM 1

If $f'(x) > 0$ for all x in an interval I, then f is increasing on I.
If $f'(x) < 0$ for all x in an interval I, then f is decreasing on I.

Critical Points

Consider this graph of a continuous function.

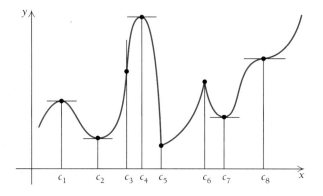

Note the following:

1. $f'(c) = 0$ at points c_1, c_2, c_4, c_7, and c_8. That is, the tangent to the graph is horizontal at these points.

2. $f'(c)$ does not exist at points c_3, c_5, and c_6. The tangent line is vertical at c_3 and there is a corner point, or sharp point, at both c_5 and c_6. (See also the discussion at the end of Section 2.4.)

DEFINITION

A *critical point* of a function is an interior point c of its domain at which the tangent to the graph is horizontal or at which the derivative does not exist. That is, c is a critical point if

$$f'(c) = 0 \quad \text{or} \quad f'(c) \text{ does not exist.}$$

Thus, in the preceding graph:

1. c_1, c_2, c_4, c_7, and c_8 are critical points because $f'(c) = 0$ for each point.

2. c_3, c_5, and c_6 are critical points because $f'(c)$ does not exist at each point.

Note also that a function can change from increasing to decreasing *only* at a critical point.

Finding Relative Maximum and Minimum Values

Consider the graph below. Note the "peaks" and "valleys" at the interior points c_1, c_2, and c_3.

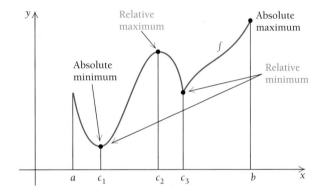

The function value $f(c_2)$ is called a **relative maximum.** Each of the function values $f(c_1)$ and $f(c_3)$ is called a **relative minimum.**

> **Definition**
>
> Suppose that f is a function whose value $f(c)$ exists at input c in an interval I in the domain of f. Then:
>
>> $f(c)$ is a *relative minimum* if there exists an open interval I_1 containing c in the domain such that $f(c) \leq f(x)$, for all x in I_1; and
>>
>> $f(c)$ is a *relative maximum* if there exists an open interval I_2 containing c in the domain such that $f(c) \geq f(x)$, for all x in I_2.

A relative maximum can be thought of as a high point that may or may not be the highest point, or *absolute maximum,* on an interval I. Similarly, a relative minimum can be thought of as a low point that may or may not be the lowest point, or *absolute minimum,* on I. For now we will consider how to find relative maximum or minimum values, stated simply as **relative extrema.**

Look again at the preceding graph. The points at which a continuous function has relative extrema are points where the derivative is 0 or where the derivative does not exist—the critical points.

THEOREM 2

If a function f has a relative extreme value $f(c)$, then c is a critical point, so

$$f'(c) = 0 \quad \text{or} \quad f'(c) \text{ does not exist.}$$

Theorem 2 is very useful, but it is important to understand it precisely. What it says is that when we are looking for points that are relative extrema, the only points we need to consider are those where the derivative is 0 or where the derivative does not exist. We can think of a critical point as a *candidate* for a relative maximum or minimum, and the candidate might or might not provide a relative extremum. That is, Theorem 2 does not say that if a point is a critical point, its function value will necessarily be a relative maximum or minimum. The existence of a critical point does *not* guarantee that a function has a relative maximum or minimum. A useful counterexample is the graph of

$$f(x) = (x - 1)^3 + 2,$$

shown below. Note that

$$f'(x) = 3(x - 1)^2,$$

and $\quad f'(1) = 3(1 - 1)^2 = 0.$

The function has a critical point at $c = 1$, but has no relative maximum or minimum at $c = 1$.

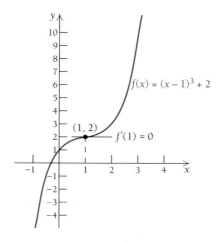

Now how can we tell when the existence of a critical point leads us to a relative extremum? The following graph leads us to a test.

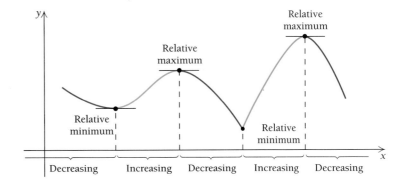

At a critical point for which there is a relative extreme value, the function is increasing on one side of the critical point and decreasing on the other side.

$f(c)$	Sign of $f'(x)$ for x in (a, c)	Sign of $f'(x)$ for x in (c, b)	Graph over the interval (a, b)
Relative minimum	−	+	
Relative maximum	+	−	
No relative maxima or minima	−	−	
No relative maxima or minima	+	+	

Derivatives tell us when a function is increasing or decreasing. This leads us to the First-Derivative Test.

THEOREM 3

The First-Derivative Test for Relative Extrema

For any continuous function f that has exactly one critical point c in an open interval (a, b):

F1. f has a relative minimum at c if $f'(x) < 0$ on (a, c) and $f'(x) > 0$ on (c, b). That is, f is decreasing to the left of c and increasing to the right of c.

F2. f has a relative maximum at c if $f'(x) > 0$ on (a, c) and $f'(x) < 0$ on (c, b). That is, f is increasing to the left of c and decreasing to the right of c.

F3. f has neither a relative maximum nor a relative minimum at c if $f'(x)$ has the same sign on (a, c) as on (c, b).

Now let us apply the First-Derivative Test to finding relative extrema and to graphing.

EXAMPLE 1   Find the relative extrema of the function f given by

$$f(x) = 2x^3 - 3x^2 - 12x + 12.$$

Then sketch the graph.

Solution   First, we must find the critical points. To do so, we find $f'(x)$:

$$f'(x) = 6x^2 - 6x - 12.$$

We then determine where $f'(x)$ does not exist or where $f'(x) = 0$. We can replace x in $f'(x) = 6x^2 - 6x - 12$ by any real number. Thus, $f'(x)$ exists for all real numbers. So the only possibilities for critical points are where $f'(x) = 0$, at which there are horizontal tangents. To find such points, we solve $f'(x) = 0$:

$$6x^2 - 6x - 12 = 0$$
$$x^2 - x - 2 = 0 \qquad \text{Dividing by 6 on both sides}$$
$$(x + 1)(x - 2) = 0 \qquad \text{Factoring}$$
$$x + 1 = 0 \quad \text{or} \quad x - 2 = 0 \qquad \text{Using the Principle of Zero Products}$$
$$x = -1 \quad \text{or} \qquad x = 2.$$

The critical points are -1 and 2. Since it is at these points that a relative maximum or minimum will exist, if there is one, we examine the intervals on each side of the critical points. We use the critical points to divide the real-number line into three intervals: $A(-\infty, -1)$, $B(-1, 2)$, and $C(2, \infty)$, as shown below.

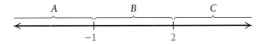

Then we analyze the sign of the derivative on each interval. If $f'(x)$ is positive for one value in the interval, then it will be positive for all numbers in the interval. Similarly, if it is negative for one value, it will be negative for all values in the interval. This is because in order for the derivative to change signs, it must become 0 or be undefined at some point. Such a point will be a critical point. Thus we merely choose a test value in each interval and make a substitution. The test values we choose are -2, 0, and 3.

A: Test -2, $f'(-2) = 6(-2)^2 - 6(-2) - 12$
$$= 24 + 12 - 12 = 24 > 0;$$

B: Test 0, $f'(0) = 6(0)^2 - 6(0) - 12 = -12 < 0;$

C: Test 3, $f'(3) = 6(3)^2 - 6(3) - 12$
$$= 54 - 18 - 12 = 24 > 0.$$

Interval	$(-\infty, -1)$	$(-1, 2)$	$(2, \infty)$
Test value	$x = -2$	$x = 0$	$x = 3$
Sign of $f'(x)$	$f'(-2) > 0$	$f'(0) < 0$	$f'(3) > 0$
Result	f is increasing	f is decreasing	f is increasing

└─── Change ─┐└─ Change ──┐
indicates a   indicates a
relative   relative
maximum.   minimum.

Therefore, by the First-Derivative Test,

f has a relative maximum at $x = -1$ given by

$$f(-1) = 2(-1)^3 - 3(-1)^2 - 12(-1) + 12$$ Substituting into the original function

$$= 19$$

and f has a relative minimum at $x = 2$ given by

$$f(2) = 2(2)^3 - 3(2)^2 - 12(2) + 12 = -8.$$

Thus there is a relative maximum at $(-1, 19)$ and a relative minimum at $(2, -8)$.

The information we have obtained can be very useful in sketching a graph of the function. We know that this polynomial is continuous, and we know where the function is increasing, where it is decreasing, and where it has relative extrema. We complete the graph by using a calculator to generate some additional function values. Many are listed in the table below. More can be generated by the student if there is a need. Some calculators can actually be programmed to generate function values by first entering a formula and then generating many outputs. The graph of the function, shown below, has been scaled to clearly show its curving nature.

x	$f(x)$
-3	-33
-2.5	-8
-2	8
-1.5	16.5
-1	19
-0.5	17
0	12
0.5	5.5
1	-1
1.5	-6
2	-8
2.5	-5.5
3	3
3.5	19
4	44

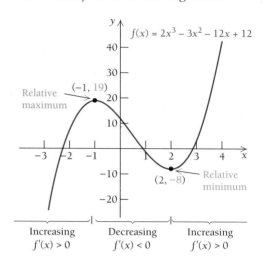

 TECHNOLOGY CONNECTION

Finding Relative Extrema on a Grapher

How can we approximate relative extrema using a grapher? Consider

$$y = -0.4x^3 + 6.2x^2 - 11.3x - 54.8.$$

We graph the function, using a viewing window that reveals the curvature.

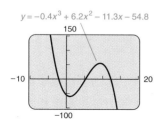

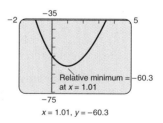

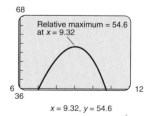

Then we look for relative extrema, using the trace feature to obtain an estimate. Estimate the relative extrema of the function f given by $f(x) = x^3 + 3x^2 - 9x - 12$. First use your grapher. Then check your work using the analytic method of Example 1.

Keep the following in mind as you find relative extrema.

The *derivative f'* is used to find the critical points of f. The test values, in the intervals defined by the critical points, are substituted into the *derivative f'*, and the function values are found using the *original* function f. Use the derivative f' to find information about the shape of the graph of f.

EXAMPLE 2   Find the relative extrema of the function f given by

$$f(x) = 2x^3 - x^4.$$

Then sketch the graph.

Solution   First, we must determine the critical points. To do so, we find $f'(x)$:

$$f'(x) = 6x^2 - 4x^3.$$

Next, we find where $f'(x)$ does not exist or where $f'(x) = 0$. We can replace x in $f'(x) = 6x^2 - 4x^3$ by any real number. Thus, $f'(x)$ exists for all real numbers. So the only possibilities for critical points are where $f'(x) = 0$, that is, where there are horizontal tangents. To find such points, we solve $f'(x) = 0$:

$$6x^2 - 4x^3 = 0$$
$$2x^2(3 - 2x) = 0 \qquad \text{Factoring}$$
$$2x^2 = 0 \quad \text{or} \quad 3 - 2x = 0$$
$$x^2 = 0 \quad \text{or} \qquad 3 = 2x$$
$$x = 0 \quad \text{or} \qquad x = \tfrac{3}{2}.$$

The critical points are 0 and $\tfrac{3}{2}$. We use these points to divide the real-number line into three intervals: $A(-\infty, 0)$, $B\left(0, \tfrac{3}{2}\right)$, and $C\left(\tfrac{3}{2}, \infty\right)$, as shown below.

We now analyze the sign of the derivative on each interval. We begin by choosing a test value in each interval and making a substitution. We usually choose as test values numbers for which it is easy to compute outputs of the derivative—in this case, -1, 1, and 2.

$$A: \text{Test } -1, \quad f'(-1) = 6(-1)^2 - 4(-1)^3$$
$$= 6 + 4 = 10 > 0;$$
$$B: \text{Test } 1, \quad f'(1) = 6(1)^2 - 4(1)^3$$
$$= 6 - 4 = 2 > 0;$$
$$C: \text{Test } 2, \quad f'(2) = 6(2)^2 - 4(2)^3$$
$$= 24 - 32 = -8 < 0.$$

Interval	$(-\infty, 0)$	$\left(0, \frac{3}{2}\right)$	$\left(\frac{3}{2}, \infty\right)$
Test value	$x = -1$	$x = 1$	$x = 2$
Sign of $f'(x)$	$f'(-1) > 0$	$f'(1) > 0$	$f'(2) < 0$
Result	f is increasing	f is increasing	f is decreasing

└── No change ──┘└── Change ──┘
indicates a
relative maximum.

Therefore, by the First-Derivative Test, f has neither a relative maximum nor a relative minimum at $x = 0$ since the function is increasing on both sides of 0, and f has a relative maximum at $x = \frac{3}{2}$ given by

$$f\left(\tfrac{3}{2}\right) = 2\left(\tfrac{3}{2}\right)^3 - \left(\tfrac{3}{2}\right)^4 = \tfrac{27}{16}.$$ Remember to substitute into
the original function.

Thus there is a relative maximum at $\left(\frac{3}{2}, \frac{27}{16}\right)$.

We use the information obtained to sketch the graph. Other function values are listed in the table below. (More can be generated by the student.) The graph follows.

x	$f(x)$, approximately
-1	-3
-0.75	-1.16
-0.5	-0.31
-0.25	-0.04
0	0
0.25	0.03
0.5	0.19
0.75	0.53
1	1
1.25	1.46
1.5	1.69
1.75	1.34
2	0

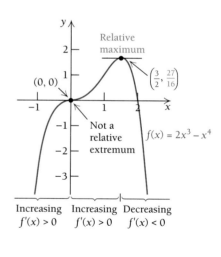

TECHNOLOGY
CONNECTION

Graph $f(x) = x^4 - 8x^3 + 18x^2$. Estimate the relative extrema. Then check your work using the analytic method of Example 2.

EXAMPLE 3   Find the relative extrema of the function f given by

$$f(x) = (x - 2)^{2/3} + 1.$$

Then sketch the graph.

Solution   First, we determine the critical points. To do so, we find $f'(x)$:

$$f'(x) = \frac{2}{3}(x-2)^{-1/3} = \frac{2}{3\sqrt[3]{x-2}}.$$

Next, we find where $f'(x)$ does not exist or where $f'(x) = 0$. Note that the derivative $f'(x)$ does not exist at 2, although $f(x)$ does. Thus the number 2 is a critical point. The equation $f'(x) = 0$ has no solution, so the only critical point is 2. We use 2 to divide the real-number line into two intervals: $A(-\infty, 2)$ and $B(2, \infty)$, as shown below.

We analyze the derivative on each interval. We begin by choosing a test value in each interval and making a substitution. We choose test points 0 and 3. It is not necessary to find an exact value of the derivative; we need only determine the sign. Sometimes we can do this by just examining the formula for the derivative:

$$A: \text{ Test } 0, \quad f'(0) = \frac{2}{3\sqrt[3]{0-2}} < 0;$$

$$B: \text{ Test } 3, \quad f'(3) = \frac{2}{3\sqrt[3]{3-2}} > 0.$$

Interval	$(-\infty, 2)$	$(2, \infty)$
Test value	$x = 0$	$x = 3$
Sign of $f'(x)$	$f'(0) < 0$	$f'(3) > 0$
Result	f is decreasing	f is increasing

Change
indicates a
relative minimum.

Since we have a change from decreasing to increasing, we conclude from the First-Derivative Test that

f has a relative minimum at $x = 2$ given by

$$f(2) = (2-2)^{2/3} + 1 = 1.$$

Thus there is a relative minimum at $(2, 1)$.

We use the information obtained to sketch the graph. Other function values are listed in the table on the following page. (More can be generated by the student.) The graph follows.

TECHNOLOGY
CONNECTION

Graph the function f given
by $f(x) = 2 - (x - 1)^{2/3}$,
using a viewing window of
$[-5, 5, -1, 8]$.
    Graph the derivative.
What happens to the graph
of the derivative at the
critical point(s)? Find the
relative extrema.
    Graph $f(x) = x - |x|$
and its derivative. What are
the critical points and the
relative extrema?

x	$f(x)$, approximately
-1	3.08
-0.5	2.84
0	2.59
0.5	2.31
1	2
1.5	1.63
2	1
2.5	1.63
3	2
3.5	2.31
4	2.59

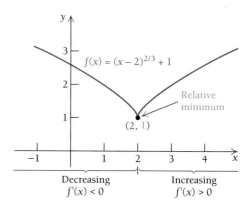

3.1 Exercise Set

Find the relative extrema of the function, if they exist.
List your answers in terms of ordered pairs. Then sketch
a graph of the function.

1. $f(x) = x^2 - 4x + 5$

2. $f(x) = x^2 - 6x - 3$

3. $f(x) = 5 + x - x^2$

4. $f(x) = 2 - 3x - 2x^2$

5. $f(x) = 1 + 6x + 3x^2$

6. $f(x) = 0.5x^2 - 2x - 11$

7. $f(x) = x^3 - x^2 - x + 2$

8. $f(x) = x^3 + \frac{1}{2}x^2 - 2x + 5$

9. $f(x) = x^3 - 3x + 6$

10. $f(x) = x^3 - 3x^2$

11. $f(x) = 3x^2 - 2x^3$

12. $f(x) = x^3 - 3x$

13. $f(x) = 2x^3$

14. $f(x) = 1 - x^3$

15. $f(x) = x^3 - 6x^2 + 10$

16. $f(x) = 12 + 9x - 3x^2 - x^3$

17. $f(x) = x^3 - x^4$

18. $f(x) = x^4 - 2x^3$

19. $f(x) = x^4 - 8x^2 + 3$

20. $f(x) = x^4 - 2x^2 + 5$

21. $f(x) = 1 - x^{2/3}$

22. $f(x) = (x + 3)^{2/3} - 5$

23. $f(x) = \dfrac{-8}{x^2 + 1}$

24. $f(x) = \dfrac{5}{x^2 + 1}$

25. $f(x) = \dfrac{4x}{x^2 + 1}$

26. $f(x) = \dfrac{x^2}{x^2 + 1}$

27. $f(x) = \sqrt[3]{x}$

28. $f(x) = (x + 1)^{1/3}$

APPLICATIONS

◆ **Business and Economics**

29. *Advertising.* A firm estimates that it will sell N units
    of a product after spending a dollars on advertising,

where

$$N(a) = -a^2 + 300a + 6, \quad 0 \le a \le 300,$$

and a is in thousands of dollars. Find the relative extrema and sketch a graph of the function.

◆ **Life and Physical Sciences**

30. *Temperature during an illness.* The temperature of a person during an illness is given by

$$T(t) = -0.1t^2 + 1.2t + 98.6, \quad 0 \le t \le 12,$$

where T = the temperature (°F) at time t, in days. Find the relative extrema and sketch a graph of the function.

◆ **General Interest**

31. *Path of the Olympic arrow.* The Olympic flame at the 1992 summer Olympics was lit by a flaming arrow. As the arrow moved d feet horizontally from the archer, its height h, in feet, was approximated by the function

$$h = -0.002d^2 + 0.8d + 6.6.$$

Find the relative maximum and sketch a graph of the function.

32. Use the graph in Exercise 31 to describe the words "the path of the Olympic arrow." Include the height from which it was launched, its maximum height, and the horizontal distance from the archer at which it hit the ground.

33. ◆ Consider this graph.

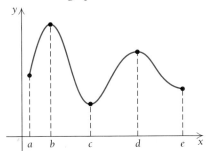

Using the graph and the intervals noted, explain how to relate the concept of the function being increasing or decreasing to the first derivative.

34. ◆ Consider this graph.

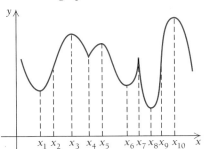

Explain the idea of a critical point. Then determine which points on the graph are critical points and why.

TECHNOLOGY CONNECTION

Graph the function. Then estimate any relative extrema. Consider input–output tables.

35. $f(x) = x^4 + 4x^3 - 36x^2 - 160x + 400$

36. $f(x) = -x^6 - 4x^5 + 54x^4 + 160x^3 - 641x^2 - 828x + 1200$

37. $f(x) = x\sqrt{9 - x^2}$

38. $f(x) = \sqrt[3]{|4 - x^2|} + 1$

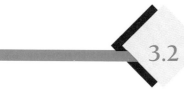

3.2 Using Second Derivatives to Find Maximum and Minimum Values and Sketch Graphs

OBJECTIVES

- Find the relative extrema of a function using the Second-Derivative Test.
- Sketch the graph of a continuous function.

Concavity: Increasing and Decreasing Derivatives

The graphs of two functions are shown in Fig. 1. The graph in Fig. 1(a) is turning up and the graph in Fig. 1(b) is turning down. Let us see if we can relate this to their derivatives.

Consider the graph of f (Fig. 1a). Take a ruler, or straightedge, and draw tangent lines as you move along the curve from left to right. What happens to the slopes of the tangent lines? Do the same for the graph of g (Fig. 1b). Look for a pattern.

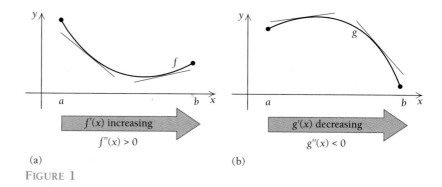

(a)   (b)

FIGURE 1

In Fig. 1(a), the slopes are increasing. That is, f' is increasing on the interval. We know that f' is increasing if we know that f'' is positive, since the relationship between f' and f'' is like the relationship between f and f'. Note also that all the tangent lines are below the graph. In Fig. 1(b), the slopes are decreasing. That is, we know that g' is decreasing if we know that g'' is negative. All the tangent lines are above the graph.

DEFINITION

Suppose that f is a function whose derivative f' exists at every point in an open interval I. Then:

1. f is *concave up* on the interval I if f' is increasing on I.
2. f is *concave down* on the interval I if f' is decreasing on I.

For example, the graph in Fig. 1(a) is concave up and the graph in Fig. 1(b) is concave down. We then have the following theorem, which allows us to use second derivatives to determine concavity.

THEOREM 4

A Test for Concavity

1. If $f''(x) > 0$ on an interval I, then the graph of f is turning up. (f' is increasing, so f is concave up on I.)

2. If $f''(x) < 0$ on an interval I, then the graph of f is turning down. (f' is decreasing, so f is concave down on I.)

A helpful memory device follows.

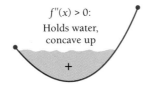
$f''(x) > 0$:
Holds water,
concave up

$f''(x) < 0$:
Loses water,
concave down

 TECHNOLOGY CONNECTION

Graph the function

$$f(x) = -\tfrac{1}{3}x^3 + 6x^2 - 11x - 50$$

and its second derivative

$$f''(x) = -2x + 12$$

using the same viewing window $[-10, 25, -100, 150]$.

Over what intervals is the graph of f concave up?

Over what intervals is the graph of f concave down?

Over what intervals is the graph of f'' positive?

Over what intervals is the graph of f'' negative?

What can you conjecture?

Now graph the first derivative

$$f'(x) = -x^2 + 12x - 11$$

and the second derivative

$$f''(x) = -2x + 12$$

using the same viewing window $[-10, 25, -200, 50]$.

Over what intervals is the first derivative f' increasing?

Over what intervals is the first derivative f' decreasing?

Over what intervals is the graph of f'' positive?

Over what intervals is the graph of f'' negative?

What can you conjecture?

Finding Relative Extrema Using Second Derivatives

In the following discussion, we see how we can use second derivatives to determine whether a function has a relative extreme value on an open interval.

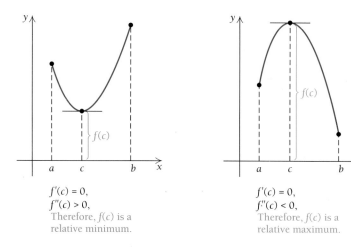

$f'(c) = 0,$
$f''(c) > 0,$
Therefore, $f(c)$ is a relative minimum.

$f'(c) = 0,$
$f''(c) < 0,$
Therefore, $f(c)$ is a relative maximum.

THEOREM 5

The Second-Derivative Test for Relative Extrema

Suppose that f is a function for which $f'(x)$ exists for every x in an open interval (a, b) contained in its domain, and that there is a critical point c in (a, b) for which $f'(c) = 0$. Then:

1. $f(c)$ is a relative minimum if $f''(c) > 0$.
2. $f(c)$ is a relative maximum if $f''(c) < 0$.

The test fails if $f''(c) = 0$. The First-Derivative Test would then have to be used.

Note that $f''(c) = 0$ does not tell us that there is no relative extremum. It just tells us that we do not know at this point and that we must use some other means, such as the First-Derivative Test, to determine whether there is.

Consider the following graphs. In each one, f' and f'' are both 0 at $c = 2$, but the first function has an extremum and the second function has *no* extremum. Also note that if $f'(c)$ does not exist, then $f''(c)$ does not exist and the Second-Derivative Test cannot be used.

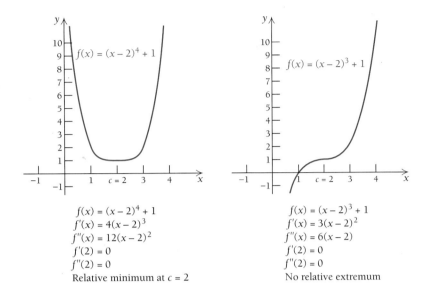

$$f(x) = (x - 2)^4 + 1$$
$$f'(x) = 4(x - 2)^3$$
$$f''(x) = 12(x - 2)^2$$
$$f'(2) = 0$$
$$f''(2) = 0$$
Relative minimum at $c = 2$

$$f(x) = (x - 2)^3 + 1$$
$$f'(x) = 3(x - 2)^2$$
$$f''(x) = 6(x - 2)$$
$$f'(2) = 0$$
$$f''(2) = 0$$
No relative extremum

EXAMPLE 1   Find the relative extrema of the function f given by

$$f(x) = x^3 + 3x^2 - 9x - 13.$$

Solution   We find both the first and second derivatives, $f'(x)$ and $f''(x)$:

$$f'(x) = 3x^2 + 6x - 9,$$
$$f''(x) = 6x + 6.$$

Then we solve $f'(x) = 0$:

$$3x^2 + 6x - 9 = 0$$
$$x^2 + 2x - 3 = 0 \qquad \text{Dividing by 3 on both sides}$$
$$(x + 3)(x - 1) = 0 \qquad \text{Factoring}$$
$$x + 3 = 0 \quad \text{or} \quad x - 1 = 0 \qquad \text{Using the Principle of Zero Products}$$
$$x = -3 \quad \text{or} \qquad x = 1.$$

We now use the Second-Derivative Test with the numbers -3 and 1:

$$f''(-3) = 6(-3) + 6 = -12 < 0 \longrightarrow \qquad \text{Relative maximum}$$
$$f''(1) = 6(1) + 6 = 12 > 0 \longrightarrow \qquad \text{Relative minimum}$$

We then find second coordinates by substituting in the original function:

$$f(-3) = (-3)^3 + 3(-3)^2 - 9(-3) - 13 = 14;$$
$$f(1) = (1)^3 + 3(1)^2 - 9(1) - 13 = -18.$$

TECHNOLOGY
CONNECTION

Consider the function f given by

$$f(x) = x^3 - 3x^2 - 9x - 1.$$

Use your grapher to estimate the relative extrema. Then find the first and second derivatives. Graph both on the same set of axes. Determine, using the trace feature, where the first derivative is zero. At those x-values, verify the relative extrema by checking the sign of the second derivative. Then check your work using the analytic method of Example 1.

Thus there is a relative maximum at $(-3, 14)$ and a relative minimum at $(1, -18)$. The relative extrema are shown in the graph.

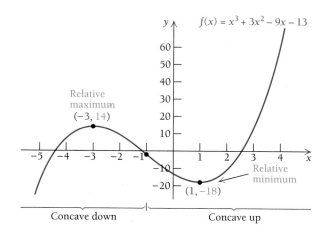

EXAMPLE 2   Find the relative extrema of the function f given by

$$f(x) = 3x^5 - 20x^3.$$

Solution   We find both the first and second derivatives, $f'(x)$ and $f''(x)$:

$$f'(x) = 15x^4 - 60x^2,$$
$$f''(x) = 60x^3 - 120x.$$

Then we solve for $f'(x) = 0$:

$$15x^4 - 60x^2 = 0$$
$$15x^2(x^2 - 4) = 0$$
$$15x^2(x + 2)(x - 2) = 0 \qquad \text{Factoring}$$
$$15x^2 = 0 \quad \text{or} \quad x + 2 = 0 \qquad \text{or} \quad x - 2 = 0 \qquad \text{Using the Principle of Zero Products}$$
$$x = 0 \quad \text{or} \qquad x = -2 \quad \text{or} \qquad x = 2.$$

We now use the Second-Derivative Test with the numbers -2, 2, and 0:

$$f''(-2) = 60(-2)^3 - 120(-2) = -240 < 0 \longrightarrow \qquad \text{Relative maximum;}$$
$$f''(2) = 60(2)^3 - 120(2) = 240 > 0 \longrightarrow \qquad \text{Relative minimum;}$$
$$f''(0) = 60(0)^3 - 120(0) = 0 \longrightarrow \qquad \text{Second-Derivative Test fails;}$$
$$\text{Use the First-Derivative Test.}$$

We then find second coordinates by substituting in the original function:

$$f(-2) = 3(-2)^5 - 20(-2)^3 = 64;$$
$$f(2) = 3(2)^5 - 20(2)^3 = -64;$$
$$f(0) = 3(0)^5 - 20(0)^3 = 0.$$

Thus there is a relative maximum at $(-2, 64)$ and a relative minimum at $(2, -64)$. Because the function decreases to the left and to the right of $x = 0$, we know by the First-Derivative Test that it has no relative extremum at $(0, 0)$. The extrema are shown in the graph.

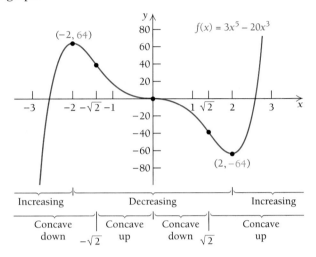

Points of Inflection

A **point of inflection,** or an **inflection point,** is a point across which the direction of concavity changes. For example, in Figs. 2–4, point P is an inflection point.

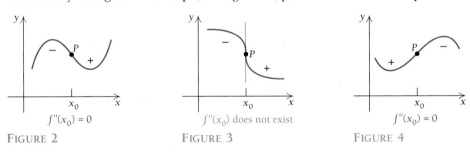

FIGURE 2   FIGURE 3   FIGURE 4

There are points of inflection in other examples that we have considered in this section. In Example 1, the graph has a point of inflection at $(-1, -2)$, and in Example 2, the graph has points of inflection at $(-\sqrt{2}, 28\sqrt{2})$, $(0, 0)$, and $(\sqrt{2}, -28\sqrt{2})$.

As we move to the right along the curve in Fig. 2, the concavity changes from concave down, $f''(x) < 0$, on the left of P to concave up, $f''(x) > 0$, on the right of P. Since, as we move through P, $f''(x)$ changes sign from $-$ to $+$, the value of $f''(x_0)$ at P must be 0, as in Fig 2; or $f''(x_0)$ does not exist, as in Fig. 3. A similar change in concavity occurs at P in Fig. 4.

TECHNOLOGY
CONNECTION

Consider the function f
given by $f(x) = 3x^5 - 5x^3$.
Graph the second derivative
$f''(x) = 60x^3 - 30x$.
Estimate any points of
inflection.

THEOREM 6

Finding Points of Inflection

If a function f has a point of inflection, it occurs at a point x_0, where

$$f''(x_0) = 0 \quad \text{or} \quad f''(x_0) \text{ does not exist.}$$

Thus we find candidates for points of inflection by looking for numbers x_0 for which $f''(x_0) = 0$ or for which $f''(x_0)$ does not exist. Then if $f''(x)$ changes sign as x moves through x_0, we have a point of inflection at $x = x_0$. If $f''(x) = k$, where $k \neq 0$, then x is not a candidate for a point of inflection.

Curve Sketching

What we have learned thus far in this chapter will greatly enhance our ability to sketch curves. We use the following strategy, writing it first in abbreviated form.

Strategy for Sketching Graphs

a) Derivatives

b) Critical points of f

c) Increasing, decreasing, relative extrema

d) Inflection points

e) Concavity

f) Sketch

Below we expand on the strategy.

Strategy for Sketching Graphs

a) *Derivatives.* Find $f'(x)$ and $f''(x)$.

b) *Critical points of f.* Find the critical points of f by solving $f'(x) = 0$ and finding where $f'(x)$ does not exist. These numbers yield candidates for relative maxima or minima. Find the function values at these points.

c) *Increasing, decreasing, relative extrema.* Use the critical points of f from step (b) to define intervals. Determine whether f is increasing or decreasing on the intervals. Do this by selecting test values and substituting into $f'(x)$. Use this information and/or the second derivative to determine the relative maxima and minima.

d) *Inflection points.* Determine candidates for inflection points by finding where $f''(x) = 0$ or where $f''(x)$ does not exist. Find the function values at these points.

e) *Concavity.* Use the candidates for inflection points from step (d) to define intervals. Determine the concavity by checking to see where f' is increasing and where f' is decreasing. Do this by selecting test values and substituting into $f''(x)$.

f) *Sketch the graph.* Sketch the graph using the information from steps (a) through (e), plotting extra points (computing them with your calculator) if the need arises.

EXAMPLE 3   Find the relative maxima and minima of the function f given by

$$f(x) = x^3 - 3x + 2$$

and sketch the graph.

Solution

a) *Derivatives.* Find $f'(x)$ and $f''(x)$:

$$f'(x) = 3x^2 - 3,$$
$$f''(x) = 6x.$$

b) *Critical points of f.* Find the critical points of f by finding where $f'(x)$ does not exist and by solving $f'(x) = 0$. Now $f'(x) = 3x^2 - 3$ exists for all values of x, so the only critical points are where

$$3x^2 - 3 = 0$$
$$3x^2 = 3$$
$$x^2 = 1$$
$$x = \pm 1.$$

Now $f(-1) = 4$ and $f(1) = 0$. These give the points $(-1, 4)$ and $(1, 0)$ on the graph.

c) *Increasing, decreasing, relative extrema.* Find the intervals on which f is increasing and the intervals on which f is decreasing. The critical points are -1 and 1. We use these points to divide the real-number line into three intervals: $A(-\infty, -1)$, $B(-1, 1)$, and $C(1, \infty)$. We choose a test point in each interval and make a substitution. The test values we select are -2, 0, and 3:

A: Test -2, $f'(-2) = 3(-2)^2 - 3 = 9 > 0$;
B: Test 0, $f'(0) = 3(0)^2 - 3 = -3 < 0$;
C: Test 3, $f'(3) = 3(3)^2 - 3 = 24 > 0$.

Interval	$(-\infty, -1)$	$(-1, 1)$	$(1, \infty)$
Test value	$x = -2$	$x = 0$	$x = 3$
Sign of $f'(x)$	$f'(-2) > 0$	$f'(0) < 0$	$f'(3) > 0$
Result	f is increasing	f is decreasing	f is increasing

Change indicates a relative maximum.   Change indicates a relative minimum.

Therefore, by the First-Derivative Test, there is a relative maximum at $(-1, 4)$ and a relative minimum at $(1, 0)$. That these are relative extrema can also be verified by the Second-Derivative Test: $f''(-1) < 0$ tells us that $(-1, 4)$ is a relative maximum, and $f''(1) > 0$ tells us that $(1, 0)$ is a relative minimum.

d) *Inflection points.* Find possible inflection points by finding where $f''(x)$ does not exist and by solving $f''(x) = 0$. Now $f''(x) = 6x$ exists for all values of x, so we try to solve $f''(x) = 0$:

$$6x = 0$$
$$x = 0.$$

Now $f(0) = 2$. This gives us another point, $(0, 2)$, that lies on the graph.

e) *Concavity.* Find the intervals on which f is concave up and concave down. We do this by determining where f' is increasing and decreasing using the numbers found in step (d). There is only one such number, 0. We use this point to divide the real-number line into two intervals: $A(-\infty, 0)$ and $B(0, \infty)$. We choose a test point in each interval and make a substitution into f''. The test points we select are -4 and 4:

A: Test -4, $f''(-4) = 6(-4) = -24 < 0$;
B: Test 4, $f''(4) = 6(4) = 24 > 0$.

Interval	$(-\infty, 0)$	$(0, \infty)$
Test value	$x = -4$	$x = 4$
Sign of $f''(x)$	$f''(-4) < 0$	$f''(4) > 0$
Result	f' is decreasing	f' is increasing

Change indicates a point of inflection.

Since f' is decreasing on the interval $(-\infty, 0)$ and f' is increasing on the interval $(0, \infty)$, f is concave down on $(-\infty, 0)$ and concave up on $(0, \infty)$. The graph changes concavity across $(0, 2)$, so it is a point of inflection.

f) *Sketch the graph.* Sketch the graph using the information in the following table. Calculate some extra function values if desired. The graph follows.

x	$f(x)$
-3	-16
-2	0
-1	4
0	2
1	0
2	4
3	20

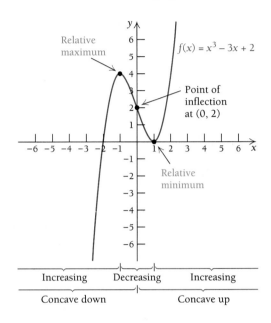

EXAMPLE 4 Find the relative maxima and minima of the function f given by

$$f(x) = x^4 - 2x^2$$

and sketch the graph.

Solution

a) *Derivatives.* Find $f'(x)$ and $f''(x)$:

$$f'(x) = 4x^3 - 4x,$$
$$f''(x) = 12x^2 - 4.$$

b) *Critical points of f.* Since $f'(x) = 4x^3 - 4x$ exists for all values of x, the only critical points of f are where

$$4x^3 - 4x = 0$$
$$4x(x^2 - 1) = 0$$
$$4x = 0 \quad \text{or} \quad x^2 - 1 = 0$$
$$x = 0 \quad \text{or} \quad x^2 = 1$$
$$x = \pm 1.$$

Now $f(0) = 0$, $f(-1) = -1$, and $f(1) = -1$. These give the points $(0, 0)$, $(-1, -1)$, and $(1, -1)$ on the graph.

c) *Increasing, decreasing, relative extrema.* Use the critical points of f—namely, -1, 0, and 1—to divide the real-number line into four intervals: $A(-\infty, -1)$, $B(-1, 0)$, $C(0, 1)$, and $D(1, \infty)$. We choose a test value in each interval and make a substitution. The test values we select are -2, $-\frac{1}{2}$, $\frac{1}{2}$, and 2:

A: Test -2, $f'(-2) = 4(-2)^3 - 4(-2) = -24 < 0$;
B: Test $-\frac{1}{2}$, $f'\left(-\frac{1}{2}\right) = 4\left(-\frac{1}{2}\right)^3 - 4\left(-\frac{1}{2}\right) = \frac{3}{2} > 0$;
C: Test $\frac{1}{2}$, $f'\left(\frac{1}{2}\right) = 4\left(\frac{1}{2}\right)^3 - 4\left(\frac{1}{2}\right) = -\frac{3}{2} < 0$;
D: Test 2, $f'(2) = 4(2)^3 - 4(2) = 24 > 0$.

Interval	$(-\infty, -1)$	$(-1, 0)$	$(0, 1)$	$(1, \infty)$
Test value	$x = -2$	$x = -\frac{1}{2}$	$x = \frac{1}{2}$	$x = 2$
Sign of $f'(x)$	$f'(-2) < 0$	$f'\left(-\frac{1}{2}\right) > 0$	$f'\left(\frac{1}{2}\right) < 0$	$f'(2) > 0$
Result	f is decreasing	f is increasing	f is decreasing	f is increasing

Change indicates a relative minimum. Change indicates a relative maximum. Change indicates a relative minimum.

Thus, by the First-Derivative Test, there is a relative maximum at $(0, 0)$ and two relative minima at $(-1, -1)$ and $(1, -1)$. That these are relative extrema can also be verified by the Second-Derivative Test: $f''(0) < 0$, $f''(-1) > 0$, and $f''(1) > 0$.

d) *Inflection points.* Find where $f''(x)$ does not exist and where $f''(x) = 0$. Since $f''(x)$ exists for all real numbers, we just solve $f''(x) = 0$:

$$12x^2 - 4 = 0$$
$$4(3x^2 - 1) = 0$$
$$3x^2 - 1 = 0$$
$$3x^2 = 1$$
$$x^2 = \frac{1}{3}$$
$$x = \pm\sqrt{\frac{1}{3}}$$
$$= \pm\frac{1}{\sqrt{3}}.$$

Now

$$f\left(\frac{1}{\sqrt{3}}\right) = \left(\frac{1}{\sqrt{3}}\right)^4 - 2\left(\frac{1}{\sqrt{3}}\right)^2 = \frac{1}{9} - \frac{2}{3} = -\frac{5}{9}$$

and

$$f\left(-\frac{1}{\sqrt{3}}\right) = -\frac{5}{9}.$$

This gives the points

$$\left(-\frac{1}{\sqrt{3}}, -\frac{5}{9}\right) \quad \text{and} \quad \left(\frac{1}{\sqrt{3}}, -\frac{5}{9}\right)$$

on the graph—$(-0.6, -0.6)$ and $(0.6, -0.6)$, approximately.

e) *Concavity.* Find the intervals on which f is concave up and concave down. We do this by determining where f' is increasing and decreasing using the numbers found in step (d). Those numbers divide the real-number line into three intervals:

$$A\left(-\infty, -\frac{1}{\sqrt{3}}\right), \qquad B\left(-\frac{1}{\sqrt{3}}, \frac{1}{\sqrt{3}}\right), \quad \text{and} \quad C\left(\frac{1}{\sqrt{3}}, \infty\right).$$

Next, we choose a test value in each interval and make a substitution into f''. The test points we select are -1, 0, and 1:

A: Test -1, $f''(-1) = 12(-1)^2 - 4 = 8 > 0$;

B: Test 0, $f''(0) = 12(0)^2 - 4 = -4 < 0$;

C: Test 1, $f''(1) = 12(1)^2 - 4 = 8 > 0$.

Interval	$(-\infty, -1/\sqrt{3})$	$(-1/\sqrt{3}, 1/\sqrt{3})$	$(1/\sqrt{3}, \infty)$
Test value	$x = -1$	$x = 0$	$x = 1$
Sign of $f''(x)$	$f''(-1) > 0$	$f''(0) < 0$	$f''(1) > 0$
Result	f' is increasing	f' is decreasing	f' is increasing

Change indicates a point of inflection.   Change indicates a point of inflection.

Therefore, f' is increasing on the interval $(-\infty, -1/\sqrt{3})$, decreasing on the interval $(-1/\sqrt{3}, 1/\sqrt{3})$, and increasing on the interval $(1/\sqrt{3}, \infty)$. The function f is concave up on $(-\infty, -1/\sqrt{3})$ and $(1/\sqrt{3}, \infty)$ and concave down on $(-1/\sqrt{3}, 1/\sqrt{3})$. The graph changes concavity across $(-1/\sqrt{3}, -5/9)$ and $(1/\sqrt{3}, -5/9)$, so these are points of inflection.

TECHNOLOGY
CONNECTION

Consider $f(x) = x^3(x - 2)^3$.
How many relative extrema
do you anticipate finding?
Where do you think they
will be?

 Graph

$$f(x) = x^3(x - 2)^3$$

using $[-1, 3, -2, 6]$ as a
viewing window, with
xScl = 0.5 and yScl = 0.5.
Use the Product Rule to
find the first derivative of
the function. Graph the first
derivative. How many
points of inflection do you
see? Where are they?

f) *Sketch.* Sketch the graph using the information in the following table. By solving $x^4 - 2x^2 = 0$, we can find the x-intercepts easily. They are $(-\sqrt{2}, 0)$, $(0, 0)$, and $(\sqrt{2}, 0)$. This also aids the graphing. Extra function values can be calculated if desired. The graph is shown below.

x	$f(x)$, approximately
-2	8
-1.5	0.56
-1	-1
-0.5	-0.44
0	0
0.5	-0.44
1	-1
1.5	0.56
2	8

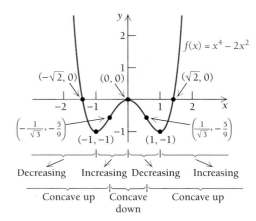

3.2 Exercise Set

Find the relative extrema of the function. List your
answers in terms of ordered pairs. Use the Second-
Derivative Test, where possible. Then sketch the graph.

1. $f(x) = 2 - x^2$

2. $f(x) = 3 - x^2$

3. $f(x) = x^2 + x - 1$

4. $f(x) = x^2 - x$

5. $f(x) = -4x^2 + 3x - 1$

6. $f(x) = 7 - 8x + 5x^2$

7. $f(x) = 2x^3 - 3x^2 - 36x + 28$

8. $f(x) = 3x^3 - 36x - 3$

9. $f(x) = \frac{8}{3}x^3 - 2x + \frac{1}{3}$

10. $f(x) = 80 - 9x^2 - x^3$

11. $f(x) = -x^3 + 3x^2 - 4$

12. $f(x) = -x^3 + 3x - 2$

13. $f(x) = 3x^4 - 16x^3 + 18x^2$

14. $f(x) = 3x^4 + 4x^3 - 12x^2 + 5$

15. $f(x) = (x + 1)^{2/3}$

16. $f(x) = (x - 1)^{2/3}$

17. $f(x) = x^4 - 6x^2$

18. $f(x) = 2x^2 - x^4$

19. $f(x) = x^3 - 2x^2 - 4x + 3$

20. $f(x) = x^3 - 6x^2 + 9x + 1$

21. $f(x) = 3x^4 + 4x^3$

22. $f(x) = x^4 - 2x^3$

23. $f(x) = x^3 - 6x^2 - 135x$

24. $f(x) = x^3 - 3x^2 - 144x - 140$

25. $f(x) = \dfrac{x}{x^2 + 1}$

26. $f(x) = \dfrac{8x}{x^2 + 1}$

27. $f(x) = \dfrac{3}{x^2 + 1}$

28. $f(x) = \dfrac{-4}{x^2 + 1}$

29. $f(x) = (x - 1)^3$

30. $f(x) = (x + 2)^3$

31. $f(x) = x^2(1 - x)^2$

32. $f(x) = x^2(3 - x)^2$

33. $f(x) = 20x^3 - 3x^5$

34. $f(x) = 5x^3 - 3x^5$

35. $f(x) = x\sqrt{4 - x^2}$

36. $f(x) = -x\sqrt{1 - x^2}$

37. $f(x) = (x - 1)^{1/3} - 1$

38. $f(x) = 2 - x^{1/3}$

Find all points of inflection, if they exist.

39. $f(x) = x^3 + 3x + 1$

40. $f(x) = x^3 - 6x^2 + 12x - 6$

41. $f(x) = \frac{4}{3}x^3 - 2x^2 + x$

42. $f(x) = x^4 - 4x^3 + 10$

APPLICATIONS

◆ **Business and Economics**

Total revenue, cost, and profit. Using the same set of axes, sketch the graphs of the total-revenue, total-cost, and total-profit functions.

43. $R(x) = 50x - 0.5x^2, \quad C(x) = 4x + 10$

44. $R(x) = 50x - 0.5x^2, \quad C(x) = 10x + 3$

◆ **Life and Physical Sciences**

45. *Coughing velocity.* A person coughs when a foreign object is in the windpipe. The velocity of the cough depends on the size of the object. Suppose a person has a windpipe with a 20-mm radius. If a foreign object has a radius r, in millimeters, then the velocity V, in millimeters/second, needed to remove the object by a cough is given by

$$V(r) = k(20r^2 - r^3), \quad 0 \leq r \leq 20,$$

where k is some positive constant. For what size object is the maximum velocity needed to remove the object?

46. *Temperature in January.* Suppose that the temperature T, in degrees Fahrenheit, during a 24-hr day in January is given by

$$T(x) = 0.0027(x^3 - 34x + 240), \quad 0 \leq x \leq 24,$$

where x = the number of hours since midnight. Estimate the relative minimum temperature and when it occurs.

SYNTHESIS

In each of Exercises 47 and 48, determine which graph is the derivative of the other.

47.

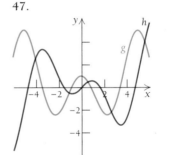

48.

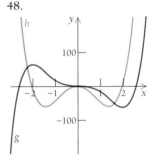

Social science: Three aspects of love. Researchers at Yale University have suggested that the following graphs* may represent three different aspects of love.

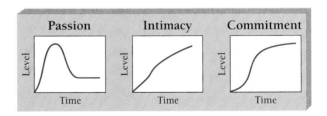

49. ◆ Analyze each of these graphs in terms of the concepts you have learned in Sections 3.1 and 3.2: relative extrema, concavity, increasing, decreasing, and so on.

50. ◆ Do you agree with the researchers regarding the shape of these graphs? Explain your reasons.

*From "A Triangular Theory of Love," by R. J. Sternberg, 1986, *Psychological Review*, 93(2), 119–135. Copyright 1986 by the American Psychological Association, Inc. Reprinted by permission.

TECHNOLOGY CONNECTION

Graph the function. Then estimate any relative extrema.

51. $f(x) = 3x^{2/3} - 2x$

52. $f(x) = 4x - 6x^{2/3}$

53. $f(x) = x^2(x - 2)^3$

54. $f(x) = x^2(1 - x)^3$

55. $f(x) = x - \sqrt{x}$

56. $f(x) = (x - 1)^{2/3} - (x + 1)^{2/3}$

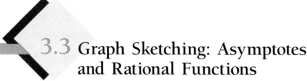

3.3 Graph Sketching: Asymptotes and Rational Functions

O B J E C T I V E

• Graph rational functions.

Vertical and Horizontal Asymptotes

Thus far we have considered a strategy for graphing a continuous function using the tools of calculus. We now want to consider some discontinuous functions, most of which are rational functions. Our graphing skills will now have to take into account the discontinuities of the graph and certain lines called *asymptotes*.

Let us reconsider the definition of a rational function.

DEFINITION

A *rational function* is a function f that can be described by

$$f(x) = \frac{P(x)}{Q(x)},$$

where $P(x)$ and $Q(x)$ are polynomials having no common factor other than 1 and -1, and with $Q(x)$ not the zero polynomial. The domain of f consists of all inputs x for which $Q(x) \neq 0$.

Polynomial functions are themselves special kinds of rational functions, since $Q(x)$ can be the polynomial 1. Here we are interested in graphing rational functions in which the denominator is not a constant.

Figure 1 shows the graph of the rational function

$$f(x) = \frac{x^2 - 1}{x^2 + x - 6} = \frac{(x - 1)(x + 1)}{(x - 2)(x + 3)}.$$

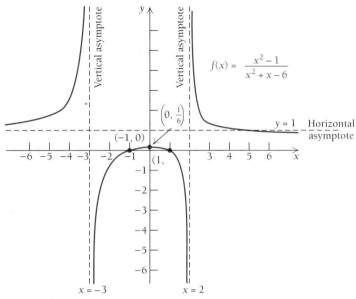

$$f(x) = \frac{x^2 - 1}{x^2 + x - 6}$$

FIGURE 1

Let us make some observations about this graph.

First, note that as x gets closer to 2 from the left, the function values get smaller and smaller negatively, approaching $-\infty$. As x gets closer to 2 from the right, the function values get larger and larger positively. Thus,

$$\lim_{x \to 2^-} f(x) = -\infty \quad \text{and} \quad \lim_{x \to 2^+} f(x) = \infty.$$

For this graph, we can think of the line $x = 2$ as a "limiting line" called a *vertical asymptote*. Similarly, the line $x = -3$ is a vertical asymptote.

The line $x = a$ is a *vertical asymptote* if any of the following limit statements is true:

$$\lim_{x \to a^-} f(x) = \infty \quad \text{or} \quad \lim_{x \to a^-} f(x) = -\infty \quad \text{or}$$

$$\lim_{x \to a^+} f(x) = \infty \quad \text{or} \quad \lim_{x \to a^+} f(x) = -\infty.$$

The graph of a rational function *never* crosses a vertical asymptote. If a is an input that makes the denominator 0, then the line $x = a$ is a vertical asymptote. (Remember that the numerator and the denominator have no common factor other than -1 or 1.)

Figure 2 shows the four ways in which a vertical asymptote can occur.

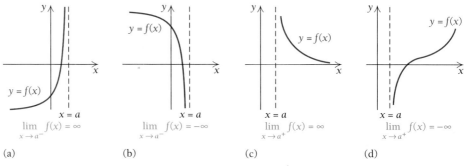

(a)

(b)

(c)

(d)

FIGURE 2

Look again at the graph in Fig. 1. Note that function values get closer and closer to 1 as x approaches $-\infty$, meaning that $f(x) \to 1$ as $x \to -\infty$. Also, function values get closer and closer to 1 as x approaches ∞, meaning that $f(x) \to 1$ as $x \to \infty$. Thus,

$$\lim_{x \to -\infty} f(x) = 1 \quad \text{and} \quad \lim_{x \to \infty} f(x) = 1.$$

The line $y = 1$ is called a *horizontal asymptote.*

> The line $y = b$ is a *horizontal asymptote* if either or both of the following limit statements is true:
>
> $$\lim_{x \to -\infty} f(x) = b \quad \text{or} \quad \lim_{x \to \infty} f(x) = b.$$
>
> The graph of a rational function may or may not cross a horizontal asymptote. Horizontal asymptotes occur when the degree of the numerator is less than or equal to the degree of the denominator.

In Figs. 3–5, we see three ways in which horizontal asymptotes can occur.

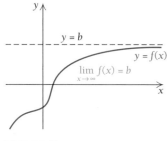

FIGURE 3

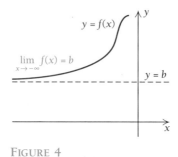

FIGURE 4

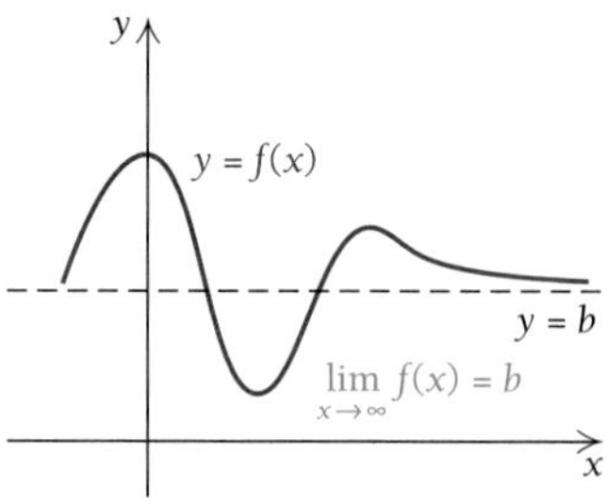

FIGURE 5

TECHNOLOGY CONNECTION

Asymptotes and Graphers

Our discussion now allows us to attach the term "vertical asymptote" to those mysterious vertical lines that appear with the use of some graphers. For example, consider the graph of $f(x) = 8/(x^2 - 4)$, using the viewing window $[-6, 6, -8, 8]$. Vertical asymptotes occur at $x = -2$ and $x = 2$. These lines are not part of the graph.

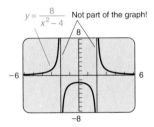

$y = \dfrac{8}{x^2 - 4}$ Not part of the graph!

Graph each of the following. Try to visually locate the vertical asymptotes. Then verify your results using the methods of Examples 1 and 2. You may need to try different viewing windows.

$$f(x) = \frac{x^2 + 7x + 10}{x^2 + 3x - 28},$$

$$f(x) = \frac{x^2 + 5}{x^3 - x^2 - 6x}$$

Occurrences of Asymptotes

It is important in graphing rational functions to determine where the asymptotes, if any, occur. Vertical asymptotes are easy to locate when a denominator can be factored. The x-inputs that make a denominator 0 give us the vertical asymptotes.

EXAMPLE 1   Determine the vertical asymptotes:

$$f(x) = \frac{3x - 2}{x(x - 5)(x + 3)}.$$

Solution   The vertical asymptotes are the lines $x = 0$, $x = 5$, and $x = -3$.   ◆

EXAMPLE 2   Determine the vertical asymptotes:

$$f(x) = \frac{x - 2}{x^3 - x}.$$

Solution   We factor the denominator:

$$x^3 - x = x(x^2 - 1)$$
$$= x(x - 1)(x + 1).$$

The vertical asymptotes are $x = 0$, $x = -1$, and $x = 1$.   ◆

EXAMPLE 3   Find the horizontal asymptotes:

$$f(x) = \frac{2x + 3}{x^3 - 2x^2 + 4}.$$

Solution   Since the degree of the numerator is less than the degree of the denominator, there is a horizontal asymptote. We then divide the numerator and the denominator by x^3 and find the limits as x approaches $-\infty$ and ∞; that is, as $|x|$ gets larger and larger. (You might want to review the part of Section 2.2 on limits involving infinity.)

$$f(x) = \frac{2x + 3}{x^3 - 2x^2 + 4} = \frac{\dfrac{2}{x^2} + \dfrac{3}{x^3}}{1 - \dfrac{2}{x} + \dfrac{4}{x^3}}$$

As x gets smaller and smaller negatively, $|x|$ gets larger and larger. Similarly, as x gets larger and larger positively, $|x|$ gets larger and larger. Thus, as $|x|$ becomes very large, each expression with x or some power of x in the denominator takes on values ever closer to 0. Thus the numerator approaches 0 and the denominator approaches 1;

hence the entire expression takes on values ever closer to 0. We have

$$f(x) \approx \frac{0 + 0}{1 - 0 + 0},$$

so $\lim\limits_{x \to -\infty} f(x) = 0$ and $\lim\limits_{x \to \infty} f(x) = 0,$

and the x-axis, the line $y = 0$, is a horizontal asymptote.   ◆

EXAMPLE 4   Find the horizontal asymptotes:

$$f(x) = \frac{3x^2 + 2x - 4}{2x^2 - x + 1}.$$

Solution   The numerator and the denominator have the same degree, so there is a horizontal asymptote. We divide the numerator and the denominator by x^2 and find the limit as $|x|$ gets larger and larger:

$$f(x) = \frac{3x^2 + 2x - 4}{2x^2 - x + 1}$$

$$= \frac{3 + \dfrac{2}{x} - \dfrac{4}{x^2}}{2 - \dfrac{1}{x} + \dfrac{1}{x^2}}.$$

As $|x|$ gets very large, the numerator approaches 3 and the denominator approaches 2. Therefore, the function gets very close to $\frac{3}{2}$. Thus,

$$\lim\limits_{x \to -\infty} f(x) = \frac{3}{2} \quad\text{and}\quad \lim\limits_{x \to \infty} f(x) = \frac{3}{2}.$$

The line $y = \frac{3}{2}$ is a horizontal asymptote.   ◆

TECHNOLOGY CONNECTION

Graph each of the following. Try to visually locate the horizontal asymptotes. Use the trace feature to move the cursor along the graph as x approaches positive infinity and then as x approaches negative infinity. You may need to try different viewing windows. Verify your results using the methods of Examples 3 and 4.

$$f(x) = \frac{x^2 + 5}{x^3 - x^2 - 6x},$$

$$f(x) = \frac{9x^4 - 7x^2 - 9}{3x^4 + 7x^2 + 9},$$

$$f(x) = \frac{135x^5 - x^2}{x^7},$$

$$f(x) = \frac{3x^2 - 4x + 3}{6x^2 + 2x - 5}$$

When the degree of the numerator is less than the degree of the denominator, the x-axis, or the line $y = 0$, is a horizontal asymptote.

When the degree of the numerator is the same as the degree of the denominator, the line $y = a/b$ is a horizontal asymptote, where a is the leading coefficient of the numerator and b is the leading coefficient of the denominator.

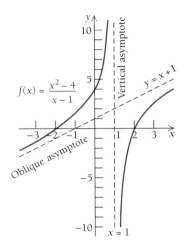

$f(x) = \dfrac{x^2 - 4}{x - 1}$

Vertical asymptote

$y = x + 1$

Oblique asymptote

$x = 1$

Oblique Asymptotes

There are asymptotes that are neither vertical nor horizontal. For example, in the graph of

$$f(x) = \frac{x^2 - 4}{x - 1},$$

shown at left, the line $x = 1$ is a vertical asymptote. The line $y = x + 1$ is called an *oblique asymptote*. Note that as $|x|$ gets larger and larger, the curve gets closer and closer to $y = x + 1$.

> The line $y = mx + b$ is an *oblique asymptote* of the rational function $f(x) = P(x)/Q(x)$ if $f(x)$ can be expressed as
>
> $$f(x) = (mx + b) + g(x),$$
>
> where $g(x) \to 0$ as $|x| \to \infty$. Oblique asymptotes occur when the degree of the numerator is one more than the degree of the denominator. A graph can cross an oblique asymptote.

How can we find an oblique asymptote? One way is by division.

TECHNOLOGY CONNECTION

Graph each of the following. Try to visually locate the oblique asymptotes. Use the trace feature to move the cursor along the graph as x approaches positive infinity and then as x approaches negative infinity. Graph what appear to be the oblique asymptotes. You may need to try different viewing windows. Verify your results using the method of Example 5.

$$f(x) = \frac{3x^2 - 7x + 8}{x - 2},$$

$$f(x) = \frac{5x^3 + 2x + 1}{x^2 - 4}$$

EXAMPLE 5   Find the oblique asymptotes:

$$f(x) = \frac{x^2 - 4}{x - 1}.$$

Solution   When we divide the numerator by the denominator, we obtain a quotient of $x + 1$ and a remainder of -3. Thus,

$$f(x) = \frac{x^2 - 4}{x - 1} = (x + 1) + \frac{-3}{x - 1}.$$

$$
\begin{array}{r}
x + 1 \\
x - 1 \overline{) x^2 \quad\quad - 4} \\
\underline{x^2 - x} \\
x - 4 \\
\underline{x - 1} \\
-3
\end{array}
$$

Now we can see that when $|x|$ gets very large, $-3/(x - 1)$ approaches 0. Thus, for very large $|x|$, the expression $x + 1$ is the dominant part of

$$(x + 1) + \frac{-3}{x - 1}.$$

Thus, $y = x + 1$ is an oblique asymptote.   ◆

Intercepts

If they exist, the **x-intercepts** of a function occur at those values of x for which $y = f(x) = 0$ and give us points at which the graph crosses the x-axis. If it exists, the **y-intercept** of a function occurs at the value of y for which $x = 0$ and gives us the point at which the graph crosses the y-axis.

EXAMPLE 6   Find the intercepts of

$$f(x) = \frac{x^3 - x^2 - 6x}{x^2 - 3x + 2}.$$

Solution   We factor the numerator and the denominator:

$$f(x) = \frac{x(x + 2)(x - 3)}{(x - 1)(x - 2)}.$$

To find the x-intercepts, we solve the equation $f(x) = 0$. Such values occur when the numerator is 0 but the denominator is not. Thus we solve the equation

$$x(x + 2)(x - 3) = 0.$$

The x-values making the numerator 0 are 0, -2, and 3. Since none of these makes the denominator 0, they yield the x-intercepts of the function: $(0, 0)$, $(-2, 0)$, and $(3, 0)$.

To find the y-intercept, we let $x = 0$:

$$f(0) = \frac{0^3 - 0^2 - 6(0)}{0^2 - 3(0) + 2} = 0.$$

In this case, the y-intercept is also an x-intercept, $(0, 0)$.   ◆

TECHNOLOGY
CONNECTION

Graph each of the following. Use the trace and zoom features to find the intercepts.

$$f(x) = \frac{x(x - 3)(x + 5)}{(x + 2)(x - 4)},$$

$$f(x) = \frac{x^3 + 2x^2 - 3x}{x^2 + 5}$$

Sketching Graphs

We can now refine our strategy for graphing.

Strategy for Sketching Graphs

a) *Intercepts.* Find the x-intercept(s) and the y-intercept of the graph.

b) *Asymptotes.* Find the vertical, horizontal, and oblique asymptotes.

c) *Derivatives.* Find $f'(x)$ and $f''(x)$.

d) *Undefined values and critical points of f.* Find the inputs for which the function is not defined, giving denominators of 0. Find also the critical points of f.

e) *Increasing, decreasing, relative extrema.* Use the points found in step (d) to determine intervals on which the function f is increasing or decreasing. Use this information and/or the second derivative to determine the relative maxima and minima. A relative extremum can occur only at a point c for which $f(c)$ exists.

f) *Inflection points.* Determine candidates for inflection points by finding points x_0 where $f''(x_0)$ does not exist or where $f''(x_0) = 0$. Find the function values at these points. If a function value $f(x)$ does not exist, then the function does not have an inflection point at x.

g) *Concavity.* Use the values c from step (f) as endpoints of intervals. Determine the concavity by checking to see where f' is decreasing and where f' is increasing. Do this by selecting test points and substituting into $f''(x)$.

h) *Sketch the graph.* Use the information from steps (a) through (g) to sketch the graph, plotting extra points (computing them with your calculator) as the need arises.

EXAMPLE 7   Sketch the graph of $f(x) = \dfrac{8}{x^2 - 4}$.

Solution

a) *Intercepts.* The x-intercepts occur at the points where the numerator is 0 but the denominator is not. Since in this case the numerator is the constant 8, there are no x-intercepts. To find the y-intercept, we compute $f(0)$:

$$f(0) = \frac{8}{0^2 - 4} = \frac{8}{-4} = -2.$$

This gives us one point on the graph, $(0, -2)$.

b) *Asymptotes.*

Vertical: The denominator $x^2 - 4 = (x + 2)(x - 2)$. It is 0 for x-values of -2 and 2. Thus the graph has the lines $x = -2$ and $x = 2$ as vertical asymptotes. We plot them using dashed lines.

Horizontal: The degree of the numerator is less than the degree of the denominator, so the x-axis, or the line $y = 0$, is a horizontal asymptote. It is already plotted as an axis.

Oblique: There is no oblique asymptote since the degree of the numerator is not one more than the degree of the denominator.

c) *Derivatives.* Find $f'(x)$ and $f''(x)$. Using the Quotient Rule, we get

$$f'(x) = \frac{-16x}{(x^2 - 4)^2} \quad \text{and} \quad f''(x) = \frac{16(3x^2 + 4)}{(x^2 - 4)^3}.$$

d) *Undefined values and critical points of f.* The domain of the original function is all real numbers except -2 and 2, where the vertical asymptotes occur. We find the critical points of f by looking for values of x where $f'(x) = 0$ or where $f'(x)$ does not exist. Now $f'(x) = 0$ for values of x for which $-16x = 0$, but the denominator is not 0. The only such number is 0 itself. The derivative $f'(x)$ does not exist at -2 and 2. Thus the undefined values and the critical points are -2, 0, and 2.

e) *Increasing, decreasing, relative extrema.* Use the undefined values and the critical points to determine the intervals on which f is increasing and the intervals on which f is decreasing. The points to consider are -2, 0, and 2. These divide the real-number line into four intervals. We choose a test point in each interval and make a substitution into the derivative f':

$$A:\ \text{Test } -3,\quad f'(-3) = \frac{-16(-3)}{[(-3)^2 - 4]^2} = \frac{48}{25} > 0;$$

$$B:\ \text{Test } -1,\quad f'(-1) = \frac{-16(-1)}{[(-1)^2 - 4]^2} = \frac{16}{9} > 0;$$

$$C:\ \text{Test } 1,\quad f'(1) = \frac{-16(1)}{[(1)^2 - 4]^2} = \frac{-16}{9} < 0;$$

$$D:\ \text{Test } 3,\quad f'(3) = \frac{-16(3)}{[(3)^2 - 4]^2} = \frac{-48}{25} < 0.$$

Interval	$(-\infty, -2)$	$(-2, 0)$	$(0, 2)$	$(2, \infty)$
Test value	$x = -3$	$x = -1$	$x = 1$	$x = 3$
Sign of $f'(x)$	$f'(-3) > 0$	$f'(-1) > 0$	$f'(1) < 0$	$f'(3) < 0$
Result	f is increasing	f is increasing	f is decreasing	f is decreasing

└─ No change ─┘   └─ Change ─┘   └─ No change ─┘
indicates a
relative maximum
at 0.

Now $f(0) = -2$, so there is a relative maximum at $(0, -2)$.

f) *Inflection points.* Determine candidates for inflection points by finding where $f''(x)$ does not exist and where $f''(x) = 0$. Now $f(x)$ does not exist at -2 and 2, so $f''(x)$ does not exist at -2 and 2. We then determine where $f''(x) = 0$, or

$$16(3x^2 + 4) = 0.$$

But $16(3x^2 + 4) > 0$ for all real numbers x, so there are no points of inflection since $f(-2)$ and $f(2)$ do not exist.

g) *Concavity.* Use the values found in step (f) as endpoints of intervals. Determine the concavity by checking to see where f' is increasing and decreasing. The points -2 and 2 divide the real-number line into three intervals. We choose test points in each interval and make a substitution into f'':

A: Test -3, $\quad f''(-3) = \dfrac{16[3(-3)^2 + 4]}{[(-3)^2 - 4]^3} > 0;$

B: Test 0, $\quad f''(0) = \dfrac{16[3(0)^2 + 4]}{[(0)^2 - 4]^3} < 0;$

C: Test 3, $\quad f''(3) = \dfrac{16[3(3)^2 + 4]}{[(3)^2 - 4]^3} > 0.$

Interval	$(-\infty, -2)$	$(-2, 2)$	$(2, \infty)$
Test value	$x = -3$	$x = 0$	$x = 3$
Sign of $f''(x)$	$f''(-3) > 0$	$f''(0) < 0$	$f''(3) > 0$
Result	f' is increasing	f' is decreasing	f' is increasing

Change does not indicate a point of inflection since $f(-2)$ does not exist.

Change does not indicate a point of inflection since $f(2)$ does not exist.

The function is concave up on the intervals $(-\infty, -2)$ and $(2, \infty)$. The function is concave down on the interval $(-2, 2)$.

h) *Sketch.* Sketch the graph using the information in the following table, plotting extra points by computing values from your calculator as needed. The graph follows.

x	$f(x)$, approximately
-5	0.38
-4	0.67
-3	1.6
-1	-2.67
0	-2
1	-2.67
3	1.6
4	0.67
5	0.38

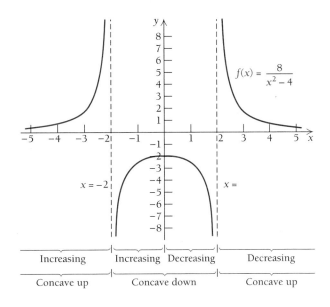

EXAMPLE 8   Sketch the graph of $f(x) = \dfrac{x^2 + 4}{x}$.

Solution

a) *Intercepts.* The equation $f(x) = 0$ has no real-number solution. Thus there are no x-intercepts. The number 0 is not in the domain of the function. Thus there is no y-intercept.

b) *Asymptotes.*

   Vertical: The denominator is x. Since its replacement by 0 makes the denominator 0, the line $x = 0$ is a vertical asymptote.

   Horizontal: The degree of the numerator is greater than the degree of the denominator, so there are no horizontal asymptotes.

   Oblique: The degree of the numerator is one greater than the degree of the denominator, so there is an oblique asymptote. We do the division and express the function in the form

$$f(x) = x + \frac{4}{x}. \qquad \begin{array}{r} x \\ x\overline{)x^2 + 4} \\ \underline{x^2 } \\ 4 \end{array}$$

   As $|x|$ gets larger, the term $4/x$ approaches 0, so the line $y = x$ is an oblique asymptote.

c) *Derivatives.* Find $f'(x)$ and $f''(x)$:

$$f'(x) = 1 - 4x^{-2}$$
$$= 1 - \frac{4}{x^2},$$
$$f''(x) = 8x^{-3}$$
$$= \frac{8}{x^3}.$$

d) *Undefined values and critical points of f.* The number 0 is not in the domain of f. The derivative exists for all values of x except 0. Thus, to find critical points, we solve $f'(x) = 0$, looking for solutions other than 0:

$$1 - \frac{4}{x^2} = 0$$
$$1 = \frac{4}{x^2}$$
$$x^2 = 4$$
$$x = \pm 2.$$

Thus, -2 and 2 are critical points. Now $f(0)$ does not exist, but $f(-2) = -4$ and $f(2) = 4$. These give the points $(-2, -4)$ and $(2, 4)$ on the graph.

e) *Increasing, decreasing, relative extrema.* Use the points found in step (d) to find intervals on which f is increasing and intervals on which f is decreasing. The points to consider are -2, 0, and 2. These divide the real-number line into four intervals. We choose test points in each interval and make a substitution into f':

$$A:\ \text{Test } -3, \quad f'(-3) = 1 - \frac{4}{(-3)^2} = \frac{5}{9} > 0;$$

$$B:\ \text{Test } -1, \quad f'(-1) = 1 - \frac{4}{(-1)^2} = -3 < 0;$$

$$C:\ \text{Test } 1, \quad f'(1) = 1 - \frac{4}{1^2} = -3 < 0;$$

$$D:\ \text{Test } 3, \quad f'(3) = 1 - \frac{4}{3^2} = \frac{5}{9} > 0.$$

Interval	$(-\infty, -2)$	$(-2, 0)$	$(0, 2)$	$(2, \infty)$
Test value	$x = -3$	$x = -1$	$x = 1$	$x = 3$
Sign of $f'(x)$	$f'(-3) > 0$	$f'(-1) < 0$	$f'(1) < 0$	$f'(3) > 0$
Result	f is increasing	f is decreasing	f is decreasing	f is increasing

Change — indicates a relative maximum at -2.   No change   Change — indicates a relative minimum at 2.

Now $f(-2) = -4$ and $f(2) = 4$. There is a relative maximum at $(-2, -4)$ and a relative minimum at $(2, 4)$.

f) *Inflection points.* Determine candidates for inflection points by finding where $f''(x_0)$ does not exist or where $f''(x_0) = 0$. Now $f''(0)$ does not exist, but because $f(0)$ does not exist, there cannot be an inflection point at 0. Then look for values of x for which $f''(x) = 0$:

$$\frac{8}{x^3} = 0.$$

But this equation has no solution. Thus there are no points of inflection.

g) *Concavity.* Use the values found in step (f) as endpoints of intervals. Determine the concavity by checking to see where f' is increasing and decreasing. The number 0 divides the real-number line into two intervals, $(-\infty, 0)$ and $(0, \infty)$. We could choose a test value in each interval and make a substitution into f''. But note that for any $x < 0$, $x^3 < 0$, so

$$f''(x) = \frac{8}{x^3} < 0,$$

and for any $x > 0$, $x^3 > 0$, so

$$f''(x) = \frac{8}{x^3} > 0.$$

Thus, f is concave down on the interval $(-\infty, 0)$ and concave up on the interval $(0, \infty)$.

h) *Sketch.* Sketch the graph using the preceding information and any computed values of f as the need arises. The graph follows.

x	$f(x)$, approximately
-6	-6.67
-5	-5.8
-4	-5
-3	-4.3
-2	-4
-1	-5
-0.5	-8.5
0.5	8.5
1	5
2	4
3	4.3
4	5
5	5.8
6	6.67

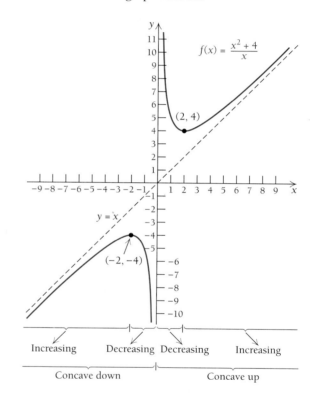

3.3 Exercise Set

Sketch a graph of the function.

1. $f(x) = \dfrac{4}{x}$

2. $f(x) = -\dfrac{5}{x}$

3. $f(x) = \dfrac{-2}{x-5}$

4. $f(x) = \dfrac{1}{x-5}$

5. $f(x) = \dfrac{1}{x - 3}$

6. $f(x) = \dfrac{1}{x + 2}$

7. $f(x) = \dfrac{-2}{x + 5}$

8. $f(x) = \dfrac{-3}{x - 3}$

9. $f(x) = \dfrac{2x + 1}{x}$

10. $f(x) = \dfrac{3x - 1}{x}$

11. $f(x) = x + \dfrac{9}{x}$

12. $f(x) = x + \dfrac{2}{x}$

13. $f(x) = \dfrac{2}{x^2}$

14. $f(x) = \dfrac{-1}{x^2}$

15. $f(x) = \dfrac{x}{x - 3}$

16. $f(x) = \dfrac{x}{x + 2}$

17. $f(x) = \dfrac{1}{x^2 + 3}$

18. $f(x) = \dfrac{-1}{x^2 + 2}$

19. $f(x) = \dfrac{x - 1}{x + 2}$

20. $f(x) = \dfrac{x - 2}{x + 1}$

21. $f(x) = \dfrac{x^2 - 4}{x + 3}$

22. $f(x) = \dfrac{x^2 - 9}{x + 1}$

23. $f(x) = \dfrac{x - 1}{x^2 - 2x - 3}$

24. $f(x) = \dfrac{x + 2}{x^2 + 2x - 15}$

25. $f(x) = \dfrac{2x^2}{x^2 - 16}$

26. $f(x) = \dfrac{x^2 + x - 2}{2x^2 + 1}$

27. $f(x) = \dfrac{1}{x^2 - 1}$

28. $f(x) = \dfrac{10}{x^2 + 4}$

29. $f(x) = \dfrac{x^2 + 1}{x}$

30. $f(x) = \dfrac{x^3}{x^2 - 1}$

APPLICATIONS

◆ **Business and Economics**

31. *Depreciation.* Suppose that the value V of a certain product decreases, or depreciates, with time t, in months, where

$$V(t) = 50 - \dfrac{25t^2}{(t + 2)^2}.$$

a) Find $V(0)$, $V(5)$, $V(10)$, and $V(70)$.
b) Find the relative maximum value of the product over the interval $[0, \infty)$.
c) Sketch a graph of V.
d) Find $\lim_{t \to \infty} V(t)$.

e) Does there seem to be a value below which V will never fall?

32. *Average cost.* The total-cost function for Acme, Inc., to produce x units of a product is given by

$$C(x) = 3x^2 + 80.$$

a) The *average cost* is given by $A(x) = C(x)/x$. Find $A(x)$.
b) Graph the average cost.
c) Find the oblique asymptote for the graph of $y = A(x)$ and interpret it.

33. *Cost of pollution control.* Cities and companies find the cost of pollution control to increase tremendously with respect to the percentage of pollutants to be removed from a situation. Suppose that the cost C of removing $p\%$ of the pollutants from a chemical dumping site is given by

$$C(p) = \dfrac{\$48,000}{100 - p}.$$

a) Find $C(0)$, $C(20)$, $C(80)$, and $C(90)$.
b) Find $\lim_{p \to 100^-} C(p)$.
c) Sketch a graph of C.
d) Can the company afford to remove 100% of the pollutants?

◆ **Life and Physical Sciences**

34. *Medication in the bloodstream.* After an injection, the amount of a medication A in the bloodstream decreases after time t, in hours. Suppose under certain conditions that A is given by

$$A(t) = \dfrac{A_0}{t^2 + 1},$$

where $A_0 = $ the initial amount of the medication given. Assume that an initial amount of 100 cc is injected.

a) Find $A(0)$, $A(1)$, $A(2)$, $A(7)$, and $A(10)$.
b) Find $\lim_{t \to \infty} A(t)$.
c) Find the relative maximum value of the injection over the interval $[0, \infty)$.
d) Sketch a graph of the function.
e) According to this function, does the medication ever completely leave the bloodstream?

SYNTHESIS

35. ◈ Using graphs and limits, explain the idea of an

asymptote to the graph of a function. Describe three types of asymptotes.

36. ◆ Explain why a vertical asymptote cannot be part of the graph of a function.

TECHNOLOGY CONNECTION

Graph the function.

37. $f(x) = \dfrac{x}{\sqrt{x^2 + 1}}$

38. $f(x) = x^2 + \dfrac{1}{x^2}$

39. $f(x) = \dfrac{x^3 + 2x^2 - 15x}{x^2 - 5x - 14}$

40. $f(x) = \dfrac{x^3 + 4x^2 + x - 6}{x^2 - x - 2}$

41. $f(x) = \left| \dfrac{1}{x} - 2 \right|$

42. $f(x) = \dfrac{x^3 + 2x^2 - 3x}{x^2 - 25}$

43. Graph the function

$$f(x) = \dfrac{x^2 - 3}{2x - 4}$$

using various viewing windows. Also, use the trace and zoom features. From this work only:

a) Find all the x-intercepts.
b) Find the y-intercept.
c) Find all the asymptotes.

3.4 Using Derivatives to Find Absolute Maximum and Minimum Values

OBJECTIVES

- Find absolute extrema using Maximum–Minimum Principle 1.
- Find absolute extrema using Maximum–Minimum Principle 2.

Absolute Maximum and Minimum Values

A relative minimum may or may not be an absolute minimum, meaning the smallest value of the function over its entire domain. Similarly, a relative maximum may or may not be an absolute maximum, meaning the greatest value of a function over its entire domain.

The function in the following graph has relative minima at interior points c_1 and c_3 of the closed interval $[a, b]$.

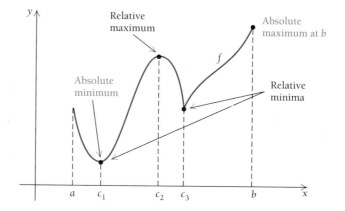

The relative minimum at c_1 also happens to be an absolute minimum. The function has a relative maximum at c_2 but it is not an absolute maximum. The absolute maximum occurs at the endpoint b.

DEFINITION

Suppose that f is a function whose value $f(c)$ exists at input c in an interval I in the domain of f. Then:

$f(c)$ is an *absolute minimum* if $f(c) \leq f(x)$ for all x in I.

$f(c)$ is an *absolute maximum* if $f(x) \leq f(c)$ for all x in I.

Finding Absolute Maximum and Minimum Values on Closed Intervals

We first consider a continuous function on a closed interval. To do so, look at these graphs and try to determine the points on the closed interval at which the absolute maxima and minima (extrema) occur.

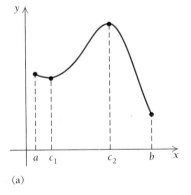
(a)

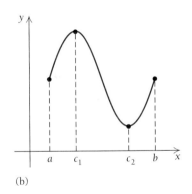
(b)

You may have discovered two theorems. Each of the functions did indeed have an absolute maximum and an absolute minimum value. This leads us to one of the theorems.

THEOREM 7

The Extreme-Value Theorem

A continuous function f defined on a closed interval $[a, b]$ must have an absolute maximum value and an absolute minimum value at points in $[a, b]$.

Look carefully at the preceding graphs and consider the critical points and the endpoints. In part (a), the graph starts at $f(a)$ and falls to $f(c_1)$. Then it rises from $f(c_1)$ to $f(c_2)$. From there it falls to $f(b)$. In part (b), the graph starts at $f(a)$ and rises to $f(c_1)$. Then it falls from $f(c_1)$ to $f(c_2)$. From there it rises to $f(b)$. It seems reasonable that whatever the maximum and minimum values are, they occur among the function values $f(a), f(c_1), f(c_2)$, and $f(b)$. This leads us to another theorem regarding *absolute extrema*.

THEOREM 8

Maximum–Minimum Principle 1

Suppose that f is a continuous function over a closed interval $[a, b]$. To find the absolute maximum and minimum values of the function on $[a, b]$:

a) First find $f'(x)$.

b) Then determine the critical points of f in $[a, b]$. That is, find all points c for which

$$f'(c) = 0 \quad \text{or} \quad f'(c) \text{ does not exist.}$$

c) List the critical points of f and the endpoints of the interval:

$$a, c_1, c_2, \ldots, c_n, b.$$

d) Find the function values at the points in part (c):

$$f(a), f(c_1), f(c_2), \ldots, f(c_n), f(b).$$

The largest of these is the *absolute maximum* of f on the interval $[a, b]$. The smallest of these is the *absolute minimum* of f on the interval $[a, b]$.

EXAMPLE 1 Find the absolute maximum and minimum values of

$$f(x) = x^3 - 3x + 2$$

on the interval $\left[-2, \frac{3}{2}\right]$.

Solution In each of Examples 1–6 of this section, we show the related graph (as here) so that you can see the absolute extrema. The procedures we use do not require the drawing of a graph.

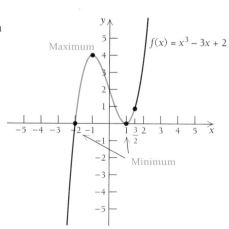

a) Find $f'(x)$:

$$f'(x) = 3x^2 - 3.$$

b) Find the critical points. The derivative exists for all real numbers. Thus we merely solve $f'(x) = 0$:

$$3x^2 - 3 = 0$$
$$3x^2 = 3$$
$$x^2 = 1$$
$$x = \pm 1.$$

c) List the critical points and the endpoints. These points are -2, -1, 1, and $\frac{3}{2}$.

d) Find the function values at the points in step (c):

$$f(-2) = (-2)^3 - 3(-2) + 2 = -8 + 6 + 2 = 0; \longrightarrow \text{Minimum}$$
$$f(-1) = (-1)^3 - 3(-1) + 2 = -1 + 3 + 2 = 4; \longrightarrow \text{Maximum}$$
$$f(1) = (1)^3 - 3(1) + 2 = 1 - 3 + 2 = 0; \longrightarrow \text{Minimum}$$
$$f(\tfrac{3}{2}) = (\tfrac{3}{2})^3 - 3(\tfrac{3}{2}) + 2 = \tfrac{27}{8} - \tfrac{9}{2} + 2 = \tfrac{7}{8}.$$

The largest of these values, 4, is the maximum. It occurs at $x = -1$. The smallest of these values is 0. It occurs twice: at $x = -2$ and $x = 1$. Thus on the interval $\left[-2, \frac{3}{2}\right]$ the

$$\text{absolute maximum} = 4 \text{ at } x = -1$$

and the

$$\text{absolute minimum} = 0 \text{ at } x = -2 \text{ and } x = 1.$$ ◆

Note that an absolute maximum or minimum value can occur at more than one point.

EXAMPLE 2   Find the absolute maximum and minimum values of

$$f(x) = x^3 - 3x + 2$$

on the interval $\left[-3, -\frac{3}{2}\right]$.

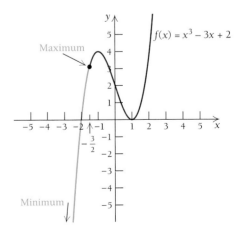

Solution   As in Example 1, the derivative is 0 at -1 and 1. But neither -1 nor 1 is in the interval $\left[-3, -\frac{3}{2}\right]$, so there are no critical points in this interval. Thus the maximum and minimum values occur at the endpoints:

$$f(-3) = (-3)^3 - 3(-3) + 2$$

$$= -27 + 9 + 2 = -16; \longrightarrow \text{Minimum}$$

$$f\left(-\tfrac{3}{2}\right) = \left(-\tfrac{3}{2}\right)^3 - 3\left(-\tfrac{3}{2}\right) + 2$$

$$= -\tfrac{27}{8} + \tfrac{9}{2} + 2 = \tfrac{25}{8} = 3\tfrac{1}{8}. \longrightarrow \text{Maximum}$$

Thus, on the interval $\left[-3, -\frac{3}{2}\right]$, the

$$\text{absolute maximum} = 3\tfrac{1}{8} \text{ at } x = -\tfrac{3}{2}$$

and the

$$\text{absolute minimum} = -16 \text{ at } x = -3. \qquad \blacklozenge$$

TECHNOLOGY CONNECTION

Use your grapher to estimate the absolute maximum and minimum values of $f(x) = x^3 - x^2 - x + 2$ first on the interval $[-2, 1]$ and then on the interval $[-1, 2]$. Use the trace feature. You may need different viewing windows. Then check your work using the methods of Examples 1 and 2.

Finding Absolute Maximum and Minimum Values on Other Intervals

When there is only one critical point c in I, we may not need to check endpoint values to determine whether the function has an absolute maximum or minimum at that point.

THEOREM 9

Maximum–Minimum Principle 2

Suppose that f is a function such that $f'(x)$ exists for *every* x in an interval I, and that there is *exactly one* (critical) point c, interior to I, for which $f'(c) = 0$. Then

   $f(c)$ is the absolute maximum value on I if $f''(c) < 0$

or

   $f(c)$ is the absolute minimum value on I if $f''(c) > 0$.

This theorem holds no matter what the interval I is—whether open, closed, or extending to infinity. If $f''(c) = 0$, either we must use Maximum–Minimum Principle 1 or we must know more about the behavior of the function on the given interval.

EXAMPLE 3   Find the absolute maximum and minimum values of

$$f(x) = 4x - x^2.$$

Solution   When no interval is specified, we consider the entire domain of the function. In this case, the domain is the set of all real numbers.

a) Find $f'(x)$:

$$f'(x) = 4 - 2x.$$

b) Find the critical points. The derivative exists for all real numbers. Thus we merely solve $f'(x) = 0$:

$$4 - 2x = 0$$
$$-2x = -4$$
$$x = 2.$$

c) Since there is only one critical point, we can apply Maximum–Minimum Principle 2 using the second derivative:

$$f''(x) = -2.$$

The second derivative is constant. Thus, $f''(2) = -2$, and since this is negative, we have the

$$\text{absolute maximum} = f(2) = 4 \cdot 2 - 2^2 = 8 - 4 = 4 \text{ at } x = 2.$$

The function has no minimum, as the graph below indicates.

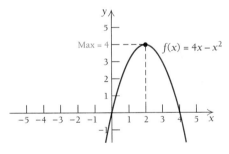

EXAMPLE 4   Find the absolute maximum and minimum values of $f(x) = 4x - x^2$ on the interval $[0, 4]$.

Solution   By the reasoning in Example 3, we know that the absolute maximum of f on $(-\infty, \infty)$ is $f(2)$, or 4. Since 2 is in the interval $[0, 4]$, we know that the absolute maximum of f on $[0, 4]$ will occur at 2. To find the absolute minimum, we need to check the endpoints:

$$f(0) = 4 \cdot 0 - 0^2 = 0$$

and $$f(4) = 4 \cdot 4 - 4^2 = 0.$$

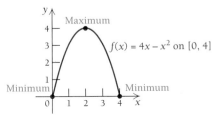

We see in the figure that the minimum is 0. It occurs twice, at $x = 0$ and $x = 4$. Thus the

absolute maximum $= 4$ at $x = 2$

and the

absolute minimum $= 0$ at $x = 0$ and $x = 4$.   ◆

We have thus far restricted the use of Maximum–Minimum Principle 2 to intervals with one critical point. Suppose a closed interval contains two critical points. Then we could break the interval up into two subintervals, consider maximums and minimums on those subintervals, and compare. But we would need to consider values at the endpoints, and since we would, in effect, be using Maximum–Minimum Principle 1, we may as well use it at the outset.

A Strategy for Finding Maximum and Minimum Values

The following general strategy can be used when finding maximum and minimum values of continuous functions.

A Strategy for Finding Absolute Maximum and Minimum Values

To find absolute maximum and minimum values of a continuous function on an interval:

a) Find $f'(x)$.

b) Find the critical points.

c) If the interval is closed and there is more than one critical point, use Maximum–Minimum Principle 1.

d) If the interval is closed and there is exactly one critical point, use either Maximum–Minimum Principle 1 or Maximum–Minimum Principle 2. If the function is easy to differentiate, use Maximum–Minimum Principle 2.

e) If the interval is not closed, does not have endpoints, or does not contain its endpoints, such as $(-\infty, \infty)$, $(0, \infty)$, or (a, b), and the function has only one critical point, use Maximum–Minimum Principle 2. In such a case, if the function has a maximum, it will have no minimum; and if it has a minimum, it will have no maximum.

The case of finding absolute maximum and minimum values when more than one critical point occurs in an interval described in step (e) above must be dealt with by a detailed graph or by techniques beyond the scope of this book.

EXAMPLE 5   Find the absolute maximum and minimum values of

$$f(x) = (x - 2)^3 + 1.$$

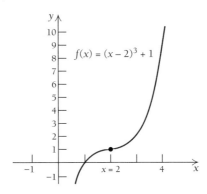

Solution

a) Find $f'(x)$:

$$f'(x) = 3(x - 2)^2.$$

b) Find the critical points. The derivative exists for all real numbers. Thus we solve $f'(x) = 0$:

$$3(x - 2)^2 = 0$$
$$(x - 2)^2 = 0$$
$$x - 2 = 0$$
$$x = 2.$$

c) Since there is only one critical point and there are no endpoints, we can try to apply Maximum–Minimum Principle 2 using the second derivative:

$$f''(x) = 6(x - 2).$$

Now

$$f''(2) = 6(2 - 2) = 0,$$

so Maximum–Minimum Principle 2 fails. We cannot use Maximum–Minimum Principle 1 because there are no endpoints. But note that $f'(x) = 3(x - 2)^2$ is never negative. Thus $f(x)$ is increasing everywhere except at $x = 2$, so there is no maximum and no minimum. For $x < 2$, $x - 2 < 0$, so $f''(x) = 6(x - 2) < 0$. Similarly, for $x > 2$, $x - 2 > 0$, so $f''(x) = 6(x - 2) > 0$. Thus, at $x = 2$, the function has a *point of inflection*.   ◆

EXAMPLE 6   Find the absolute maximum and minimum values of

$$f(x) = 5x + \frac{35}{x}$$

on the interval $(0, \infty)$.

Solution

a) Find $f'(x)$. We first express $f(x)$ as

$$f(x) = 5x + 35x^{-1}.$$

Then

$$f'(x) = 5 - 35x^{-2}$$

$$= 5 - \frac{35}{x^2}.$$

b) Find the critical points. Now $f'(x)$ exists for all values of x in $(0, \infty)$. Thus the only critical points are those for which $f'(x) = 0$:

$$5 - \frac{35}{x^2} = 0$$

$$5 = \frac{35}{x^2}$$

$$5x^2 = 35 \qquad \text{Multiplying by } x^2, \text{ since } x \neq 0$$

$$x^2 = 7$$

$$x = \pm\sqrt{7}.$$

e) The interval is not closed and is $(0, \infty)$. The only critical point is $\sqrt{7}$. Therefore, we can apply Maximum–Minimum Principle 2 using the second derivative,

$$f''(x) = 70x^{-3}$$

$$= \frac{70}{x^3},$$

to determine whether we have a maximum or a minimum. Now $f''(x)$ is positive for all values of x in $(0, \infty)$, so $f''(\sqrt{7}) > 0$, and the

$$\text{absolute minimum} = f(\sqrt{7})$$

$$= 5 \cdot \sqrt{7} + \frac{35}{\sqrt{7}}$$

$$= 5\sqrt{7} + \frac{35}{\sqrt{7}} \cdot \frac{\sqrt{7}}{\sqrt{7}}$$

$$= 5\sqrt{7} + \frac{35\sqrt{7}}{7}$$

$$= 5\sqrt{7} + 5\sqrt{7}$$

$$= 10\sqrt{7}$$

at $x = \sqrt{7}$.

TECHNOLOGY
CONNECTION

Use your grapher to estimate the absolute maximum and minimum values of $f(x) = 10x + 1/x$ on the interval $(0, \infty)$. Then check your work using the method of Example 6.

The function has no maximum value.

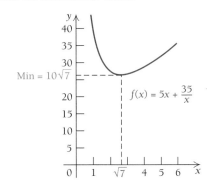

3.4 Exercise Set

The curves on the graph below show the gasoline mileage obtained when traveling at a constant speed for an average-size car and for a compact car.

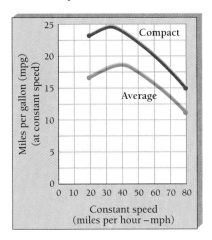

Constant speed
(miles per hour —mph)

1. Consider the graph for the average-size car over the interval [20, 80].

   a) Estimate the speed at which the absolute maximum gasoline mileage is obtained.
   b) Estimate the speed at which the absolute minimum gasoline mileage is obtained.
   c) What is the mileage obtained at 70 mph?
   d) What is the mileage obtained at 55 mph?
   e) What percent increase in mileage is there by traveling at 55 mph rather than at 70 mph?

2. Answer the questions in Exercise 1 for the compact car.

Find the absolute maximum and minimum values of the function, if they exist, on the indicated interval.

3. $f(x) = 5 + x - x^2$; [0, 2]

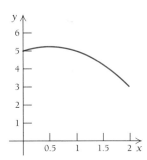

4. $f(x) = 4 + x - x^2$; [0, 2]

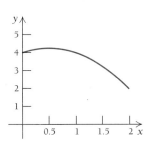

5. $f(x) = x^3 - x^2 - x + 2$; $[0, 2]$

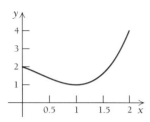

6. $f(x) = x^3 + \frac{1}{2}x^2 - 2x + 5$; $[0, 1]$

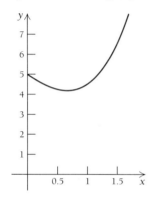

7. $f(x) = x^3 - x^2 - x + 2$; $[-1, 0]$
8. $f(x) = x^3 + \frac{1}{2}x^2 - 2x + 5$; $[-2, 0]$
9. $f(x) = 3x - 2$; $[-1, 1]$
10. $f(x) = 2x + 4$; $[-1, 1]$
11. $f(x) = 7 - 4x$; $[-2, 5]$
12. $f(x) = -2 + 8x$; $[-10, 10]$
13. $f(x) = -5$; $[-1, 1]$
14. $g(x) = 24$; $[4, 13]$
15. $f(x) = x^2 - 6x - 3$; $[-1, 5]$
16. $f(x) = x^2 - 4x + 5$; $[-1, 3]$
17. $f(x) = 3 - 2x - 5x^2$; $[-3, 3]$
18. $f(x) = 1 + 6x - 3x^2$; $[0, 4]$
19. $f(x) = x^3 - 3x^2$; $[0, 5]$
20. $f(x) = x^3 - 3x + 6$; $[-1, 3]$
21. $f(x) = x^3 - 3x$; $[-5, 1]$
22. $f(x) = 3x^2 - 2x^3$; $[-5, 1]$
23. $f(x) = 1 - x^3$; $[-8, 8]$

24. $f(x) = 2x^3$; $[-10, 10]$
25. $f(x) = 12 + 9x - 3x^2 - x^3$; $[-3, 1]$
26. $f(x) = x^3 - 6x^2 + 10$; $[0, 4]$
27. $f(x) = x^4 - 2x^3$; $[-2, 2]$
28. $f(x) = x^3 - x^4$; $[-1, 1]$
29. $f(x) = x^4 - 2x^2 + 5$; $[-2, 2]$
30. $f(x) = x^4 - 8x^2 + 3$; $[-3, 3]$
31. $f(x) = (x + 3)^{2/3} - 5$; $[-4, 5]$
32. $f(x) = 1 - x^{2/3}$; $[-8, 8]$

33. $f(x) = x + \dfrac{1}{x}$; $[1, 20]$

34. $f(x) = x + \dfrac{4}{x}$; $[-8, -1]$

35. $f(x) = \dfrac{x^2}{x^2 + 1}$; $[-2, 2]$

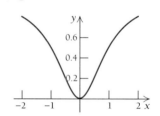

36. $f(x) = \dfrac{4x}{x^2 + 1}$; $[-3, 3]$

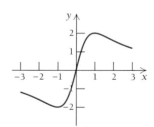

37. $f(x) = (x + 1)^{1/3}$; $[-2, 26]$
38. $f(x) = \sqrt[3]{x}$; $[8, 64]$

Find the absolute maximum and minimum values of the function, if they exist, on the indicated interval. When no interval is specified, use the real line $(-\infty, \infty)$.

39. $f(x) = x(70 - x)$
40. $f(x) = x(50 - x)$
41. $f(x) = 2x^2 - 40x + 400$

42. $f(x) = 2x^2 - 20x + 100$

43. $f(x) = x - \frac{4}{3}x^3; \quad (0, \infty)$

44. $f(x) = 16x - \frac{4}{3}x^3; \quad (0, \infty)$

45. $f(x) = 17x - x^2$

46. $f(x) = 27x - x^2$

47. $f(x) = \frac{1}{3}x^3 - 3x; \quad [-2, 2]$

48. $f(x) = \frac{1}{3}x^3 - 5x; \quad [-3, 3]$

49. $f(x) = -0.001x^2 + 4.8x - 60$

50. $f(x) = -0.01x^2 + 1.4x - 30$

51. $f(x) = -\frac{1}{3}x^3 + 6x^2 - 11x - 50; \quad (0, 3)$

52. $f(x) = -x^3 + x^2 + 5x - 1; \quad (0, \infty)$

53. $f(x) = 15x^2 - \frac{1}{2}x^3; \quad [0, 30]$

54. $f(x) = 4x^2 - \frac{1}{2}x^3; \quad [0, 8]$

55. $f(x) = 2x + \dfrac{72}{x}; \quad (0, \infty)$

56. $f(x) = x + \dfrac{3600}{x}; \quad (0, \infty)$

57. $f(x) = x^2 + \dfrac{432}{x}; \quad (0, \infty)$

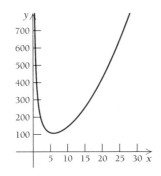

58. $f(x) = x^2 + \dfrac{250}{x}; \quad (0, \infty)$

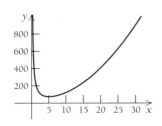

59. $f(x) = 2x^4 - x; \quad [-1, 1]$

60. $f(x) = 2x^4 + x; \quad [-1, 1]$

61. $f(x) = \sqrt[3]{x}; \quad [0, 8]$

62. $f(x) = \sqrt{x}; \quad [0, 4]$

63. $f(x) = (x + 1)^3$

64. $f(x) = (x - 1)^3$

65. $f(x) = 2x - 3; \quad [-1, 1]$

66. $f(x) = 9 - 5x; \quad [-10, 10]$

67. $f(x) = 2x - 3$

68. $f(x) = 9 - 5x$

69. $f(x) = x^{2/3}; \quad [-1, 1]$

70. $g(x) = x^{2/3}$

71. $f(x) = \frac{1}{3}x^3 - x + \frac{2}{3}$

72. $f(x) = \frac{1}{3}x^3 - \frac{1}{2}x^2 - 2x + 1$

73. $f(x) = \frac{1}{3}x^3 - 2x^2 + x; \quad [0, 4]$

74. $g(x) = \frac{1}{3}x^3 + 2x^2 + x; \quad [-4, 0]$

75. $t(x) = x^4 - 2x^2$

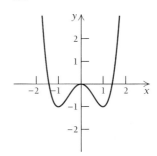

76. $f(x) = 2x^4 - 4x^2 + 2$

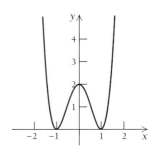

APPLICATIONS

◆ **Business and Economics**

77. *Monthly productivity.* An employee's monthly productivity M, in number of units produced, is found to be

a function of the number t of years of service. For a certain product, a productivity function is given by

$$M(t) = -2t^2 + 100t + 180, \quad 0 \le t \le 40.$$

Find the maximum productivity and the year in which it is achieved.

78. **Advertising.** A firm estimates that it will sell N units of a product after spending a dollars on advertising, where

$$N(a) = -a^2 + 300a + 6, \quad 0 \le a \le 300,$$

and a is in thousands of dollars. Find the maximum number of units that can be sold and the amount that must be spent on advertising in order to achieve that maximum.

79. **Maximizing profit.** A firm determines that its total profit in dollars from the production and sale of x units of a product is given by

$$P(x) = \frac{1500}{x^2 - 6x + 10}.$$

Find the number of units x for which the total profit is a maximum.

◆ **Life and Physical Sciences**

80. **Blood pressure.** For a dosage of x cubic centimeters (cc) of a certain drug, the resulting blood pressure B is approximated by

$$B(x) = 0.05x^2 - 0.3x^3, \quad 0 \le x \le 0.16.$$

Find the maximum blood pressure and the dosage at which it occurs.

81. **Minimizing automobile accidents.** At travel speed (constant velocity) x, in miles per hour, there are y accidents at nighttime for every 100 million miles of travel, where y is given by

$$y = -6.1x^2 + 752x + 22,620.$$

At what travel speed does the greatest number of accidents occur?

82. **Temperature during an illness.** The temperature T of a person during an illness is given by

$$T(t) = -0.1t^2 + 1.2t + 98.6, \quad 0 \le t \le 12,$$

where $T = $ the temperature (°F) at time t, in days. Find the maximum value of the temperature and when it occurs.

Find the absolute maximum and minimum values of the function, if they exist, over the indicated interval.

83. $g(x) = x\sqrt{x + 3}; \quad [-3, 3]$

84. $h(x) = x\sqrt{1 - x}; \quad [0, 1]$

85. **Business: Total cost.** Several costs in a business environment can be separated into two components: those that increase with volume and those that decrease with volume. Although the quality of customer service becomes more expensive as it is increased, part of the increased cost is offset by customer goodwill. A firm has determined that its cost of service is given by the following function of "quality units,"

$$C(x) = (2x + 4) + \left(\frac{2}{x - 6}\right), \quad x > 6.$$

Find the number of "quality units" that the firm should use in order to minimize its total cost of service.

86. Let

$$y = (x - a)^2 + (x - b)^2.$$

For what value of x is y a minimum?

87. ◈ Explain the usefulness of the first derivative in finding the absolute extrema of a function.

88. ◈ Explain the usefulness of the second derivative in finding the absolute extrema of a function.

 TECHNOLOGY CONNECTION

Use your grapher to graph each function over the given interval. Visually estimate where absolute maximum and minimum values occur. Then use the trace feature to refine your estimate.

89. $f(x) = x^4 - 4x^3 + 10; \quad [0, 4]$

90. $f(x) = x^{2/3}(x - 5); \quad [1, 4]$

91. $f(x) = \frac{3}{4}(x^2 - 1)^{2/3}$

92. $f(x) = x\left(\frac{x}{2} - 5\right)^4$

Contractions during pregnancy. The following table and graph give the size of a pregnant woman's contractions as a function of time.

Time (t), in minutes	Pressure, in millimeters of mercury
0	10
1	8
2	9.5
3	15
4	12
5	14
6	14.5

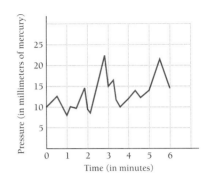

Use a grapher that does linear regression.

93. Fit a linear equation to the data. Predict the size of the contraction after 7 min.

94. Fit a quartic polynomial to the data. Predict the size of the contraction after 7 min. Find the smallest contraction over the interval [0, 10].

3.5 Maximum–Minimum Problems

OBJECTIVE

• Solve maximum–minimum problems using calculus.

One very important application of differential calculus is the solving of maximum–minimum problems, that is, finding the absolute maximum or minimum value of some varying quantity Q and the point at which that maximum or minimum occurs.

EXAMPLE 1   A hobby store has 20 ft of fencing to fence off a rectangular area for an electric train in one corner of its display room. The two sides up against the wall require no fence. What dimensions of the rectangle will maximize the area? What is the maximum area?

Solution   At first glance, we might think that it does not matter what dimensions we use: They will all yield the same area. This is not the case. Let us first make a drawing and express the area in terms of one variable. If we let $x =$ the length of one side and $y =$ the length of the other, then since the sum of the lengths must be 20 ft, we have

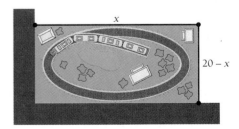

$$x + y = 20 \quad \text{and} \quad y = 20 - x.$$

Thus the area is given by

$$A = xy$$
$$= x(20 - x)$$
$$= 20x - x^2.$$

We are trying to find the maximum value of

$$A = 20x - x^2 \quad \text{on the interval} \quad (0, 20).$$

We consider the interval $(0, 20)$ because x is the length of one side and cannot be negative or 0. Since there is only 20 ft of fencing, x cannot be greater than 20. Also, x cannot be 20 because then the length of y would be 0.

a) We first find $A'(x)$, where $A(x) = 20x - x^2$:

$$A'(x) = 20 - 2x.$$

b) This derivative exists for all values of x in $(0, 20)$. Thus the only critical points are where

$$A'(x) = 20 - 2x = 0$$
$$-2x = -20$$
$$x = 10.$$

Since there is only one critical point in the interval, we can use the second derivative to determine whether we have a maximum. Note that

$$A''(x) = -2,$$

which is a constant. Thus, $A''(10)$ is negative, so $A(10)$ is a maximum. Now

$$A(10) = 10(20 - 10)$$
$$= 10 \cdot 10$$
$$= 100.$$

Thus the maximum area of 100 ft^2 is obtained using 10 ft for the length of one side and $20 - 10$, or 10 ft for the other. Note that $A(5) = 75$, $A(16) = 64$, and $A(12) = 96$; so length does affect area.   ◆

TECHNOLOGY CONNECTION

Exploratory Exercises

Complete this table using your grapher.

x	$y = 20 - x$	$A = x(20 - x)$
0		
4		
6.5		
8		
10		
12		
13.2		
20		

Graph $A(x) = x(20 - x)$ on the interval $[0, 20]$. Use the zoom and trace features to estimate a maximum value and where it would occur.

Here is a general strategy for solving maximum–minimum problems. While it may not guarantee success, it should certainly improve your chances.

A Strategy For Solving Maximum–Minimum Problems

1. Read the problem carefully. If relevant, draw a picture.
2. Label the picture with appropriate variables and constants, noting what varies and what stays fixed.
3. Translate the problem to an equation involving a quantity Q to be maximized or minimized. Try to represent Q in terms of the variables of step (2).
4. Try to express Q as a function of *one* variable. Use the procedures developed in Sections 3.1, 3.2, and 3.4 to determine the maximum or minimum values and the points at which they occur.

EXAMPLE 2   From a thin piece of cardboard 8 in. by 8 in., square corners are cut out so that the sides can be folded up to make a box. What dimensions will yield a box of maximum volume? What is the maximum volume?

Solution   We might again think at first that it does not matter what the dimensions are, but our experience with Example 1 should lead us to think otherwise. We make a drawing.

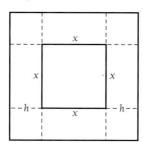
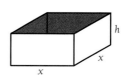

When squares of length h on a side are cut out of the corners, we are left with a square base with sides of length x. The volume of the resulting box is

$$V = lwh = x \cdot x \cdot h.$$

We want to express V in terms of one variable. Note that the overall length of a side of the cardboard is 8 in. We see from the figure that

$$h + x + h = 8,$$
or $x + 2h = 8.$

Solving for h, we get

$$2h = 8 - x$$
$$h = \tfrac{1}{2}(8 - x) = \tfrac{1}{2} \cdot 8 - \tfrac{1}{2}x = 4 - \tfrac{1}{2}x.$$

Thus,

$$V = x \cdot x \cdot \left(4 - \tfrac{1}{2}x\right) = x^2\left(4 - \tfrac{1}{2}x\right) = 4x^2 - \tfrac{1}{2}x^3.$$

TECHNOLOGY CONNECTION

Exploratory Exercises

Complete this table using your grapher.

x	$h = 4 - \tfrac{1}{2}x$	$V = 4x^2 - \tfrac{1}{2}x^3$
0		
1		
2		
3		
4		
4.6		
5		
6		
6.8		
7		
8		

Graph $V(x) = 4x^2 - \tfrac{1}{2}x^3$ on the interval $[0, 8]$. Use the zoom and trace features to estimate a maximum value and where it would occur.

We are trying to find the maximum value of

$$V(x) = 4x^2 - \tfrac{1}{2}x^3 \quad \text{on the interval} \quad (0, 8).$$

We first find $V'(x)$:

$$V'(x) = 8x - \tfrac{3}{2}x^2.$$

Now $V'(x)$ exists for all x in the interval $(0, 8)$, so we set it equal to 0 to find the critical values:

$$V'(x) = 8x - \tfrac{3}{2}x^2 = 0$$
$$x\left(8 - \tfrac{3}{2}x\right) = 0$$
$$x = 0 \quad \text{or} \quad 8 - \tfrac{3}{2}x = 0$$
$$x = 0 \quad \text{or} \quad -\tfrac{3}{2}x = -8$$
$$x = 0 \quad \text{or} \quad x = -\tfrac{2}{3}(-8) = \tfrac{16}{3}.$$

The only critical point in $(0, 8)$ is $\tfrac{16}{3}$. Thus we can use the second derivative,

$$V''(x) = 8 - 3x,$$

to determine whether we have a maximum. Since

$$V''\left(\tfrac{16}{3}\right) = 8 - 3 \cdot \tfrac{16}{3}$$
$$= -8,$$

$V''\left(\tfrac{16}{3}\right)$ is negative, so $V\left(\tfrac{16}{3}\right)$ is a maximum, and

$$V\left(\tfrac{16}{3}\right) = 4 \cdot \left(\tfrac{16}{3}\right)^2 - \tfrac{1}{2}\left(\tfrac{16}{3}\right)^3 = \tfrac{1024}{27} = 37\tfrac{25}{27}.$$

The maximum volume is $37\tfrac{25}{27}$ in^3. The dimensions that yield this maximum volume are

$$x = \tfrac{16}{3} = 5\tfrac{1}{3} \text{ in.,} \quad \text{by } x = 5\tfrac{1}{3} \text{ in.,} \quad \text{by } h = 4 - \tfrac{1}{2}\left(\tfrac{16}{3}\right) = 1\tfrac{1}{3} \text{ in.} \qquad \blacklozenge$$

In the following problem, an open-top container of fixed volume is to be constructed. We want to determine the dimensions that will allow it to be built with the least amount of material. Such a problem could be important from an ecological standpoint.

EXAMPLE 3   A container firm is designing an open-top rectangular box, with a square base, that will hold 108 cubic centimeters (cc). What dimensions yield the minimum surface area? What is the minimum surface area?

Solution   We make a drawing. The surface area of the box is given by

$$S = x^2 + 4xy.$$

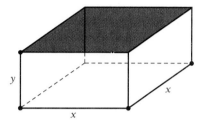

The volume must be 108 cc, and is given by

$$V = x^2y = 108.$$

To express S in terms of one variable, we solve $x^2y = 108$ for y:

$$y = \frac{108}{x^2}.$$

Then

$$S = x^2 + 4x\left(\frac{108}{x^2}\right) = x^2 + \frac{432}{x}.$$

Since S is defined only for positive numbers and the problem dictates that the length x be positive, we are minimizing S on the interval $(0, \infty)$. We first find dS/dx:

$$\frac{dS}{dx} = 2x - \frac{432}{x^2}.$$

Since dS/dx exists for all x in $(0, \infty)$, the only critical points occur where $dS/dx = 0$. Thus we solve the following equation:

$$2x - \frac{432}{x^2} = 0$$

$$x^2\left(2x - \frac{432}{x^2}\right) = x^2 \cdot 0 \qquad \text{We multiply by } x^2 \text{ to clear the fractions.}$$

$$2x^3 - 432 = 0$$

$$2x^3 = 432$$

$$x^3 = 216$$

$$x = 6.$$

This is the only critical point, so we can use the second derivative to determine whether we have a minimum:

$$\frac{d^2S}{dx^2} = 2 + \frac{864}{x^3}.$$

Note that this is positive for all positive values of x. Thus we have a minimum at $x = 6$. When $x = 6$, it follows that $y = 3$:

$$y = \frac{108}{6^2}$$

$$= \frac{108}{36}$$

$$= 3.$$

Thus the surface area is minimized when $x = 6$ cm (centimeters) and $y = 3$ cm. The minimum surface area is

$$S = 6^2 + 4 \cdot 6 \cdot 3$$

$$= 108 \text{ cm}^2.$$

By coincidence, this is the same number as the fixed volume.   ◆

Business and Economics Applications: Marginal Analysis

EXAMPLE 4   *Business: Maximizing revenue.* A stereo manufacturer determines that in order to sell x units of a new stereo, the price per unit must be

$$p = 1000 - x.$$

The manufacturer also determines that the total cost of producing x units is given by

$$C(x) = 3000 + 20x.$$

a) Find the total revenue $R(x)$.

b) Find the total profit $P(x)$.

c) How many units must the company produce and sell in order to maximize profit?

d) What is the maximum profit?

e) What price per unit must be charged in order to make this maximum profit?

Solution

a) $R(x) = $ Total revenue $= $ (Number of units) $\cdot$ (Price per unit)

$$= \quad x \quad \cdot \quad p$$

$$= x(1000 - x) = 1000x - x^2$$

b) $P(x) = R(x) - C(x) = (1000x - x^2) - (3000 + 20x)$

$$= -x^2 + 980x - 3000$$

c) To find the maximum value of $P(x)$, we first find $P'(x)$:

$$P'(x) = -2x + 980.$$

This is defined for all real numbers (actually we are interested in numbers x in $[0, \infty)$ only, since we cannot produce a negative number of stereos). Thus we solve:

$$P'(x) = -2x + 980 = 0$$
$$-2x = -980$$
$$x = 490.$$

Since there is only one critical point, we can try to use the second derivative to determine whether we have a maximum. Note that

$$P''(x) = -2, \quad \text{a constant.}$$

Thus, $P''(490)$ is negative, so $P(490)$ is a maximum.

d) The maximum profit is given by

$$P(490) = -(490)^2 + 980 \cdot 490 - 3000$$
$$= \$237{,}100.$$

Thus the stereo manufacturer makes a maximum profit of \$237,100 by producing and selling 490 stereos.

e) The price per unit needed to make the maximum profit is

$$p = 1000 - 490 = \$510. \qquad \blacklozenge$$

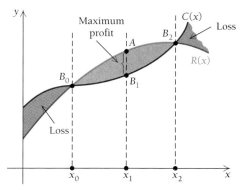

FIGURE 1

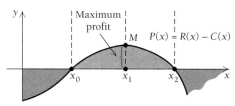

FIGURE 2

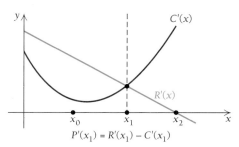

$P'(x_1) = R'(x_1) - C'(x_1)$

FIGURE 3

Let us take a general look at the total-profit function and its related functions.

Figure 1 shows an example of total-cost and total-revenue functions. We can estimate what the maximum profit might be by looking for the widest gap between $R(x)$ and $C(x)$. Points B_0 and B_2 are break-even points.

Figure 2 shows an example of total-profit function. Note that when production is too low ($< x_0$), there is a loss because of high fixed or initial costs and low revenue. When production is too high ($> x_2$), there is also a loss due to high marginal costs and low marginal revenues (as shown in Fig. 3).

The business operates at a profit everywhere between x_0 and x_2. Note that maximum profit occurs at a critical point x_1 of $P(x)$. If we assume that $P'(x)$ exists for all x in some interval, usually $[0, \infty)$, this critical point occurs at some number x such that

$$P'(x) = 0$$

and

$$P''(x) < 0.$$

Since $P(x) = R(x) - C(x)$, it follows that

$$P'(x) = R'(x) - C'(x)$$

and

$$P''(x) = R''(x) - C''(x).$$

Thus the maximum profit occurs at some number x such that

$$P'(x) = R'(x) - C'(x) = 0$$

and

$$P''(x) = R''(x) - C''(x) < 0,$$

or

$$R'(x) = C'(x)$$

and

$$R''(x) < C''(x).$$

In summary, we have the following theorem.

THEOREM 10

Maximum profit is achieved when marginal revenue equals marginal cost and the rate of change of marginal revenue is less than the rate of change of marginal cost:

$$R'(x) = C'(x) \quad \text{and} \quad R''(x) < C''(x).$$

EXAMPLE 5 *Business: Determining a ticket price.* Fight promoters ride a thin line between profit and loss, especially in determining the price to charge for admission to closed-circuit television showings in local theaters. By keeping records, a theater determines that if the admission price is $20, it averages 1000 people in attendance. But for every increase of $1, it loses 100 customers from the average number. Every customer spends an average of $1.80 on concessions. What admission price should the theater charge in order to maximize total revenue?

Solution   Let $x =$ the amount by which the price of \$20 should be increased. (If x is negative, the price is decreased.) We first express the total revenue R as a function of x. Note that

$$R(x) = \text{(Revenue from tickets)} + \text{(Revenue from concessions)}$$
$$= \text{(Number of people)} \cdot \text{(Ticket price)} + \$1.80\text{(Number of people)}$$
$$= (1000 - 100x)(20 + x) + 1.80(1000 - 100x)$$
$$= 20{,}000 - 2000x + 1000x - 100x^2 + 1800 - 180x$$
$$R(x) = -100x^2 - 1180x + 21{,}800.$$

We are trying to find the maximum value of R over the set of all real numbers. To find x such that $R(x)$ is a maximum, we first find $R'(x)$:

$$R'(x) = -200x - 1180.$$

This derivative exists for all real numbers x. Thus the only critical points are where $R'(x) = 0$, so we solve that equation:

$$-200x - 1180 = 0$$
$$-200x = 1180$$
$$x = -5.9 = -\$5.90.$$

Since this is the only critical point, we can use the second derivative,

$$R''(x) = -200,$$

to determine whether we have a maximum. Since $R''(-5.9)$ is negative, $R(-5.9)$ is a maximum. Therefore, in order to maximize revenue, the theater should charge

$$\$20 + (-\$5.90), \quad \text{or} \quad \$14.10 \text{ per ticket.}$$

That is, this reduced ticket price will attract more people to the movie theater,

$$1000 - 100(-5.9), \quad \text{or} \quad 1590,$$

and will result in maximum revenue.   ◆

Minimizing Inventory Costs

A retail outlet of a business is concerned about inventory costs. Suppose, for example, that an appliance store sells 2500 television sets per year. It *could* operate by ordering all the sets at once. But then the owners would face the carrying costs (insurance, building space, and so on) of storing them all. Thus they might make several smaller orders, say 5, so that the largest number they would ever have to store is 500. On the other hand, each time they reorder, there are costs for paperwork, delivery charges, manpower, and so on. It would seem, therefore, that there must be some balance between carrying costs and reorder costs. Let us see how calculus can help to determine what that balance might be. We are trying to minimize the following function:

$$\text{Total inventory costs} = \left(\begin{array}{c}\text{Yearly carrying}\\\text{costs}\end{array}\right) + \left(\begin{array}{c}\text{Yearly reorder}\\\text{costs}\end{array}\right).$$

The *lot size x* refers to the largest amount ordered each reordering period. If x is ordered each period, then during that time there is somewhere between 0 and x units in stock. To have a representative expression for the amount in stock at any one time in the period, we can use the average, $x/2$. This represents the average amount held in stock over the course of the year.

Refer to the following graphs. If the lot size is 2500, then during the period between orders, there is somewhere between 0 and 2500 units in stock. On the average, there is 2500/2, or 1250 units in stock. If the lot size is 1250, then during the period between orders, there is somewhere between 0 and 1250 units in stock. On the average, there is 1250/2, or 625 units in stock.

How can inventory costs be minimized?

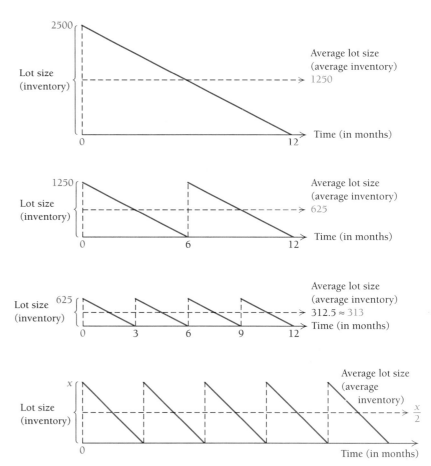

EXAMPLE 6 *Business: Minimizing inventory costs.* A retail appliance store sells 2500 television sets per year. It costs $10 to store one set for a year. To reorder, there is a fixed cost of $20, plus $9 for each set. How many times per year should the store reorder, and in what lot size, in order to minimize inventory costs?

Solution   Let x = the lot size. Inventory costs are given by

$$C(x) = (\text{Yearly carrying costs}) + (\text{Yearly reorder costs}).$$

We consider each separately.

a) *Yearly carrying costs.* The average amount held in stock is $x/2$, and it costs $10 per set for storage. Thus

$$\text{Yearly carrying costs} = \left(\begin{array}{c}\text{Yearly cost}\\ \text{per item}\end{array}\right) \cdot \left(\begin{array}{c}\text{Average number}\\ \text{of items}\end{array}\right)$$

$$= 10 \cdot \frac{x}{2}.$$

b) *Yearly reorder costs.* We know that x = the lot size, and suppose that there are N reorders each year. Then $Nx = 2500$, and $N = 2500/x$. Thus,

$$\text{Yearly reorder costs} = \left(\begin{array}{c}\text{Cost of each}\\ \text{order}\end{array}\right) \cdot \left(\begin{array}{c}\text{Number of}\\ \text{reorders}\end{array}\right)$$

$$= (20 + 9x)\frac{2500}{x}.$$

c) Hence

$$C(x) = 10 \cdot \frac{x}{2} + (20 + 9x)\frac{2500}{x}$$

$$= 5x + \frac{50{,}000}{x} + 22{,}500.$$

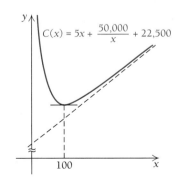

$$C(x) = 5x + \frac{50{,}000}{x} + 22{,}500$$

100

TECHNOLOGY CONNECTION

Many graphers and software packages have the ability to make tables and/or spreadsheets of function values. In reference to Example 6, without a knowledge of calculus, one might make a rough estimate of the lot size that will minimize total inventory costs by using a table like the following. Complete the table and make an estimate.

Lot size, x	Number of reorders, $\dfrac{2500}{x}$	Average inventory, $\dfrac{x}{2}$	Carrying costs, $10 \cdot \dfrac{x}{2}$	Cost of each order, $20 + 9x$	Reorder costs, $(20 + 9x)\dfrac{2500}{x}$	Total inventory costs, $C(x) = 10 \cdot \dfrac{x}{2} + (20 + 9x)\dfrac{2500}{x}$
2500	1	1250	$12,500	$22,500	$22,520	$35,020
1250	2	625	6,250	11,270	22,540	
500	5	250	2,500	4,520		
250	10	125				
167	15	84				
125	20					
100	25					
90	28					
50	50					

Use your grapher to graph $C(x)$ over the interval $[0, 2500]$. Estimate a minimum value and where it occurs.

d) We want to find a minimum value of C on the interval $[1, 2500]$. (See the graph above.) We first find $C'(x)$:

$$C'(x) = 5 - \frac{50{,}000}{x^2}.$$

e) $C'(x)$ exists for all x in $[1, 2500]$, so the only critical points are those x such that $C'(x) = 0$. We solve $C'(x) = 0$:

$$5 - \frac{50{,}000}{x^2} = 0$$

$$5 = \frac{50{,}000}{x^2}$$

$$5x^2 = 50{,}000$$

$$x^2 = 10{,}000$$

$$x = \pm 100.$$

Since there is only one critical point in the interval $[1, 2500]$, that is, $x = 100$, we can use the second derivative to see whether we have a maximum or a minimum:

$$C''(x) = \frac{100,000}{x^3}.$$

$C''(x)$ is positive for all x in $[1, 2500]$, so we do have a minimum at $x = 100$. Thus in order to minimize inventory costs, the store should order sets $(2500/100)$, or 25 times per year. The lot size is 100.   ◆

What happens in such problems when the answer is not a whole number? For those functions, we consider the two whole numbers closest to the answer and substitute them into $C(x)$. The value that yields the smaller $C(x)$ is the lot size.

EXAMPLE 7   *Business: Minimizing inventory costs.* Let us repeat Example 6 using all the data given, but change the \$10 storage cost to \$20. How many times per year should the store reorder television sets, and in what lot size, in order to minimize inventory costs?

Solution   Comparing this with Example 6, we find that the inventory cost function becomes

$$C(x) = 20 \cdot \frac{x}{2} + (20 + 9x)\frac{2500}{x}$$

$$= 10x + \frac{50,000}{x} + 22,500.$$

Then we find $C'(x)$, set it equal to 0, and solve for x:

$$C'(x) = 10 - \frac{50,000}{x^2} = 0$$

$$10 = \frac{50,000}{x^2}$$

$$10x^2 = 50,000$$

$$x^2 = 5000$$

$$x = \sqrt{5000} \approx 70.7.$$

Since it does not make sense to reorder 70.7 sets each time, we consider the two numbers closest to 70.7, which are 70 and 71. Now

$$C(70) \approx \$23,914.29 \quad \text{and} \quad C(71) \approx \$23,914.23.$$

It follows that the lot size that will minimize cost is 71, although the difference, \$0.06, is not significant. (*Note:* Such a procedure will not work for all types of functions, but will work for the type we are considering here. The number of times that an order should be placed is $2500/71 \approx 35$, so there is still some estimating involved.)   ◆

The lot size that minimizes total inventory costs is often referred to as the *economic ordering quantity*. There are three assumptions made in using the preceding method to determine the economic ordering quantity. The first is that the demand for the product is the same throughout the year. For television sets this may be reasonable, but for seasonal items such as clothing or skis, this assumption may not be reasonable. The second assumption is that the time between the placing of an order and its receipt will be consistent throughout the year. The third assumption is that the various costs involved, such as storage, shipping charges, and so on, do not vary. This may not be reasonable in a time of inflation, although one may account for them by anticipating what they might be and using average costs. Nevertheless, the model described above can be useful, and it allows us to analyze a seemingly difficult problem using calculus.

3.5 Exercise Set

1. Of all numbers whose sum is 50, find the two that have the maximum product. That is, maximize $Q = xy$, where $x + y = 50$.

2. Of all numbers whose sum is 70, find the two that have the maximum product. That is, maximize $Q = xy$, where $x + y = 70$.

3. In Exercise 1, can there be a minimum product? Explain.

4. In Exercise 2, can there be a minimum product? Explain.

5. Of all numbers whose difference is 4, find the two that have the minimum product.

6. Of all numbers whose difference is 6, find the two that have the minimum product.

7. Maximize $Q = xy^2$, where x and y are positive numbers, such that $x + y^2 = 1$.

8. Maximize $Q = xy^2$, where x and y are positive numbers, such that $x + y^2 = 4$.

9. Minimize $Q = x^2 + y^2$, where $x + y = 20$.

10. Minimize $Q = x^2 + y^2$, where $x + y = 10$.

11. Maximize $Q = xy$, where x and y are positive numbers, such that $\frac{4}{3}x^2 + y = 16$.

12. Maximize $Q = xy$, where x and y are positive numbers, such that $x + \frac{4}{3}y^2 = 1$.

13. A rancher wants to build a rectangular fence next to a river, using 120 yd of fencing. What dimensions of the rectangle will maximize the area? What is the maximum area? (Note that the rancher need not fence in the side next to the river.)

14. A rancher wants to enclose two rectangular areas near a river, one for sheep and one for cattle. There is 240 yd of fencing available. What is the largest total area that can be enclosed?

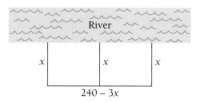

15. A carpenter is building a rectangular room with a fixed perimeter of 54 ft. What are the dimensions of the largest room that can be built? What is its area?

16. Of all rectangles that have a perimeter of 34 ft, find the dimensions of the one with the largest area. What is its area?

17. From a thin piece of cardboard 30 in. by 30 in., square corners are cut out so that the sides can be folded up to make a box. What dimensions will yield

a box of maximum volume? What is the maximum volume?

18. From a thin piece of cardboard 20 in. by 20 in., square corners are cut out so that the sides can be folded up to make a box. What dimensions will yield a box of maximum volume? What is the maximum volume?

19. A container company is designing an open-top, square-based, rectangular box that will have a volume of 62.5 in^3. What dimensions yield the minimum surface area? What is the minimum surface area?

20. A soup company is constructing an open-top, square-based, rectangular metal tank that will have a volume of 32 ft^3. What dimensions yield the minimum surface area? What is the minimum surface area?

APPLICATIONS

◆   Business and Economics

Maximizing profit. Find the maximum profit and the number of units that must be produced and sold in order to yield the maximum profit.

21. $R(x) = 50x - 0.5x^2$, $C(x) = 4x + 10$

22. $R(x) = 50x - 0.5x^2$, $C(x) = 10x + 3$

23. $R(x) = 2x$, $C(x) = 0.01x^2 + 0.6x + 30$

24. $R(x) = 5x$, $C(x) = 0.001x^2 + 1.2x + 60$

25. $R(x) = 9x - 2x^2$, $C(x) = x^3 - 3x^2 + 4x + 1$; $R(x)$ and $C(x)$ are in thousands of dollars, and x is in thousands of units.

26. $R(x) = 100x - x^2$, $C(x) = \frac{1}{3}x^3 - 6x^2 + 89x + 100$; $R(x)$ and $C(x)$ are in thousands of dollars, and x is in thousands of units.

27. *Maximizing profit.* Raggs, Ltd., a clothing firm, determines that in order to sell x suits, the price per suit must be

$$p = 150 - 0.5x.$$

It also determines that the total cost of producing x suits is given by

$$C(x) = 4000 + 0.25x^2.$$

a) Find the total revenue $R(x)$.
b) Find the total profit $P(x)$.

c) How many suits must the company produce and sell in order to maximize profit?
d) What is the maximum profit?
e) What price per suit must be charged in order to make this maximum profit?

28. *Maximizing profit.* An appliance firm is marketing a new refrigerator. It determines that in order to sell x refrigerators, the price per refrigerator must be

$$p = 280 - 0.4x.$$

It also determines that the total cost of producing x refrigerators is given by

$$C(x) = 5000 + 0.6x^2.$$

a) Find the total revenue $R(x)$.
b) Find the total profit $P(x)$.
c) How many refrigerators must the company produce and sell in order to maximize profit?
d) What is the maximum profit?
e) What price per refrigerator must be charged in order to make this maximum profit?

29. *Maximizing revenue.* A university is trying to determine what price to charge for football tickets. At a price of $6 per ticket, it averages 70,000 people per game. For every increase of $1, it loses 10,000 people from the average number. Every person at the game spends an average of $1.50 on concessions. What price per ticket should be charged in order to maximize revenue? How many people will attend at that price?

30. *Maximizing revenue.* Suppose that you are the owner of a 30-unit motel. All units are occupied when you charge $20 a day per unit. For every increase of x dollars in the daily rate, there are x units vacant. Each occupied room costs $2 per day to service and maintain. What should you charge per unit in order to maximize profit?

31. *Maximizing yield.* An apple farm yields an average of 30 bushels of apples per tree when 20 trees are planted on an acre of ground. Each time 1 more tree is planted per acre, the yield decreases 1 bu per tree due to the extra congestion. How many trees should be planted in order to get the highest yield?

32. *Maximizing revenue.* When a theater owner charges $3 for admission, there is an average attendance of 100 people. For every $0.10 increase in admission, there is a loss of 1 customer from the average number.

What admission should be charged in order to maximize revenue?

33. *Minimizing costs.* A rectangular box with a volume of 320 ft³ is to be constructed with a square base and top. The cost per square foot for the bottom is 15¢, for the top is 10¢, and for the sides is 2.5¢. What dimensions will minimize the cost?

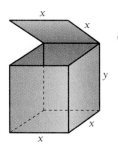

34. ◈ A merchant who was purchasing a display sign from a salesclerk said, "I want a sign 10 ft by 10 ft." The salesclerk responded, "That's just what we'll give you; only to make it more aesthetically pleasing, why don't we change it to 7 ft by 13 ft?" Comment.

35. *Maximizing profit.* The amount of money deposited in a financial institution in savings accounts is directly proportional to the interest rate that the financial institution pays on the money. Suppose that a financial institution can loan *all* the money it takes in on its savings accounts at an interest rate of 18%. What interest rate should it pay on its savings accounts in order to maximize profit?

36. *Maximizing area.* A page in this book is 73.125 in². On the average, there is a 0.75-in. margin at the top and at the bottom of each page and a 0.5-in. margin on each of the sides. What should the outside dimensions of each page be so that the printed area is a maximum? Measure the outside dimensions to see whether the actual dimensions maximize the printed area.

37. *Minimizing inventory costs.* A sporting goods store sells 100 pool tables per year. It costs $20 to store one pool table for one year. To reorder, there is a fixed cost of $40, plus $16 for each pool table. How many times per year should the store order pool tables, and in what lot size, in order to minimize inventory costs?

38. *Minimizing inventory costs.* A pro shop in a bowling center sells 200 bowling balls per year. It costs $4 to store one bowling ball for one year. To reorder, there is a fixed cost of $1, plus $0.50 for each bowling ball. How many times per year should the shop order bowling balls, and in what lot size, in order to minimize inventory costs?

39. *Minimizing inventory costs.* A retail outlet for Boxowitz Calculators sells 360 calculators per year. It costs $8 to store one calculator for one year. To reorder, there is a fixed cost of $10, plus $8 for each calculator. How many times per year should the store order calculators, and in what lot size, in order to minimize inventory costs?

40. *Minimizing inventory costs.* A sporting goods store in southern California sells 720 surfboards per year. It costs $2 to store one surfboard for one year. To reorder, there is a fixed cost of $5, plus $2.50 for each surfboard. How many times per year should the store order surfboards, and in what lot size, in order to minimize inventory costs?

41. *Minimizing inventory costs.* Repeat Exercise 39 using all the data given, but change the $8 storage charge to $9.

42. *Minimizing inventory costs.* Repeat Exercise 40 using all the data given, but change the $5 fixed cost to $4.

◆   **General Interest**

43. *Maximizing volume.* The postal service places a limit of 84 in. on the combined length and girth (distance around) of a package to be sent parcel post. What dimensions of a rectangular box with square cross-section will contain the largest volume that can be mailed? (*Hint:* There are two different girths.)

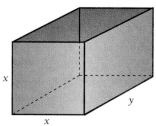

44. *Minimizing cost.* A rectangular play area is to be fenced off in a person's yard and is to contain 48 yd². The neighbor agrees to pay half the cost of the fence

on the side of the play area that lines the lot. What dimensions will minimize the cost of the fence?

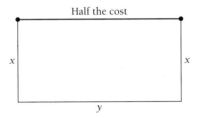

Half the cost

45. *Maximizing light.* A Norman window is a rectangle with a semicircle on top. Suppose that the perimeter of a particular Norman window is to be 24 ft. What should its dimensions be in order to allow the maximum amount of light to enter through the window?

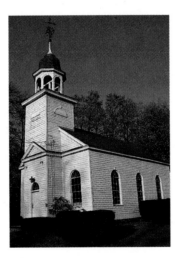

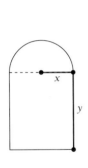

46. *Maximizing light.* Repeat Exercise 45, but assume that the semicircle is to be stained glass, which transmits only half as much light as the semicircle in Exercise 45.

$A = xy \qquad (x+1)(4+1.5) = 73.125$

$x = \dfrac{73.125}{4+1.5} - 1$

SYNTHESIS

47. For what positive number is the sum of its reciprocal and five times its square a minimum?

48. For what positive number is the sum of its reciprocal and four times its square a minimum?

49. *Business: Minimizing inventory costs—a general solution.* A store sells Q units of a product per year. It costs a dollars to store one unit for one year. To reorder, there is a fixed cost of b dollars, plus c dollars for each unit. How many times per year should the store reorder, and in what lot size, in order to minimize inventory costs?

50. *Business: Minimizing inventory costs.* Use the general solution found in Exercise 49 to find how many times per year a store should reorder, and in what lot size, when $Q = 2500$, $a = \$10$, $b = \$20$, and $c = \$9$.

51. A 24-in. piece of string is cut in two pieces. One piece is used to form a circle and the other to form a square. How should the string be cut so that the sum of the areas is a minimum? a maximum?

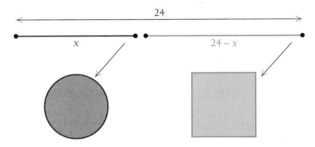

52. *Business: Minimizing costs.* A power line is to be constructed from a power station at point A to an island at point C, which is 1 mi directly out in the water from a point B on the shore. Point B is 4 mi downshore from the power station at A.

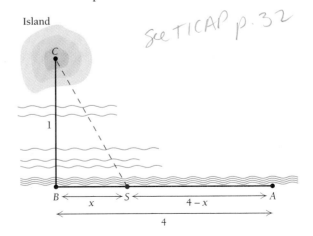

see TICAP p. 32

It costs $5000 per mile to lay the power line under water and $3000 per mile to lay the line under ground. At what point S downshore from A should the line come to the shore in order to minimize cost? Note that S could very well be B or A. (*Hint:* The length of CS is $\sqrt{1 + x^2}$.)

53. **Life science: Flights of homing pigeons.** It is known that homing pigeons tend to avoid flying over water in the daytime, perhaps because the downdrafts of air over water make flying difficult. Suppose a homing pigeon is released on an island at point C, which is 3 mi directly out in the water from a point B on shore. Point B is 8 mi downshore from the pigeon's home loft at point A. Assume that a pigeon requires 1.28 times the rate of energy over land to fly over water. Toward what point S downshore from A should the pigeon fly in order to minimize the total energy required to get to home loft A? Assume that

(Total energy)

  = (Energy rate over water) · (Distance over water)

  + (Energy rate over land) · (Distance over land).

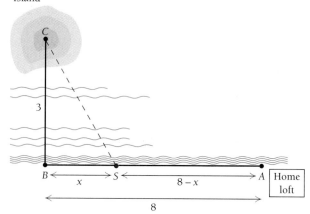

Island

54. **Business: Minimizing distance.** A road is to be built between two cities C_1 and C_2, which are on opposite sides of a river of uniform width r. Because of the river, a bridge must be built. C_1 is a units from the river, and C_2 is b units from the river; $a \leq b$. Where should the bridge be located in order to minimize the

total distance between the cities? Give a general solution using the constants a, b, p, and r in the figure shown here.

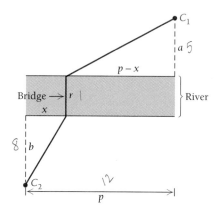

55. **Business: Minimizing cost.** The total-cost function for producing x units of a certain product is given by

$$C(x) = 8x + 20 + \frac{x^3}{100}.$$

a) Find the marginal cost $C'(x)$.
b) Find the average cost $A(x) = C(x)/x$.
c) Find the marginal average cost $A'(x)$.
d) Find the minimum of $A(x)$ and the value x_0 at which it occurs. Find the marginal cost at x_0.
e) Compare $A(x_0)$ and $C'(x_0)$.

56. **Business: Minimizing cost.** Consider $A(x) = C(x)/x$.

a) Find $A'(x)$ in terms of $C'(x)$ and $C(x)$.
b) Show that $A(x)$ has a minimum at that value of x_0 such that

$$C'(x_0) = A(x_0) = \frac{C(x_0)}{x_0}.$$

This shows that when marginal cost and average cost are the same, a product is being produced at the least average cost.

57. Minimize $Q = x^3 + 2y^3$, where x and y are positive numbers, such that $x + y = 1$.

58. Minimize $Q = 3x + y^3$, where $x^2 + y^2 = 2$.

59. ◈ Explain how marginal revenue, cost, and profit can be used to find maximum profit.

3.6 Differentials

Delta Notation

Recall the difference quotient

$$\frac{f(x + h) - f(x)}{h},$$

illustrated in this graph.

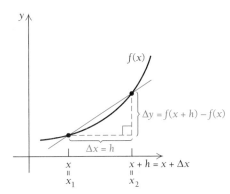

The difference quotient is used to define the derivative of a function at x. The number h is considered to be a *change* in x. Another notation for such a change is Δx, read "delta x" and called **delta notation.** The expression Δx is *not* the product of Δ and x, but is its own entity; that is, it is a new type of variable that represents the *change* in the value of x from a *first* value to a *second.* Thus,

$$\Delta x = (x + h) - x = h.$$

If subscripts are used for the first and second values of x, we have

$$\Delta x = x_2 - x_1, \quad \text{or} \quad x_2 = x_1 + \Delta x.$$

Δx can be positive or negative.

EXAMPLE 1

a) If $x_1 = 4$ and $\Delta x = 0.7$, then $x_2 = 4.7$.

b) If $x_1 = 4$ and $\Delta x = -0.7$, then $x_2 = 3.3$.   ◆

We generally omit the subscripts and use x and $x + \Delta x$.

Now suppose we have a function given by $y = f(x)$. A change in x from x to $x + \Delta x$ yields a change in y from $f(x)$ to $f(x + \Delta x)$. The change in y is given by

$$\Delta y = f(x + \Delta x) - f(x).$$

EXAMPLE 2   For $y = x^2$, $x = 4$, and $\Delta x = 0.1$, find Δy.

Solution

$$\Delta y = (4 + 0.1)^2 - 4^2$$
$$= (4.1)^2 - 4^2 = 16.81 - 16 = 0.81$$

◆

EXAMPLE 3   For $y = x^3$, $x = 2$, and $\Delta x = -0.1$, find Δy.

Solution

$$\Delta y = [2 + (-0.1)]^3 - 2^3$$
$$= (1.9)^3 - 2^3 = 6.859 - 8 = -1.141$$

◆

If delta notation is used, the difference quotient

$$\frac{f(x + h) - f(x)}{h}$$

becomes

$$\frac{f(x + \Delta x) - f(x)}{\Delta x} = \frac{\Delta y}{\Delta x}.$$

We can then express the derivative as

$$\frac{dy}{dx} = \lim_{\Delta x \to 0} \frac{\Delta y}{\Delta x}.$$

Note that the delta notation resembles the Leibniz notation.

For values of Δx close to 0, we have the approximation

$$\frac{dy}{dx} \approx \frac{\Delta y}{\Delta x}, \quad \text{or} \quad f'(x) \approx \frac{\Delta y}{\Delta x}.$$

Multiplying both sides of the second expression by Δx gives us

$$\Delta y \approx f'(x) \, \Delta x.$$

We can see this in the following graph.

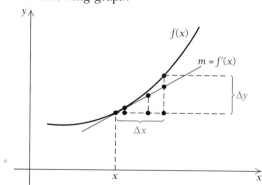

Recall that the derivative is a limit of slopes $\Delta y/\Delta x$ of secant lines. Thus as Δx gets smaller, the ratio $\Delta y/\Delta x$ gets closer to dy/dx. Note too that over small intervals, the tangent line is a good approximation to the function. Thus it is reasonable to assume that average rates of change $\Delta y/\Delta x$ of the function are approximately the same as the slope of the tangent line.

Let us use the fact that

$$\Delta y \approx f'(x) \, \Delta x$$

to make certain approximations, such as square roots.

EXAMPLE 4   Approximate $\sqrt{27}$ using $\Delta y \approx f'(x) \, \Delta x$.

Solution   We first think of the number closest to 27 that is a perfect square. This is 25. What we will do is approximate how y, or $\sqrt{x}$, changes when 25 changes by $\Delta x = 2$. Let

$$y = f(x) = \sqrt{x}.$$

Then

$$\Delta y = \sqrt{x + \Delta x} - \sqrt{x} = \sqrt{x + \Delta x} - y,$$

so

$$y + \Delta y = \sqrt{x + \Delta x}.$$

Now

$$\Delta y \approx f'(x) \, \Delta x = \frac{1}{2} x^{-1/2} \, \Delta x = \frac{1}{2\sqrt{x}} \, \Delta x.$$

Let $x = 25$ and $\Delta x = 2$. Then

$$\Delta y \approx f'(x) \, \Delta x = \frac{1}{2\sqrt{25}} \cdot 2$$

$$= \frac{1}{\sqrt{25}} = \frac{1}{5} = 0.2.$$

Thus,

$$\sqrt{27} = \sqrt{x + \Delta x} = y + \Delta y \approx \sqrt{25} + 0.2 = 5 + 0.2 = 5.2.$$

To five decimal places, $\sqrt{27} = 5.19615$. Thus our approximation is fairly accurate. ◆

Suppose we have a total-cost function $C(x)$. When $\Delta x = 1$, we have

$$\Delta C \approx C'(x).$$

Whether this is a good approximation depends on the function and on the values of x. Let us consider an example.

EXAMPLE 5   Consider the total-cost function

$$C(x) = 2x^3 - 12x^2 + 30x + 200.$$

a) Find ΔC and $C'(x)$ when $x = 2$ and $\Delta x = 1$.

b) Find ΔC and $C'(x)$ when $x = 100$ and $\Delta x = 1$.

Solution

a) We have

$$\Delta C = C(2 + 1) - C(2) = C(3) - C(2) = \$236 - \$228 = \$8.$$

Recall that $C(2)$ is the total cost of producing 2 units, and $C(3)$ is the total cost of producing 3 units, so $C(3) - C(2)$, or $\$8$, is the cost of the third unit. Now

$$C'(x) = 6x^2 - 24x + 30, \quad \text{so} \quad C'(2) = \$6.$$

b) We have

$$\Delta C = C(100 + 1) - C(100) = C(101) - C(100) = \$58,220.$$

Note that this is the cost of the 101st unit. Now

$$C'(100) = \$57,630.$$

Note that in part (a), the approximation of $\$6$ is $\$2$ less than the "correct" value of $\$8$; this is a percentage difference of $\$2/\$8$, or 25%, from the correct value. In part (b), the approximation of $\$57,630$ is $\$590$ less than the "correct" value of $\$58,220$; this is a percentage difference of $\$590/\$58,220$, or about 1% from the correct value, so it is a very close approximation.   ◆

We purposely used $\Delta x = 1$ in Example 5 to illustrate the following.

$$C'(x) \approx C(x + 1) - C(x)$$

Marginal cost is (approximately) the cost of the $(x + 1)$st, or next, unit.

This is the historical definition that economists have given to marginal cost. Similarly, the following is true.

$$R'(x) \approx R(x + 1) - R(x)$$

Marginal revenue is (approximately) the revenue from the sale of the $(x + 1)$st, or next, unit.

And

$$P'(x) \approx P(x + 1) - P(x)$$

Marginal profit is (approximately) the profit from the production and sale of the $(x + 1)$st, or next, unit.

Differentials

Up to now we have not defined the symbols dy and dx as separate entities, and we have treated dy/dx as one symbol. We now define dy and dx. These symbols are called **differentials.**

DEFINITION

For $y = f(x)$, we define

dx, called the *differential of x*, by $dx = \Delta x$

and

dy, called the *differential of y*, by $dy = f'(x)\, dx$.

We can illustrate dx and dy as shown below. Note that $dx = \Delta x$, but $dy \neq \Delta y$, though $dy \approx \Delta y$, for small values of dx.

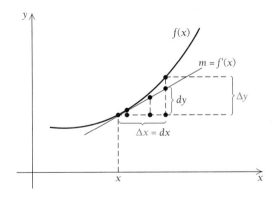

EXAMPLE 6 For $y = x(4 - x)^3$, (a) find dy and (b) find dy when $x = 5$ and $dx = 0.2$.

Solution

a) First, we find dy/dx:

$$\frac{dy}{dx} = -x[3(4-x)^2] + (4-x)^3 \qquad \text{Using the Product Rule and the Extended Power Rule}$$

$$= (4-x)^2[-3x + (4-x)]$$

$$= (4-x)^2[-4x+4] \qquad \text{Simplifying}$$

$$= -4(4-x)^2(x-1).$$

Then

$$dy = -4(4-x)^2(x-1)\ dx.$$

Note that the expression for dy contains *two* variables, x and dx.

b) When $x = 5$ and $dx = 0.2$,

$$dy = -4(4-5)^2(5-1)0.2 = -4(-1)^2(4)(0.2) = -3.2. \qquad \blacklozenge$$

3.6 Exercise Set

Find Δy and $f'(x)\ \Delta x$.

1. For $y = f(x) = x^2$, $x = 2$, and $\Delta x = 0.01$.

2. For $y = f(x) = x^3$, $x = 2$, and $\Delta x = 0.01$.

3. For $y = f(x) = x + x^2$, $x = 3$, and $\Delta x = 0.04$.

4. For $y = f(x) = x - x^2$, $x = 3$, and $\Delta x = 0.02$.

5. For $y = f(x) = 1/x^2$, $x = 1$, and $\Delta x = 0.5$.

6. For $y = f(x) = 1/x$, $x = 1$, and $\Delta x = 0.2$.

7. For $y = f(x) = 3x - 1$, $x = 4$, and $\Delta x = 2$.

8. For $y = f(x) = 2x - 3$, $x = 8$, and $\Delta x = 0.5$.

9. For the total-cost function

$$C(x) = 0.01x^2 + 0.6x + 30,$$

find ΔC and $C'(x)$ when $x = 70$ and $\Delta x = 1$.

10. For the total-cost function

$$C(x) = 0.01x^2 + 1.6x + 100,$$

find ΔC and $C'(x)$ when $x = 80$ and $\Delta x = 1$.

11. For the total-revenue function

$$R(x) = 2x,$$

find ΔR and $R'(x)$ when $x = 70$ and $\Delta x = 1$.

12. For the total-revenue function

$$R(x) = 3x,$$

find ΔR and $R'(x)$ when $x = 80$ and $\Delta x = 1$.

13. a) Using $C(x)$ of Exercise 9 and $R(x)$ of Exercise 11, find the total profit $P(x)$.

   b) Find ΔP and $P'(x)$ when $x = 70$ and $\Delta x = 1$.

14. a) Using $C(x)$ of Exercise 10 and $R(x)$ of Exercise 12, find the total profit $P(x)$.

   b) Find ΔP and $P'(x)$ when $x = 80$ and $\Delta x = 1$.

Approximate using $\Delta y \approx f'(x)\ \Delta x$.

15. $\sqrt{19}$ **16.** $\sqrt{10}$ **17.** $\sqrt{102}$

18. $\sqrt{103}$ **19.** $\sqrt[3]{10}$ **20.** $\sqrt[3]{28}$

Find dy.

21. $y = (2x^3 + 1)^{3/2}$ **22.** $y = x^3(2x+5)^2$

23. $y = \sqrt[5]{x + 27}$ **24.** $y = \dfrac{x^3 + x + 2}{x^2 + 3}$

25. $y = x^4 - 2x^3 + 5x^2 + 3x - 4$

26. $y = (7 - x)^8$

27. In Exercise 25, find dy when $x = 2$ and $dx = 0.1$.

28. In Exercise 26, find dy when $x = 1$ and $dx = 0.01$.

APPLICATIONS

◆ **Business and Economics**

29. *Average cost.* The average cost of a company to produce x units of a product is given by the function

$$A(x) = \frac{13x + 100}{x}.$$

By approximately how much does the average cost change as production goes from 100 units to 101 units?

30. *Supply.* A supply function for a certain product is given by

$$S(p) = 0.08p^3 + 2p^2 + 10p + 11.$$

Approximately how many more units will a seller supply when the price changes from $18.00 per unit to $18.20 per unit?

31. *Advertising.* A firm estimates that it will sell N units of a product after spending a dollars on advertising, where

$$N(a) = -a^2 + 300a + 6,$$

and a is in thousands of dollars. Approximately how many more products will a company sell by increasing its advertising expenditure from $100 thousand to $101 thousand?

◆ **Life and Physical Sciences**

32. *Healing wound.* The circular area of a healing wound is given by

$$A = \pi r^2,$$

where r is the radius, in centimeters. By approximately how much does the area decrease when the radius is decreased from 2 cm to 1.9 cm? Use 3.14 for π.

33. *Tumor growth.* The spherical volume of a tumor is given by

$$V = \tfrac{4}{3}\pi r^3,$$

where r is the radius, in centimeters. By approximately how much does the volume increase when the radius is increased from 1 cm to 1.2 cm? Use 3.14 for π.

34. *Medical dosage.* The function

$$N(t) = \frac{0.8t + 1000}{5t + 4}$$

gives the bodily concentration $N(t)$, in parts per million, of a dosage of medication after time t, in hours. By approximately how much does the concentration change as time changes from 2.8 hr to 2.9 hr?

◆ **General Interest**

35. Suppose a rope surrounds the earth at the equator. The rope is lengthened by 10 ft. By about how much is the rope raised above the earth?

SYNTHESIS

36. ◈ Look up the idea of a differential in a book on the history of mathematics. Write a short paragraph.

37. ◈ Explain the uses of the differential.

3.7 Implicit Differentiation and Related Rates*

OBJECTIVES

• Differentiate implicitly.
• Solve related-rate problems.

Implicit Differentiation

Consider the equation

$$y^3 = x.$$

---

*This section can be omitted without loss of continuity.

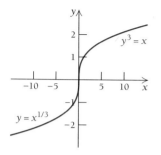

This equation *implies* that y is a function of x, for if we solve for y, we get

$$y = \sqrt[3]{x}$$
$$= x^{1/3}.$$

We know from our work in this chapter that

$$\frac{dy}{dx} = \frac{1}{3}x^{-2/3}. \tag{1}$$

A method known as **implicit differentiation** allows us to find dy/dx *without* solving for y. We use the Chain Rule, treating y as a function of x. We use the Extended Power Rule and differentiate both sides of

$$y^3 = x$$

with respect to x:

$$\frac{d}{dx}y^3 = \frac{d}{dx}x.$$

The derivative on the left side is found using the Extended Power Rule:

$$3y^2\frac{dy}{dx} = 1.$$

Then

$$\frac{dy}{dx} = \frac{1}{3y^2}, \quad \text{or} \quad \frac{1}{3}y^{-2}.$$

We can show that this indeed gives us the same answer as Eq. (1) by replacing y by $x^{1/3}$:

$$\frac{dy}{dx} = \frac{1}{3}y^{-2} = \frac{1}{3}(x^{1/3})^{-2} = \frac{1}{3}x^{-2/3}.$$

Often, it is difficult or impossible to solve for y, obtaining an explicit expression in terms of x. For example, the equation

$$y^3 + x^2y^5 - x^4 = 27$$

determines y as a function of x, but it would be difficult to solve for y. We can nevertheless find a formula for the derivative of y *without* solving for y. This involves computing $\frac{d}{dx}y^n$ for various integers n, and hence involves the Extended Power Rule in the form

$$\frac{d}{dx}y^n = ny^{n-1} \cdot \frac{dy}{dx}.$$

EXAMPLE 1   For

$$y^3 + x^2 y^5 - x^4 = 27:$$

a) Find dy/dx using implicit differentiation.

b) Find the slope of the tangent line to the curve at the point $(0, 3)$.

Solution

a) We differentiate the term $x^2 y^5$ using the Product Rule. Note that whenever an expression involving y is differentiated, dy/dx must be a factor of the answer. When an expression involving just x is differentiated, there is no factor dy/dx.

$$\frac{d}{dx}(y^3 + x^2 y^5 - x^4) = \frac{d}{dx}(27)$$

$$\frac{d}{dx}y^3 + \frac{d}{dx}x^2 y^5 - \frac{d}{dx}x^4 = 0$$

$$3y^2 \cdot \frac{dy}{dx} + x^2 \cdot 5y^4 \cdot \frac{dy}{dx} + 2x \cdot y^5 - 4x^3 = 0.$$

Then

$$3y^2 \cdot \frac{dy}{dx} + 5x^2 y^4 \cdot \frac{dy}{dx} = 4x^3 - 2xy^5 \qquad \text{Only those terms involving } dy/dx \text{ should appear on one side.}$$

$$(3y^2 + 5x^2 y^4)\frac{dy}{dx} = 4x^3 - 2xy^5$$

$$\frac{dy}{dx} = \frac{4x^3 - 2xy^5}{3y^2 + 5x^2 y^4} \qquad \text{Solving for } dy/dx. \text{ Leave the answer in terms of } x \text{ and } y.$$

b) To find the slope of the tangent line to the curve at $(0, 3)$, we replace x by 0 and y by 3:

$$\frac{dy}{dx} = \frac{4 \cdot 0^3 - 2 \cdot 0 \cdot 3^5}{3 \cdot 3^2 + 5 \cdot 0^2 \cdot 3^4} = 0.$$

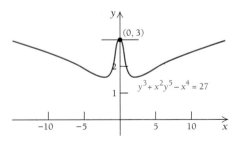

The demand function for a product (see Sections 1.5 and 2.6) is often given implicitly.

EXAMPLE 2 For the following demand equation, differentiate implicitly to find dp/dx:

$$x = \sqrt{200 - p^3}.$$

Solution

$$\frac{d}{dx} x = \frac{d}{dx} \sqrt{200 - p^3}$$

$$1 = \frac{1}{2} (200 - p^3)^{-1/2} \cdot (-3p^2) \cdot \frac{dp}{dx}$$

$$1 = \frac{-3p^2}{2\sqrt{200 - p^3}} \cdot \frac{dp}{dx}$$

$$\frac{2\sqrt{200 - p^3}}{-3p^2} = \frac{dp}{dx} \qquad \blacklozenge$$

Related Rates

Suppose that y is a function of x, say

$$y = f(x),$$

and x varies with time t (as a function of time t). Since y depends on x and x depends on t, y also depends on t. That is, y is also a function of time t. The Chain Rule gives the following:

$$\frac{dy}{dt} = \frac{dy}{dx} \cdot \frac{dx}{dt}.$$

Thus the rate of change of y is *related* to the rate of change of x. Let us see how this comes up in problems. It helps to keep in mind that any variable can be thought of as a function of time t, even though a specific expression in terms of t may not be given.

EXAMPLE 3 *Business: Service area.* A restaurant supplier services the restaurants in a circular area in such a way that the radius r is increasing at the rate of 2 mi per year at the moment when r goes through the value $r = 5$ mi. At that moment, how fast is the area increasing?

Solution The area A and the radius r are always related by the equation for the area of a circle:

$$A = \pi r^2.$$

We take the derivative of both sides with respect to t:

$$\frac{dA}{dt} = 2\pi r \cdot \frac{dr}{dt}.$$

At the moment in question, $dr/dt = 2$ mi/yr (miles per year) and $r = 5$ mi, so

$$\frac{dA}{dt} = 2\pi(5 \text{ mi})\left(2\frac{\text{mi}}{\text{yr}}\right)$$

$$= 20\pi \frac{\text{mi}^2}{\text{yr}}$$

$$\approx 63 \text{ square miles per year.} \qquad \blacklozenge$$

EXAMPLE 4   *Business: Rates of change of revenue, cost, and profit.* For a company making stereos, the total revenue from the sale of x stereos is given by

$$R(x) = 1000x - x^2,$$

and the total cost is given by

$$C(x) = 3000 + 20x.$$

Suppose that the company is producing and selling stereos at the rate of 10 stereos per day at the moment when the 400th stereo is produced. At that same moment, what is the rate of change of (a) total revenue? (b) total cost? (c) total profit?

Solution

a) $\dfrac{dR}{dt} = 1000 \cdot \dfrac{dx}{dt} - 2x \cdot \dfrac{dx}{dt}$ Differentiating with respect to time

$\qquad\quad = 1000 \cdot 10 - 2(400)10$ Substituting 10 for dx/dt and 400 for x

$\qquad\quad = \$2000 \text{ per day}$

b) $\dfrac{dC}{dt} = 20 \cdot \dfrac{dx}{dt}$ Differentiating with respect to time

$\qquad\quad = 20(10)$

$\qquad\quad = \$200 \text{ per day}$

c) Since $P = R - C$,

$$\frac{dP}{dt} = \frac{dR}{dt} - \frac{dC}{dt}$$

$$= \$2000 \text{ per day} - \$200 \text{ per day}$$

$$= \$1800 \text{ per day.} \qquad \blacklozenge$$

3.7 Exercise Set

Differentiate implicitly to find dy/dx. Then find the slope of the curve at the given point.

1. $xy - x + 2y = 3;$ $\left(-5, \dfrac{2}{3}\right)$

2. $xy + y^2 - 2x = 0;$ $(1, -2)$

3. $x^2 + y^2 = 1;$ $\left(\dfrac{1}{2}, \dfrac{\sqrt{3}}{2}\right)$

4. $x^2 - y^2 = 1;$ $(\sqrt{3}, \sqrt{2})$

5. $x^2y - 2x^3 - y^3 + 1 = 0;$ $(2, -3)$

6. $4x^3 - y^4 - 3y + 5x + 1 = 0;$ $(1, -2)$

Differentiate implicitly to find dy/dx.

7. $2xy + 3 = 0$ 8. $x^2 + 2xy = 3y^2$

9. $x^2 - y^2 = 16$ 10. $x^2 + y^2 = 25$

11. $y^5 = x^3$ 12. $y^3 = x^5$

13. $x^2y^3 + x^3y^4 = 11$ 14. $x^3y^2 - x^5y^3 = -19$

For the given demand equation, differentiate implicitly to find dp/dx.

15. $p^2 + p + 2x = 40$

16. $xp^3 = 24$

17. $(p + 4)(x + 3) = 48$

18. $1000 - 300p + 25p^2 = x$

19. Two variable quantities A and B are found to be related by the equation

$$A^3 + B^3 = 9.$$

What is the rate of change dA/dt at the moment when $A = 2$ and $dB/dt = 3$?

20. Two variable quantities G and H, nonnegative, are found to be related by the equation

$$G^2 + H^2 = 25.$$

What is the rate of change dH/dt when $dG/dt = 3$ and $G = 0$? $G = 1$? $G = 3$?

APPLICATIONS

◆ **Business and Economics**

Rates of change of total revenue, cost, and profit. Find the rates of change of total revenue, cost, and profit for each of the following.

21. $R(x) = 50x - 0.5x^2,$
    $C(x) = 4x + 10,$
    when $x = 30$ and $dx/dt = 20$ units per day

22. $R(x) = 50x - 0.5x^2,$
    $C(x) = 10x + 3,$
    when $x = 10$ and $dx/dt = 5$ units per day

23. $R(x) = 2x,$
    $C(x) = 0.01x^2 + 0.6x + 30,$
    when $x = 20$ and $dx/dt = 8$ units per day

24. $R(x) = 280x - 0.4x^2,$
    $C(x) = 5000 + 0.6x^2,$
    when $x = 200$ and $dx/dt = 300$ units per day

◆ **Life and Physical Sciences**

25. **Rate of change of a tumor.** The volume of a tumor is given by

$$V = \tfrac{4}{3}\pi r^3.$$

The radius is increasing at the rate of 0.03 centimeter per day (cm/day) at the moment when $r = 1.2$ cm. How fast is the volume changing at that moment?

26. **Rate of change of a healing wound.** The area of a healing wound is given by

$$A = \pi r^2.$$

The radius is decreasing at the rate of 1 millimeter per day (-1 mm/day) at the moment when $r = 25$ mm. How fast is the area decreasing at that moment?

Poiseuille's Law. The flow of blood in a blood vessel is faster toward the center of the vessel and slower toward the outside. The speed of the blood V is given by

$$V = \frac{p}{4Lv}(R^2 - r^2),$$

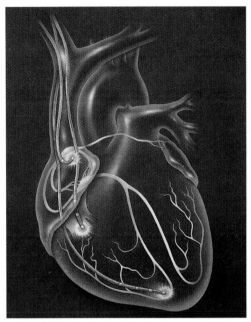

The flow of blood in a blood vessel can be modeled by Poiseuille's Law.

far from the heart. Suppose a blood
at a rate of

$$\frac{dR}{dt} = -0.0015 \text{ mm/min}$$

at a place in the blood vessel where the radius $R = 0.0075$ mm. Find the rate of change dV/dt at that location.

28. Assume that r is a constant as well as p, L, and v.

a) Find the rate of change dV/dt in terms of R and dR/dt when $L = 1$ mm, $p = 100$, and $v = 0.05$.

b) When shoveling snow in cold air, a person with a history of heart trouble can develop angina (chest pains) due to contracting blood vessels. To counteract this, he or she may take a nitroglycerin tablet, which dilates the blood vessels. Suppose that after a nitroglycerin tablet is taken, a blood vessel dilates at a rate of

$$\frac{dR}{dt} = 0.0025 \text{ mm/min}$$

at a place in the blood vessel where the radius $R = 0.02$ mm. Find the rate of change dV/dt.

where R = the radius of the blood vessel, r = the distance of the blood from the center of the vessel, and p, L, and v are physical constants related to pressure, length, and viscosity of the blood vessels, respectively. Use this formula for Exercises 27 and 28.

◆ General Interest

29. Two cars start from the same point at the same time. One travels north at 25 mph, and the other travels east at 60 mph. How fast is the distance between them increasing at the end of 1 hr? (Hint: $D^2 = x^2 + y^2$. To find D after 1 hr, solve $D^2 = 25^2 + 60^2$.)

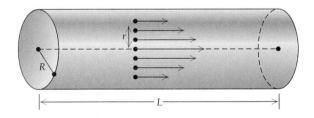

27. Assume that r is a constant as well as p, L, and v.

a) Find the rate of change dV/dt in terms of R and dR/dt when $L = 1$ mm, $p = 100$, and $v = 0.05$.

b) A person goes out into the cold to shovel snow. Cold air has the effect of contracting blood vessels

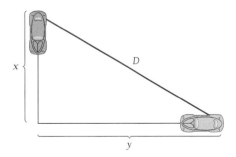

30. A ladder 26 ft long leans against a vertical wall. If the lower end is being moved away from the wall at the rate of 5 ft/sec, how fast is the height of the top

decreasing (this will be a negative rate) when the lower end is 10 ft from the wall?

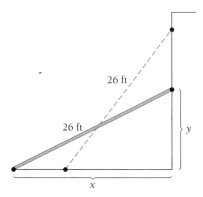

SYNTHESIS

Differentiate implicitly to find dy/dx.

31. $\sqrt{x} + \sqrt{y} = 1$

32. $\dfrac{1}{x^2} + \dfrac{1}{y^2} = 5$

33. $y^3 = \dfrac{x - 1}{x + 1}$

34. $y^2 = \dfrac{x^2 - 1}{x^2 + 1}$

35. $x^{3/2} + y^{2/3} = 1$

36. $(x - y)^3 + (x + y)^3 = x^5 + y^5$

Differentiate implicitly to find dy/dx and d^2y/dx^2.

37. $xy + x - 2y = 4$

38. $y^2 - xy + x^2 = 5$

39. $x^2 - y^2 = 5$

40. $x^3 - y^3 = 8$

41. ◈ Explain the usefulness of implicit differentiation.

42. ◈ Look up the word "implicit" in a dictionary. Explain how that definition can be related to the concept of a function that is defined "implicitly."

TECHNOLOGY CONNECTION

Some graphers are able to graph equations whether or not they have been solved for y. You simply enter the equation. For others, it may be necessary to solve for y or do so in parts. Use your grapher to graph each equation.

43. $x^2 + y^2 = 4$ (*Note*: Depending on the grapher, you may need to sketch the graph in two parts:

$$y = \sqrt{4 - x^2} \quad \text{and} \quad y = -\sqrt{4 - x^2}.$$

Then graph the tangent line to the graph at the point $(-1, \sqrt{3})$.

44. $x^4 = y^2 + x^6$

Then graph the tangent line to the graph at the point $(-0.8, 0.384)$.

45. $y^4 = y^2 - x^2$

46. $x^3 = y^2(2 - x)$

47. $y^2 = x^3$

48. $y^2 + 4xy^2 - y^4 = x^4 - 4x^3 + 3x^2 + 2x^2y^2$

---

3 Chapter Summary and Review

TERMS TO KNOW

REVIEW EXERCISES

The review exercises are for test preparation. They can also be used as a lengthened practice test. Answers are at the back of the book. The answers also contain bracketed section references, which tell you where to restudy if your answer is incorrect.

Find the relative extrema of the function. List your answers in terms of ordered pairs. Then sketch a graph of the function.

1. $f(x) = 3 - 2x - x^2$

2. $f(x) = x^4 - 2x^2 + 3$

3. $f(x) = \dfrac{-8x}{x^2 + 1}$

4. $f(x) = 4 + (x - 1)^3$

5. $f(x) = x^3 + x^2 - x + 3$

6. $f(x) = 3x^{2/3}$

7. $f(x) = 2x^3 - 3x^2 - 12x + 10$

8. $f(x) = x^3 - 3x + 2$

Sketch a graph of the function.

9. $f(x) = \dfrac{-4}{x - 3}$

10. $f(x) = \dfrac{1}{2}x + \dfrac{1}{x}$

11. $f(x) = \dfrac{x^2 - 2x + 2}{x - 1}$

12. $f(x) = \dfrac{x}{x - 2}$

Find the absolute maximum and minimum values of the function, if they exist, over the indicated interval. Where no interval is specified, use the real line.

13. $f(x) = \dfrac{1}{3}x^3 + 3x^2 + 9x + 2$

14. $f(x) = x^2 - 10x + 8$; $[-2, 6]$

15. $f(x) = 4x^3 - 6x^2 - 24x + 5$; $[-2, 3]$

16. $f(x) = 5x - 7$; $[-1, 1]$

17. $f(x) = x^2 - \dfrac{2}{x}$; $(-\infty, 0)$

18. $f(x) = 5x - 7$

19. $f(x) = 3x^4 + 2x^3 - 3x^2 + 1$; $[-3, 0]$

20. $f(x) = 5x^2 + \dfrac{5}{x^2}$; $(0, \infty)$

21. $f(x) = 5x^4 - x^5$; $[-1, 5]$

22. $f(x) = -x^2 + 5x + 7$

23. Of all numbers whose sum is 60, find the two that have the maximum product.

24. Find the minimum value of $Q = x^2 - 2y^2$, where $x - 2y = 1$.

25. *Business: Maximizing profit.* If $R(x) = 52x - 0.5x^2$ and $C(x) = 22x - 1$, find the maximum profit and the number of units that must be produced and sold in order to yield this maximum profit.

26. A rectangular box with a square base and a cover is to contain 2500 ft^3. If the cost per square foot for the bottom is \$2, for the top is \$3, and for the sides is \$1, what should the dimensions be in order to minimize the cost?

27. *Business: Minimizing inventory cost.* A store in California sells 360 multispeed bicycles per year. It costs \$8 to store one bicycle for one year. To reorder, there is a fixed cost of \$10, plus \$2 for each bicycle. How many times per year should the store order bicycles, and in what lot size, in order to minimize inventory costs?

Given $y = f(x) = x^3 - x$.

28. Find Δy and $f'(x)\, \Delta x$, given that $x = 3$ and $\Delta x = -0.5$.

29. a) Find dy.
    b) Find dy when $x = 2$ and $dx = 0.01$.

30. Approximate $\sqrt{69}$ using $\Delta y \approx f'(x)\, \Delta x$.

31. Differentiate the following implicitly to find dy/dx. Then find the slope of the curve at the given point.

$$2x^3 + 2y^3 = -9xy;\quad (-1, -2)$$

32. A ladder 25 ft long leans against a vertical wall. If the lower end is being moved away from the wall at the rate of 6 ft/sec, how fast is the height of the top decreasing when the lower end is 7 ft from the wall?

33. *Business: Total revenue, cost, and profit.* Find the rates of change of total revenue, cost, and profit for

$$R(x) = 120x - 0.5x^2 \quad \text{and} \quad C(x) = 15x + 6,$$

when $x = 100$ and $dx/dt = 30$ units per day.

SYNTHESIS

34. Find the absolute maximum and minimum values, if they exist, over the indicated interval.

$$f(x) = (x - 3)^{2/5};\quad (-\infty, \infty)$$

35. Differentiate implicitly to find dy/dx:

$$(x - y)^4 + (x + y)^4 = x^6 + y^6.$$

36. Find the relative maxima and minima of

$$y = x^4 - 8x^3 - 270x^2.$$

TECHNOLOGY CONNECTION

Use your grapher to estimate the relative extrema of each function. Use the trace feature to obtain your estimate.

37. $f(x) = 3.8x^5 - 18.6x^3$

38. $f(x) = \sqrt[3]{|9 - x^2|} - 1$

3 Chapter Test

Find the relative extrema of the function. List your answers in terms of ordered pairs. Then sketch a graph of the function.

1. $f(x) = x^2 - 4x - 5$

2. $f(x) = 2x^4 - 4x^2 + 1$

3. $f(x) = (x - 2)^{2/3} - 4$

4. $f(x) = \dfrac{16}{x^2 + 4}$

5. $f(x) = x^3 + x^2 - x + 1$

6. $f(x) = 4 + 3x - x^3$

7. $f(x) = (x + 2)^3$

8. $f(x) = x\sqrt{9 - x^2}$

Sketch a graph of the function.

9. $f(x) = \dfrac{2}{x - 1}$

10. $f(x) = \dfrac{-8}{x^2 - 4}$

11. $f(x) = \dfrac{x^2 - 1}{x}$

12. $f(x) = \dfrac{x + 2}{x - 3}$

Find the absolute maximum and minimum values of the function, if they exist, over the indicated interval. Where no interval is specified, use the real line.

13. $f(x) = x(6 - x)$

14. $f(x) = x^3 + x^2 - x + 1$; $\left[-2, \frac{1}{2}\right]$

15. $f(x) = -x^2 + 8.6x + 10$

16. $f(x) = -2x + 5$; $[-1, 1]$

17. $f(x) = -2x + 5$

18. $f(x) = 3x^2 - x - 1$

19. $f(x) = x^2 + \dfrac{128}{x}$; $(0, \infty)$

20. Of all numbers whose difference is 8, find the two that have the minimum product.

21. Minimize $Q = x^2 + y^2$, where $x - y = 10$.

22. Business: Maximum profit. Find the maximum profit and the number of units that must be produced and sold in order to yield the maximum profit.

$$R(x) = x^2 + 110x + 60,$$
$$C(x) = 1.1x^2 + 10x + 80$$

23. From a thin piece of cardboard 60 in. by 60 in., square corners are cut out so the sides can be folded up to make a box. What dimensions will yield a box of maximum volume? What is the maximum volume?

24. Business: Minimizing inventory costs. A sporting goods store sells 1225 tennis rackets per year. It costs $2 to store one tennis racket for one year. To reorder, there is a fixed cost of $1, plus $0.50 for each tennis racket. How many times per year should the sporting goods store order tennis rackets, and in what lot size, in order to minimize inventory costs?

25. For $y = f(x) = x^2 - 3$, $x = 5$, and $\Delta x = 0.1$, find Δy and $f'(x)\, \Delta x$.

26. Approximate $\sqrt{104}$ using $\Delta y \approx f'(x)\, \Delta x$.

27. For $y = \sqrt{x^2 + 3}$, (a) find dy and (b) find dy when $x = 2$ and $dx = 0.01$.

28. Differentiate the following implicitly to find dy/dx.

Then find the slope of the curve at the given point.

$$x^3 + y^3 = 9; \quad (1, 2)$$

29. A board 13 ft long leans against a vertical wall. If the lower end is being moved away from the wall at the rate of 0.4 ft/sec, how fast is the upper end coming down when the lower end is 12 ft from the wall?

SYNTHESIS

30. Find the absolute maximum and minimum values of the function, if they exist, over the indicated interval.

$$f(x) = \frac{x^2}{1 + x^3}; \quad [0, \infty)$$

31. *Business: Minimizing average cost.* The total cost of producing x units of a product is given by

$$C(x) = 100x + 100\sqrt{x} + \frac{\sqrt{x^3}}{100}.$$

a) Find the average cost $A(x)$.
b) Find the minimum value of $A(x)$.

 TECHNOLOGY CONNECTION

32. Use your grapher to estimate the relative extrema of the function. Use the trace feature to obtain your estimate.

$$f(x) = 5x^3 - 30x^2 + 45x + 5\sqrt{x}$$

 EXTENDED TECHNOLOGY APPLICATION

Maximum Sustainable Harvest

In certain situations, biologists are able to determine what is called a *reproduction curve*. This is a function

$$y = f(P)$$

such that if P is the population at a certain time t, then the population one year later, at time $t + 1$, is $f(P)$. Such a curve is shown below.

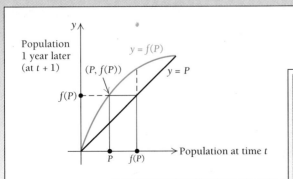

The line $y = P$ is significant for two reasons. First, if $y = P$ is a description of f, then we know that the population stays the same from year to year. But the graph of f above lies mostly above the line, indicating that, in this case, the population is increasing.

Too many deer in a forest can deplete the food supply and eventually cause the population to decrease for lack of food. Often in such cases and with some controversy, hunters are allowed to "harvest" some of the deer. Then with a greater food supply, the remaining deer population might actually prosper and increase.

We know that a population P will grow to a population $f(P)$ in one year. If this were a population of fur-bearing animals and the population were increasing, then one could "harvest" the amount

$$f(P) - P$$

each year without depleting the initial population P. If the population were remaining the same or decreasing, then such a harvest would deplete the population.

Suppose we want to know the value of P_0 that would allow the harvest to be the largest. If we could determine that P_0, then we could let the population grow until it reached that level, and then begin harvesting year after year the amount $f(P_0) - P_0$.

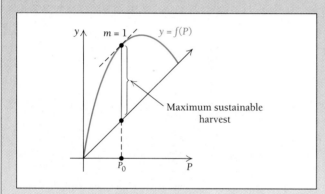

Let the harvest function H be given by

$$H(P) = f(P) - P.$$

Then

$$H'(P) = f'(P) - 1.$$

Now, if we assume that $H'(P)$ exists for all values of P and that there is only one critical point, it follows that the *maximum sustainable harvest* occurs at that value P_0 such that

$$H'(P_0) = f'(P_0) - 1 = 0$$

and

$$H''(P_0) = f''(P_0) < 0.$$

Or, equivalently, we have the following.

THEOREM

The *maximum sustainable harvest* occurs at P_0 such that

$$f'(P_0) = 1 \quad \text{and} \quad f''(P_0) < 0,$$

and is given by

$$H(P_0) = f(P_0) - P_0.$$

EXERCISES

For each reproduction curve in Exercises 1–3, do the following.

a) Graph the reproduction function, the function $y = P$, and the harvest function using the same set of axes or viewing window.
b) Find the population at which the maximum sustainable harvest occurs. Use both a graphical solution and a calculus solution.
c) Find the maximum sustainable harvest.

1. $f(P) = P(10 - P)$, where P is measured in thousands.

2. $f(P) = -0.025P^2 + 4P$, where P is measured in thousands. This is the reproduction curve in the Hudson bay area for the snowshoe hare, a fur-bearing animal.

b) Find the population at which the maximum sustainable harvest occurs. Use just a graphical solution.
c) Find the maximum sustainable harvest.

4. $f(P) = 40\sqrt{P}$, where P is measured in thousands. Assume that this is the reproduction curve for the brown trout population in a large lake.

5. $f(P) = 0.237P\sqrt{2000 - P^2}$, where P is measured in thousands.

6. The following table lists data regarding the reproduction of a certain animal.

Population, P (in thousands)	Population, $f(P)$, one year later
10	9.7
20	23.1
30	37.4
40	46.2
50	42.6

3. $f(P) = -0.01P^2 + 2P$, where P is measured in thousands. This is the reproduction curve in the Hudson Bay area for the lynx, a fur-bearing animal.

For each reproduction curve in Exercises 4 and 5, do the following.

a) Graph the reproduction function, the function $y = P$, and the harvest function using the same set of axes or viewing window.

a) Use the regression feature on your grapher and fit a cubic polynomial to these data.
b) Graph the reproduction function, the function $y = P$, and the harvest function using the same set of axes or viewing window.
c) Find the population at which the maximum sustainable harvest occurs. Use just a graphical solution.

4 Exponential and Logarithmic Functions

INTRODUCTION

In this chapter, we will consider two kinds of functions that are closely related: *exponential functions* and *logarithmic functions*. We will also learn to find derivatives of such functions. Both are rich in application, with exponential functions applying particularly to problems of population growth.

AN APPLICATION

In 1970, the average salary of major-league baseball players was $29,303. By 1994, the average salary had grown to $1,200,000. If we assume exponential growth, what will the average salary be in 2000?

THE MATHEMATICS

The average salary S is given by

$$\underbrace{S(t) = \$29,303e^{0.155t}}_{\uparrow},$$

This is an *exponential function*.

where t = the number of years since 1970.

This problem appears as Exercise 17 in Exercise Set 4.3.

4.1 Exponential Functions

OBJECTIVES

- Graph exponential functions.
- Differentiate exponential functions.

Graphs of Exponential Functions

No doubt you are aware of the high rise in the salaries of professional athletes. The following graph, like one that might appear in a newspaper, shows the average salary of major-league baseball players for various years since 1967. The average salary in 1994 was $1,200,000. A curve drawn along the graph would approximate the graph of an exponential function. We now consider such graphs.

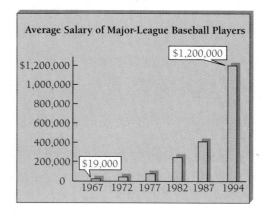

In Chapter 1, we reviewed definitions of such expressions as a^x, where x is a rational number. For example,

$$a^{2.34}, \quad \text{or} \quad a^{234/100},$$

means "raise a to the 234th power and then take the 100th root."

What about expressions with irrational exponents, such as $2^{\sqrt{2}}$, 2^π, or $2^{-\sqrt{3}}$? An irrational number is a number named by an infinite, nonrepeating decimal. Let us consider 2^π. We know that π is irrational with infinite, nonrepeating decimal expansion:

$$3.141592653\ldots.$$

This means that π is approached as a limit by the rational numbers

$$3, \ 3.1, \ 3.14, \ 3.141, \ 3.1415, \ \ldots,$$

so it seems reasonable that 2^π should be approached as a limit by the rational powers

$$2^3, \ 2^{3.1}, \ 2^{3.14}, \ 2^{3.141}, \ 2^{3.1415}, \ \ldots.$$

Estimating each power with a calculator, we get the following:

$$8, \ 8.574188, \ 8.815241, \ 8.821353, \ 8.824411, \ \ldots.$$

In general, a^x is approximated by the values of a^r for rational numbers r near x; a^x is the limit of a^r as r approaches x through rational values. In summary, for $a > 0$, the definition of a^x for rational numbers x can be extended to arbitrary real numbers x in such a way that the usual laws of exponents, such as

$$a^x \cdot a^y = a^{x+y}, \qquad a^x \div a^y = a^{x-y}, \qquad (a^x)^y = a^{xy}, \quad \text{and} \quad a^{-x} = \frac{1}{a^x},$$

still hold. Moreover, the function so obtained, $f(x) = a^x$, is continuous.

DEFINITION

An *exponential function f* is given by

$$f(x) = a^x,$$

where x is any real number, $a > 0$, and $a \neq 1$. The number a is called the *base*.

The following are examples of exponential functions:

$$f(x) = 2^x, \qquad f(x) = \left(\tfrac{1}{2}\right)^x, \qquad f(x) = (0.4)^x.$$

Note that in contrast to power functions like $y = x^2$ or $y = x^3$, the variable in an exponential function is in the exponent, not the base. Exponential functions have extensive application. Let us consider their graphs.

EXAMPLE 1   Graph: $y = f(x) = 2^x$.

Solution

a) First we find some function values.

x	0	$\frac{1}{2}$	1	2	3	-1	-2
$y = f(x)$ (or 2^x)	1	1.4	2	4	8	$\frac{1}{2}$	$\frac{1}{4}$

Note: For:

$$x = 0, \quad y = 2^0 = 1;$$
$$x = \tfrac{1}{2}, \quad y = 2^{1/2} = \sqrt{2} \approx 1.4;$$
$$x = 1, \quad y = 2^1 = 2;$$
$$x = 2, \quad y = 2^2 = 4;$$
$$x = 3, \quad y = 2^3 = 8;$$
$$x = -1, \quad y = 2^{-1} = \tfrac{1}{2};$$
$$x = -2, \quad y = 2^{-2} = \frac{1}{2^2} = \tfrac{1}{4}.$$

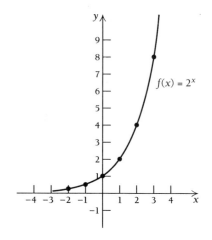

b) Next we plot the points and connect them with a smooth curve. The graph is continuous, increasing, and concave up.   ◆

EXAMPLE 2 Graph: $y = f(x) = \left(\tfrac{1}{2}\right)^x$.

Solution

a) We first find some function values. Before we do so, note that

$$y = f(x) = \left(\tfrac{1}{2}\right)^x = (2^{-1})^x = 2^{-x}.$$

This will ease our work.

x	0	$\tfrac{1}{2}$	1	2	-1	-2	-3
y	1	0.7	$\tfrac{1}{2}$	$\tfrac{1}{4}$	2	4	8

Note: For:

$$x = 0, \quad y = 2^{-0} = 1;$$
$$x = \frac{1}{2}, \quad y = 2^{-1/2} = \frac{1}{2^{1/2}}$$
$$= \frac{1}{\sqrt{2}} \approx \frac{1}{1.4} \approx 0.7;$$
$$x = 1, \quad y = 2^{-1} = \tfrac{1}{2};$$
$$x = 2, \quad y = 2^{-2} = \tfrac{1}{4};$$
$$x = -1, \quad y = 2^{-(-1)} = 2;$$
$$x = -2, \quad y = 2^{-(-2)} = 4;$$
$$x = -3, \quad y = 2^{-(-3)} = 8.$$

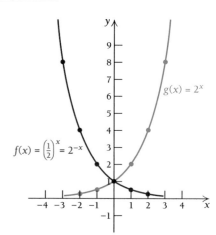

b) We plot these points and connect them with a smooth curve, as shown by the red curve in the figure. The graph is continuous, decreasing, and concave up. The graph of $g(x) = 2^x$, the blue curve, is shown for comparison.   ◆

The following are some properties of the exponential function.

TECHNOLOGY
CONNECTION

Check the graphs of the functions in Examples 1 and 2. Then graph

$$f(x) = 3^x \quad \text{and} \quad g(x) = \left(\tfrac{2}{3}\right)^x$$

and look for patterns.

1. The function $f(x) = a^x$, where $a > 1$, is a positive, increasing, continuous function. As x gets smaller, a^x approaches 0. The graph is concave up, as shown below. (If you studied Section 3.3, you also know that the x-axis is an asymptote.)

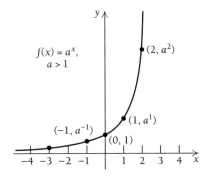

2. The function $f(x) = a^x$, where $0 < a < 1$, is a positive, decreasing, continuous function. As x gets larger, a^x approaches 0. The graph is concave up, as shown below. (The x-axis is an asymptote.)

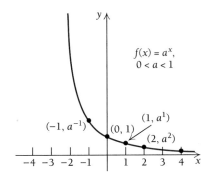

When $a = 1$, $f(x) = a^x = 1^x = 1$, and is a constant function. This is why we do not allow 1 to be the base of an exponential function.

The Derivative of a^x; the Number e

Let us consider finding the derivative of the exponential function

$$f(x) = a^x.$$

The derivative is given by

$$f'(x) = \lim_{h \to 0} \frac{f(x + h) - f(x)}{h} \qquad \text{Definition of the derivative}$$

$$= \lim_{h \to 0} \frac{a^{x+h} - a^x}{h} \qquad \text{Substituting } a^{x+h} \text{ for } f(x + h) \text{ and } a^x \text{ for } f(x)$$

$$= \lim_{h \to 0} \frac{a^x \cdot a^h - a^x \cdot 1}{h}$$

$$= \lim_{h \to 0} a^x \cdot \left(\frac{a^h - 1}{h} \right)$$

$$= a^x \cdot \lim_{h \to 0} \frac{a^h - 1}{h}. \qquad \text{Since the variable is } h \text{ and } h \to 0, \text{ we treat } a^x \text{ as a constant, and the limit of a constant times a function is the constant times the limit.}$$

We get

$$f'(x) = a^x \cdot \lim_{h \to 0} \frac{a^h - 1}{h}. \tag{1}$$

In particular, for $g(x) = 2^x$,

$$g'(x) = 2^x \cdot \lim_{h \to 0} \frac{2^h - 1}{h}.$$

Note that the limit does not depend on the value of x at which we are evaluating the derivative. In order for $g'(x)$ to exist, we must determine whether

$$\lim_{h \to 0} \frac{2^h - 1}{h} \quad \text{exists.}$$

h	$\dfrac{2^h - 1}{h}$
0.5	0.8284
0.25	0.7568
0.175	0.7369
0.0625	0.7084
0.03125	0.7007
0.00111	0.6934
0.000001	0.6931

Let us investigate this question.

We choose a sequence of numbers h approaching 0 and compute $(2^h - 1)/h$, listing the results in a table, as shown at left. It seems reasonable to assume that $(2^h - 1)/h$ has a limit as h approaches 0 and that its approximate value is 0.7; thus,

$$g'(x) \approx (0.7)2^x.$$

In other words, the derivative is a constant times the function value 2^x. Similarly, for $t(x) = 3^x$,

$$t'(x) = 3^x \cdot \lim_{h \to 0} \frac{3^h - 1}{h}.$$

h	$\dfrac{3^h - 1}{h}$
0.5	1.4641
0.25	1.2643
0.175	1.2113
0.0625	1.1372
0.03125	1.1177
0.00111	1.0993
0.000001	1.0986

Again, we can find an approximation for the limit that does not depend on the value of x at which we are evaluating the derivative. Consider the table shown at left. Again, it seems reasonable to assume that $(3^h - 1)/h$ has a limit as h approaches 0 and that its approximate value is 1.1; thus

$$t'(x) \approx (1.1)3^x.$$

In other words, the derivative is a constant times the function value 3^x.

The graphs of $g(x) = 2^x$ and $g'(x) \approx (0.7)2^x$ follow. Note that the graph of g' lies *below* the graph of g. Also shown are the graphs of $t(x) = 3^x$ and $t'(x) \approx (1.1)3^x$. Note that the graph of t' lies *above* the graph of t.

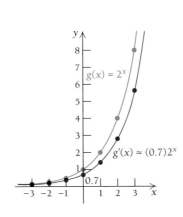

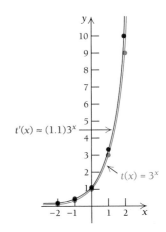

We might expect that there is exactly one base a between 2 and 3 for which a^x and its derivative have the same graph. This conjecture can be proved (though we will not do it here).

Graph $f(x) = e^x$ using the viewing window $[-5, 5, -1, 10]$, with xScl $= 1$ and yScl $= 1$. Trace along the graph to observe ordered pairs (x, y). Then find the value of the first derivative at each point.

Compare the answer to the y-value for the original function $f(x) = e^x$. Repeat this process for three other values of x. What do you observe?

DEFINITION

The number e is the unique positive real number for which

$$\lim_{h \to 0} \frac{e^h - 1}{h} = 1.$$

It follows that for the exponential function $f(x) = e^x$,

$$f'(x) = e^x \cdot \lim_{h \to 0} \frac{e^h - 1}{h} = e^x \cdot 1 = e^x.$$

That is, the derivative of e^x is e^x.

Compound Interest and an Approximation for e

In order to find an approximation for e, we use the compound-interest formula

$$A = P\left(1 + \frac{i}{n}\right)^{nt}$$

that we developed in Chapter 1, where A = the amount that an initial investment P will be worth after t years at interest rate i, compounded n times per year.

Suppose that \$1 is an initial investment at 100% interest for 1 year (though obviously no financial institution would pay this). The formula becomes

$$A = \left(1 + \frac{1}{n}\right)^{n}.$$

Suppose we were to have the compounding periods n increase indefinitely. Let us investigate the behavior of the function A. We obtain the following table of values and graph.

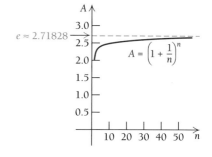

n	$A = \left(1 + \dfrac{1}{n}\right)^{n}$
1	\$2.00000
2	\$2.25000
3	\$2.37037
4	\$2.44141
12	\$2.61304
52	\$2.69260
365	\$2.71457
8,760	\$2.71813
525,600	\$2.71828

The amount in the investment grows at interest compounded continuously and approaches a limit. That limit and an approximation of e are given in the following theorem.

THEOREM 1

$$e = \lim_{n \to \infty} \left(1 + \frac{1}{n}\right)^{n}$$
$$\approx 2.718281828459$$

That is, e is that number which

$$\left(1 + \frac{1}{n}\right)^n$$

approaches as n gets larger without bound.

Finding Derivatives of Functions Involving e

We have established that for the function $f(x) = e^x$, we also have $f'(x) = e^x$, or, simply, the following.

THEOREM 2

$$\frac{d}{dx}e^x = e^x$$

Note that this says that the derivative (the slope of the tangent line) is the same as the function value at any x. Let us find some other derivatives.

EXAMPLE 3

$$\frac{d}{dx}3e^x = 3\frac{d}{dx}e^x = 3e^x \qquad\qquad\blacklozenge$$

EXAMPLE 4

$$\frac{d}{dx}(x^2e^x) = x^2 \cdot e^x + 2x \cdot e^x \qquad \text{By the Product Rule}$$

$$= e^x(x^2 + 2x), \quad \text{or} \quad xe^x(x + 2) \qquad \text{Factoring} \qquad\blacklozenge$$

EXAMPLE 5

$$\frac{d}{dx}\left(\frac{e^x}{x^3}\right) = \frac{x^3 \cdot e^x - 3x^2 \cdot e^x}{x^6} \qquad \text{By the Quotient Rule}$$

$$= \frac{x^2e^x(x - 3)}{x^6} \qquad \text{Factoring}$$

$$= \frac{e^x(x - 3)}{x^4} \qquad \text{Simplifying} \qquad\blacklozenge$$

Suppose we have a more complicated function in the exponent, such as

$$h(x) = e^{x^2 - 5x}.$$

TECHNOLOGY CONNECTION

Check the results of Examples 3–5 using your grapher. This assumes you have a grapher that graphs f and f'. Then differentiate $f(x) = e^x/x^2$ and check your answer using your grapher.

This is a composition of functions. In general, we have

$$h(x) = e^{f(x)} = g[f(x)], \quad \text{where} \quad g(x) = e^x.$$

Now $g'(x) = e^x$. Then by the Chain Rule (Section 2.8), we have

$$h'(x) = g'[f(x)] \cdot f'(x) = e^{f(x)} \cdot f'(x).$$

For the case above, $f(x) = x^2 - 5x$, so $f'(x) = 2x - 5$. Then

$$h'(x) = g'[f(x)] \cdot f'(x) = e^{f(x)} \cdot f'(x) = e^{x^2-5x}(2x - 5).$$

The next rule, which we have proven using the Chain Rule, allows us to find derivatives of functions like the one above.

THEOREM 3

$$\frac{d}{dx} e^{f(x)} = f'(x) \, e^{f(x)}$$

The derivative of e to some power is the derivative of the power times e to the power.

The following gives us a way to remember this rule.

$$g(x) = e^{x^2-5x} \quad ①$$

$$g'(x) = (2x - 5)e^{x^2-5x}$$

① Take the derivative of the exponent.

② Multiply the derivative of the exponent by the original function.

EXAMPLE 6

$$\frac{d}{dx} e^{5x} = 5e^{5x}$$

◆

EXAMPLE 7

$$\frac{d}{dx} e^{-x^2+4x-7} = (-2x + 4)e^{-x^2+4x-7}$$

◆

EXAMPLE 8

$$\frac{d}{dx} e^{\sqrt{x^2-3}} = \frac{d}{dx} e^{(x^2-3)^{1/2}} = \frac{1}{2}(x^2 - 3)^{-1/2} \cdot 2x \cdot e^{(x^2-3)^{1/2}}$$

$$= x(x^2 - 3)^{-1/2} \cdot e^{\sqrt{x^2-3}} = \frac{xe^{\sqrt{x^2-3}}}{\sqrt{x^2 - 3}}$$

◆

Graphs of e^x, e^{-x}, and $1 - e^{-kx}$

We use a calculator with an $\boxed{e^x}$ key to find approximate values of e^x and e^{-x}. We can also use a power key $\boxed{y^x}$ with $y = 2.7183$ or some other approximation for e. With these, we can draw graphs of the functions. Note that the graph of e^{-x} is a reflection, or "mirror image," of the graph of e^x across the y-axis.

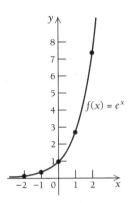
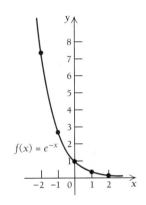

Functions of the type $f(x) = 1 - e^{-kx}$ also have important applications.

EXAMPLE 9   Graph $f(x) = 1 - e^{-2x}$ for nonnegative values of x.

Solution   We obtain these values using a calculator with an $\boxed{e^x}$ key. For example,

$$f(1) = 1 - e^{-2(1)}$$
$$= 1 - e^{-2}$$
$$= 1 - 0.135335$$
$$\approx 0.86.$$

The graph is shown below.

TECHNOLOGY CONNECTION

Graph $h(x) = e^{-x}$ and $Q(x) = 1 - e^{-x}$ and compare. Graph $f(x) = x^2$ and $g(x) = 2^x$ and compare. Graph f and f' and compare. Graph g and g' and compare.

x	$f(x)$
0	0
$\frac{1}{2}$	0.63
1	0.86
2	0.98
3	0.998

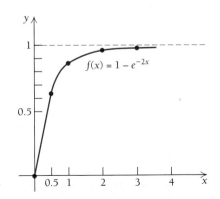

In general, the graph of $f(x) = 1 - e^{-kx}$, for $k > 0$ and $x \geq 0$, is increasing, which we expect since $f'(x) = ke^{-kx}$ is always positive. We also see that $f(x) \to 1$ as $x \to \infty$; that is, $\lim_{x \to \infty} (1 - e^{-kx}) = 1$.

A *word of caution!* Functions of the type a^x (for example, 2^x, 3^x, and e^x) are different from functions of the type x^a (for example, x^2, x^3, $x^{1/2}$). For a^x, the variable is in the exponent. For x^a, the variable is in the base. The derivative of a^x is not xa^{x-1}. In particular, we have the following:

$$\frac{d}{dx} e^x \neq xe^{x-1}, \quad \text{but} \quad \frac{d}{dx} e^x = e^x.$$

4.1 Exercise Set

Graph.

1. $y = 4^x$
2. $y = 5^x$
3. $y = (0.4)^x$
4. $y = (0.2)^x$
5. $x = 4^y$
6. $x = 5^y$

Differentiate.

7. $f(x) = e^{3x}$
8. $f(x) = e^{2x}$
9. $f(x) = 5e^{-2x}$
10. $f(x) = 4e^{-3x}$
11. $f(x) = 3 - e^{-x}$
12. $f(x) = 2 - e^{-x}$
13. $f(x) = -7e^x$
14. $f(x) = -4e^x$
15. $f(x) = \frac{1}{2}e^{2x}$
16. $f(x) = \frac{1}{4}e^{4x}$
17. $f(x) = x^4 e^x$
18. $f(x) = x^5 e^x$
19. $f(x) = \dfrac{e^x}{x^4}$
20. $f(x) = \dfrac{e^x}{x^5}$
21. $f(x) = e^{-x^2+7x}$
22. $f(x) = e^{-x^2+8x}$
23. $f(x) = e^{-x^2/2}$
24. $f(x) = e^{x^2/2}$
25. $y = e^{\sqrt{x-7}}$
26. $y = e^{\sqrt{x-4}}$
27. $y = \sqrt{e^x - 1}$
28. $y = \sqrt{e^x + 1}$
29. $y = xe^{-2x} + e^{-x} + x^3$
30. $y = e^x + x^3 - xe^x$
31. $y = 1 - e^{-x}$
32. $y = 1 - e^{-3x}$
33. $y = 1 - e^{-kx}$
34. $y = 1 - e^{-mx}$

Graph.

35. $f(x) = e^{2x}$
36. $f(x) = e^{(1/2)x}$
37. $f(x) = e^{-2x}$
38. $f(x) = e^{-(1/2)x}$
39. $f(x) = 3 - e^{-x}$, for nonnegative values of x

40. $f(x) = 2(1 - e^{-x})$, for nonnegative values of x
41. Find the tangent line to the graph of $f(x) = e^x$ at the point $(0, 1)$.
42. Find the tangent line to the graph of $f(x) = 2e^{-3x}$ at the point $(0, 2)$.

APPLICATIONS

◆ **Business and Economics**

43. *Marginal cost.* A company's total cost, in millions of dollars, is given by

$$C(t) = 100 - 50e^{-t},$$

where t = time.

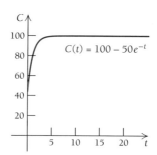

Find each of the following.

a) The marginal cost $C'(t)$
b) $C'(0)$
c) $C'(4)$

44. *Marginal cost.* A company's total cost, in millions of dollars, is given by

$$C(t) = 200 - 40e^{-t},$$

where t = time.

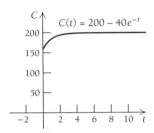

Find each of the following.

a) The marginal cost $C'(t)$
b) $C'(0)$
c) $C'(5)$

45. *Marginal demand.* The demand function for a certain kind of VCR is given by the function

$$x = D(p) = 480e^{-0.003p}.$$

a) How many VCRs will be sold when the price is $120? $180? $340?
b) Sketch a graph of $x = D(p)$ for $0 \leq p \leq 400$.
c) Find the marginal demand $D'(p)$.

46. *Marginal supply.* The supply function for the VCR in Exercise 45 is given by the function

$$x = S(p) = 150e^{0.004p}.$$

a) How many VCRs will the seller allow to be sold when the price is $120? $180? $340?
b) Sketch a graph of $x = S(p)$ for $0 \leq p \leq 400$.
c) Find the marginal supply $S'(p)$.

◆ **Life and Physical Sciences**

47. *Medication concentration.* The concentration C, in parts per million, of a medication in the body t hours after ingestion is given by the function

$$C(t) = 10t^2 e^{-t}.$$

a) Find the concentration after 0 hr; 1 hr; 2 hr; 3 hr; 10 hr.
b) Sketch a graph of the function for $0 \leq t \leq 10$.
c) Find the rate of change of the concentration $C'(t)$.
d) Find the maximum value of the concentration and where it occurs.

◆ **Social Sciences**

48. *Ebbinghaus learning model.* Suppose you are given the task of learning 100% of a block of knowledge. Human nature would tell us that we would retain only a percentage P of the knowledge t weeks after we have learned it. The *Ebbinghaus learning model* asserts that P is given by

$$P(t) = Q + (100\% - Q)e^{-kt},$$

where Q is the percentage that we would never forget and k is a constant that depends on the knowledge learned. Suppose that $Q = 40\%$ and $k = 0.7$.

a) Find the percentage retained after 0 weeks; 1 week; 2 weeks; 6 weeks; 10 weeks.
b) Find $\lim_{t \to \infty} P(t)$.
c) Sketch a graph of P.
d) Find the rate of change of P with respect to time t, $P'(t)$.

SYNTHESIS

Differentiate.

49. $y = (e^{3x} + 1)^5$

50. $y = (e^{x^2} - 2)^4$

51. $y = \dfrac{e^{3t} - e^{7t}}{e^{4t}}$

52. $y = \sqrt[3]{e^{3t} + t}$

53. $y = \dfrac{e^x}{x^2 + 1}$

54. $y = \dfrac{e^x}{1 - e^x}$

55. $f(x) = e^{\sqrt{x}} + \sqrt{e^x}$

56. $f(x) = \dfrac{1}{e^x} + e^{1/x}$

57. $f(x) = e^{x/2} \cdot \sqrt{x - 1}$

58. $f(x) = \dfrac{xe^{-x}}{1 + x^2}$

59. $f(x) = \dfrac{e^x - e^{-x}}{e^x + e^{-x}}$

60. $f(x) = e^{e^x}$

Each of the following is an expression for e. Find the function values that are approximations for e. Round to five decimal places.

61. $e = \lim_{t \to 0} f(t); f(t) = (1 + t)^{1/t}$. Find $f(1)$, $f(0.5)$, $f(0.2)$, $f(0.1)$, and $f(0.001)$.

62. $e = \lim_{t \to 1} g(t); g(t) = t^{1/(t-1)}$. Find $g(0.5)$, $g(0.9)$, $g(0.99)$, $g(0.999)$, and $g(0.9998)$.

63. Find the maximum value of $f(x) = x^2 e^{-x}$ on $[0, 4]$.

64. Find the minimum value of $f(x) = xe^x$ on $[-2, 0]$.

65. ◆ A student made the following error on a test:

$$\frac{d}{dx}e^x = xe^{x-1}.$$

Explain the error and how to correct it.

66. ◆ Describe the differences in the graphs of $f(x) = 3^x$ and $g(x) = x^3$.

TECHNOLOGY CONNECTION

Graph each of the following and find the relative extrema.

67. $f(x) = x^2 e^{-x}$ 68. $f(x) = e^{-x^2}$

Graph each of the following functions f and its derivative f'.

69. $f(x) = e^x$

70. $f(x) = e^{-x}$

71. $f(x) = 2e^{0.3x}$

72. $f(x) = 1000e^{-0.08x}$

73. Graph

$$f(x) = \left(1 + \frac{1}{x}\right)^x.$$

Consider especially large values of x to see how e is approached as a limit. Use a program for creating a table of function values if one is available.

4.2 Logarithmic Functions

OBJECTIVES

- Convert between exponential and logarithmic equations.
- Solve exponential equations.
- Solve problems involving exponential and logarithmic functions.
- Differentiate functions involving natural logarithms.

Graphs of Logarithmic Functions

Suppose we want to solve the equation

$$10^x = 1000.$$

We are trying to find that power of 10 that will give 1000. We can see that the answer is 3. The number 3 is called the "logarithm, base 10, of 1000."

DEFINITION

A *logarithm* is as follows:

$$y = \log_a x \quad \text{means} \quad x = a^y, \quad a > 0, a \neq 1.$$

The number a is called the *logarithmic base*.

Thus for logarithms base 10, $\log_{10} x$ is that number y such that $x = 10^y$. Therefore, a logarithm can be thought of as an exponent. We can convert from a logarithmic equation to an exponential equation, and conversely, as follows.

Logarithmic equation	Exponential equation
$\log_a M = N$	$a^N = M$
$\log_{10} 100 = 2$ $\log_{10} 0.01 = -2$ $\log_{49} 7 = \frac{1}{2}$	$10^2 = 100$ $10^{-2} = 0.01$ $49^{1/2} = 7$

In order to graph a logarithmic equation, we can graph its equivalent exponential equation.

EXAMPLE 1   Graph: $y = \log_2 x$.

Solution   We first write the equivalent exponential equation:

$$x = 2^y.$$

We select values for y and find the corresponding values of 2^y. Then we plot points, remembering that x is still the first coordinate, and connect the points with a smooth curve.

x, or 2^y	y
1	0
2	1
4	2
8	3
$\frac{1}{2}$	-1
$\frac{1}{4}$	-2

① Select y.

② Compute x.

The graphs of $f(x) = 2^x$ and $g(x) = \log_2 x$ are shown on the following page using the same set of axes. Note that we can obtain the graph of g by reflecting the graph of f across the line $y = x$. Graphs obtained in this manner are known as *inverses* of each other.

TECHNOLOGY
CONNECTION

Graph $f(x) = 2^x$. Then use the trace feature to find the coordinates of points on the graph. How can each ordered pair help you make a hand drawing of a graph of $g(x) = \log_2 x$?

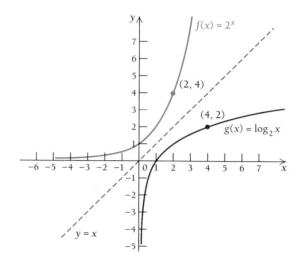

Although we cannot develop inverses in detail here, it is of interest to note that they "undo" each other. For example,

$$f(3) = 2^3 = 8 \qquad \text{The input 3 gives the output 8.}$$

and $\qquad g(8) = \log_2 8 = 3. \qquad \text{The input 8 gets us back to 3.}$

Basic Properties of Logarithms

The following are some basic properties of logarithms. The proofs follow from properties of exponents.

THEOREM 4

Properties of Logarithms

For any positive numbers M, N, and a, $a \neq 1$, and any real number k:

P1. $\log_a MN = \log_a M + \log_a N$

P2. $\log_a \dfrac{M}{N} = \log_a M - \log_a N$

P3. $\log_a M^k = k \cdot \log_a M$

P4. $\log_a a = 1$

P5. $\log_a a^k = k$

P6. $\log_a 1 = 0$

Proof of P1 and P2:   Let $X = \log_a M$ and $Y = \log_a N$. Writing the equivalent exponential equations, we then have

$$M = a^X \quad \text{and} \quad N = a^Y.$$

Then by the properties of exponents (see Section 1.1), we have

$$MN = a^X \cdot a^Y = a^{X+Y},$$

so $$\log_a MN = X + Y$$
$$= \log_a M + \log_a N.$$

Also

$$\frac{M}{N} = a^X \div a^Y = a^{X-Y},$$

so $$\log_a \frac{M}{N} = X - Y$$
$$= \log_a M - \log_a N.$$

Proof of P3:   Let $X = \log_a M$. Then

$$M = a^X,$$

so $$M^k = (a^X)^k, \quad \text{or} \quad M^k = a^{Xk}.$$

Thus,

$$\log_a M^k = Xk = kX = k \cdot \log_a M.$$

Proof of P4: $\log_a a = 1$ because $a^1 = a$.

Proof of P5: $\log_a (a^k) = k$ because $(a^k) = a^k$.

Proof of P6: $\log_a 1 = 0$ because $a^0 = 1$.

Let us illustrate these properties.

EXAMPLE 2   Given

$$\log_a 2 = 0.301 \quad \text{and} \quad \log_a 3 = 0.477,$$

find each of the following.

a) $\log_a 6$ $\log_a 6 = \log_a (2 \cdot 3) = \log_a 2 + \log_a 3$ **By P1**
$$= 0.301 + 0.477$$
$$= 0.778$$

b) $\log_a \frac{2}{3}$ $\log_a \frac{2}{3} = \log_a 2 - \log_a 3$ **By P2**
$$= 0.301 - 0.477$$
$$= -0.176$$

c) $\log_a 81$ $\log_a 81 = \log_a 3^4 = 4 \log_a 3$ By P3
$$= 4(0.477)$$
$$= 1.908$$

d) $\log_a \frac{1}{3}$ $\log_a \frac{1}{3} = \log_a 1 - \log_a 3$ By P2
$$= 0 - 0.477 \qquad \text{By P6}$$
$$= -0.477$$

e) $\log_a \sqrt{a}$ $\log_a \sqrt{a} = \log_a a^{1/2} = \frac{1}{2}$ By P5

f) $\log_a 2a$ $\log_a 2a = \log_a 2 + \log_a a$ By P1
$$= 0.301 + 1 \qquad \text{By P4}$$
$$= 1.301$$

g) $\log_a 5$ *No way to find using these properties.*
   $(\log_a 5 \neq \log_a 2 + \log_a 3)$

h) $\dfrac{\log_a 3}{\log_a 2}$ $\dfrac{\log_a 3}{\log_a 2} = \dfrac{0.477}{0.301} \approx 1.58$
   We simply divided and used none of the properties.   ◆

Common Logarithms

The number $\log_{10} x$ is called the **common logarithm** of x and is abbreviated $\log x$; that is:

> **DEFINITION**
>
> For any positive number x,
>
> $$\log x = \log_{10} x.$$

Thus, when we write $\log x$ with no base indicated, base 10 is understood. Note the following comparison of common logarithms and powers of 10.

$1000 = 10^3$	The common	$\log 1000 = 3$
$100 = 10^2$	logarithms	$\log 100 = 2$
$10 = 10^1$	at the right	$\log 10 = 1$
$1 = 10^0$	follow from	$\log 1 = 0$
$0.1 = 10^{-1}$	the powers at	$\log 0.1 = -1$
$0.01 = 10^{-2}$	the left.	$\log 0.01 = -2$
$0.001 = 10^{-3}$		$\log 0.001 = -3$

TECHNOLOGY
CONNECTION

Graph $f(x) = 10^x$, $y = x$,
and $g(x) = \log_{10} x$ using the
same set of axes. Then find
$f(3)$, $f(0.699)$, $g(5)$, and
$g(1000)$.

Since $\log 100 = 2$ and $\log 1000 = 3$, it seems reasonable that $\log 500$ is somewhere between 2 and 3. Tables were generally used for such approximations, but with the advent of the calculator, that method of finding logarithms is being used more and more infrequently. Using a calculator with a $\boxed{\text{log}}$ key, we find that $\log 500 = 2.6990$, rounded to four decimal places.

Before calculators and computers became so readily available, common logarithms were used extensively to do certain kinds of computations. In fact, computation is the reason logarithms were developed. Since the standard notation we use for numbers is based on 10, it is logical that base-10, or common, logarithms were used for computations. Today, computations with common logarithms are mainly of historical interest; the logarithmic functions, base e, are of modern importance.

Natural Logarithms

The number e, which is approximately 2.718282, was developed in Section 4.1, and has extensive application in many fields. The number $\log_e x$ is called the **natural logarithm** of x and is abbreviated $\ln x$; that is:

DEFINITION

For any positive number x,

$$\ln x = \log_e x.$$

The following is a restatement of the basic properties of logarithms in terms of natural logarithms.

THEOREM 5

P1. $\ln MN = \ln M + \ln N$

P2. $\ln \dfrac{M}{N} = \ln M - \ln N$

P3. $\ln a^k = k \cdot \ln a$

P4. $\ln e = 1$

P5. $\ln e^k = k$

P6. $\ln 1 = 0$

Let us illustrate these properties.

EXAMPLE 3 Given

$$\ln 2 = 0.6931 \quad \text{and} \quad \ln 3 = 1.0986,$$

find each of the following.

a) $\ln 6$ $\quad \ln 6 = \ln(2 \cdot 3) = \ln 2 + \ln 3 \quad$ By P1

$$= 0.6931 + 1.0986$$

$$= 1.7917$$

b) $\ln 81$ $\quad \ln 81 = \ln(3^4)$

$$= 4 \ln 3 \quad \text{By P3}$$

$$= 4(1.0986)$$

$$= 4.3944$$

c) $\ln \frac{2}{3}$ $\quad \ln \frac{2}{3} = \ln 2 - \ln 3 \quad$ By P2

$$= 0.6931 - 1.0986$$

$$= -0.4055$$

d) $\ln \frac{1}{3}$ $\quad \ln \frac{1}{3} = \ln 1 - \ln 3 \quad$ By P2

$$= 0 - 1.0986 \quad \text{By P6}$$

$$= -1.0986$$

e) $\ln 2e$ $\quad \ln 2e = \ln 2 + \ln e \quad$ By P1

$$= 0.6931 + 1 \quad \text{By P4}$$

$$= 1.6931$$

f) $\ln \sqrt{e^3}$ $\quad \ln \sqrt{e^3} = \ln e^{3/2}$

$$= \tfrac{3}{2} \quad \text{By P5} \qquad \blacklozenge$$

Finding Natural Logarithms Using a Calculator

You should have a calculator with a $\boxed{\text{ln}}$ key. You can find natural logarithms directly using this key.

EXAMPLE 4 Find each logarithm on your calculator. Round to six decimal places.

a) $\ln 5.24 = 1.656321$

b) $\ln 0.00001277 = -11.268412$ $\qquad \blacklozenge$

Exponential Equations

If an equation contains a variable in an exponent, we call the equation **exponential.** We can use logarithms to manipulate or solve exponential equations.

CONNECTION

Solve the equation $e^t = 40$ by solving the system of equations

$y_1 = e^x$ and $y_2 = 40.$

EXAMPLE 5   Solve $e^t = 40$ for t.

Solution

$$\ln e^t = \ln 40 \qquad \text{Taking the natural logarithm on both sides}$$
$$t = \ln 40 \qquad \text{By P5}$$
$$t = 3.688879$$
$$t \approx 3.7$$

It should be noted that this is an approximation for t even though an equals sign is often used.   ◆

EXAMPLE 6   Solve $e^{-0.04t} = 0.05$ for t.

Solution

$$\ln e^{-0.04t} = \ln 0.05 \qquad \text{Taking the natural logarithm on both sides}$$
$$-0.04t = \ln 0.05 \qquad \text{By P5}$$
$$t = \frac{\ln 0.05}{-0.04}$$
$$t = \frac{-2.995732}{-0.04}$$
$$t \approx 75$$

For purposes of space and explanation, we have rounded the value of $\ln 0.05$ to -2.995732 in an intermediate step. When using your calculator, you should find

$$\frac{\ln 0.05}{-0.04}$$

directly, without rounding, as say,

$$\frac{-2.995732274}{-0.04}.$$

Then divide, and round at the end. Answers at the back of the book have been found in this manner. Remember, the number of places in a table or on a calculator may affect the accuracy of the answer. Usually, your answer should agree to at least three digits.   ◆

Graphs of Natural Logarithmic Functions

There are two ways in which we might obtain the graph of $y = f(x) = \ln x$. One is by writing its equivalent equation $x = e^y$. Then we select values for y and use a calculator to find the corresponding values of e^y. We then plot points, remembering that x is still the first coordinate. The graph follows.

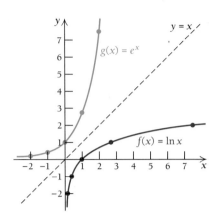

x, or e^y	y
0.1	−2
0.4	−1
1.0	0
2.7	1
7.4	2
20.1	3

① Select y.

② Compute x.

The figure above shows the graph of $g(x) = e^x$ for comparison. Note again that the functions are inverses of each other. That is, the graph of $y = \ln x$, or $x = e^y$, is a reflection, or mirror image, across the line $y = x$ of the graph of $y = e^x$. Any ordered pair (a, b) on the graph of g yields an ordered pair (b, a) on f.

The second method of graphing $y = \ln x$ is to use a calculator to find function values. For example, when $x = 2$, then $y = \ln 2 = 0.6931 \approx 0.7$. This gives the pair $(2, 0.7)$ on the graph, which follows.

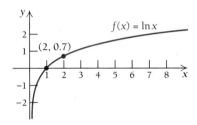

The following properties hold.

THEOREM 6

$\ln x$ exists only for positive numbers x. (See both the preceding and the following figures.)

$\ln x < 0$ for $0 < x < 1$.

$\ln x = 0$ when $x = 1$.

$\ln x > 0$ for $x > 1$.

TECHNOLOGY
CONNECTION

Consider $f(x) = \ln x$.
Graph f and f' using the
viewing window
$[0, 10, -3, 3]$, with
$xScl = 1$ and $yScl = 1$.
Make an input–output table
for the functions. Compare
a given x-value with its
corresponding y-value for f
and f'. What do you
observe?

Derivatives of Natural Logarithmic Functions

Consider $f(x) = \ln x$. We can show that $f'(x) = 1/x$ (the slope of the tangent line at x is just the reciprocal of x). We are trying to find the derivative of

$$f(x) = \ln x. \tag{1}$$

We first write its equivalent exponential equation:

$$e^{f(x)} = x. \qquad \text{$\ln x = \log_e x = f(x)$, so $e^{f(x)} = x$,}$$
$$\text{by the definition of logarithms.} \tag{2}$$

Now we differentiate on both sides of this equation:

$$\frac{d}{dx} e^{f(x)} = \frac{d}{dx} x$$

$$f'(x) \cdot e^{f(x)} = 1 \qquad \text{By the Chain Rule}$$

$$f'(x) \cdot x = 1 \qquad \text{Substituting x for $e^{f(x)}$ from Eq. (2)}$$

$$f'(x) = \frac{1}{x}.$$

Thus we have the following.

> **THEOREM 7**
>
> For any positive number x,
>
> $$\frac{d}{dx} \ln x = \frac{1}{x}.$$

This is true only for positive values of x, since $\ln x$ is defined only for positive numbers. (For negative numbers x, this derivative formula becomes

$$\frac{d}{dx} \ln |x| = \frac{1}{x},$$

but we will seldom consider such a case in this text.)

Let us find some derivatives.

EXAMPLE 7

$$\frac{d}{dx} 3 \ln x = 3 \frac{d}{dx} \ln x$$

$$= \frac{3}{x}$$

EXAMPLE 8

$$\frac{d}{dx}(x^2 \ln x + 5x) = x^2 \cdot \frac{1}{x} + 2x \cdot \ln x + 5 \qquad \text{Using the Product Rule on } x^2 \ln x$$

$$= x + 2x \cdot \ln x + 5, \qquad \text{Simplifying}$$
$$\text{or } x(1 + 2 \ln x) + 5$$

◆

EXAMPLE 9

$$\frac{d}{dx}\frac{\ln x}{x^3} = \frac{x^3 \cdot (1/x) - (3x^2)(\ln x)}{x^6} \qquad \text{By the Quotient Rule}$$

$$= \frac{x^2 - 3x^2 \ln x}{x^6}$$

$$= \frac{x^2(1 - 3 \ln x)}{x^6} \qquad \text{Factoring}$$

$$= \frac{1 - 3 \ln x}{x^4} \qquad \text{Simplifying}$$

◆

Suppose we want to differentiate a more complicated function, such as

$$h(x) = \ln (x^2 - 8x).$$

This is a composition of functions. In general, we have

$$h(x) = \ln f(x) = g[f(x)], \quad \text{where} \quad g(x) = \ln x.$$

Now $g'(x) = 1/x$. Then by the Chain Rule (Section 2.8), we have

$$h'(x) = g'[f(x)] \cdot f'(x) = \frac{1}{f(x)} \cdot f'(x).$$

For the above case, $f(x) = x^2 - 8x$, so $f'(x) = 2x - 8$. Then

$$h'(x) = \frac{1}{x^2 - 8x} \cdot (2x - 8) = \frac{2x - 8}{x^2 - 8x}.$$

The following rule, which we have proven using the Chain Rule, allows us to find derivatives of functions like the one above.

THEOREM 8

$$\frac{d}{dx} \ln f(x) = f'(x) \cdot \frac{1}{f(x)} = \frac{f'(x)}{f(x)}$$

The derivative of the natural logarithm of a function is the derivative of the function divided by the function.

The following gives us a way of remembering this rule.

$$h(x) = \ln \underbrace{(x^2 - 8x)}_{①}$$

① Differentiate the "inside" function.

$$h'(x) = \underbrace{\frac{2x - 8}{x^2 - 8x}}_{②}$$

② Divide by the "inside" function.

EXAMPLE 10

$$\frac{d}{dx} \ln 3x = \frac{3}{3x} = \frac{1}{x}$$

Note that we could have done this another way, using Property 1:

$$\ln 3x = \ln 3 + \ln x;$$

then

$$\frac{d}{dx} \ln 3x = \frac{d}{dx} \ln 3 + \frac{d}{dx} \ln x = 0 + \frac{1}{x} = \frac{1}{x}. \qquad \blacklozenge$$

EXAMPLE 11

$$\frac{d}{dx} \ln (x^2 - 5) = \frac{2x}{x^2 - 5} \qquad \blacklozenge$$

EXAMPLE 12

$$\frac{d}{dx} \ln (\ln x) = \frac{1}{x} \cdot \frac{1}{\ln x} = \frac{1}{x \ln x} \qquad \blacklozenge$$

**TECHNOLOGY
CONNECTION**

Check the results of Examples 10–13 using your grapher. Then differentiate

$$y = \ln\left(\frac{x^5 - 2}{x}\right)$$

and check your answer using your grapher.

EXAMPLE 13

$$\frac{d}{dx} \ln \left(\frac{x^3 + 4}{x}\right) = \frac{d}{dx} \left[\ln (x^3 + 4) - \ln x\right] \qquad \text{By P2. This avoids using the Quotient Rule.}$$

$$= \frac{3x^2}{x^3 + 4} - \frac{1}{x}$$

$$= \frac{3x^2}{x^3 + 4} \cdot \frac{x}{x} - \frac{1}{x} \cdot \frac{x^3 + 4}{x^3 + 4}$$

$$= \frac{(3x^2)x - (x^3 + 4)}{x(x^3 + 4)}$$

$$= \frac{3x^3 - x^3 - 4}{x(x^3 + 4)} = \frac{2x^3 - 4}{x(x^3 + 4)} \qquad \blacklozenge$$

Applications

Conduct your own memory experiment. Study this photograph carefully. Then put it aside and write down as many items as you can. Wait a half-hour and again write down as many as you can. Do this five more times. Make a graph of the number of items you remember versus the time. Does the graph appear to be logarithmic?

EXAMPLE 14 *Social science: Forgetting.* In a psychological experiment, students were shown a set of nonsense syllables, such as POK, RTZ, PDQ, and so on, and asked to recall them every second thereafter. The percentage $R(t)$ who retained the syllables after t seconds was found to be given by the logarithmic learning model

$$R(t) = 80 - 27 \ln t, \quad \text{for } t \geq 1.$$

Strictly speaking, the function is not continuous, but in order to use calculus, we "fill in" the graph with a smooth curve, considering $R(t)$ to be defined for any number $t \geq 1$. This is not unreasonable, since we could find the percentage who retained the syllables after $t = 3.417$ sec, instead of merely after integer values such as 1, 2, 3, 4, and so on.

a) What percentage retained the syllables after 1 sec?

b) Find $R'(t)$, the rate of change of R with respect to t.

c) Find the maximum and minimum values, if they exist.

TECHNOLOGY
CONNECTION

TECHNOLOGY
CONNECTION

Graph $y = 80 - 27 \ln x$
using the viewing window
$[1, 14, -1, 100]$, with
$xScl = 1$ and $yScl = 1$.
Trace along the graph. Describe the meaning of each
coordinate in an ordered
pair.

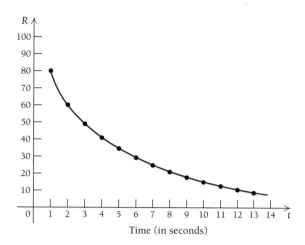

Time (in seconds)

Solution

a) $R(1) = 80 - 27 \cdot \ln 1 = 80 - 27 \cdot 0 = 80\%$

b) $R'(t) = \dfrac{-27}{t} = -\dfrac{27}{t}$

c) $R'(t)$ exists for all values of t in the interval $[1, \infty)$. Note that for $t \geq 1$, $-27/t < 0$. Thus there are no critical points and R is decreasing. Then R has a maximum value at the endpoint 1. This maximum value is $R(1)$, or 80%. There is no minimum value.   ◆

EXAMPLE 15   *Business: An advertising model.* A company begins a radio advertising campaign in New York City to market a new product. The percentage of the "target market" that buys a product is normally a function of the length of the advertising campaign. The radio station estimates this percentage as $1 - e^{-0.04t}$ for this type of product, where $t =$ the number of days of the campaign. The target market is estimated to be 1,000,000 people and the price per unit is $0.50. The costs of advertising are $1000 per day. Find the length of the advertising campaign that will result in the maximum profit.

Solution   That the percentage of the target market that buys the product can be modeled by $f(t) = 1 - e^{-0.04t}$ is justified if we look at its graph. The function increases from 0 (0%) toward 1 (100%). The longer the advertising campaign, the larger the percentage of the market that has bought the product.

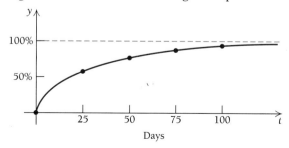

Days

The total-profit function, here expressed in terms of time t, is given by

$$\text{Profit} = \text{Revenue} - \text{Cost}$$
$$P(t) = R(t) - C(t).$$

a) Find $R(t)$:

$$R(t) = (\text{Price per unit}) \cdot (\text{Target market}) \cdot (\text{Percentage buying})$$
$$R(t) = 0.5(1{,}000{,}000)(1 - e^{-0.04t}) = 500{,}000 - 500{,}000e^{-0.04t}.$$

b) Find $C(t)$:

$$C(t) = (\text{Advertising costs per day}) \cdot (\text{Number of days})$$
$$C(t) = 1000t.$$

c) Find $P(t)$ and take its derivative:

$$P(t) = R(t) - C(t)$$
$$P(t) = 500{,}000 - 500{,}000e^{-0.04t} - 1000t$$
$$P'(t) = (-0.04)(-500{,}000e^{-0.04t}) - 1000$$
$$P'(t) = 20{,}000e^{-0.04t} - 1000.$$

d) Set the first derivative equal to 0 and solve:

$$20{,}000e^{-0.04t} - 1000 = 0$$
$$20{,}000e^{-0.04t} = 1000$$
$$e^{-0.04t} = \frac{1000}{20{,}000} = 0.05$$
$$\ln e^{-0.04t} = \ln 0.05$$
$$-0.04t = \ln 0.05$$
$$t = \frac{\ln 0.05}{-0.04}$$
$$t = \frac{-2.995732}{-0.04}$$
$$t \approx 75.$$

e) We have only one critical point, so we can use the second derivative to determine whether we have a maximum:

$$P''(t) = -0.04(20{,}000e^{-0.04t})$$
$$= -800e^{-0.04t}.$$

Since exponential functions are positive, $e^{-0.04t} > 0$ for all numbers t. Thus, since $-800e^{-0.04t} < 0$ for all numbers t, $P''(t)$ is less than 0 for $t = 75$ and we have a maximum.

The length of the advertising campaign must be 75 days in order to result in maximum profit. ◆

TECHNOLOGY
CONNECTION

Graph the function P in Example 15 and verify that there is maximum profit if the length of the advertising campaign is about 75 days.

4.2 Exercise Set

Write an equivalent exponential equation.

1. $\log_2 8 = 3$

2. $\log_3 81 = 4$

3. $\log_8 2 = \frac{1}{3}$

4. $\log_{27} 3 = \frac{1}{3}$

5. $\log_a K = J$

6. $\log_a J = K$

7. $\log_b T = v$

8. $\log_c Y = t$

Write an equivalent logarithmic equation.

9. $e^M = b$

10. $e^t = p$

11. $10^2 = 100$

12. $10^3 = 1000$

13. $10^{-1} = 0.1$

14. $10^{-2} = 0.01$

15. $M^p = V$

16. $Q^n = T$

Given $\log_b 3 = 1.099$ and $\log_b 5 = 1.609$, find each of the following.

17. $\log_b 15$

18. $\log_b \frac{3}{5}$

19. $\log_b \frac{1}{5}$

20. $\log_b \sqrt{b^3}$

21. $\log_b 5b$

22. $\log_b 75$

Given $\ln 4 = 1.3863$ and $\ln 5 = 1.6094$, find each of the following. Do not use a calculator.

23. $\ln 20$

24. $\ln \frac{5}{4}$

25. $\ln \frac{1}{4}$

26. $\ln 4e$

27. $\ln \sqrt{e^8}$

28. $\ln 100$

Find the logarithm. Round to six decimal places.

29. $\ln 5894$

30. $\ln 99{,}999$

31. $\ln 0.0182$

32. $\ln 0.00087$

33. $\ln 8100$

34. $\ln 0.011$

Solve for t.

35. $e^t = 100$

36. $e^t = 1000$

37. $e^t = 60$

38. $e^t = 90$

39. $e^{-t} = 0.1$

40. $e^{-t} = 0.01$

41. $e^{-0.02t} = 0.06$

42. $e^{0.07t} = 2$

Differentiate.

43. $y = -6 \ln x$

44. $y = -4 \ln x$

45. $y = x^4 \ln x - \frac{1}{2}x^2$

46. $y = x^5 \ln x - \frac{1}{4}x^4$

47. $y = \dfrac{\ln x}{x^4}$

48. $y = \dfrac{\ln x}{x^5}$

49. $y = \ln \dfrac{x}{4}$ $\left(Hint: \ln \dfrac{x}{4} = \ln x - \ln 4. \right)$

50. $y = \ln \dfrac{x}{2}$

51. $f(x) = \ln (5x^2 - 7)$

52. $f(x) = \ln (7x^3 + 4)$

53. $f(x) = \ln (\ln 4x)$

54. $f(x) = \ln (\ln 3x)$

55. $f(x) = \ln \left(\dfrac{x^2 - 7}{x} \right)$

56. $f(x) = \ln \left(\dfrac{x^2 + 5}{x} \right)$

57. $f(x) = e^x \ln x$

58. $f(x) = e^{2x} \ln x$

59. $f(x) = \ln (e^x + 1)$

60. $f(x) = \ln (e^x - 2)$

61. $f(x) = (\ln x)^2$
   (*Hint:* Use the Extended Power Rule.)

62. $f(x) = (\ln x)^3$

APPLICATIONS

◆ **Business and Economics**

63. *Advertising.* A model for advertising response is given by

$$N(a) = 1000 + 200 \ln a, \quad a \geq 1,$$

where $N(a) =$ the number of units sold and $a =$ the amount spent on advertising, in thousands of dollars.

a) How many units were sold after spending $1000 ($a = 1$) on advertising?

b) Find $N'(a)$ and $N'(10)$.

c) Find the maximum and minimum values, if they exist.

64. *Advertising.* A model for advertising response is given by

$$N(a) = 2000 + 500 \ln a, \quad a \geq 1,$$

where $N(a) =$ the number of units sold and $a =$ the amount spent on advertising, in thousands of dollars.

a) How many units were sold after spending $1000 ($a = 1$) on advertising?

b) Find $N'(a)$ and $N'(10)$.

c) Find the maximum and minimum values, if they exist.

65. *An advertising model.* Solve Example 15 given that the costs of advertising are $2000 per day.

66. *An advertising model.* Solve Example 15 given that the costs of advertising are $4000 per day.

67. *Growth of a stock.* The value V of a stock is modeled by

$$V(t) = \$58(1 - e^{-1.1t}) + \$20,$$

where V = the value of the stock after time t, in months.

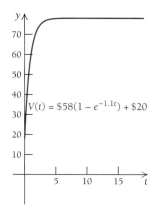

a) Find $V(1)$ and $V(12)$.

b) Find $V'(t)$.

c) After how many months will the value of the stock be $75?

68. *Marginal revenue.* The demand function for a certain product is given by

$$x = D(p) = 800e^{-0.125p}.$$

Recall that total revenue is given by $R(p) = p D(p)$.

a) Find $R(p)$.

b) Find the marginal revenue, $R'(p)$.

c) At what price per unit p will the revenue be maximum?

◆ Life and Physical Sciences

69. *Acceptance of a new medicine.* The percentage P of doctors who accept a new medicine is given by

$$P(t) = 100(1 - e^{-0.2t}),$$

where t = the time, in months.

a) Find $P(1)$ and $P(6)$.

b) Find $P'(t)$.

c) How many months will it take for 90% of the doctors to accept the new medicine?

70. *The Reynolds number.* For many kinds of animals, the Reynolds number R is given by

$$R = A \ln r - Br,$$

where A and B are positive constants and r = the radius of the aorta. Find the maximum value of R.

◆ Social Sciences

71. *Forgetting.* Students in college botany took a final exam. They took equivalent forms of the exam in monthly intervals thereafter. The average score $S(t)$, in percent, after t months was found to be given by

$$S(t) = 68 - 20 \ln (t + 1), \quad t \ge 0.$$

a) What was the average score when they initially took the test, $t = 0$?

b) What was the average score after 4 months?

c) What was the average score after 24 months?

d) What percentage of the initial score did they retain after 2 years (24 months)?

e) Find $S'(t)$.

f) Find the maximum and minimum values, if they exist.

72. *Forgetting.* Students in college zoology took a final exam. They took equivalent forms of the exam in monthly intervals thereafter. The average score $S(t)$, in percent, after t months was found to be given by

$$S(t) = 78 - 15 \ln (t + 1), \quad t \ge 0.$$

a) What was the average score when they initially took the test, $t = 0$?

b) What was the average score after 4 months?

c) What was the average score after 24 months?

d) What percentage of the initial score did they retain after 2 years (24 months)?

e) Find $S'(t)$.

f) Find the maximum and minimum values, if they exist.

73. *Walking speed.* Bornstein and Bornstein found in a study that the average walking speed v of a person living in a city of population p, in thousands, is

given by

$$v(p) = 0.37 \ln p + 0.05,$$

where v is in feet per second.

a) The population of Seattle is 531,000. What is the average walking speed of a person living in Seattle? [*Hint:* Find $v(531)$.]

b) The population of New York is 7,900,000. What is the average walking speed of a person living in New York?

c) Find $v'(p)$. Interpret $v'(p)$.

74. *The Hullian learning model.* A typist learns to type W words per minute after t weeks of practice, where W is given by

$$W(t) = 100(1 - e^{-0.3t}).$$

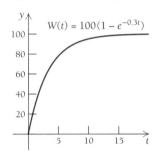

a) Find $W(1)$ and $W(8)$.

b) Find $W'(t)$.

c) After how many weeks will the typist's speed be 95 words per minute?

SYNTHESIS

Differentiate.

75. $y = (\ln x)^{-4}$

76. $y = (\ln x)^n$

77. $f(t) = \ln (t^3 + 1)^5$

78. $f(t) = \ln (t^2 + t)^3$

79. $f(x) = [\ln (x + 5)]^4$

80. $f(x) = \ln [\ln (\ln 3x)]$

81. $f(t) = \ln [(t^3 + 3)(t^2 - 1)]$

82. $f(t) = \ln \dfrac{1 - t}{1 + t}$

83. $y = \ln \dfrac{x^5}{(8x + 5)^2}$

84. $y = \ln \sqrt{5 + x^2}$

85. $f(t) = \dfrac{\ln t^2}{t^2}$

86. $f(x) = \dfrac{1}{5}x^5\left(\ln x - \dfrac{1}{5}\right)$

87. $y = \dfrac{x^{n+1}}{n + 1}\left(\ln x - \dfrac{1}{n + 1}\right)$

88. $y = \dfrac{x \ln x - x}{x^2 + 1}$

89. $y = \ln (t + \sqrt{1 + t^2})$

90. $f(x) = \ln \dfrac{1 + \sqrt{x}}{1 - \sqrt{x}}$

91. $f(x) = \ln [\ln x]^3$

92. $f(x) = \dfrac{\ln x}{1 + (\ln x)^2}$

93. Find $\lim\limits_{h \to 0} \dfrac{\ln (1 + h)}{h}$.

94. ▦ Which is larger, e^π or π^e?

95. ▦ Find $\sqrt[e]{e}$. Compare it to other expressions of the type $\sqrt[x]{x}$, $x > 0$. What can you conclude?

Solve for t.

96. $P = P_0 e^{-kt}$

97. $P = P_0 e^{kt}$

▦ Use input–output tables to find the limit.

98. $\lim\limits_{x \to 1} \ln x$

99. $\lim\limits_{x \to \infty} \ln x$

Verify each of the following.

100. $\log x = \dfrac{\ln x}{\ln 10} \approx 0.4343 \ln x$

101. $\ln x = \dfrac{\log x}{\log e} \approx 2.3026 \log x$

102. ◈ Explain how the graph of $y = \ln x$ could be used to find the graph of $y = e^x$.

103. ◈ Consider the true statement

$$\dfrac{d}{dx} \ln 4x = \dfrac{1}{x}.$$

Explain what this means graphically.

Graph each of the following functions f and its derivative f'.

104. $f(x) = \ln x$

105. $f(x) = x \ln x$

106. $f(x) = x^2 \ln x$

107. $f(x) = \dfrac{\ln x}{x^2}$

Find the minimum value of the function.

108. $f(x) = x \ln x$

109. $f(x) = x^2 \ln x$

4.3 Applications: The Uninhibited Growth Model, $dP/dt = kP$

OBJECTIVES

* Find functions that satisfy $dP/dt = kP$.
* Convert between growth rate and doubling time.
* Solve application problems involving exponential growth.

Exponential Growth

Consider the function

$$f(x) = 2e^{3x}.$$

Differentiating, we get

$$f'(x) = 3 \cdot 2e^{3x}$$
$$= 3 \cdot f(x).$$

Graphically, this says that the derivative, or slope of the tangent line, is simply the constant 3 times the function value.

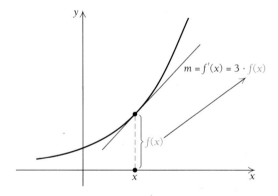

In general, we have the following.

THEOREM 9

A function $y = f(x)$ satisfies the equation

$$\frac{dy}{dx} = ky \qquad [f'(x) = k \cdot f(x)]$$

if and only if

$$y = ce^{kx} \qquad [f(x) = ce^{kx}]$$

for some constant c.

No matter what the variables, you should be able to write the function that satisfies what is called a *differential equation.*

EXAMPLE 1   Find the function that satisfies the equation

$$\frac{dA}{dt} = kA.$$

Solution   The function is $A = ce^{kt}$, or $A(t) = ce^{kt}$.   ◆

EXAMPLE 2   Find the function that satisfies the equation

$$\frac{dP}{dt} = kP.$$

Solution   The function is $P = ce^{kt}$, or $P(t) = ce^{kt}$.   ◆

EXAMPLE 3   Find the function that satisfies the equation

$$f'(Q) = k \cdot f(Q).$$

Solution   The function is $f(Q) = ce^{kQ}$.   ◆

The equation

$$\frac{dP}{dt} = kP, \quad k > 0 \qquad [P'(t) = k \cdot P(t), \quad k > 0]$$

is the basic model of uninhibited population growth, whether it be a population of humans, a bacteria culture, or money invested at interest compounded continuously. Neglecting special inhibiting and stimulating factors, we know that a population normally reproduces itself at a rate proportional to its size, and this is exactly what the equation $dP/dt = kP$ says. The solution of the equation is

$$P(t) = ce^{kt}, \tag{1}$$

where t = the time. At $t = 0$, we have some "initial" population $P(0)$ that we will represent by P_0. We can rewrite Eq. (1) in terms of P_0 as

$$P_0 = P(0) = ce^{k \cdot 0} = ce^0 = c \cdot 1 = c.$$

Thus, $P_0 = c$, so we can express $P(t)$ as

$$P(t) = P_0 e^{kt}.$$

Its graph shows how uninhibited growth produces a "population explosion."

What will the world population be in 2000?

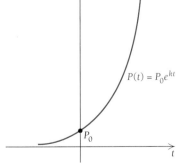

TECHNOLOGY CONNECTION

Exploratory Exercises: Growth

Use a sheet of $8\frac{1}{2}$-in. by 11-in. paper that has been cut into two equal pieces. Then cut these into four equal pieces. Then cut these into eight equal pieces, and so on, performing five cutting steps.

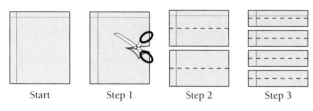

Start   Step 1   Step 2   Step 3

	t	$0.004 \cdot 2^t$
Start	0	$0.004 \cdot 2^0$, or 0.004
Step 1	1	$0.004 \cdot 2^1$, or 0.008
Step 2	2	$0.004 \cdot 2^2$, or 0.016
Step 3	3	
Step 4	4	
Step 5	5	

a) Place all the pieces in a stack and measure the thickness.

b) A piece of paper is typically 0.004 in. thick. Check the calculation in part (a) by completing the table.

c) Graph the function $f(t) = 0.004(2)^t$.

d) Compute the thickness of the paper (in miles) after 25 steps.

The constant k is called the **rate of exponential growth**, or simply the **growth rate.** This is not the rate of change of the population size, which is

$$\frac{dP}{dt} = kP,$$

but the constant by which P must be multiplied in order to get its rate of change. It is thus a different use of the word *rate*. It is like the *interest rate* paid by a bank. If the interest rate is 7%, or 0.07, we do not mean that your bank balance P is growing at the rate of 0.07 dollars per year, but at the rate of $0.07P$ dollars per year. We therefore express the rate as 7% per year, rather than 0.07 dollars per year. We could say that the rate is 0.07 dollars *per dollar* per year. When interest is compounded continuously, the interest rate is a true exponential growth rate.

EXAMPLE 4   *Business: Interest compounded continuously.* Suppose an amount P_0 is invested in a savings account where interest is compounded continuously at 7% per year. That is, the balance P grows at the rate given by

$$\frac{dP}{dt} = 0.07P.$$

a) Find the function that satisfies the equation. List it in terms of P_0 and 0.07.

b) Suppose $100 is invested. What is the balance after 1 year?

c) After what period of time will an investment of $100 double itself?

Solution

a) $P(t) = P_0 e^{0.07t}$

b) $P(1) = 100 e^{0.07(1)} = 100 e^{0.07}$
$$= 100(1.072508)$$
$$\approx \$107.25$$

c) We are looking for that time T for which $P(T) = \$200$. The number T is called the **doubling time.** To find T, we solve the equation

$$200 = 100 e^{0.07 \cdot T}$$
$$2 = e^{0.07T}.$$

We use natural logarithms to solve this equation:

$$\ln 2 = \ln e^{0.07T}$$
$$\ln 2 = 0.07T \qquad \text{By P5: } \ln e^k = k$$
$$\frac{\ln 2}{0.07} = T$$
$$\frac{0.693147}{0.07} = T$$
$$9.9 \approx T.$$

Thus, \$100 will double itself in 9.9 years.   ◆

Let us consider another way of developing the formula
$$P(t) = P_0 e^{kt}$$
by starting with
$$A = P\left(1 + \frac{i}{n}\right)^{nt}$$
and compounding interest continuously. Let $P = P_0$ and $i = k$ to obtain
$$A = P_0\left(1 + \frac{k}{n}\right)^{nt}.$$

We are interested in what happens as n gets very large, that is, as n approaches ∞. To determine this limit, we first let
$$\frac{k}{n} = \frac{1}{h}.$$

Then
$$hk = n \quad \text{and} \quad h = \frac{n}{k},$$

which shows that since k is a positive constant, as n gets large, so does h. To find a

Under ideal conditions, the growth rate of this rabbit population might be 11.7% per day. When will this population of rabbits double?

formula for continuously compounded interest, we evaluate the following limit:

$$P(t) = \lim_{n \to \infty} \left[P_0 \left(1 + \frac{k}{n} \right)^{nt} \right]$$
  Letting the number of compounding periods become infinite

$$= P_0 \lim_{h \to \infty} \left[\left(1 + \frac{1}{h} \right)^{hkt} \right]$$
  The limit of a constant times a function is the constant times the limit. We also substitute $1/h$ for k/n and hk for n. Also, $h \to \infty$ because $n \to \infty$.

$$= P_0 \left[\lim_{h \to \infty} \left(1 + \frac{1}{h} \right)^{h} \right]^{kt}$$
  The limit of a power is the power of the limit: a form of L2 in Section 2.2.

$$= P_0 [e]^{kt}.$$
  Theorem 1

We can find a general expression relating the growth rate k and the doubling time T by solving the following equation:

$$2P_0 = P_0 e^{kT}$$
$$2 = e^{kT} \qquad \text{Dividing by } P_0$$
$$\ln 2 = \ln e^{kT}$$
$$\ln 2 = kT.$$

THEOREM 10

The *growth rate k* and the *doubling time T* are related by

$$kT = \ln 2 = 0.693147,$$

or
$$k = \frac{\ln 2}{T} = \frac{0.693147}{T},$$

and
$$T = \frac{\ln 2}{k} = \frac{0.693147}{k}.$$

Note that this relationship between k and T does not depend on P_0.

EXAMPLE 5 *Business: Doubling time.* At one time in Canada, a bank advertised that it would double your money in 6.6 years. What was the interest rate on such an account, assuming interest were compounded continuously?

Solution

$$k = \frac{\ln 2}{T} = \frac{0.693147}{6.6} \approx 0.105 = 10.5\%$$

The interest rate was 10.5%.

EXAMPLE 6 *Life science: World population growth.* The population of the world was 5.2 billion in 1990. On the basis of data available at that time, it was estimated that the population P was growing exponentially at the rate of 1.6% per year. That is, $dP/dt = 0.016P$, where $t =$ the time, in years, from 1990. (To facilitate computations, we assume that the population was 5.2 billion at the beginning of 1990.)

a) Find the function that satisfies the equation. Assume that $P_0 = 5.2$ and $k = 0.016$.

b) Estimate the world population in 2000 ($t = 10$).

c) After what period of time will the population be double that in 1990?

Solution

a) $P(t) = 5.2e^{0.016t}$

b) $P(10) = 5.2e^{0.016(10)} = 5.2e^{0.16} = 5.2(1.173511)$
$$\approx 6.1 \text{ billion}$$

c) $T = \dfrac{\ln 2}{k} = \dfrac{0.693147}{0.016} \approx 43.3 \text{ yr}$

Thus, according to this model, the 1990 population will double in the year 2033. (No wonder ecologists are alarmed!) ◆

The Rule of 70

The relationship between doubling time T and interest rate k is the basis of a rule often used in the investment world, called the *Rule of 70*. To estimate how long it will take to double your money at varying rates of return, divide 70 by the rate of return. To see how this works, let the interest rate $k = r\%$. Then

$$T = \frac{\ln 2}{k} = \frac{0.693147}{r\%} = \frac{0.693147}{r \times 0.01}$$

$$= \frac{0.693147}{r \times 0.01} \cdot \frac{100}{100} = \frac{69.3147}{r} \approx \frac{70}{r}.$$

Modeling Other Phenomena

EXAMPLE 7 *Life science: Alcohol absorption and the risk of having an accident.* Extensive research has provided data relating the risk R (in percent) of having an automobile accident to the blood alcohol level b (in percent). Note that these data are not a perfect fit (see the part of the graph between $b = 0$ and $b = 0.05$), but we can approximate the data with an exponential function. The modeling assumption is that the rate of change of the risk R with respect to the blood alcohol level b is given by

$$\frac{dR}{db} = kR.$$

a) Find the function that satisfies the equation. Assume that $R_0 = 1\%$.

Some myths about alcohol. It's a fact—the blood alcohol concentration (BAC) in the human body is measurable. And there's no cure for its effect on the central nervous system except time. It takes time for the body's metabolism to recover. That means a cup of coffee, a cold shower, and fresh air can't erase the effect of several drinks.

There are variables, of course: a person's body weight, how many drinks have been consumed in a given time, how much has been eaten, and so on. These account for different BAC levels. But the myth that some people can "handle their liquor" better than others is a gross rationalization—especially when it comes to driving. Some people can act more sober than others. But an automobile doesn't act; it reacts.

b) Find k using the data point $R(0.14) = 20$. (This is how one might fit an exponential equation to the data.)

c) Rewrite $R(b)$ in terms of k.

d) At what blood alcohol level will the risk of having an accident be 100%? Round to the nearest hundredth.

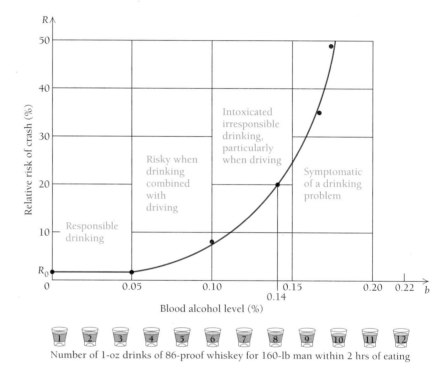

Number of 1-oz drinks of 86-proof whiskey for 160-lb man within 2 hrs of eating

Solution

a) Since both R and b are percents, we omit the % symbol for ease of computation. The solution is

$$R(b) = e^{kb}, \quad \text{since } R_0 = 1.$$

b) We solve this equation for k, using natural logarithms:

$$20 = e^{k(0.14)} = e^{0.14k}$$

$$\ln 20 = \ln e^{0.14k}$$

$$\ln 20 = 0.14k$$

$$\frac{\ln 20}{0.14} = k$$

$$\frac{2.995732}{0.14} = k$$

$$21.4 \approx k. \qquad \text{Rounded to the nearest tenth}$$

c) $R(b) = e^{21.4b}$

d) We substitute 100 for $R(b)$ and solve the equation for b:

$$100 = e^{21.4b}$$

$$\ln 100 = \ln e^{21.4b}$$

$$\ln 100 = 21.4b$$

$$\frac{\ln 100}{21.4} = b$$

$$\frac{4.605170}{21.4} = b$$

$$0.22 \approx b. \qquad \text{Rounded to the nearest hundredth}$$

The calculations done in this problem can be performed more conveniently on your calculator if you do not stop to round. For example, in part (b), we find $\ln 20$ and divide by 0.14, obtaining $21.39808767\ldots$. We then use that value for k in part (d). Answers will be found that way in the exercises. You may note some variance in the last one or two decimal places if you round as you go.

Thus when the blood alcohol level is 0.22%, according to this model, the risk of an accident is 100%. From the graph, we see that this would occur for a 160-lb man after 12 1-oz drinks of 86-proof whiskey. "Theoretically," the model tells us that after 12 drinks of whiskey, one is "sure" to have an accident. This might be questioned in reality, since a person who has had 12 drinks might not be able to drive at all.   ◆

Models of Limited Growth

The growth model $P(t) = P_0 e^{kt}$ has many applications, as we have seen thus far in this section. It seems reasonable that there can be factors that prevent a population from exceeding some limiting value L—perhaps a limitation on food, living space, or other natural resources. One model of such growth is

$$P(t) = \frac{L}{1 + be^{-kt}},$$

which is called the *logistic equation*.

EXAMPLE 8   *Life science: Limited population growth.* A ship carrying 1000 passengers has the misfortune to be shipwrecked on a small island from which the passengers are never rescued. The natural resources of the island limit the growth of the population to a *limiting value* of 5780, to which the population gets closer and closer but which it never reaches. The population of the island after time t, in years, is given by the logistic equation

$$P(t) = \frac{5780}{1 + 4.78e^{-0.4t}}.$$

a) Find the population after 0 years; 1 year; 2 years; 5 years; 10 years; 20 years.

b) Find the rate of change $P'(t)$.

c) Sketch a graph of the function.

Solution

a) We use a calculator to find the function values, listing them in a table, as shown at left.

t	$P(t)$, approximately
0	1000
1	1375
2	1836
5	3510
10	5315
20	5771

b) We find the rate of change $P'(t)$ as follows, using the Quotient Rule:

$$P'(t) = \frac{(1 + 4.78e^{-0.4t}) \cdot 0 - [(-0.4)(4.78)e^{-0.4t}](5780)}{(1 + 4.78e^{-0.4t})^2}$$

$$= \frac{11{,}051.36e^{-0.4t}}{(1 + 4.78e^{-0.4t})^2}.$$

c) We consider the curve-sketching procedure discussed in Section 3.2. The derivative $P'(t)$ exists for all real numbers t; thus there are no critical points for which the derivative does not exist. The equation $P'(t) = 0$ holds only when the numerator is 0. But the numerator is positive for all real numbers t. Thus the function has no critical points, and hence no relative extrema.

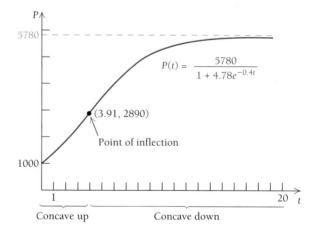

$$P(t) = \frac{5780}{1 + 4.78e^{-0.4t}}$$

(3.91, 2890)

Point of inflection

Concave up   Concave down

Since $e^{-0.4t}$ is positive for all real numbers t, $P'(t)$ is positive for all real numbers t. Thus, P is increasing over the entire real line. Though we will not develop it here, the second derivative can be used to show that the graph has a point of inflection at (3.91, 2890). This function is concave up on the interval (0, 3.91) and concave down on the interval (3.91, ∞). The graph is the S-shaped curve shown above. ◆

Another model of limited growth is provided by the function

$$P(t) = L(1 - e^{-kt}),$$

which is graphed below. This function also increases over the entire interval $[0, ∞)$.

This island may have achieved its limited population growth.

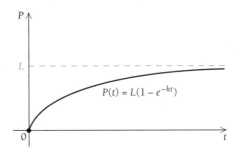

4.3 Exercise Set

1. Find the function that satisfies the equation $dQ/dt = kQ$.
2. Find the function that satisfies the equation $dR/dt = kR$.

APPLICATIONS

◆ **Business and Economics**

3. *Compound interest.* Suppose P_0 is invested in a savings account in which interest is compounded continuously at 6.5% per year. That is, the balance P grows at the rate given by

$$\frac{dP}{dt} = 0.065P.$$

   a) Find the function that satisfies the equation. List it in terms of P_0 and 0.065.
   b) Suppose $1000 is invested. What is the balance after 1 yr? after 2 yr?
   c) When will an investment of $1000 double itself?

4. *Compound interest.* Suppose P_0 is invested in a savings account in which interest is compounded continuously at 8% per year. That is, the balance P grows at the rate given by

$$\frac{dP}{dt} = 0.08P.$$

   a) Find the function that satisfies the equation. List it in terms of P_0 and 0.08.
   b) Suppose $20,000 is invested. What is the balance after 1 yr? after 2 yr?
   c) When will an investment of $20,000 double itself?

5. *Annual interest rate.* A bank advertises that it compounds interest continuously and that it will double your money in 10 yr. What is its annual interest rate?

6. *Annual interest rate.* A bank advertises that it compounds interest continuously and that it will double your money in 12 yr. What is its annual interest rate?

7. *Franchise expansion.* A national hamburger firm is selling franchises throughout the country. The president estimates that the number of franchises N will increase at the rate of 10% per year, that is,

$$\frac{dN}{dt} = 0.10N.$$

   a) Find the function that satisfies the equation. Assume that the number of franchises at $t = 0$ is 50.
   b) How many franchises will there be in 20 yr?
   c) After what period of time will the initial number of 50 franchises double?

8. *Franchise expansion.* Pizza, Unltd., a national pizza firm, is selling franchises throughout the country. The president estimates that the number of franchises N will increase at the rate of 15% per year, that is,

$$\frac{dN}{dt} = 0.15N.$$

   a) Find the function that satisfies the equation. Assume that the number of franchises at $t = 0$ is 40.
   b) How many franchises will there be in 20 yr?
   c) After what period of time will the initial number of 40 franchises double?

9. *Oil demand.* The growth rate of the demand for oil in the United States is 10% per year. When will the demand be double that of 1996?

10. **Coal demand.** The growth rate of the demand for coal in the world is 4% per year. When will the demand be double that of 1996?

11. **Value of a Van Gogh painting.** The Van Gogh painting "Irises," shown below, sold for $84,000 in 1947, but sold again for $53,900,000 in 1987. Assuming the exponential model:

a) Find the value k ($V_0 = \$84,000$), and write the function.
b) Estimate the value of the painting in the year 1997.
c) What is the doubling time for the value of the painting?
d) How long after 1947 will the value of the painting be $1 billion?

Van Gogh's "Irises," a 28-by-32-in. oil on canvas.

12. **Cost of a double-dip ice cream cone.** In 1970, the cost of a double-dip ice cream cone was 52¢. In 1978, it was 66¢. Assuming the exponential model:

a) Find the value k ($P_0 = 52$), and write the function.
b) Estimate the cost of a cone in 1998.
c) What is the doubling time for the cost of a double-dip ice cream cone?

13. **Consumer price index.** The *consumer price index* compares the costs of goods and services over various years, where 1967 is used as a base ($t = 0$). The same goods and services that cost $100 in 1967 cost $184.50 in 1977. Assuming the exponential model:

a) Find the value k ($P_0 = \$100$), and write the function.
b) Estimate what the same goods and services will cost in 1997.

c) After what period of time did the same goods and services cost double that of 1967?

14. **Job opportunities.** It is estimated that there were 714,000 accountants employed in 1972 and 935,000 in 1985. Assuming the exponential model:

a) Find the value k ($P_0 = 714,000$), and write the function.
b) Estimate the number of accountants needed in 1996.
c) After what period of time will the need for accountants be double that of 1972?

15. **Cellular phones.** The number of cellular phones has grown exponentially in recent years. In 1985, there were 203,600 cellular phones in use. This number increased to 19.3 million in 1994. Find the growth rate k and write an exponential function. Then use this function to predict the number of cellular phones in 2000.

16. **Total revenue.** Intel, a computer chip manufacturer, reported $1265 million in total revenue in 1986. In 1994, the total revenue was $11,521 million. Assuming the exponential model, find the growth rate k and write the function. Then estimate the total revenue in 2006.

17. **Value of Manhattan Island.** Peter Minuit of the Dutch West India Company purchased Manhattan Island from the Indians in 1626 for $24 worth of merchandise. Assuming an exponential rate of inflation of 8%, how much will Manhattan be worth in 1998?

18. **Cost of a first-class postage stamp.** The cost of a first-class postage stamp in 1962 was 4¢. In 1995, it was 32¢. This was exponential growth. What was the growth rate? What will be the cost of a first-class postage stamp in 1998? in 2004?

19. **Average salary of major-league baseball players.** In 1970, the average salary of major-league baseball players was $29,303. By 1994, the average salary had grown to $1,200,000. This was exponential growth. What was the growth rate? What will the average salary be in 1997? in 2000?

20. **Cost of a Hershey bar.** The cost of a Hershey bar in 1962 was $0.05, and was increasing at an exponential growth rate of 9.7%. What will the cost of a Hershey bar be in 1996? in 1998?

21. **Effect of advertising.** A company introduces a new product on a trial run in a city. They advertised the product on television and found that the percentage P

of people who bought the product after t ads were run satisfied the function

$$P(t) = \frac{100\%}{1 + 49e^{-0.13t}}.$$

a) What percentage buy the product without seeing the ad ($t = 0$)?
b) What percentage buy the product after the ad is run 5 times? 10 times? 20 times? 30 times? 50 times? 60 times?
c) Find the rate of change $P'(t)$.
d) Sketch a graph of the function.

◆   **Life and Physical Sciences**

22. *Population growth.* The growth rate of the population of Alaska is 2.8% per year (one of the highest of the fifty states). What is the doubling time?

23. *Population growth.* The growth rate of the population of Central America is 3.5% per year (one of the highest in the world). What is the doubling time?

24. *Population growth.* The population of western Europe was 430 million in 1961. It was estimated that the population was growing exponentially at the rate of 1% per year, that is,

$$\frac{dP}{dt} = 0.01P.$$

a) Find the function that satisfies the equation. Assume that $P_0 = 430$ and $k = 0.01$.
b) Estimate the population of Europe in 1996.
c) After what period of time will the population be double that of 1961?

25. *Population growth.* The population of the United States was 241 million in 1986. It was estimated that the population P was growing exponentially at the rate of 0.9% per year, that is,

$$\frac{dP}{dt} = 0.009P,$$

where t = the time, in years.

a) Find the function that satisfies the equation. Assume that $P_0 = 241$ and $k = 0.009$.
b) Estimate the population of the United States in 1998 ($t = 12$).
c) After what period of time will the population be double that in 1986?

26. *Blood alcohol level.* In Example 7 (on alcohol absorption), at what blood alcohol level will the risk of an accident be 90%?

27. *Blood alcohol level.* In Example 7, at what blood alcohol level will the risk of an accident be 80%?

28. *Female Olympic athletes.* In 1985, the number of female athletes participating in summer Olympic-type games was 500. In 1996, it is estimated that 3600 will participate in the Summer Olympics in Atlanta. Assuming the exponential model:

a) Find the value of k ($P_0 = 500$), and write the function.
b) Estimate the number of female athletes participating in 2000.

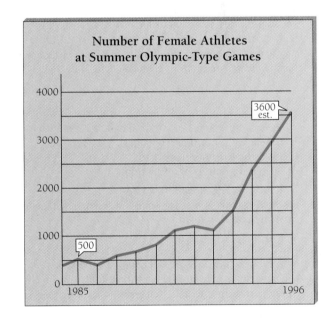

29. *Limited population growth.* A lake is stocked with 400 fish of a new variety. The size of the lake, the availability of food, and the number of other fish restrict growth in the lake to a *limiting value* of 2500. The population of fish in the lake after time t, in months, is given by

$$P(t) = \frac{2500}{1 + 5.25e^{-0.32t}}.$$

a) Find the population after 0 months; 1 month; 5 months; 10 months; 15 months; 20 months.
b) Find the rate of change $P'(t)$.
c) Sketch a graph of the function.

30. *Bicentennial growth of the United States.* The population of the United States in 1776 was about 2,508,000. In its bicentennial year, the population was about 216,000,000. Assuming the exponential model, what was the growth rate of the United States through its bicentennial year?

◆ **Social Sciences**

31. *Diffusion of information.* Pharmaceutical firms spend an immense amount of money in order to test a new medication. After the drug is approved by the Federal Drug Administration, it still takes time for physicians to fully accept and make use of the medication. The use approaches a *limiting value* of 100%, or 1, after time t, in months. Suppose that for a new cancer medication the percentage P of physicians using the product after t months is given by

$$P(t) = 100\%(1 - e^{-0.4t}).$$

a) What percentage of doctors have accepted the medication after 0 months? 1 month? 2 months? 3 months? 5 months? 12 months? 16 months?
b) Find the rate of change $P'(t)$.
c) Sketch a graph of the function.

32. *Hullian learning model.* The Hullian learning model asserts that the probability p of mastering a task after t learning trials is given by

$$p(t) = 1 - e^{-kt},$$

where k is a constant that depends on the task to be learned. Suppose a new task is introduced in a factory on an assembly line. For that particular task, the constant $k = 0.28$.

a) What is the probability of learning the task after 1 trial? 2 trials? 5 trials? 11 trials? 16 trials? 20 trials?
b) Find the rate of change $p'(t)$.
c) Sketch a graph of the function.

SYNTHESIS

Business: Effective annual yield. Suppose $100 is invested at 7%, compounded continuously, for 1 year. We know from Example 4 that the balance will be $107.25. This is

the same as if $100 were invested at 7.25% and compounded once a year (simple interest). The 7.25% is called the *effective annual yield.* In general, if P_0 is invested at k% compounded continuously, then the effective annual yield is that number i satisfying $P_0(1 + i) = P_0 e^k$. Then $1 + i = e^k$, or

$$\text{Effective annual yield} = i = e^k - 1.$$

33. An amount is invested at 7.3% per year compounded continuously. What is the effective annual yield?

34. An amount is invested at 8% per year compounded continuously. What is the effective annual yield?

35. The effective annual yield on an investment compounded continuously is 9.42%. At what rate was it invested?

36. The effective annual yield on an investment compounded continuously is 6.61%. At what rate was it invested?

37. Find an expression relating the growth rate k and the *tripling time* T_3.

38. Find an expression relating the growth rate k and the *quadrupling time* T_4.

39. Gather data concerning population growth in your city. Estimate the population in 1996; in 2010.

40. A quantity Q_1 grows exponentially with a doubling time of 1 year. A quantity Q_2 grows exponentially with a doubling time of 2 years. If the initial amounts of Q_1 and Q_2 are the same, when will Q_1 be twice the size of Q_2?

41. A growth rate of 100% per day corresponds to what exponential growth rate per hour?

42. Show that any two measurements of an exponentially growing population will determine k. That is, show that if y has the values y_1 at t_1 and y_2 at t_2, then

$$k = \frac{\ln (y_2/y_1)}{t_2 - t_1}.$$

43. ◈ Complete this table relating growth rate k and doubling time T.

Growth rate (k), % per year	1%	2%			14%
Doubling time (T), in years			15	10	

Graph $T = \ln 2/k$. Is this a linear relationship? Explain.

44. ◈ Explain the Rule of 70 to a fellow student.

TECHNOLOGY CONNECTION

45. The concept of *regression* can be used to fit an exponential equation to a set of data. Again, the development of this topic will not be considered until Section 7.5. The grapher may give a value of a constant r, called a *correlation coefficient*. The closer r is to 1, the better the data fits the proposed equation. Use a grapher that does regression.

Year (t)	CD sales, in millions
1985 ($t = 0$)	20
1989 ($t = 4$)	210
1993 ($t = 8$)	495.4

a) Fit an exponential function $N(t) = Be^{kt}$ to the following data regarding CD music recording sales in various years. Be sure to use all the data.

b) Graph the function.

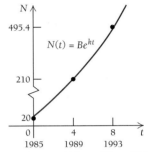

c) How many CDs will be sold in 1999?
d) In what year were 1 billion CDs sold?

4.4 Applications: Decay

OBJECTIVES

- Find a function that satisfies $dP/dt = -kP$.
- Convert between decay rate and half-life.
- Solve application problems involving exponential decay.

In the equation of population growth $dP/dt = kP$, the constant k is actually given by

$$k = (\text{Birth rate}) - (\text{Death rate}).$$

Thus a population "grows" only when the *birth rate* is greater than the *death rate*. When the birth rate is less than the death rate, k will be negative so the population will be decreasing, or "decaying," at a rate proportional to its size. For convenience in our computations, we will express such a negative value as $-k$, where $k > 0$. The equation

$$\frac{dP}{dt} = -kP, \quad \text{where } k > 0,$$

shows P to be *decreasing* as a function of time, and the solution

$$P(t) = P_0 e^{-kt}$$

shows it to be decreasing exponentially. This is called **exponential decay.** The amount present initially at $t = 0$ is again P_0.

TECHNOLOGY
CONNECTION

Using the same set of axes,
graph $y_1 = e^{2x}$ and $y_2 = e^{-2x}$. Compare the graphs.
   Then, using the same
set of axes, graph $y_1 = 100e^{-0.06x}$ and $y_2 = 100e^{0.06x}$.
Compare these graphs.

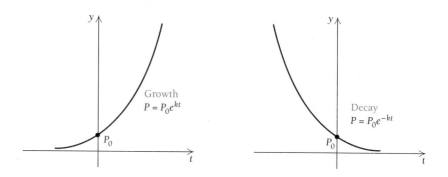

Radioactive Decay

Radioactive elements decay exponentially; that is, they disintegrate at a rate that is proportional to the amount present.

EXAMPLE 1 *Life science: Decay.* Strontium-90 has a decay rate of 2.8% per year. The rate of change of an amount N is given by

$$\frac{dN}{dt} = -0.028N.$$

a) Find the function that satisfies the equation in terms of N_0 (the amount present at $t = 0$).

b) Suppose 1000 grams of strontium-90 is present at $t = 0$. How much will remain after 100 years?

c) After how long will half of the 1000 grams remain?

Solution

a) $N(t) = N_0 e^{-0.028t}$

b)
$$\begin{aligned}
N(100) = 1000e^{-0.028(100)} &= 1000e^{-2.8} \\
&= 1000(0.060810) \\
&\approx 60.8 \text{ grams}
\end{aligned}$$

c) We are asking, "At what time T will $N(T) = 500$?" The number T is called the **half-life.** To find T, we solve the equation

$$500 = 1000e^{-0.028T}$$

$$\frac{1}{2} = e^{-0.028T}$$

$$\ln \frac{1}{2} = \ln e^{-0.028T}$$

$$\ln 1 - \ln 2 = -0.028T$$

$$0 - \ln 2 = -0.028T.$$

Then

$$\frac{-\ln 2}{-0.028} = T$$

$$\frac{\ln 2}{0.028} = T$$

$$\frac{0.693147}{0.028} = T$$

$$25 \approx T.$$

Thus the half-life of strontium-90 is 25 years.   ◆

We can find a general expression relating the decay rate k and the half-life T by solving the equation

$$\tfrac{1}{2}P_0 = P_0 e^{-kT}$$

$$\tfrac{1}{2} = e^{-kT}$$

$$\ln \tfrac{1}{2} = \ln e^{-kT}$$

$$\ln 1 - \ln 2 = -kT$$

$$0 - \ln 2 = -kT$$

$$-\ln 2 = -kT$$

$$\ln 2 = kT.$$

Again, we have the following.

THEOREM 11

The *decay rate* k and the *half-life* T are related by

$$kT = \ln 2 = 0.693147,$$

or

$$k = \frac{\ln 2}{T}$$

and

$$T = \frac{\ln 2}{k}.$$

Thus the half-life T depends only on the decay rate k. In particular, it is independent of the initial population size.

The effect of half-life is shown in the radioactive decay curve that follows.

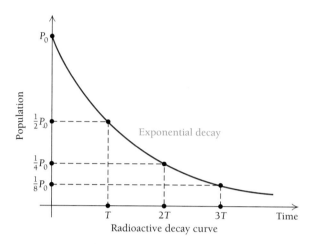

Radioactive decay curve

How can scientists determine that an animal bone has lost 30% of its carbon-14? The assumption is that the percentage of carbon-14 in the atmosphere and in living plants and animals is the same. When a plant or animal dies, the amount of carbon-14 decays exponentially. The scientist burns the animal bone and uses a Geiger counter to determine the percentage of the smoke that is carbon-14. The amount by which this varies from the percentage in the atmosphere indicates how much carbon-14 has been lost.

The process of carbon-14 dating was developed by the American chemist Willard F. Libby in 1952. It is known that the radioactivity in a living plant is 16 disintegrations per gram per minute. Since the half-life of carbon-14 is 5750 years, a dead plant with an activity of 8 disintegrations per gram per minute is 5750 years old, one with an activity of 4 disintegrations per gram per minute is 11,500 years old, and so on. Carbon-14 dating can be used to measure the age of objects from 30,000 to 40,000 years old. Beyond such an age, it is too difficult to measure the radioactivity, and some other method would have to be used.

Note that the exponential function gets close to, but never reaches, 0 as t gets larger. Thus, in theory, a radioactive substance never completely decays.

EXAMPLE 2   *Life science: Half-life.* Plutonium, a common product and ingredient of nuclear reactors, is of great concern to those who are against the building of nuclear reactors. Its decay rate is 0.003% per year. What is its half-life?

Solution

$$T = \frac{\ln 2}{k} = \frac{0.693147}{0.00003} \approx 23{,}100 \text{ years}$$ ◆

EXAMPLE 3   *Life science: Carbon dating.* The radioactive element carbon-14 has a half-life of 5750 years. The percentage of carbon-14 present in the remains of plants and animals can be used to determine age. Archaeologists found that the linen wrapping from one of the Dead Sea Scrolls had lost 22.3% of its carbon-14. How old was the linen wrapping?

Solution

a) Find the decay rate k:

$$k = \frac{\ln 2}{T} = \frac{0.693147}{5750} \approx 0.0001205, \quad \text{or} \quad 0.01205\% \text{ per year.}$$

b) Find the exponential equation for the amount $N(t)$ that remains from an initial amount N_0 after t years:

$$N(t) = N_0 e^{-0.0001205t}.$$

(*Note:* This equation can be used for any subsequent carbon-dating problem.)

In 1947, a Bedouin youth looking for a stray goat climbed into a cave at Kirbet Qumran on the shores of the Dead Sea near Jericho and came upon earthenware jars containing an incalculable treasure of ancient manuscripts. Shown here are fragments of those so-called Dead Sea Scrolls, a portion of some 600 or so texts found so far and which concern the Jewish books of the Bible. Officials date them before 70 A.D., making them the oldest Biblical manuscripts by 1000 years.

c) If the Dead Sea Scrolls lost 22.3% of their carbon-14 from an initial amount P_0, then 77.7% (P_0) is the amount present. To find the age t of the scrolls, we solve the following equation for t:

$$77.7\% \ N_0 = N_0 e^{-0.0001205t}$$
$$0.777 = e^{-0.0001205t}$$
$$\ln 0.777 = \ln e^{-0.0001205t}$$
$$\ln 0.777 = -0.0001205t$$
$$-0.2523149 = -0.0001205t$$
$$\frac{0.2523149}{0.0001205} = t$$
$$2094 \approx t.$$

Thus the linen wrapping of the Dead Sea Scrolls is about 2094 years old. ◆

Present Value

A representative of a financial institution is often asked to solve a problem like the following.

EXAMPLE 4   *Business: Present value.* Following the birth of a child, a parent wants to make an initial investment of P_0 that will grow to $10,000 by the child's 20th birthday. Interest is compounded continuously at 8%. What should the initial investment be?

Solution   Using the equation $P = P_0 e^{kt}$, we find P_0 such that

$$10,000 = P_0 e^{0.08 \cdot 20},$$

or

$$10,000 = P_0 e^{1.6}.$$

Now

$$\frac{10,000}{e^{1.6}} = P_0,$$

or

$$10,000 e^{-1.6} = P_0,$$

and, using a calculator, we have

$$P_0 = 10,000 e^{-1.6}$$
$$= 10,000(0.201897)$$
$$= \$2018.97.$$

Thus the parent must deposit $2018.97, which will grow to $10,000 by the child's 20th birthday. ◆

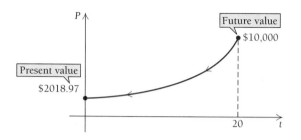

Economists call $2018.97 the *present value* of $10,000 due 20 years from now at 8%, compounded continuously. The process of computing present value is called *discounting*.

Computing present value can be interpreted as exponential decay from the future back to the present.

In general, the present value P_0 of an amount P due t years later is found by solving the following equation for P_0:

$$P_0 e^{kt} = P$$

$$P_0 = \frac{P}{e^{kt}} = Pe^{-kt}.$$

THEOREM 12

The *present value* P_0 of an amount P due t years later at interest rate k, compounded continuously, is given by

$$P_0 = Pe^{-kt}.$$

Newton's Law of Cooling

Consider the following situation. A hot cup of soup, at a temperature of 200°, is placed in a room whose temperature is 70°. The temperature of the soup cools over time t, in minutes, according to the mathematical model, or equation, called **Newton's Law of Cooling.**

Newton's Law of Cooling

The temperature T of a cooling object drops at a rate that is proportional to the difference $T - C$, where C is the constant temperature of the surrounding medium. Thus,

$$\frac{dT}{dt} = -k(T - C). \tag{1}$$

The function that satisfies Eq. (1) is

$$T = T(t) = ae^{-kt} + C. \tag{2}$$

We can check this by differentiating:

$$\frac{dT}{dt} = -kae^{-kt}$$

$$= -k(ae^{-kt})$$

$$= -k(T - C).$$

EXAMPLE 5   *Physical science: Cooling.* The temperature of a cup of freshly brewed coffee is 200° and the room temperature is 70°. The temperature of the coffee cools to 190° in 5 minutes.

a) What is the temperature after 10 minutes?

b) How long does it take the cup of coffee to cool to 90°?

Solution

a) We first find the value of a in Eq. (2). At $t = 0$, $T = 200°$. We solve the following equation for a:

$$200 = ae^{-k \cdot 0} + 70$$

$$200 = a \cdot 1 + 70$$

$$130 = a.$$

Now we find k using the fact that at $t = 5$, $T = 190°$. We solve the following equation for k:

$$190 = 130e^{-k \cdot 5} + 70$$

$$120 = 130e^{-5k}$$

$$\frac{12}{13} = e^{-5k}$$

$$\ln \frac{12}{13} = \ln e^{-5k}$$

$$\ln \frac{12}{13} = -5k$$

$$-5k = -0.080043 \qquad \text{Divide 12 by 13 and take the}$$
$$\text{natural logarithm.}$$

$$k \approx 0.016.$$

Now we find the temperature at $t = 10$:

$$T(10) = 130e^{-0.016(10)} + 70$$

$$= 130e^{-0.16} + 70$$

$$= 130(0.852144) + 70$$

$$\approx 181°.$$

b) To find how long it will take for the cup of coffee to cool to 90°, we solve for t:

$$90 = 130e^{-0.016t} + 70$$
$$20 = 130e^{-0.016t}$$
$$\frac{2}{13} = e^{-0.016t}.$$

Thus,

$$\ln \frac{2}{13} = \ln e^{-0.016t}$$
$$-1.871802 = -0.016t$$
$$t \approx 117 \text{ min.}$$

The graph of $T(t) = ae^{-kt} + C$ shows that $\lim_{t \to \infty} T(t) = C$. The temperature of the object decreases toward the temperature of the surrounding medium.

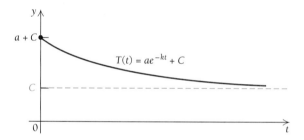

Mathematically, this model tells us that the temperature never reaches C, but in practice this happens eventually. At least, the temperature of the cooling object gets so close to that of the surrounding medium that no device could detect a difference. Let us now see how Newton's Law of Cooling can be used in solving a crime.

EXAMPLE 6   *Life science: When was the murder committed?* The police discover the body of a calculus professor. Critical to solving the crime is determining when the murder was committed. The police call the coroner, who arrives at 12 noon. He immediately takes the temperature of the body and finds it to be 94.6°. He waits 1 hour, takes the temperature again, and finds it to be 93.4°. He also notes that the temperature of the room is 70°. When was the murder committed?

Solution   We first find a in the equation $T(t) = ae^{-kt} + C$. Assuming that the temperature of the body was normal when the murder occurred, we have $T = 98.6°$ at $t = 0$. Thus,

$$98.6 = ae^{-k \cdot 0} + 70,$$

so $a = 28.6$.

Thus, T is given by $T(t) = 28.6e^{-kt} + 70$.

We want to find the number of hours N since the murder was committed. To find N, we must first determine k. From the two temperature readings the coroner made, we have

$$94.6 = 28.6e^{-kN} + 70, \quad \text{or} \quad 24.6 = 28.6e^{-kN}; \tag{3}$$

$$93.4 = 28.6e^{-k(N+1)} + 70, \quad \text{or} \quad 23.4 = 28.6e^{-k(N+1)}. \tag{4}$$

Dividing Eq. (3) by Eq. (4), we get

$$\frac{24.6}{23.4} = \frac{28.6e^{-kN}}{28.6e^{-k(N+1)}}$$

$$= e^{-kN+k(N+1)} = e^{-kN+kN+k} = e^{k}.$$

We solve this equation for k:

$$\ln \frac{24.6}{23.4} = \ln e^{k} \qquad \text{Taking the natural logarithm on both sides}$$

$$0.05 \approx k.$$

Now we substitute back into Eq. (3) and solve for N:

$$24.6 = 28.6e^{-0.05N}$$

$$\frac{24.6}{28.6} = e^{-0.05N}$$

$$\ln \frac{24.6}{28.6} = \ln e^{-0.05N}$$

$$-0.150660 = -0.05N$$

$$3 \approx N.$$

Since the coroner arrived at 12 noon, the murder was committed at about 9:00 A.M.

◆

4.4 Exercise Set

APPLICATIONS

◆ **Business and Economics**

1. *Present value.* Following the birth of a child, a parent wants to make an initial investment P_0 that will grow to $5000 by the child's 20th birthday. Interest is compounded continuously at 6%. What should the initial investment be?

2. *Present value.* Following the birth of a child, a parent wants to make an initial investment P_0 that will grow to $5000 by the child's 20th birthday. Interest is com-
pounded continuously at 7.2%. What should the initial investment be?

3. *Present value.* Find the present value of $60,000 due 8 years later at 6.4%, compounded continuously.

4. *Present value.* Find the present value of $50,000 due 16 years later at 8.2%, compounded continuously.

5. *Salvage value.* A business estimates that the salvage value V of a piece of machinery after t years is given by

$$V(t) = \$40,000e^{-t}.$$

a) What did the machinery cost initially?

b) What is the salvage value after 2 yr?

6. *Supply and demand.* The supply and demand for the sale of stereos by a sound company are given by

$$x = S(p) = \ln p \quad \text{and} \quad x = D(p) = \ln \frac{163,000}{p},$$

where $S(p) =$ the number of stereos that the company is willing to sell at price p and $D(p) =$ the quantity that the public is willing to buy at price p. Find the equilibrium point. (For reference, see Section 1.5.)

7. *Present value.* A person knows that a trust fund will yield \$80,000 in 13 years. A CPA is preparing a financial statement for this client and wants to take into account the present value of the trust fund in computing the client's net worth. Interest is compounded continuously at 8.3%. What is the present value of the trust fund?

8. *Consumer price index.* The consumer price index compares the costs of goods and services over various years, where 1967 is used as a base. The same goods and services that cost \$100 in 1967 cost \$42 in 1940. Assuming the exponential-decay model:

a) Find the value k, and write the equation. Round to the nearest hundredth.

b) Estimate what the same goods and services cost in 1900.

◆   **Life and Physical Sciences**

9. *Decay rate.* The decay rate of iodine-131 is 9.6% per day. What is its half-life?

10. *Decay rate.* The decay rate of krypton-85 is 6.3% per year. What is its half-life?

11. *Half-life.* The half-life of polonium is 3 min. What is its decay rate?

12. *Half-life.* The half-life of lead is 22 yr. What is its decay rate?

13. *Half-life.* Of an initial amount of 1000 g of polonium, how much will remain after 20 min? See Exercise 11 for the value of k.

14. *Half-life.* Of an initial amount of 1000 g of lead, how much will remain after 100 yr? See Exercise 12 for the value of k.

15. *Carbon dating.* How old is a piece of wood that has lost 90% of its carbon-14?

16. *Carbon dating.* How old is an ivory tusk that has lost 40% of its carbon-14?

17. *Carbon dating.* How old is a Chinese artifact that has lost 60% of its carbon-14?

18. *Carbon dating.* How old is a skeleton that has lost 50% of its carbon-14?

19. In a *chemical reaction,* substance A decomposes at a rate proportional to the amount of A present.

a) Write an equation relating A to the amount left of an initial amount A_0 after time t.

b) It is found that 8 g of A will reduce to 4 g in 3 hr. After how long will there be only 1 g left?

20. In a *chemical reaction,* substance A decomposes at a rate proportional to the amount of A present.

a) Write an equation relating A to the amount left of an initial amount A_0 after time t.

b) It is found that 10 lb of A will reduce to 5 lb in 3.3 hr. After how long will there be only 1 lb left?

21. *Weight loss.* The initial weight of a starving animal is W_0. Its weight W after t days is given by

$$W = W_0 e^{-0.008t}.$$

a) What percentage of its weight does it lose each day?

b) What percentage of its initial weight remains after 30 days?

22. *Weight loss.* The initial weight of a starving animal is W_0. Its weight W after t days is given by

$$W = W_0 e^{-0.009t}.$$

a) What percentage of its weight does it lose each day?

b) What percentage of its initial weight remains after 30 days?

Beer–Lambert Law

A beam of light enters a medium such as water or smoky air with initial intensity I_0. Its intensity is decreased depending on the thickness (or concentration) of the medium. The intensity I at a depth (or concentration) of x units is given by

$$I = I_0 e^{-\mu x}.$$

The constant μ ("mu"), called the *coefficient of absorption,* varies with the medium.

---

23. *Light through sea water* has $\mu = 1.4$ when x is measured in meters.

    a) What percentage of I_0 remains at a depth of sea water that is 1 m? 2 m? 3 m?
    b) Plant life cannot exist below 10 m. What percentage of I_0 remains at 10 m?

24. *Light through smog.* Particulate concentrations of pollution reduce sunlight. In a smoggy area, $\mu = 0.01$ and $x =$ the concentration of particulates measured in micrograms per cubic meter. What percentage of an initial amount I_0 of sunlight passes through smog that has a concentration of 100 micrograms per cubic meter?

25. *Cooling.* The temperature of a hot liquid is 100° and the room temperature is 75°. The liquid cools to 90° in 10 min.

    a) Find the value of the constant a in Newton's Law of Cooling.
    b) Find the value of the constant k. Round to the nearest hundredth.
    c) What is the temperature after 20 min?
    d) How long does it take the liquid to cool to 80°?

26. *Cooling.* The temperature of a hot liquid is 100°. The liquid is placed in a refrigerator where the temperature is 40°, and it cools to 90° in 5 min.

    a) Find the value of the constant a in Newton's Law of Cooling.
    b) Find the value of the constant k. Round to the nearest hundredth.
    c) What is the temperature after 10 min?
    d) How long does it take the liquid to cool to 41°?

27. *Cooling body.* The coroner arrives at the scene of a murder at 11 P.M. She takes the temperature of the body and finds it to be 85.9°. She waits 1 hour, takes the temperature again, and finds it to be 83.4°. She notes that the room temperature is 60°. When was the murder committed?

28. *Cooling body.* The coroner arrives at the scene of a murder at 2 A.M. He takes the temperature of the body and finds it to be 61.6°. He waits 1 hour, takes the temperature again, and finds it to be 57.2°. The body is in a meat freezer, where the temperature is 10°. When was the murder committed?

29. *Population decrease of Cincinnati.* The population of Cincinnati was 453,000 in 1970 and 385,000 in 1980. Assuming the population is decreasing according to

the exponential-decay model:

    a) Find the value k, and write the equation.
    b) Estimate the population of Cincinnati in 2000.
    c) After how long (theoretically) will Cincinnati have just 1 person?

30. *Population decrease of Panama.* The population of Panama was 1,464,000 in 1970 and 1,260,000 in 1980. Assuming the population is decreasing according to the exponential-decay model:

    a) Find the value k, and write the equation.
    b) Estimate the population of Panama in 2000.
    c) After how long will the population of Panama be 100,000?

31. *Population of the USSR in an earlier year.* The population of the USSR was 258 million in 1980 and was growing at the rate of 1% per year. What was the population in 1970? in 1940?

32. *Atmospheric pressure.* Atmospheric pressure P at altitude a is given by

$$P = P_0 e^{-0.00005a},$$

where $P_0 =$ the pressure at sea level. Assume that $P_0 = 14.7$ lb/in^2 (pounds per square inch).

    a) Find the pressure at an altitude of 1000 ft.
    b) Find the pressure at an altitude of 20,000 ft.
    c) At what altitude is the pressure 1.47 lb/in^2?

33. *Satellite power.* The power supply of a satellite is a radioisotope. The power output P, in watts, decreases at a rate proportional to the amount present; P is given by

$$P = 50e^{-0.004t},$$

where $t =$ the time, in days.

    a) How much power will be available after 375 days?
    b) What is the half-life of the power supply?
    c) The satellite's equipment cannot operate on fewer than 10 watts of power. How long can the satellite stay in operation?
    d) How much power did the satellite have to begin with?

34. *Carbon dating.* Recently, while digging in Chaco Canyon, New Mexico, archeologists found corn pollen that was 4000 years old. This was evidence that Indians had begun cultivating crops in the Southwest centuries earlier than scientists had thought. What percent of the carbon-14 had been lost from the pollen?

◆ **Social Sciences**

35. Decline of LP records. The sales S of LP records have declined considerably in the past 10 years because of the advent of the CD and cassette tapes. There were 205 million LP records sold in 1983 and 1.2 million records sold in 1993.

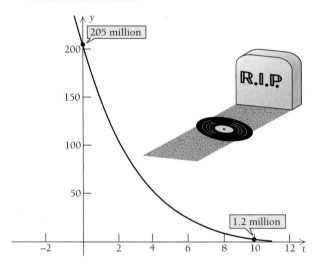

Assuming the sales are decreasing according to the exponential-decay model:

a) Find the value k, and write the equation.
b) Estimate the sales of LP records in the year 2000.
c) In what year (theoretically) will only 1 LP record be sold?

36. Decline in beef consumption. The annual consumption of beef B per person was about 80 lb in 1985 and about 67 lb in 1996. Assuming consumption is decreasing according to the exponential-decay model:

a) Find the value k, and write the equation.
b) Estimate the consumption of beef in the year 2000.
c) In what year (theoretically) will the consumption of beef be 20 lb per person?

SYNTHESIS

37. Economics: Supply and demand. The demand and supply functions for a certain type of VCR are given by

$$x = D(p) = 480e^{-0.003p} \quad \text{and} \quad x = S(p) = 150e^{0.004p}.$$

Find the equilibrium point.

38. ◈ **Newton's Law of Cooling.** Consider the following exploratory situation. Draw a glass of hot tap water. Place a thermometer in the glass and check the temperature. Check the temperature every 30 min thereafter. Plot your data on this graph, and connect the points with a smooth curve.

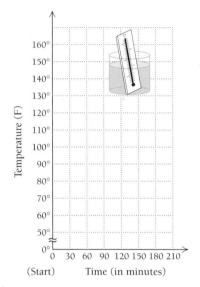

a) What was the temperature when you began?
b) At what temperature does there seem to be a leveling off of the graph?
c) What is the difference between your answers to parts (a) and (b)?
d) How does the temperature in part (b) compare with the room temperature?
e) Find an equation that "fits" the data. Use this equation to check values of other data points. How do they compare?
f) Is it ever "theoretically" possible for the temperature of the water to be the same as the room temperature? Explain.

39. ◈ An interest rate decreases from 8% to 7.2%. Explain why this increases the present value of an amount due 10 yr later.

4.5 The Derivatives of a^x and $\log_a x$

OBJECTIVES

• Convert exponential expressions to powers of e.
• Differentiate functions involving a^x and $\log_a x$.

The Derivative of a^x

To find the derivative of a^x, for any base a, we first express a^x as a power of e. In order to do this, we first prove the following additional property of a logarithm.

> **P7.** $b^{\log_b x} = x$

To prove this, let

$$y = \log_b x.$$

Then, by the definition of a logarithm,

$$b^y = x.$$

Substituting $\log_b x$ for y, we have

$$b^{\log_b x} = x.$$

We can now express a^x as a power of e. Using Property 7, where $b = e$ and $x = a$, we have

$$a = e^{\ln a}. \qquad \text{Remember: } \ln a = \log_e a.$$

Raising both sides to the power x, we get

$$a^x = (e^{\ln a})^x$$
$$= e^{x \cdot \ln a}. \qquad \text{Multiplying exponents}$$

Thus we have the following.

> **THEOREM 13**
>
> $$a^x = e^{x \cdot \ln a}$$

EXAMPLE 1   Express as a power of e.

a) 3^2 $\qquad 3^2 = e^{2 \cdot \ln 3}$
$\qquad\qquad\quad = e^{2(1.098612)}$
$\qquad\qquad\quad \approx e^{2.1972}$

b) 10^x $10^x = e^{x \cdot \ln 10}$

$$= e^{x(2.302585)}$$

$$\approx e^{2.3026x} \qquad \blacklozenge$$

Now we can differentiate.

EXAMPLE 2

$$\frac{d}{dx} 2^x = \frac{d}{dx} e^{x \cdot \ln 2} \qquad \text{Theorem 13}$$

$$= \left[\frac{d}{dx} (x \cdot \ln 2) \right] \cdot e^{x \cdot \ln 2}$$

$$= (\ln 2)(e^{\ln 2})^x$$

$$= (\ln 2) 2^x$$

We completed this by taking the derivative of $x \ln 2$ and replacing $e^{x \ln 2}$ by 2^x. Note that $\ln 2 \approx 0.7$, so the above verifies our earlier approximation of the derivative of 2^x as $(0.7)2^x$ (see Section 4.1). $\blacklozenge$

In general,

$$\frac{d}{dx} a^x = \frac{d}{dx} e^{x \cdot \ln a} \qquad \text{Theorem 13}$$

$$= \left[\frac{d}{dx} (x \cdot \ln a) \right] \cdot e^{x \cdot \ln a}$$

$$= (\ln a) \, a^x.$$

Thus we have the following.

THEOREM 14

$$\frac{d}{dx} a^x = (\ln a) \, a^x$$

EXAMPLE 3

$$\frac{d}{dx} 3^x = (\ln 3) \, 3^x \qquad \blacklozenge$$

EXAMPLE 4

$$\frac{d}{dx} (1.4)^x = (\ln 1.4)(1.4)^x \qquad \blacklozenge$$

Compare these formulas:

$$\frac{d}{dx} a^x = (\ln a)\, a^x \quad \text{and} \quad \frac{d}{dx} e^x = e^x.$$

It is the simplicity of the latter formula that is a reason for the use of base e in calculus. The many applications of e in natural phenomena provide other reasons.

One other result also follows from what we have done. If

$$f(x) = a^x,$$

we know that

$$f'(x) = a^x\, (\ln a).$$

In Section 4.1, we also showed that

$$f'(x) = a^x \left(\lim_{h \to 0} \frac{a^h - 1}{h} \right).$$

Since $a^x > 0$, we have the following.

THEOREM 15

$$\ln a = \lim_{h \to 0} \frac{a^h - 1}{h}$$

The Derivative of $\log_a x$

Just as the derivative of a^x is expressed in terms of $\ln a$, so too is the derivative of $\log_a x$. To find this derivative, we first express $\log_a x$ in terms of $\ln a$ using Property 7:

$$a^{\log_a x} = x.$$

Then

$$\ln (a^{\log_a x}) = \ln x$$

$$(\log_a x) \cdot \ln a = \ln x \qquad \text{By P3, treating } \log_a x \text{ as an exponent}$$

and

$$\log_a x = \boxed{\frac{1}{\ln a}} \cdot \ln x.$$

$$\underset{\text{constant}}{}$$

The derivative of $\log_a x$ follows.

THEOREM 16

$$\frac{d}{dx} \log_a x = \frac{1}{\ln a} \cdot \frac{1}{x}$$

Comparing this with

$$\frac{d}{dx} \ln x = \frac{1}{x},$$

we again see a reason for the use of base e in calculus.

EXAMPLE 5

$$\frac{d}{dx} \log_3 x = \frac{1}{\ln 3} \cdot \frac{1}{x}$$

◆

EXAMPLE 6

$$\frac{d}{dx} \log x = \frac{1}{\ln 10} \cdot \frac{1}{x} \qquad \log x = \log_{10} x$$

◆

EXAMPLE 7

$$\frac{d}{dx} x^2 \log x = x^2 \frac{1}{\ln 10} \cdot \frac{1}{x} + 2x \log x \qquad \text{By the Product Rule}$$

$$= \frac{x}{\ln 10} + 2x \log x, \quad \text{or} \quad x\left(\frac{1}{\ln 10} + 2 \log x\right)$$

◆

TECHNOLOGY
CONNECTION

Check the results of Examples 3–7 using your grapher. Then differentiate $y = \log_2 x$ and check the result on your grapher.

4.5 Exercise Set

Express as a power of e.

1. 5^4

2. 2^3

3. $(3.4)^{10}$

4. $(5.3)^{20}$

5. 4^k

6. 5^R

7. 8^{kT}

8. 10^{kR}

Differentiate.

9. $y = 6^x$

10. $y = 7^x$

11. $f(x) = 10^x$

12. $f(x) = 100^x$

13. $f(x) = x(6.2)^x$

14. $f(x) = x(5.4)^x$

15. $y = x^3 10^x$

16. $y = x^4 5^x$

17. $y = \log_4 x$

18. $y = \log_5 x$

19. $f(x) = 2 \log x$

20. $f(x) = 5 \log x$

21. $f(x) = \log \dfrac{x}{3}$

22. $f(x) = \log \dfrac{x}{5}$

23. $y = x^3 \log_8 x$ 24. $y = x \log_6 x$

APPLICATIONS

◆   Business and Economics

25. *Recycling aluminum cans.* It is known that one fourth of all aluminum cans distributed will be recycled each year. A beverage company distributes 250,000 cans. The number still in use after time t, in years, is given by

$$N(t) = 250{,}000\left(\tfrac{1}{4}\right)^t.$$

Find $N'(t)$.

26. *Double declining-balance depreciation.* An office machine is purchased for $5200. Under certain assumptions, its salvage value V depreciates according to a method called double declining balance, basically 80% each year, and is given by

$$V(t) = \$5200(0.80)^t,$$

where t is the time, in years. Find $V'(t)$.

◆   Life and Physical Sciences

Earthquake magnitude. The magnitude R (measured on the Richter scale) of an earthquake of intensity I is defined as

$$R = \log \frac{I}{I_0},$$

where I_0 is a minimum intensity used for comparison. When one earthquake is 10 times as intense as another, its magnitude on the Richter scale is 1 higher. If one earthquake is 100 times as intense as another, its magnitude on the Richter scale is 2 higher, and so on. Thus an earthquake whose magnitude is 7 on the Richter scale is 10 times as intense as an earthquake whose magnitude is 6. Earthquakes can be interpreted as multiples of the minimum intensity I_0.

27. In 1986, there was an earthquake near Cleveland, Ohio. It had an intensity of $10^5 \cdot I_0$. What was its magnitude on the Richter scale?

28. On October 17, 1989, there was an earthquake in San Francisco, California, during the World Series. It had an intensity of $10^{6.9} \cdot I_0$. What was its magnitude on the Richter scale?

This photograph shows part of the damage in the San Francisco, California, area earthquake in 1989.

29. *Earthquake intensity.* The intensity I of an earthquake is given by

$$I = I_0 10^R,$$

where R = the magnitude on the Richter scale and I_0 = the minimum intensity, where $R = 0$, used for comparison.

a) Find I, in terms of I_0, for an earthquake of magnitude 7 on the Richter scale.
b) Find I, in terms of I_0, for an earthquake of magnitude 8 on the Richter scale.
c) Compare your answers to parts (a) and (b).
d) Find the rate of change dI/dR.

30. *Intensity of sound.* The intensity of a sound is given by

$$I = I_0 10^{0.1L},$$

where L = the loudness of the sound as measured in decibels and I_0 = the minimum intensity detectable by the human ear.

a) Find I, in terms of I_0, for the loudness of a power mower, which is 100 decibels.
b) Find I, in terms of I_0, for the loudness of just audible sound, which is 10 decibels.
c) Compare your answers to parts (a) and (b).
d) Find the rate of change dI/dL.

31. *Earthquake magnitude.* The magnitude R (measured on the Richter scale) of an earthquake of intensity I is

defined as

$$R = \log \frac{I}{I_0},$$

where I_0 = the minimum intensity (used for comparison). (The exponential form of this definition is given in Exercise 29.) Find the rate of change dR/dI.

32. Loudness of sound. The loudness L of a sound of intensity I is defined as

$$L = 10 \log \frac{I}{I_0},$$

where I_0 = the minimum intensity detectable by the human ear and L = the loudness measured in decibels. (The exponential form of this definition is given in Exercise 30.) Find the rate of change dL/dI.

33. Response to drug dosage. The response y to a dosage x of a drug is given by

$$y = m \log x + b.$$

The response may be hard to measure with a number.

The patient might perspire more, have an increase in temperature, or faint. Find the rate of change dy/dx.

SYNTHESIS

34. Find $\lim\limits_{h \to 0} \dfrac{3^h - 1}{h}$.

Use the Chain Rule and other formulas given in this section to differentiate each of the following. In some cases, it may help to take the logarithm on both sides of the equation before differentiating.

35. $f(x) = 3^{2x}$

36. $y = 2^{x^4}$

37. $y = x^x, x > 0$

38. $y = \log_3 (x^2 + 1)$

39. $f(x) = x^{e^x}, x > 0$

40. $y = a^{f(x)}$

41. $y = \log_a f(x), f(x)$ positive

42. $y = [f(x)]^{g(x)}, f(x)$ positive

43. ◈ Describe in your own words how to justify the formula for finding the derivative of $f(x) = a^x$.

44. ◈ Describe in your own words how to justify the formula for finding the derivative of $f(x) = \log_a x$.

4.6 An Economic Application: Elasticity of Demand

OBJECTIVES

- Find the elasticity of a demand function.
- Find the maximum of a total-revenue function.
- Characterize demand in terms of elasticity.

Suppose that x represents a quantity of goods sold and p is the price per unit of the goods. Recall that x and p are related by the demand function

$$x = D(p).$$

Suppose that there is a change Δp in the price per unit of a product. The percent change in price is

$$\frac{\Delta p}{p}.$$

A change in the price of a product produces a change Δx in the quantity sold. The percent change in quantity is

$$\frac{\Delta x}{x}.$$

The ratio of the percent change in quantity to the percent change in price is

$$\frac{(\Delta x/x)}{(\Delta p/p)},$$

which can be expressed as

$$\frac{p}{x} \cdot \frac{\Delta x}{\Delta p}. \tag{1}$$

For continuous functions,

$$\lim_{\Delta p \to 0} \frac{\Delta x}{\Delta p} = \frac{dx}{dp},$$

and the limit as Δp approaches 0 of the expression in Eq. (1) becomes

$$\frac{p}{x} \cdot \frac{dx}{dp} = \frac{p}{D(p)} \cdot D'(p) = \frac{p\, D'(p)}{D(p)}.$$

DEFINITION

The *elasticity of demand* E is given as a function of price p by

$$E(p) = -\frac{p\, D'(p)}{D(p)}.$$

The numbers x, or $D(p)$, and p are always nonnegative. The slope of the demand curve $dx/dp = D'(p)$ is always negative, since the demand curve is decreasing, and $D(p)$ is always nonnegative and must be positive in order for the elasticity to exist. Economists have used the $-$ sign in the definition of elasticity to make $E(p)$ nonnegative and easier to work with.

EXAMPLE 1   *Economics: Demand for videotape rentals.* A videotape store works out a demand function for its videotape rentals and finds it to be

$$x = D(p) = 120 - 20p,$$

where $x =$ the number of videotapes rented per day when p is the price per rental. Find each of the following.

a) The quantity demanded when the price is $2 per rental

b) The elasticity as a function of p

c) The elasticity at $p = \$2$ and at $p = \$5$. Interpret the meaning of these values of the elasticity.

d) The value of p for which $E(p) = 1$. Interpret the meaning of this price.

e) The total-revenue function $R(p) = p\,D(p)$

f) The price p at which total revenue is a maximum

Solution

a) At $p = \$2$, $x = D(2) = 120 - 20(2) = 80$. Thus, 80 videotapes will be rented per day when the price per rental is $2.

b) To find the elasticity, we first find the derivative $D'(p)$:

$$D'(p) = -20.$$

Then we substitute -20 for $D'(p)$ and $120 - 20p$ for $D(p)$ in the expression for elasticity:

$$E(p) = -\frac{p\,D'(p)}{D(p)} = -\frac{p \cdot (-20)}{120 - 20p} = \frac{20p}{120 - 20p} = \frac{p}{6 - p}.$$

c) $E(\$2) = \dfrac{2}{6 - 2} = \dfrac{1}{2}$

At $p = \$2$, the elasticity is $\frac{1}{2}$, which is less than 1. Thus the ratio of the percent change in quantity to the percent change in price is less than 1. A small increase in price will cause a percentage decrease in the quantity that is smaller than the percentage change in price.

$$E(\$5) = \frac{5}{6 - 5} = 5$$

At $p = \$5$, the elasticity is 5, which is greater than 1. Thus the ratio of the percent change in quantity to the percent change in price is greater than 1. A small increase in price will cause a percentage decrease in the quantity that is greater than the percentage change in price.

d) We set $E(p) = 1$ and solve for p:

$$\frac{p}{6 - p} = 1$$

$$p = 6 - p \qquad \text{We multiply by } 6 - p, \text{ assuming } p \neq 6.$$

$$2p = 6$$

$$p = 3.$$

Thus when the price is $3 per rental, the ratio of the percent change in quantity to the percent change in price is 1.

e) Recall that the total revenue $R(p)$ is given by $p\,D(p)$. Then

$$R(p) = p\,D(p)$$

$$= p(120 - 20p)$$

$$= 120p - 20p^2.$$

TECHNOLOGY CONNECTION

Economics: Demand for Hand-Held Radios

A company determines that the demand function for hand-held radios is

$$x = D(p) = 300 - p,$$

where x = the number sold per day when the price is p dollars per radio. Find the elasticity E and the total revenue R. Graph all three using the same set of axes, but use only the first quadrant. Then find the price p for which the total revenue is a maximum. Use the method of Example 1. Check the answer on your grapher.

f) To find the price p that maximizes total revenue, we find $R'(p)$:

$$R'(p) = 120 - 40p.$$

Now $R'(p)$ exists for all values of p in the interval $[0, \infty)$. Thus we solve:

$$R'(p) = 120 - 40p = 0$$
$$-40p = -120$$
$$p = 3.$$

Since there is only one critical point, we can try to use the second derivative to see if we have a maximum:

$$R''(p) = -40, \quad \text{a constant.}$$

Thus, $R''(3)$ is negative, so $R(3)$ is a maximum. That is, total revenue is a maximum at $p = \$3$ per rental.   ◆

Note in Example 1 that the value of p for which $E(p) = 1$ is the same as the value of p for which the total revenue is a maximum. This is always true.

THEOREM 17

Total revenue is a maximum at the value(s) of p for which $E(p) = 1$.

We can prove this as follows. We know that

$$R(p) = p\,D(p),$$

so $$R'(p) = p \cdot D'(p) + 1 \cdot D(p)$$

$$= D(p)\left[\frac{p\,D'(p)}{D(p)} + 1\right] \qquad \text{Check this by multiplying.}$$

$$= D(p)[-E(p) + 1]$$

$$= D(p)[1 - E(p)] \tag{2}$$

$$R'(p) = 0 \quad \text{when} \quad 1 - E(p) = 0, \quad \text{or} \quad E(p) = 1.$$

We have shown that total revenue is a maximum when $E = 1$. What do we know when $E < 1$ and $E > 1$? We answer this by looking at Eq. (2) in the preceding proof.

For prices p for which $D(p) > 0$ and for which $E(p) < 1$, it follows that $1 - E(p) > 0$, so by Eq. (2), $R'(p) > 0$. This tells us that the total-revenue function is increasing.

For values of p for which $D(p) > 0$ and for which $E(p) > 1$, it follows that $1 - E(p) < 0$, so by Eq. (2), $R'(p) < 0$. This tells us that the total-revenue function is decreasing.

The following statement and the graphs shown below summarize our results.

For a particular value of the price p:

1. The demand is *inelastic* if $E(p) < 1$. An increase in price will bring an increase in revenue. If demand is inelastic, then revenue is increasing.

2. The demand has *unit elasticity* if $E(p) = 1$. The demand has unit elasticity when revenue is at a maximum.

3. The demand is *elastic* if $E(p) > 1$. An increase in price will bring a decrease in revenue. If demand is elastic, then revenue is decreasing.

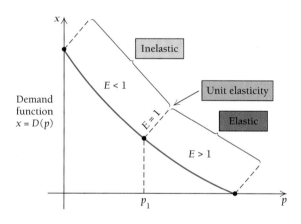

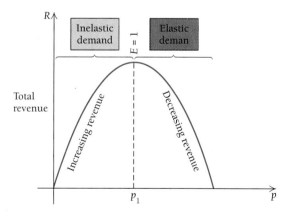

In summary, suppose that the videotape store in Example 1 raises the price per rental and that the total revenue increases. Then we say the demand is *inelastic*. If the total revenue decreases, we say the demand is *elastic*. Some price elasticities in the U.S. economy are listed in the following table.

Price elasticities in the U.S. economy

Industry	Elasticity
Elastic Demands	
Metals	1.52
Electrical engineering products	1.39
Mechanical engineering products	1.30
Furniture	1.26
Motor vehicles	1.14
Instrument engineering products	1.10
Professional services	1.09
Transportation services	1.03
Inelastic Demands	
Gas, electricity, and water	0.92
Oil	0.91
Chemicals	0.89
Beverages (all types)	0.78
Tobacco	0.61
Food	0.58
Banking and insurance services	0.56
Housing services	0.55
Clothing	0.49
Agricultural and fish products	0.42
Books, magazines, and newspapers	0.34
Coal	0.32

Source: Ahsan Mansur and John Whalley, "Numerical specification of applied general equilibrium models: Estimation, calibration, and data." In H. E. Scarf and J. B. Shoven (eds.), *Applied General Equilibrium Analysis.* (New York: Cambridge University Press, 1984), p. 109.

4.6 Exercise Set

For each demand function, find:

a) the elasticity;
b) the elasticity at the given price, stating whether the demand is elastic or inelastic; and
c) the value(s) of p for which total revenue is a maximum.

1. $x = D(p) = 400 - p$; $p = \$125$

2. $x = D(p) = 500 - p$; $p = \$38$

3. $x = D(p) = 200 - 4p$; $p = \$46$

4. $x = D(p) = 500 - 2p$; $p = \$57$

5. $x = D(p) = \dfrac{400}{p}$; $p = \$50$

6. $x = D(p) = \dfrac{3000}{p}$; $p = \$60$

7. $x = D(p) = \sqrt{500 - p}$; $p = \$400$

8. $x = D(p) = \sqrt{300 - p}$; $p = \$250$

9. $x = D(p) = 100e^{-0.25p}$; $p = \$10$

10. $x = D(p) = 200e^{-0.05p}$; $p = \$80$

11. $x = D(p) = \dfrac{100}{(p + 3)^2}$; $p = \$1$

12. $x = D(p) = \dfrac{300}{(p + 8)^2}$; $p = \$4.50$

APPLICATIONS

◆   **Business and Economics**

13. *Demand for chocolate chip cookies.* A bakery works out a demand function for its sale of chocolate chip cookies and finds it to be

$$x = D(p) = 967 - 25p,$$

where $x =$ the quantity of cookies sold when the price per cookie, in cents, is p.

a) Find the elasticity.
b) At what price is the elasticity of demand equal to 1?
c) At what prices is the elasticity of demand elastic?
d) At what prices is the elasticity of demand inelastic?
e) At what price is the revenue a maximum?
f) At a price of 20¢ per cookie, will a small increase

in price cause the total revenue to increase or decrease?

14. *Demand for oil.* Suppose you have been hired by OPEC as an economic consultant for the world demand for oil. The demand function is

$$x = D(p) = 63,000 + 50p - 25p^2, \quad 0 \le p \le 50,$$

where x is measured in millions of barrels of oil per day at a price of p dollars per barrel.

a) Find the elasticity.
b) Find the elasticity at a price of $10 per barrel, stating whether the demand is elastic or inelastic.
c) Find the elasticity at a price of $20 per barrel, stating whether the demand is elastic or inelastic.
d) Find the elasticity at a price of $30 per barrel, stating whether the demand is elastic or inelastic.
e) At what price is the revenue a maximum?
f) What quantity of oil will be sold at the price that maximizes revenue? Check current world prices to see how this compares with your answer.
g) At a price of $30 per barrel, will a small increase in price cause the total revenue to increase or decrease?

15. *Demand for videogames.* A computer software store determines the following demand function for a new videogame:

$$x = D(p) = \sqrt{200 - p^3},$$

where $x =$ the number of videogames sold per day when the price is p dollars per game.

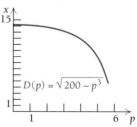

a) Find the elasticity.
b) Find the elasticity when $p = \$3$ per game.
c) At $p = \$3$, will a small increase in price cause the total revenue to increase or decrease?

16. *Demand for tomato plants.* A garden store determines the following demand function during early summer for a certain size of tomato plant:

$$x = D(p) = \frac{2p + 300}{10p + 11},$$

where x = the number of plants sold per day when the price is p dollars per plant.

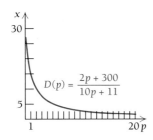

a) Find the elasticity.
b) Find the elasticity when $p = \$3$ per plant.
c) At $p = \$3$, will a small increase in price cause the total revenue to increase or decrease?

SYNTHESIS

17. *Economics: Constant elasticity curve*
   a) Find the elasticity of the demand function

   $$x = D(p) = \frac{k}{p^n},$$

where k is a positive constant and n is an integer greater than 0.
   b) Is the value of the elasticity dependent on the price per unit?
   c) Does the total revenue have a maximum? When?

18. *Economics: Exponential demand curve*
   a) Find the elasticity of the demand function

   $$x = D(p) = Ae^{-kp},$$

   where A and k are positive constants.
   b) Is the value of the elasticity dependent on the price per unit?
   c) Does the total revenue have a maximum? At what value of p?

19. Let

$$L(p) = \ln D(p).$$

   Describe the elasticity in terms of $L'(p)$.

20. ◆ Describe in your own words the concept of elasticity and explain its usefulness to economists. Do some library research or consult an economist in order to determine when and how this concept was first developed.

21. ◆ Explain how the elasticity of demand for a product can be affected by the availability of substitutes for the product.

4 Chapter Summary and Review

TERMS TO KNOW

REVIEW EXERCISES

The review exercises are for test preparation. They can also be used as a lengthened practice test. Answers are at the back of the book. The answers also contain bracketed section references, which tell you where to restudy if your answer is incorrect.

Differentiate.

1. $y = \ln x$

2. $y = e^x$

3. $y = \ln (x^4 + 5)$

4. $y = e^{2\sqrt{x}}$

5. $f(x) = \ln x^6$

6. $f(x) = e^{4x} + x^4$

7. $f(x) = \dfrac{\ln x}{x^3}$

8. $f(x) = e^{x^2} \cdot \ln 4x$

9. $f(x) = e^{4x} - \ln \dfrac{x}{4}$

10. $g(x) = x^8 - 8 \ln x$

11. $y = \dfrac{\ln e^x}{e^x}$

Given $\log_a 2 = 1.8301$ and $\log_a 7 = 5.0999$, find each of the following.

12. $\log_a 14$

13. $\log_a \frac{2}{7}$

14. $\log_a 28$

15. $\log_a 3.5$

16. $\log_a \sqrt{7}$

17. $\log_a \frac{1}{4}$

18. Find the function that satisfies $dQ/dt = kQ$. List the answer in terms of Q_0.

19. *Life science: Population growth.* The population of a certain city doubled in 16 years. What was the growth rate of the city? Round to the nearest tenth of a percent.

20. *Business: Interest compounded continuously.* Suppose $8300 is deposited in a savings and loan association in which the interest rate is 6.8%, compounded continuously. How long will it take for the $8300 to double itself? Round to the nearest tenth of a year.

21. *Business: Cost of a prime-rib dinner.* The average cost C of a prime-rib dinner was $4.65 in 1962. In 1986, it was $15.81. Assuming that the exponential-growth model applies:

    a) Find the exponential-growth rate, and write the equation.

    b) What will the cost of such a dinner be in 1997? in 2007?

22. *Business: Franchise growth.* A clothing firm is selling franchises throughout the United States and Canada.

It is estimated that the number of franchises N will increase at the rate of 12% per year, that is,

$$\frac{dN}{dt} = 0.12N,$$

where t = the time, in years.

a) Find the function that satisfies the equation, assuming that the number of franchises in 1988 ($t = 0$) is 60.

b) How many franchises will there be in 1998?

c) After how long will the number of franchises be 120? Round to the nearest tenth of a year.

23. *Life science: Decay rate.* The decay rate of a certain radioactive isotope is 13% per year. What is its half-life? Round to the nearest tenth of a year.

24. *Life science: Half-life.* The half-life of radon-222 is 3.8 days. What is its decay rate? Round to the nearest tenth of a percent.

25. *Life science: Decay rate.* A certain radioactive element has a decay rate of 7% per day, that is,

$$\frac{dA}{dt} = -0.07A,$$

where A = the amount of the element present at time t, in days.

a) Find a function that satisfies the equation if the amount of the element present at $t = 0$ is 800 g.

b) After 20 days, how much of the 800 g will remain? Round to the nearest gram.

c) After how long does half the original amount remain?

26. *Social science: The Hullian learning model.* The probability p of mastering a certain assembly line task after t learning trials is given by

$$p(t) = 1 - e^{-0.7t}.$$

a) What is the probability of learning the task after 1 trial? 2 trials? 5 trials? 10 trials? 14 trials?

b) Find the rate of change $p'(t)$.

c) Sketch a graph of the function.

27. *Business: Present value.* Find the present value of $1,000,000 due 40 yr later at 8.4%, compounded continuously.

Differentiate.

28. $y = 3^x$

29. $f(x) = \log_{15} x$

30. **Economics: Elasticity of demand.** Consider the demand function

$$x = D(p) = \frac{600}{(p + 4)^2}.$$

a) Find the elasticity.
b) Find the elasticity at $p = \$1$, stating whether the demand is elastic or inelastic.
c) Find the elasticity at $p = \$12$, stating whether the demand is elastic or inelastic.
d) At a price of $12, will a small increase in price cause the total revenue to increase or decrease?
e) Find the value of p for which the total revenue is a maximum.

SYNTHESIS

31. Differentiate: $y = \dfrac{e^{2x} + e^{-2x}}{e^{2x} - e^{-2x}}$.

32. Find the minimum value of $f(x) = x^4 \ln 4x$.

 TECHNOLOGY CONNECTION

33. Graph: $f(x) = \dfrac{e^{1/x}}{(1 + e^{1/x})^2}$.

34. Find $\displaystyle\lim_{x \to 0} \frac{e^{1/x}}{(1 + e^{1/x})^2}$.

4 Chapter Test

Differentiate.

1. $y = e^x$

2. $y = \ln x$

3. $f(x) = e^{-x^2}$

4. $f(x) = \ln \dfrac{x}{7}$

5. $f(x) = e^x - 5x^3$

6. $f(x) = 3e^x \ln x$

7. $y = \ln (e^x - x^3)$

8. $y = \dfrac{\ln x}{e^x}$

Given $\log_b 2 = 0.2560$ and $\log_b 9 = 0.8114$, find each of the following.

9. $\log_b 18$ 10. $\log_b 4.5$ 11. $\log_b 3$

12. Find the function that satisfies $dM/dt = kM$. List the answer in terms of M_0.

13. The doubling time of a certain bacteria culture is 4 hours. What is the growth rate? Round to the nearest tenth of a percent.

14. **Business: Interest compounded continuously.** An investment is made at 6.931% per year, compounded continuously. What is the doubling time? Round to the nearest year.

15. **Business: Cost of milk.** The cost C of a gallon of milk

was $0.54 in 1941. In 1985, it was $2.31. Assuming the exponential-growth model applies:

a) Find the exponential-growth rate, and write the equation.
b) Find the cost of a gallon of milk in 2000; in 2010.

16. **Life science: Drug dosage.** A dose of a drug is injected into the body of a patient. The drug amount in the body decreases at the rate of 10% per hour, that is,

$$\frac{dA}{dt} = -0.1A,$$

where $A =$ the amount in the body and $t =$ the time, in hours.

a) A dose of 3 cubic centimeters (cc) is administered. Assuming $A_0 = 3$ and $k = 0.1$, find the function that satisfies the equation.
b) How much of the initial dose of 3 cc will remain after 10 hr?
c) After how long does half the original dose remain?

17. **Life science: Decay rate.** The decay rate of zirconium is 1.1% per day. What is its half-life?

18. **Life science: Half-life.** The half-life of tellurium is

1,000,000 years. What is its decay rate? As a percent, round to six decimal places.

19. Business: Effect of advertising. A company introduces a new product on a trial run in a city. They advertised the product on television and found that the percentage P of people who bought the product after *t* ads had been run satisfied the function

$$P(t) = \frac{100\%}{1 + 24e^{-0.28t}}.$$

a) What percentage buys the product without having seen the ad ($t = 0$)?
b) What percentage buys the product after the ad is run 1 time? 5 times? 10 times? 15 times? 20 times? 30 times? 35 times?
c) Find the rate of change $P'(t)$.
d) Sketch a graph of the function.

20. Business: Present value. Find the present value of $80,000 due 12 years later at 8.6%, compounded continuously.

Differentiate.

21. $f(x) = 20^x$ **22.** $y = \log_{20} x$

23. Economics: Elasticity of demand. Consider the demand function

$$x = D(p) = 400e^{-0.2p}.$$

a) Find the elasticity.
b) Find the elasticity at $p = \$3$, stating whether the demand is elastic or inelastic.
c) Find the elasticity at $p = \$18$, stating whether the demand is elastic or inelastic.
d) At a price of $3, will a small increase in price cause the total revenue to increase or decrease?
e) Find the value of p for which the total revenue is a maximum.

SYNTHESIS

24. Differentiate: $y = x (\ln x)^2 - 2x \ln x + 2x$.

25. Find the maximum and minimum values of $f(x) = x^4 e^{-x}$ on $[0, 10]$.

  TECHNOLOGY CONNECTION

26. Graph: $f(x) = \dfrac{e^x - e^{-x}}{e^x + e^{-x}}$.

27. Find $\displaystyle \lim_{x \to 0} \frac{e^x - e^{-x}}{e^x + e^{-x}}$.

  EXTENDED TECHNOLOGY APPLICATION

The Business of Professional Sports Salaries

Professional baseball. In the opener for this chapter and in Exercise 19 of Exercise Set 4.3, you considered the possible exponential growth of the average salaries of major-league baseball players. The data used for that function are for salaries prior to 1995. During the 1994–1995 season, a players' strike hit professional baseball. When that strike was over, salaries began to fall as did attendance. Note the data and the graph shown here.

Year	Average salary of major-league baseball players
0. (1990)	$ 578,930
1. (1991)	891,188
2. (1992)	1,084,408
3. (1993)	1,120,254
4. (1994)	1,188,679
5. (1995)	1,073,579

Professional football. Now let's consider salaries in professional football.

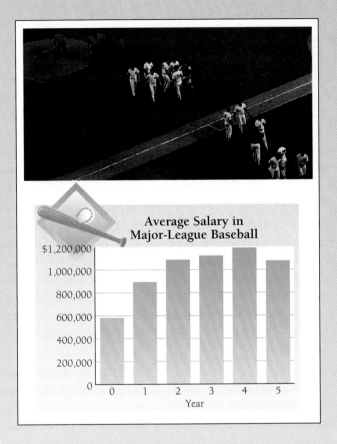

Average Salary in Major-League Baseball

Year		
$1,200,000		
1,000,000		
800,000		
600,000		
400,000		
200,000		
0	0 1 2 3 4 5	

Year

	Average salary of NFL football
Year	players
0. (1990)	$285,000
1. (1991)	386,000
2. (1992)	475,000
3. (1993)	737,000
4. (1994)	650,000
5. (1995)	?

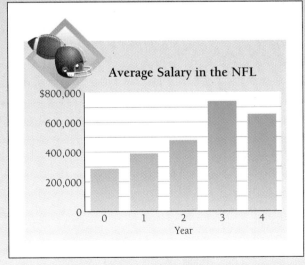

Average Salary in the NFL

$800,000	
600,000	
400,000	
200,000	
0	0 1 2 3 4

Year

EXERCISES

1. Use a grapher that does regression and fit an exponential function of the type $y = Be^{kx}$ to the data. Use all the data except that for 1995.

2. Use the function from Exercise 1 to predict the average salary of major-league baseball players in the year 2000.

3. Use a grapher that does regression and fit a cubic polynomial function of the type $y = ax^3 + bx^2 + cx + d$ to the data. Use all the data.

4. **a)** Use the function from Exercise 3 to predict the average salary of major-league baseball players in the year 2000.
   b) Find the rate of change of salaries in 1995 and in 2000.

5. Compare your answers to Exercises 2 and 4. Does each seem reasonable to you? If not, what kind of function might you consider?

We notice a similar trend to that which occurs in baseball. Up to 1993, the salaries seem to follow an exponential trend, but start to drop off in 1994, due to a salary cap agreement between owners and players. At the time of this writing, data were not available for 1995.

EXERCISES

6. Use a grapher that does regression and fit an exponential function of the type $y = Be^{kx}$ to the data. Use all the data except that for 1994.

7. Use the function from Exercise 6 to predict the average salary of NFL players in the year 2000.

8. Use a grapher that does regression and fit a cubic polynomial function of the type $y = ax^3 + bx^2 + cx + d$ to the data. Use all the data.

9. a) Use the function from Exercise 8 to predict the average salary of NFL players in the year 2000.
   b) Find the rate of change in 1995 and in 2000.

10. Compare your answers to Exercises 7 and 9. Which seems more reasonable to you?

Professional basketball. For professional basketball, we have the following data and graph.

Year	Average salary of NBA basketball players
0. (1990)	$ 748,000
1. (1991)	1,034,000
2. (1992)	1,202,000
3. (1993)	1,348,000
4. (1994)	1,558,000
5. (1995)	?

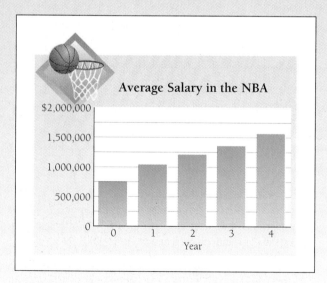

Average Salary in the NBA

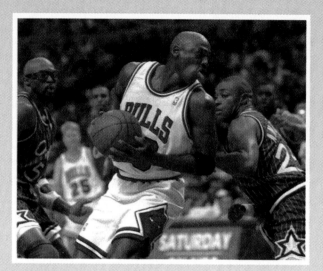

The data and the graph seem to indicate a linear, or possibly exponential, trend that is increasing. A salary cap implemented during the 1984–85 season may be a reason for a lack of exponential growth, but it has not stopped the increase in salaries, as seems to be the case in baseball and football.

EXERCISES

11. Use a grapher that does regression and fit a linear function to the data. Predict the average salary of NBA players in the year 2000.

12. Use a grapher that does regression and fit an exponential function of the type $y = Be^{kx}$ to the data. Predict the average salary of NBA players in the year 2000.

13. What kind of additional information might be helpful to determine whether you use a linear or an exponential function?

14. Use a grapher that does regression and fit a quartic polynomial function of the type $y = ax^4 + bx^3 + cx^2 + dx + e$ to the data.

15. Use the function from Exercise 14 to predict the average salary of NBA players in the year 2000.

16. Compare your answers to Exercises 11, 12, and 15 together with their graphs. Which seems more reasonable to you over the interval [0, 10]?

$\diamondsuit$ 5 Integration

INTRODUCTION

Suppose we do the reverse of differentiating; that is, suppose we try to find a function whose derivative is a given function. This process is called *antidifferentiation*, or *integration*; it is the main topic of this chapter. We will see that we can use integration to find the area under a curve over a closed interval as well as to find the accumulation of a certain quantity over an interval.

We first consider the meaning of integration and then we learn several techniques for integrating.

AN APPLICATION

The divorce rate in the United States is approximated by the function

$$D(t) = 100{,}000e^{0.025t},$$

where $D(t) =$ the number of divorces occurring at time t and $t =$ the number of years measured from 1900. Find the total number of divorces from 1980 to 1998.

THE MATHEMATICS

The total number of divorces from 1980 to 1998 is given by

$$\underbrace{\int_{80}^{98} 100{,}000e^{0.025t}\, dt.}$$

$\uparrow$

This is an *integral*.

This expression also gives the area under the graph of $D(t)$ over the interval [80, 98].

This problem appears as Exercise 56 in Exercise Set 5.5.

5.1 Integration

In Chapters 2, 3, and 4, we have considered several interpretations of the derivative, some of which are listed below.

Function	Derivative
Distance	Velocity
Revenue	Marginal revenue
Cost	Marginal cost
Population	Rate of growth of population

For population we actually considered the derivative first and then the function. Many problems can be solved by doing the reverse of differentiation, called *antidifferentiation*.

The Antiderivative

Suppose that y is a function of x and that the derivative is the constant 8. Can we find y? It is easy to see that one such function is $8x$. That is, $8x$ is a function whose derivative is 8. Is there another function whose derivative is 8? In fact, there are many. Here are some examples:

$$8x + 3, \qquad 8x - 10, \qquad 8x + 7.4, \qquad 8x + \sqrt{2}.$$

All these functions are $8x$ plus some constant. There are no other functions having a derivative of 8 other than those of the form $8x + C$. Another way of saying this is that any two functions having a derivative of 8 must differ by a constant. In general, any two functions having the same derivative differ by a constant.

> **THEOREM 1**
>
> If two functions F and G have the same derivative on an interval, then
>
> $$F(x) = G(x) + C, \quad \text{where } C \text{ is a constant.}$$

The reverse of differentiating is *antidifferentiating*, and the result of antidifferentiating is called an **antiderivative.** Above, we found antiderivatives of the function 8. There are several of them, but they are all $8x$ plus some constant.

EXAMPLE 1 Antidifferentiate (find the antiderivatives of) x^2.

Solution One antiderivative is $x^3/3$. All other antiderivatives differ from this by a constant, so we can denote them as follows:

$$\frac{x^3}{3} + C.$$

To check this, differentiate $x^3/3 + C$:

$$\frac{d}{dx}\left(\frac{x^3}{3} + C\right) = 3\left(\frac{x^2}{3}\right) = x^2.$$

So $x^3/3 + C$ is an antiderivative of x^2. ◆

The solution to Example 1 is the *general form* of the antiderivative of x^2.

Integrals and Integration

The process of antidifferentiating is often called **integration.** To indicate that the antiderivative of x^2 is $x^3/3 + C$, we write

$$\int x^2 \, dx = \frac{x^3}{3} + C, \tag{1}$$

and we note that

$$\int f(x) \, dx$$

is the symbolism we will use from now on to call for the antiderivative of a function $f(x)$. More generally, we can write

$$\int f(x) \, dx = F(x) + C, \tag{2}$$

where $F(x) + C$ is the general form of the antiderivative of $f(x)$. Equation (2) is read "the *indefinite integral* of $f(x)$ is $F(x) + C$." The constant C is called the *constant of integration* and can have any fixed value; hence the use of the word "indefinite."

The symbolism $\int f(x) \, dx$, from Leibniz, is called an *integral,* or more precisely, an **indefinite integral.** The symbol $\int$ is called an *integral sign.* The left side of Eq. (2) is often read "the integral of $f(x)$, dx" or more briefly, "the integral of $f(x)$." In this context, $f(x)$ is called the *integrand.*

EXAMPLE 2 Evaluate $\int x^9 \, dx$.

Solution

$$\int x^9 \, dx = \frac{x^{10}}{10} + C$$

The symbol on the left is read "the integral of x^9, dx." (The "dx" is often omitted in the reading.) In this case, the integrand is x^9. The constant C is called the constant of integration.   ◆

EXAMPLE 3   Evaluate $\int 5e^{4x}\,dx$.

Solution

$$\int 5e^{4x}\,dx = \frac{5}{4}e^{4x} + C$$

We can check by differentiating the antiderivative. The derivative of $\frac{5}{4}e^{4x} + C$ is $4\left(\frac{5}{4}\right)e^{4x}$, or $5e^{4x}$, which is the integrand.   ◆

To integrate (or antidifferentiate), we make use of differentiation formulas, in effect, reading them in reverse. Below are some of these, stated in reverse, as integration formulas. These can be checked by differentiating the right-hand side and noting that the result is, in each case, the integrand.

THEOREM 2

Basic Integration Formulas

1. $\displaystyle\int k\,dx = kx + C$ (k is a constant)

2. $\displaystyle\int x^r\,dx = \frac{x^{r+1}}{r+1} + C,$ provided $r \neq -1$

   (To integrate a power of x other than -1, increase the power by 1 and divide by the increased power.)

3. $\displaystyle\int x^{-1}\,dx = \int \frac{1}{x}\,dx = \int \frac{dx}{x} = \ln x + C,\quad x > 0$

   $\displaystyle\int x^{-1}\,dx = \ln |x| + C,\quad x < 0$

   (We will generally assume that $x > 0$.)

4. $\displaystyle\int be^{ax}\,dx = \frac{b}{a}e^{ax} + C$

The following rules combined with the preceding formulas allow us to find many integrals. They can be derived by reversing two familiar differentiation rules.

THEOREM 3

Rule A. $\displaystyle\int kf(x)\,dx = k\int f(x)\,dx$

(The integral of a constant times a function is the constant times the integral of the function.)

Rule B. $\displaystyle\int [f(x) + g(x)]\,dx = \int f(x)\,dx + \int g(x)\,dx$

(The integral of a sum is the sum of the integrals.)

EXAMPLE 4

$$\int (5x + 4x^3)\,dx = \int 5x\,dx + \int 4x^3\,dx \qquad \text{Rule B}$$

$$= 5\int x\,dx + 4\int x^3\,dx \qquad \text{Rule A}$$

Then

$$\int (5x + 4x^3)\,dx = 5\cdot\frac{x^2}{2} + 4\cdot\frac{x^4}{4} + C = \frac{5}{2}x^2 + x^4 + C.$$

Don't forget the constant of integration. It is not necessary to write two constants of integration. If we did, we could add them and consider C as the sum.

We can check this as follows:

$$\frac{d}{dx}\left(\frac{5}{2}x^2 + x^4 + C\right) = 2\cdot\frac{5}{2}\cdot x + 4x^3 = 5x + 4x^3. \qquad \blacklozenge$$

We can always check an integration by differentiating.

EXAMPLE 5

$$\int (7e^{6x} - \sqrt{x})\,dx = \int 7e^{6x}\,dx - \int \sqrt{x}\,dx$$

$$= \int 7e^{6x}\,dx - \int x^{1/2}\,dx$$

$$= \frac{7}{6}e^{6x} - \frac{x^{(1/2)+1}}{\frac{1}{2}+1} + C = \frac{7}{6}e^{6x} - \frac{x^{3/2}}{\frac{3}{2}} + C$$

$$= \frac{7}{6}e^{6x} - \frac{2}{3}x^{3/2} + C \qquad \blacklozenge$$

EXAMPLE 6

$$\int \left(1 - \frac{3}{x} + \frac{1}{x^4}\right) dx = \int 1\, dx - 3\int \frac{dx}{x} + \int x^{-4}\, dx$$

$$= x - 3\ln x + \frac{x^{-4+1}}{-4+1} + C$$

$$= x - 3\ln x - \frac{x^{-3}}{3} + C \qquad \blacklozenge$$

Another Look at Antiderivatives

The graphs of the antiderivatives of x^2 are the graphs of the functions

$$y = \int x^2\, dx = \frac{x^3}{3} + C$$

for the various values of the constant C.

As we can see in the figures below, x^2 is the derivative of each function; that is, the tangent line at the point

$$\left(a, \frac{a^3}{3} + C\right)$$

has slope a^2. The curves $(x^3/3) + C$ fill up the plane, with exactly one curve going through any given point (x_0, y_0).

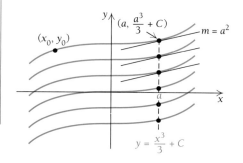
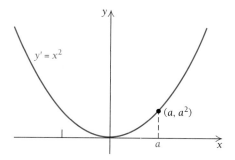

Suppose we look for an antiderivative of x^2 with a specified value at a certain point, say $f(-1) = 2$. We find that there is only one such function.

EXAMPLE 7 Find the function f such that

$$f'(x) = x^2$$

and

$$f(-1) = 2.$$

Solution

a) We find $f(x)$ by integrating:

$$f(x) = \int x^2 \, dx = \frac{x^3}{3} + C.$$

b) The condition $f(-1) = 2$ allows us to find C,

$$f(-1) = \frac{(-1)^3}{3} + C = 2,$$

and solving for C, we get

$$-\tfrac{1}{3} + C = 2$$
$$C = 2 + \tfrac{1}{3}, \quad \text{or} \quad \tfrac{7}{3}.$$

Thus, $f(x) = \dfrac{x^3}{3} + \dfrac{7}{3}$.   ◆

We are often given further information about a function, such as, in Example 7, $f(-1) = 2$ or $(-1, 2)$. The fact that $f(-1) = 2$ is called a *boundary condition*, or an *initial condition*. Sometimes the terms *boundary value*, or *initial value*, are used.

Applications

DETERMINING TOTAL COST FROM MARGINAL COST

EXAMPLE 8 *Business: Total cost from marginal cost.* A company determines that the marginal cost C' of producing the xth unit of a certain product is given by

$$C'(x) = x^3 + 2x.$$

Find the total-cost function C, assuming that fixed costs (costs when 0 units are produced) are \$45. The boundary condition, or initial condition, is that $C = \$45$ when $x = 0$; that is, $C(0) = \$45$.

Solution

a) We integrate to find $C(x)$, using K for the integration constant to avoid confusion with the cost function C:

$$C(x) = \int C'(x) \, dx = \int (x^3 + 2x) \, dx = \frac{x^4}{4} + x^2 + K.$$

b) Fixed costs are \$45; that is, $C(0) = 45$. This allows us to determine the value of K:

$$C(0) = \frac{0^4}{4} + 0^2 + K = 45$$

$$K = 45.$$

Thus, $C(x) = \dfrac{x^4}{4} + x^2 + 45$.   ◆

FINDING VELOCITY AND DISTANCE FROM ACCELERATION

Recall that the position coordinate at time t of an object moving along a number line is $s(t)$. Then

$$s'(t) = v(t) = \text{the } velocity \text{ at time } t,$$
$$v'(t) = a(t) = \text{the } acceleration \text{ at time } t.$$

EXAMPLE 9   *Physical science: Distance.* Suppose that $v(t) = 5t^4$ and $s(0) = 9$. Find $s(t)$.

Solution

a) We find $s(t)$ by integrating:

$$s(t) = \int v(t)\, dt = \int 5t^4 \, dt = t^5 + C.$$

b) We determine C by using the boundary condition $s(0) = 9$, which is the initial value, or position, of s at time $t = 0$:

$$s(0) = 0^5 + C = 9$$
$$C = 9.$$

Thus, $s(t) = t^5 + 9$.   ◆

EXAMPLE 10   *Physical science: Distance.* Suppose that $a(t) = 12t^2 - 6$, $v(0) = $ the initial velocity $= 5$, and $s(0) = $ the initial position $= 10$. Find $s(t)$.

Solution

a) We find $v(t)$ by integrating $a(t)$:

$$v(t) = \int a(t)\, dt = \int (12t^2 - 6)\, dt$$
$$= 4t^3 - 6t + C_1.$$

b) The condition $v(0) = 5$ allows us to find C_1:

$$v(0) = 4 \cdot 0^3 - 6 \cdot 0 + C_1 = 5$$
$$C_1 = 5.$$

Thus, $v(t) = 4t^3 - 6t + 5$.

c) We find $s(t)$ by integrating $v(t)$:

$$s(t) = \int v(t)\, dt = \int (4t^3 - 6t + 5)\, dt$$
$$= t^4 - 3t^2 + 5t + C_2.$$

d) The condition $s(0) = 10$ allows us to find C_2:

$$s(0) = 0^4 - 3 \cdot 0^2 + 5 \cdot 0 + C_2 = 10$$
$$C_2 = 10.$$

Thus, $s(t) = t^4 - 3t^2 + 5t + 10$.   ◆

5.1 Exercise Set

Evaluate.

1. $\int x^6 \, dx$

2. $\int x^7 \, dx$

3. $\int 2 \, dx$

4. $\int 4 \, dx$

5. $\int x^{1/4} \, dx$

6. $\int x^{1/3} \, dx$

7. $\int (x^2 + x - 1) \, dx$

8. $\int (x^2 - x + 2) \, dx$

9. $\int (t^2 - 2t + 3) \, dt$

10. $\int (3t^2 - 4t + 7) \, dt$

11. $\int 5e^{8x} \, dx$

12. $\int 3e^{5x} \, dx$

13. $\int (x^3 - x^{8/7}) \, dx$

14. $\int (x^4 - x^{6/5}) \, dx$

15. $\int \frac{1000}{x} \, dx$

16. $\int \frac{500}{x} \, dx$

17. $\int \frac{dx}{x^2} \left(\text{or } \int \frac{1}{x^2} \, dx \right)$

18. $\int \frac{dx}{x^3}$

19. $\int \sqrt{x} \, dx$

20. $\int \sqrt[3]{x^2} \, dx$

21. $\int \frac{-6}{\sqrt[3]{x^2}} \, dx$

22. $\int \frac{20}{\sqrt[5]{x^4}} \, dx$

23. $\int 8e^{-2x} \, dx$

24. $\int 7e^{-0.25x} \, dx$

25. $\int \left(x^2 - \frac{3}{2}\sqrt{x} + x^{-4/3} \right) dx$

26. $\int \left(x^4 + \frac{1}{8\sqrt{x}} - \frac{4}{5}x^{-2/5} \right) dx$

Find f such that:

27. $f'(x) = x - 3, \quad f(2) = 9$

28. $f'(x) = x - 5, \quad f(1) = 6$

29. $f'(x) = x^2 - 4, \quad f(0) = 7$

30. $f'(x) = x^2 + 1, \quad f(0) = 8$

APPLICATIONS

◆   **Business and Economics**

31. *Total cost from marginal cost.* A company determines that the marginal cost C' of producing the xth unit of a certain product is given by

$$C'(x) = x^3 - 2x.$$

Find the total-cost function C, assuming fixed costs to be $100.

32. *Total cost from marginal cost.* A company determines that the marginal cost C' of producing the xth unit of a certain product is given by

$$C'(x) = x^3 - x.$$

Find the total-cost function C, assuming fixed costs to be $200.

33. *Total revenue from marginal revenue.* A company determines that the marginal revenue R' from selling the xth unit of a certain product is given by

$$R'(x) = x^2 - 3.$$

a) Find the total-revenue function R, assuming that $R(0) = 0$.
b) Why is $R(0) = 0$ a reasonable assumption?

34. Total revenue from marginal revenue. A company determines that the marginal revenue R' from selling the xth unit of a certain product is given by

$$R'(x) = x^2 - 1.$$

a) Find the total-revenue function R, assuming that $R(0) = 0$.

b) Why is $R(0) = 0$ a reasonable assumption?

35. Demand from marginal demand. A company finds that the rate at which consumer demand quantity changes with respect to price is given by the marginal demand function

$$D'(p) = -\frac{4000}{p^2}.$$

Find the demand function if it is known that 1003 units of the product are demanded by consumers when the price is \$4 per unit.

36. Supply from marginal supply. A company finds that the rate at which a seller's quantity changes with respect to price is given by the marginal supply function

$$S'(p) = 0.24p^2 + 4p + 10.$$

Find the supply function if it is known that the seller will sell 121 units of the product when the price is \$5 per unit.

37. Efficiency of a machine operator. The rate at which a machine operator's efficiency E (expressed as a percentage) changes with respect to time t is given by

$$\frac{dE}{dt} = 30 - 10t,$$

where $t =$ the number of hours that the operator has been at work.

a) Find $E(t)$, given that the operator's efficiency after working 2 hr is 72%; that is, $E(2) = 72$.

b) Use the answer to part (a) to find the operator's efficiency after 3 hr; after 5 hr.

38. Efficiency of a machine operator. The rate at which a machine operator's efficiency E (expressed as a percentage) changes with respect to time t is given by

$$\frac{dE}{dt} = 40 - 10t,$$

where $t =$ the number of hours that the operator has been at work.

A machine operator's efficiency changes with respect to time.

a) Find $E(t)$, given that the operator's efficiency after working 2 hr is 72%; that is, $E(2) = 72$.

b) Use the answer to part (a) to find the operator's efficiency after 4 hr; after 8 hr.

◆   Life and Physical Sciences

Find $s(t)$.

39. $v(t) = 3t^2$, $s(0) = 4$ **40.** $v(t) = 2t$, $s(0) = 10$

Find $v(t)$.

41. $a(t) = 4t$, $v(0) = 20$ **42.** $a(t) = 6t$, $v(0) = 30$

Find $s(t)$.

43. $a(t) = -2t + 6$, $v(0) = 6$, and $s(0) = 10$

44. $a(t) = -6t + 7$, $v(0) = 10$, and $s(0) = 20$

45. Distance. For a freely falling object, $a(t) = -32$ ft/sec^2, $v(0) =$ initial velocity $= v_0$, and $s(0) =$ initial height $= s_0$. Find a general expression for $s(t)$ in terms of v_0 and s_0.

46. Time. A ball is thrown from a height of 10 ft, where $s(0) = 10$, at an initial velocity of 80 ft/sec, where $v(0) = 80$. How long will it take before the ball hits the ground? (See Exercise 45.)

47. Distance. A car with constant acceleration goes from 0 to 60 mph in $\frac{1}{2}$ min. How far does the car travel during that time?

48. Area of a healing wound. The area A of a healing wound is decreasing at a rate given by

$$A'(t) = -43.4t^{-2}, \quad 1 \le t \le 7,$$

where t = the time, in days, and A is in square centimeters.

a) Find $A(t)$ if $A(1) = 39.7$.
b) Find the area of the wound after 7 days.

◆   **Social Sciences**

49. *Memory.* In a certain memory experiment, the rate of memorizing is given by

$$M'(t) = 0.2t - 0.003t^2,$$

where $M(t)$ = the number of Spanish words memorized in t minutes.

a) Find $M(t)$ if it is known that $M(0) = 0$.
b) How many words are memorized in 8 min?

SYNTHESIS

Find f.

50. $f'(t) = \sqrt{t} + \dfrac{1}{\sqrt{t}}, \quad f(4) = 0$

51. $f'(t) = t^{\sqrt{3}}, \quad f(0) = 8$

Evaluate. Each of the following can be inte the rules developed in this section, but son be required beforehand.

52. $\displaystyle\int (5t + 4)^2 \, dt$

53. $\displaystyle\int (x - 1)^2 x^3 \, dx$

54. $\displaystyle\int (1 - t)\sqrt{t} \, dt$

55. $\displaystyle\int \frac{(t + 3)^2}{\sqrt{t}} \, dt$

56. $\displaystyle\int \frac{x^4 - 6x^2 - 7}{x^3} \, dx$

57. $\displaystyle\int (t + 1)^3 \, dt$

58. $\displaystyle\int \frac{1}{\ln 10} \frac{dx}{x}$

59. $\displaystyle\int b e^{ax} \, dx$

60. $\displaystyle\int (3x - 5)(2x + 1) \, dx$

61. $\displaystyle\int \sqrt[3]{64x^4} \, dx$

62. $\displaystyle\int \frac{x^2 - 1}{x + 1} \, dx$

63. $\displaystyle\int \frac{t^3 + 8}{t + 2} \, dt$

64. ◈   On a test, a student makes the statement, "The function $f(x) = x^2$ has a unique integral." Discuss.

65. ◈   Describe the graphical interpretation of an antiderivative.

5.2 Finding Area Using Antiderivatives

OBJECTIVES

• Find the area under a curve on a given closed interval.

• Interpret the area under a curve in two other ways.

We now consider the application of integration to finding areas of certain regions. Consider a function whose outputs are positive in an interval (the function might be 0 at one of the endpoints). We wish to find the area of the region between the graph of the function and the x-axis on that interval, as illustrated in the following figures.

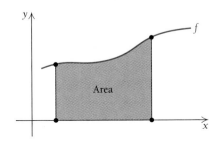

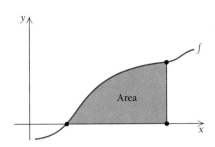

Let us first consider a constant function $f(x) = m$ on the interval from 0 to x, $[0, x]$.

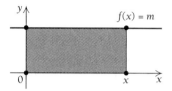

Refer to the figure above, which shows a rectangle whose area is mx. Suppose that we allow x to vary, giving us rectangles of different areas. The area of each rectangle is still mx. We now have an area function:

$$A(x) = mx.$$

Its graph is shown below.

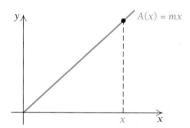

Let us consider next the linear function $f(x) = mx$ on the interval from 0 to x, $[0, x]$, as shown below.

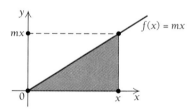

The figure formed this time is a triangle, and its area is one-half the base times the height, $\frac{1}{2} \cdot x \cdot (mx)$, or $\frac{1}{2}mx^2$. If we allow x to vary, we again get an area function:

$$A(x) = \tfrac{1}{2}mx^2.$$

Its graph is shown below.

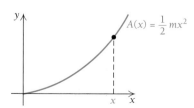

Now consider the linear function $f(x) = mx + b$ on the interval from 0 to x, $[0, x]$, as shown here.

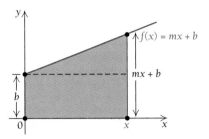

The figure formed this time is a trapezoid, and its area is one-half the height times the sum of the lengths of its parallel sides (or, noting the dashed line, the area of the triangle plus the area of the rectangle):

$$\frac{1}{2} \cdot x \cdot [b + (mx + b)],$$

or $\frac{1}{2} \cdot x \cdot (mx + 2b),$

or $\frac{1}{2}mx^2 + bx.$

If we allow x to vary, we again get an area function:

$$A(x) = \frac{1}{2}mx^2 + bx.$$

Its graph is shown below.

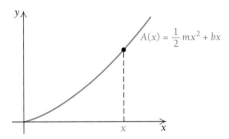

Now we consider the function $f(x) = x^2 + 1$ on the interval from 0 to x, $[0, x]$. The graph of the region in question is shown at left, but it is not so easy this time to find the area function because the graph of $f(x)$ is not a straight line. Let us look for a pattern in the following table.

$f(x)$	$A(x)$
$f(x) = 3$	$A(x) = 3x$
$f(x) = m$	$A(x) = mx$
$f(x) = 3x$	$A(x) = \frac{3}{2}x^2$
$f(x) = mx$	$A(x) = \frac{1}{2}mx^2$
$f(x) = mx + b$	$A(x) = \frac{1}{2}mx^2 + bx$

You may have conjectured that the area function $A(x)$ is an antiderivative of $f(x)$. In the following exploratory exercises, you will investigate further.

Exploratory Exercises: Finding Areas

1. Consider the constant function $f(x) = 3$.

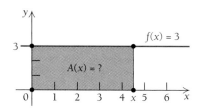

a) Find $A(x)$.
b) Find $A(1)$, $A(2)$, and $A(5)$.
c) Graph $A(x)$.
d) How are $f(x)$ and $A(x)$ related?

2. Consider the function $f(x) = 3x$.

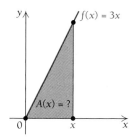

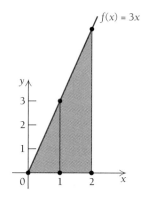

a) Find $A(x)$.
b) Find $A(1)$, $A(2)$, and $A(3.5)$.
c) Graph $A(x)$.
d) How are $f(x)$ and $A(x)$ related?

3. The region under the graph of $f(x) = x^2 + 1$, on the interval $[0, 2]$, is shown below.

a) On a sheet of thin paper, make a copy of the shaded region.
b) Cut up the shaded region in any way you wish in order to fill up squares in the grid. Make an estimate of the total area.

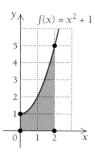

c) Using the antiderivative

$$F(x) = \frac{x^3}{3} + x,$$

find $F(2)$.
d) Compare your answers in parts (b) and (c).

4. Repeat Exercises 3(a) and (b) for the shaded region of the following graph, using the grid shown below.

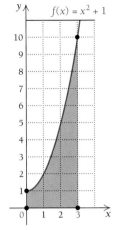

$f(x) = x^2 + 1$

c) Using the antiderivative

$$F(x) = \frac{x^3}{3} + x,$$

find $F(3)$.
d) Compare your answers in parts (b) and (c).

The conjecture concerning areas and antiderivatives (or integrals) is true. It is expressed as follows.

THEOREM 4

Let f be a positive, continuous function on an interval, and let $A(x)$ be the area of the region between the graph of f and the x-axis on the interval $[a, x]$. Then $A(x)$ is a differentiable function of x and

$A'(x) = f(x).$

Proof: The situation described in the theorem is shown in the following figure.

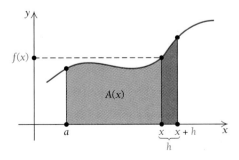

The derivative of $A(x)$ is, by the definition of a derivative,

$$A'(x) = \lim_{h \to 0} \frac{A(x + h) - A(x)}{h}.$$

Note from the figure that $A(x + h) - A(x)$ is the area of the small, orange, vertical strip. The area of this small strip is approximately that of a rectangle of base h and height $f(x)$, especially for small values of h. Thus we have

$$A(x + h) - A(x) \approx f(x) \cdot h.$$

Now

$$A'(x) = \lim_{h \to 0} \frac{A(x + h) - A(x)}{h}$$

$$= \lim_{h \to 0} \frac{f(x) \cdot h}{h}$$

$$= \lim_{h \to 0} f(x) = f(x),$$

since $f(x)$ does not involve h.

The theorem above also holds if $f(x) = 0$ at one or both endpoints of an interval.

Since the area function A is an antiderivative of f, and since any two antiderivatives differ by a constant, we easily conclude that the area function and any antiderivative differ by a constant.

We can think of the function A as given by

$$A(x) = \text{the area on the interval } [a, x],$$

where a is some fixed point and x varies as illustrated in the following figures.

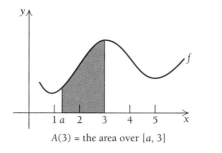

$A(3)$ = the area over $[a, 3]$

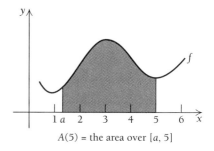

$A(5)$ = the area over $[a, 5]$

Now let us find some areas.

EXAMPLE 1   Find the area under the graph of $y = x^2 + 1$ on the interval $[-1, 2]$.

Solution

a) We first make a drawing. This includes a graph of the function as well as the region in question, as seen in Fig. 1.

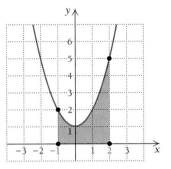

FIGURE 1

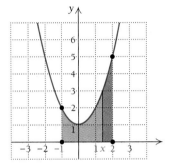

FIGURE 2

b) Next we make a drawing (Fig. 2) showing a portion of the region from -1 to x. Now $A(x)$ is the area of this portion, that is, on the interval $[-1, x]$.

c) Now

$$A(x) = \int (x^2 + 1)\, dx = \frac{x^3}{3} + x + C,$$

where C must be determined. Since we know that $A(-1) = 0$ (there is no area above the number -1), we can substitute -1 for x in $A(x)$, as follows:

$$A(-1) = \frac{(-1)^3}{3} + (-1) + C = 0$$

$$-\frac{1}{3} - 1 + C = 0$$

$$C = \frac{4}{3}.$$

This determines that $C = \frac{4}{3}$, so we have

$$A(x) = \frac{x^3}{3} + x + \frac{4}{3}.$$

Then the area on the interval $[-1, 2]$ is $A(2)$. We compute $A(2)$ as follows:

$$A(2) = \frac{2^3}{3} + 2 + \frac{4}{3}$$

$$= \frac{8}{3} + 2 + \frac{4}{3} = \frac{12}{3} + 2$$

$$= 6.$$

EXAMPLE 2   Find the area under the graph of $y = x^3$ on the interval $[0, 5]$.

Solution

a) We first make a drawing (Fig. 3) that includes a graph of the function and the region in question.

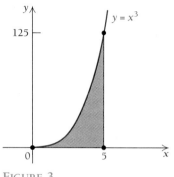

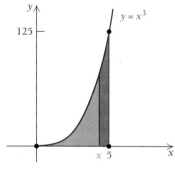

FIGURE 3   FIGURE 4

b) Next we make a drawing (Fig. 4) showing a portion of that region from 0 to x. Now $A(x)$ is the area of this portion, that is, on the interval $[0, x]$.

c) Now

$$A(x) = \int x^3 \, dx = \frac{x^4}{4} + C,$$

where C must be determined. Since we know that $A(0) = 0$, we can substitute 0 for x in $A(x)$, as follows:

$$A(0) = \frac{0^4}{4} + C = 0$$

$$C = 0.$$

This determines C. Thus,

$$A(x) = \frac{x^4}{4}.$$

Then the area on the interval $[0, 5]$ is $A(5)$. We can compute $A(5)$ as follows:

$$A(5) = \frac{5^4}{4} = \frac{625}{4} = 156\tfrac{1}{4}.$$ ◆

We have seen that the antiderivative of $f(x)$, where f is positive, on $[a, x]$ can be interpreted as an accumulating area, which (as x moves from a to b) finally gives the total area on $[a, b]$.

In a similar way, we can interpret the antiderivative of a velocity function $v(t)$ over $[0, b]$ as the area under the curve $y = v(t)$ on $[0, b]$; and the area represents the total distance, assuming that $v(t) > 0$. For example, if we have a velocity function over an interval $[0, b]$, then the area under the curve in that interval is the total distance. Suppose the velocity function is

$$v(t) = t^3.$$

In 5 hours, the total distance covered is $156\frac{1}{4}$. We can see this in Example 2, simply by changing the variable from x to t.

For a marginal cost function $C'(x)$ over the interval $[0, x]$, the area under the curve is the total cost, or the accumulated cost, of producing x units.

EXAMPLE 3   *Business: Total cost from marginal cost.* Raggs, Ltd., a clothing firm, does all it can to reduce production costs. The president of the company takes a calculus course and then manages the business in such a way that the marginal cost per suit becomes

$$C'(x) = 0.0003x^2 - 0.2x + 50.$$

Find the total cost of producing 400 suits (ignore fixed costs).

Solution

a) First we make a drawing (Fig. 5) that includes a graph of the function and the region in question.

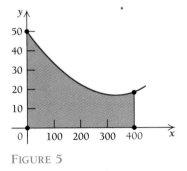

FIGURE 5

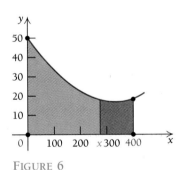

FIGURE 6

b) Next we make a drawing (Fig. 6) showing a portion of that region from 0 to x. Now $A(x)$ is the area of that portion on the interval $[0, x]$.

c) Now

$$C(x) = A(x) = \int (0.0003x^2 - 0.2x + 50)\, dx$$

$$= 0.0001x^3 - 0.1x^2 + 50x + K,$$

where K must be determined. Since we are ignoring fixed costs (that is, $K = 0$), we have

$$C(x) = 0.0001x^3 - 0.1x^2 + 50x.$$

Then the area in the interval $[0, 400]$ is $A(400)$, or $C(400)$. We can compute $C(400)$ as follows:

$$C(400) = 0.0001 \cdot 400^3 - 0.1 \cdot 400^2 + 50 \cdot 400, \quad \text{or} \quad \$10,400. \qquad \blacklozenge$$

5.2 Exercise Set

Find the area under the given curve on the interval indicated.

1. $y = 4$; $[1, 3]$

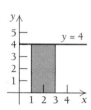

2. $y = 5$; $[1, 3]$

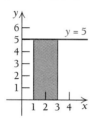

3. $y = 2x$; $[1, 3]$

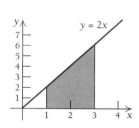

4. $y = x^2$; $[0, 3]$

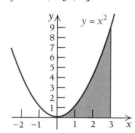

5. $y = x^2$; $[0, 5]$

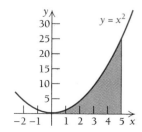

6. $y = x^3$; $[0, 2]$

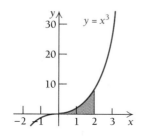

7. $y = x^3$; $[0, 1]$

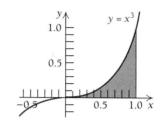

8. $y = 1 - x^2$; $[-1, 1]$

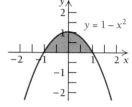

9. $y = 4 - x^2$; $[-2, 2]$

10. $y = e^x$; $[0, 2]$

11. $y = e^x$; $[0, 3]$

12. $y = \dfrac{1}{x}$; $[1, 2]$

13. $y = \dfrac{1}{x}$; $[1, 3]$

14. $y = x^2 - 4x$; $[-4, -2]$

15. $y = x^2 - 4x$; $[-4, -1]$

In each case, give two interpretations of the shaded region other than as area.

16.

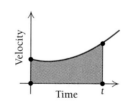

17.

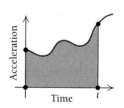

18.

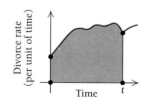

19.

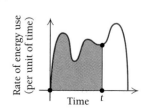

20.

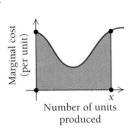

Marginal cost (per unit)

Number of units produced

21.

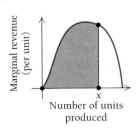

Marginal revenue (per unit)

Number of units produced

22.

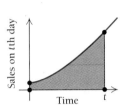

Sales on *t*th day

Time

23.

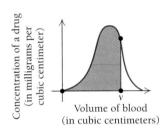

Concentration of a drug (in milligrams per cubic centimeter)

Volume of blood (in cubic centimeters)

24.

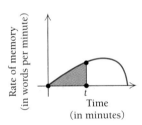

Rate of memory (in words per minute)

Time (in minutes)

APPLICATIONS

◆   **Business and Economics**

25. *Total cost from marginal cost.* A clothing firm, Raggs, Ltd., determines that the marginal cost of each suit that it produces is $50.

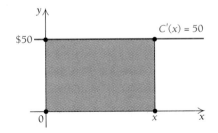

$C'(x) = 50$

$50

a) Find the total cost $C(x)$ of producing x suits, assuming that fixed costs are $0 (that is, ignore fixed costs).

b) Find the area of the shaded rectangle in the figure. Compare your answer with part (a).

c) Graph $C(x)$. Why is this an increasing function?

26. *Total cost from marginal cost.* Raggs, Ltd., of Exercise 25, installs new sewing machines. This allows the marginal cost per suit to decrease continually in such a way that

$$C'(x) = -0.1x + 50.$$

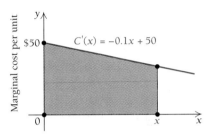

$50 $C'(x) = -0.1x + 50$

Marginal cost per unit

a) Find the total cost of producing x suits, ignoring fixed costs.

b) Find the area of the shaded trapezoid in the figure above.

c) Find the total cost of producing 400 suits.

27. *Total cost, revenue, and profit.* A sound company determines that the marginal cost of producing the xth stereo is given by

$$C'(x) = 100 - 0.2x, \quad C(0) = 0.$$

It also determines that its marginal revenue from the sale of the xth stereo is given by

$$R'(x) = 100 + 0.2x, \quad R(0) = 0.$$

a) Find the total cost of producing x stereos.

b) Find the total revenue from selling x stereos.

c) Find the total profit from the production and sale of x stereos.

d) Find the total profit from the production and sale of 1000 stereos.

28. *Total cost, revenue, and profit.* A refrigeration company determines that the marginal cost of producing the xth refrigerator is given by

$$C'(x) = 50 - 0.4x, \quad C(0) = 0.$$

It also determines that its marginal revenue from the sale of the xth refrigerator is given by

$$R'(x) = 50 + 0.4x, \quad R(0) = 0.$$

a) Find the total cost of producing x refrigerators.
b) Find the total revenue from selling x refrigerators.
c) Find the total profit from the production and sale of x refrigerators.
d) Find the total profit from the production and sale of 1000 refrigerators.

◆   Life and Physical Sciences

29. *Distance from velocity.* A particle starts out from the origin. Its velocity at time t is given by

$$v(t) = 3t^2 + 2t.$$

a) Find the distance that the particle has traveled after t hr.
b) Find the distance that the particle has traveled after 5 hr.

30. *Distance from velocity.* A particle starts out from the origin. Its velocity at time t is given by

$$v(t) = 4t^3 + 2t.$$

a) Find the distance that the particle has traveled after t hr.

b) Find the distance that the particle has traveled after 2 hr.

Find the area under the curve on the interval indicated.

31. $y = \dfrac{x^2 - 1}{x - 1}$;   [2, 3]

32. $y = \dfrac{x^5 - x^{-1}}{x^2}$;   [1, 5]

33. $y = (x - 1)\sqrt{x}$;   [4, 16]

34. $y = (x + 2)^3$;   [0, 1]

35. $y = \dfrac{\sqrt[3]{x^2} - 1}{\sqrt[3]{x}}$;   [1, 8]

36. $y = \dfrac{x^3 + 8}{x + 2}$;   [0, 1]

37. ◆   Discuss as many interpretations as you can of the integral of a positive function over a closed interval.

5.3 Integration on an Interval: The Definite Integral

OBJECTIVES

• Evaluate definite integrals and find the area under a graph.
• Solve application problems involving definite integrals.

Let f be a positive, continuous function on an interval $[a, b]$. Let F and G be any two antiderivatives of f. Then

$$F(b) - F(a) = G(b) - G(a).$$

To understand this, recall that F and G differ by a constant; that is, $F(x) = G(x) + C$. Then

$$F(b) - F(a) = [G(b) + C] - [G(a) + C] = G(b) - G(a).$$

Thus the difference $F(b) - F(a)$ has the same value for all antiderivatives of f. It is called the **definite integral** of f from a to b.

Definite integrals are generally symbolized as follows:

$$\int_a^b f(x)\, dx, \quad \text{where } a < b.$$

This is read "the integral from a to b of $f(x)$ dx." (The "dx" is often omitted from the reading.) From the preceding development, we see that to find a definite integral $\int_a^b f(x)\ dx$, we first find an antiderivative $F(x)$. The simplest one is the one for which the constant of integration is 0. From the value of F at b, we subtract the value of F at a. Then we have $F(b) - F(a)$.

DEFINITION

Let f be any positive, continuous function on the interval $[a, b]$, and let F be any antiderivative of f. Then

$$\int_a^b f(x)\ dx = F(b) - F(a).$$

Evaluating definite integrals is called *integrating*. The numbers a and b are known as the **limits of integration.**

EXAMPLE 1   Evaluate $\int_a^b x^2\ dx$.

Solution   Using the antiderivative $F(x) = x^3/3$, we have

$$\int_a^b x^2\ dx = \frac{b^3}{3} - \frac{a^3}{3}.$$ ◆

It is convenient to use an intermediate notation:

$$\int_a^b f(x)\ dx = [F(x)]_a^b = F(b) - F(a).$$

We now evaluate several definite integrals.

EXAMPLE 2

$$\int_{-1}^2 x^2\ dx = \left[\frac{x^3}{3}\right]_{-1}^2$$

$$= \frac{2^3}{3} - \frac{(-1)^3}{3}$$

$$= \frac{8}{3} - \left(-\frac{1}{3}\right) = \frac{8}{3} + \frac{1}{3} = 3$$ ◆

EXAMPLE 3

$$\int_0^3 e^x\ dx = [e^x]_0^3 = e^3 - e^0 = e^3 - 1$$ ◆

EXAMPLE 4

$$\int_1^4 (x^2 - x)\, dx = \left[\frac{x^3}{3} - \frac{x^2}{2}\right]_1^4 = \left(\frac{4^3}{3} - \frac{4^2}{2}\right) - \left(\frac{1^3}{3} - \frac{1^2}{2}\right)$$

$$= \left(\frac{64}{3} - \frac{16}{2}\right) - \left(\frac{1}{3} - \frac{1}{2}\right)$$

$$= \frac{64}{3} - 8 - \frac{1}{3} + \frac{1}{2} = 13\frac{1}{2}$$ ◆

EXAMPLE 5

$$\int_1^e \left(1 + 2x - \frac{1}{x}\right) dx = [x + x^2 - \ln x]_1^e$$

$$= (e + e^2 - \ln e) - (1 + 1^2 - \ln 1)$$

$$= (e + e^2 - 1) - (1 + 1 - 0)$$

$$= e + e^2 - 1 - 1 - 1$$

$$= e + e^2 - 3$$ ◆

For some applications that we will soon consider, it is important to keep in mind that in $\int_a^b f(x)\, dx$, we assumed that $a < b$. That is, the larger number is on the top!

The area under a curve can be expressed by a definite integral.

THEOREM 5

Let f be a positive, continuous function over the closed interval $[a, b]$. The area under the graph of f on the interval $[a, b]$ is

$$\int_a^b f(x)\, dx.$$

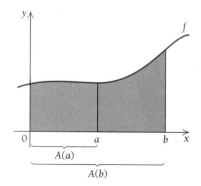

Proof: Let

$A(x) =$ the area of the region over $[0, x]$.

Then

$A'(x) = f(x),$

so $A(x)$ is an antiderivative of $f(x)$. Then

$$\int_a^b f(x)\, dx = A(b) - A(a).$$

But $A(b) - A(a)$ is the area over $[0, b]$, minus the area over $[0, a]$, which is the area over $[a, b]$.

Let us now find some areas.

EXAMPLE 6   Find the area under $y = x^2 + 1$ on $[-1, 2]$.

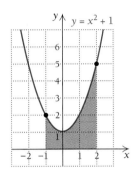

Solution

$$\int_{-1}^{2} (x^2 + 1) \, dx = \left[\frac{x^3}{3} + x \right]_{-1}^{2}$$

$$= \left(\frac{2^3}{3} + 2 \right) - \left(\frac{(-1)^3}{3} + (-1) \right)$$

$$= \left(\frac{8}{3} + 2 \right) - \left(-\frac{1}{3} - 1 \right)$$

$$= \frac{8}{3} + 2 + \frac{1}{3} + 1 = 6$$

Compare this example with Example 1 of Section 5.2.   ◆

EXAMPLE 7   Find the area under $y = x^3$ on $[0, 5]$.

Solution

$$\int_{0}^{5} x^3 \, dx = \left[\frac{x^4}{4} \right]_{0}^{5}$$

$$= \frac{5^4}{4} - \frac{0^4}{4} = \frac{625}{4}$$

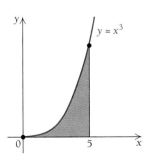

Compare this example with Example 2 of Section 5.2.   ◆

EXAMPLE 8   Find the area under $y = 1/x$ on $[1, 4]$.

Solution

$$\int_1^4 \frac{dx}{x} = [\ln x]_1^4 = \ln 4 - \ln 1$$

$$= \ln 4 \approx 1.3863$$

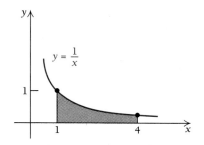

 TECHNOLOGY CONNECTION

Most graphers can draw a graph, shade the area under the graph on a closed interval, and approximate the area using procedures that we will develop in Section 5.8. For example, the following figure shows the area under $f(x) = \ln x/x^2$ over the interval $[1, 6]$. Some graphers have a shade feature to show the area and an fnint feature to approximate the area.

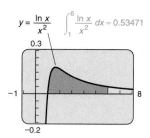

Use your grapher to approximate the area in Example 8.

EXAMPLE 9   Find the area under $y = 1/x^2$ on $[1, b]$.

Solution

$$\int_1^b \frac{dx}{x^2} = \int_1^b x^{-2}\, dx$$

$$= \left[\frac{x^{-2+1}}{-2 + 1} \right]_1^b$$

$$= \left[\frac{x^{-1}}{-1} \right]_1^b = \left[-\frac{1}{x} \right]_1^b$$

$$= \left(-\frac{1}{b} \right) - \left(-\frac{1}{1} \right)$$

$$= 1 - \frac{1}{b}$$

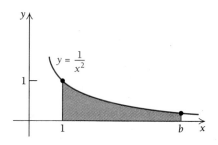

The following properties of definite integrals can be derived rather easily from the definition of a definite integral and from the properties of the indefinite integral.

PROPERTY 1

$$\int_a^b k \cdot f(x) \; dx = k \cdot \int_a^b f(x) \; dx$$

The integral of a constant times a function is the constant times the integral of the function. That is, we can "factor out" a constant from the integrand.

EXAMPLE 10

$$\int_0^5 100e^x \; dx = 100 \int_0^5 e^x \; dx$$
$$= 100[e^x]_0^5$$
$$= 100(e^5 - e^0)$$
$$= 100(e^5 - 1)$$
$$\approx 14{,}741.32$$

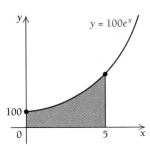

PROPERTY 2

$$\int_a^b [f(x) + g(x)] \; dx = \int_a^b f(x) \; dx + \int_a^b g(x) \; dx$$

The integral of a sum is the sum of the integrals.

PROPERTY 3

For $a < c < b$,

$$\int_a^b f(x) \; dx = \int_a^c f(x) \; dx + \int_c^b f(x) \; dx.$$

For any number c between a and b, the integral from a to b is the integral from a to c plus the integral from c to b.

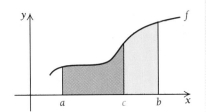

Property 3 has particular application when a function is defined piecewise in different ways over subintervals.

EXAMPLE 11 Find the area under the graph of $y = f(x)$ from -4 to 5, where

$$f(x) = \begin{cases} 9, & \text{if } x < 3, \\ x^2, & \text{if } x \geq 3. \end{cases}$$

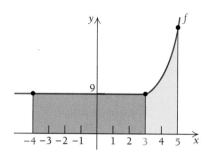

Solution

$$\int_{-4}^{5} f(x) \, dx = \int_{-4}^{3} f(x) \, dx + \int_{3}^{5} f(x) \, dx$$

$$= \int_{-4}^{3} 9 \, dx + \int_{3}^{5} x^2 \, dx$$

$$= 9 \int_{-4}^{3} dx + \int_{3}^{5} x^2 \, dx$$

$$= 9[x]_{-4}^{3} + \left[\frac{x^3}{3}\right]_{3}^{5}$$

$$= 9[3 - (-4)] + \left(\frac{5^3}{3} - \frac{3^3}{3}\right)$$

$$= 95\tfrac{2}{3}$$ ◆

An Application

EXAMPLE 12 *Business: Accumulated sales.* The sales of a company are expected to grow continuously at a rate given by the function

$$S'(t) = 100e^t,$$

where $S'(t) =$ the sales rate, in dollars per day, at time t, in days.

a) Find the accumulated sales for the first 7 days.

b) On what day will accumulated sales exceed $810,000?

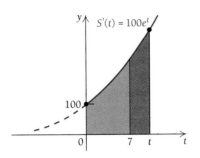

Solution

a) Accumulated sales through day 7 are

$$\int_0^7 S'(t)\, dt = \int_0^7 100e^t\, dt = 100 \int_0^7 e^t\, dt$$

$$= 100[e^t]_0^7$$
$$= 100(e^7 - e^0)$$
$$= 100(1096.633158 - 1)$$
$$= 100(1095.633158)$$
$$\approx \$109,563.32.$$

b) Accumulated sales through day k are

$$\int_0^k S'(t)\, dt = \int_0^k 100e^t\, dt = 100 \int_0^k e^t\, dt$$

$$= 100[e^t]_0^k = 100(e^k - e^0)$$
$$= 100(e^k - 1).$$

We set this equal to \$810,000 and solve for k:

$$100(e^k - 1) = 810,000$$
$$e^k - 1 = 8100$$
$$e^k = 8101.$$

We now solve this equation for k using natural logarithms:

$$\ln e^k = \ln 8101$$
$$k = 8.999743$$
$$k \approx 9.$$

Accumulated sales will exceed \$810,000 approximately on day 9.   ◆

5.3 Exercise Set

Evaluate.

1. $\int_0^1 (x - x^2)\, dx$

2. $\int_1^2 (x^2 - x)\, dx$

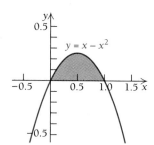

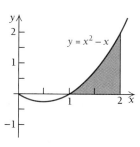

3. $\int_{-1}^1 (x^2 - x^4)\, dx$

4. $\int_0^b 2e^{3x}\, dx$

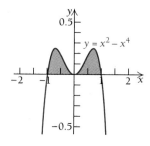

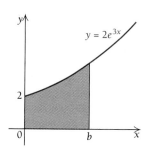

5. $\int_a^b e^t\, dt$

6. $\int_0^a (ax - x^2)\, dx$

7. $\int_a^b 3t^2\, dt$

8. $\int_a^b 4t^3\, dt$

9. $\int_1^e \left(x + \dfrac{1}{x}\right) dx$

10. $\int_1^e \left(x - \dfrac{1}{x}\right) dx$

11. $\int_0^1 \sqrt{x}\, dx$

12. $\int_0^1 3\sqrt{x}\, dx$

13. $\int_0^1 \dfrac{10}{17} t^3\, dt$

14. $\int_0^1 \dfrac{12}{13} t^2\, dt$

Find the area under the graph on the interval indicated.

15. $y = x^3$; $[0, 2]$

16. $y = x^4$; $[0, 1]$

17. $y = x^2 + x + 1$; $[2, 3]$

18. $y = 2 - x - x^2$; $[-2, 1]$

19. $y = 5 - x^2$; $[-1, 2]$

20. $y = e^x$; $[-2, 3]$

21. $y = e^x$; $[-1, 5]$

22. $y = 2x + \dfrac{1}{x^2}$; $[1, 4]$

23. $y = 2x - \dfrac{1}{x^2}$; $[1, 3]$

Find the area under the graph on $[-2, 3]$, where:

24. $f(x) = \begin{cases} x^2, & \text{if } x < 1, \\ 1, & \text{if } x \geq 1. \end{cases}$

25. $f(x) = \begin{cases} 4 - x^2, & \text{if } x < 0, \\ 4, & \text{if } x \geq 0. \end{cases}$

APPLICATIONS

◆ Business and Economics

26. *Accumulated sales.* Raggs, Ltd., estimates that its sales will grow continuously at a rate given by the function

$$S'(t) = 10e^t,$$

where $S'(t) = $ the sales rate, in dollars per day, at time t, in days.

a) Find the accumulated sales for the first 5 days.
b) Find the sales from the 2nd day through the 5th day. (This is the integral from 1 to 5.)
c) On what day will accumulated sales exceed $40,000?

27. *Accumulated sales.* A company estimates that its sales will grow continuously at a rate given by the function

$$S'(t) = 20e^t,$$

where $S'(t) = $ the sales rate, in dollars per day, at time t, in days.

a) Find the accumulated sales for the first 5 days.
b) Find the sales from the 2nd day through the 5th day. (This is the integral from 1 to 5.)
c) On what day will accumulated sales exceed $20,000?

28. **Total cost from marginal cost.** Raggs, Ltd., determines that the marginal cost per suit is given by

$$C'(x) = 0.0003x^2 - 0.2x + 50.$$

Ignoring fixed costs, find the total cost of producing the 101st suit through the 400th suit (that is, integrate from $x = 100$ to $x = 400$).

29. **Total cost from marginal cost.** Using the information in Exercise 28, find the cost of producing the 201st suit through the 400th suit (that is, integrate from $x = 200$ to $x = 400$).

30. **Total profit from marginal profit.** A company has the marginal-profit function given by

$$P'(x) = -2x + 980.$$

This means that the rate of change of total profit with respect to the number of units x produced is $P'(x)$. Find the total profit from the production and sale of the 101st unit through the 800th unit of the product.

31. **Total revenue from marginal revenue.** A fight promoter sells x tickets and has a marginal-revenue function given by

$$R'(x) = 200x - 1080.$$

This means that the rate of change of total revenue with respect to the number of tickets sold x is $R'(x)$. Find the total revenue from the sale of the 1001st ticket through the 1300th ticket.

◆   Life and Physical Sciences

32. A particle starts out from the origin. Its velocity at time t is given by

$$v(t) = 3t^2 + 2t.$$

How far does it travel from the 2nd hour through the 5th hour (from $t = 1$ to $t = 5$)?

33. A particle starts out from the origin. Its velocity at time t is given by

$$v(t) = 4t^3 + 2t.$$

How far does it travel from the start through the 3rd hour (from $t = 0$ to $t = 3$)?

34. **Total pollution.** A factory is polluting a lake in such a way that the rate of pollutants entering the lake at time t, in months, is given by

$$N'(t) = 280t^{3/2},$$

where N = the total number of pounds of pollutants in the lake at time t.

a) How many pounds of pollutants enter the lake in 16 months?
b) An environmental expert tells the factory that it will have to begin cleanup procedures after 50,000 lb of pollutants have entered the lake. After what amount of time will this occur?

35. **Bacteria growth.** A population of bacteria in a biological experiment grows at the rate of

$$P'(t) = 1200e^{0.32t},$$

where $P(t)$ = the total population at time t, in days.

a) Find the increase in the bacteria population in the first 20 days.
b) The experiment will cease when the population reaches 4 million. After what amount of time will this occur?

◆   Social Sciences

Memorizing. In the psychological process of memorizing, the rate of memorizing (say, in words per minute) initially increases with respect to time. Eventually, however, a maximum rate of memorizing is reached, after which the memory rate begins to decrease.

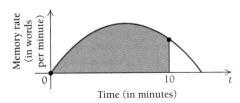

36. Suppose in a certain memory experiment that the rate of memorizing is given by

$$M'(t) = -0.009t^2 + 0.2t,$$

where $M'(t) =$ the memory rate, in words per minute. How many words are memorized in the first 10 min (from $t = 0$ to $t = 10$)?

37. Suppose in a certain memory experiment that the rate of memorizing is given by

$$M'(t) = -0.003t^2 + 0.2t,$$

where $M'(t) =$ the memory rate, in words per minute. How many words are memorized in the first 10 min (from $t = 0$ to $t = 10$)?

38. *Industrial learning curve.* A company is producing a new product. However, due to the nature of the product, it is felt that the time required to produce each unit will decrease as the workers become more familiar with production procedure. It is determined that the function for the learning process is

$$T(x) = ax^b,$$

where $T(x) =$ the average time to produce x units, $x =$ the number of units produced, $a =$ the number of hours required to produce the first unit, and $b =$ the slope of the learning curve.

a) Find an expression for the total time required to produce 100 units.

b) Suppose that $a = 100$ hr and $b = -0.322$. Find the total time required to produce 100 units.

SYNTHESIS

Evaluate.

39. $\displaystyle\int_1^2 (4x + 3)(5x - 2)\, dx$

40. $\displaystyle\int_2^5 (t + \sqrt{3})(t - \sqrt{3})\, dt$

41. $\displaystyle\int_0^1 (t + 1)^3\, dt$

42. $\displaystyle\int_1^3 \left(x - \frac{1}{x}\right)^2 dx$

43. $\displaystyle\int_1^3 \frac{t^5 - t}{t^3}\, dt$

44. $\displaystyle\int_4^9 \frac{t + 1}{\sqrt{t}}\, dt$

45. $\displaystyle\int_3^5 \frac{x^2 - 4}{x - 2}\, dx$

46. $\displaystyle\int_0^1 \frac{t^3 + 1}{t + 1}\, dt$

Find the error in each of the following. Explain.

47. $\displaystyle\int_1^2 (x^2 + x + 1)\, dx = \left[\frac{1}{3}x^3 + \frac{1}{2}x^2 + x\right]_1^2$

$$= \left(\frac{1}{3} \cdot 2^3 + \frac{1}{2} \cdot 2^2 + 2\right)$$

$$= \frac{10}{3}$$

48. $\displaystyle\int_1^2 (\ln x - e^x)\, dx = \left[\frac{1}{x} - e^x\right]_1^2$

$$= \left(\frac{1}{2} - e^2\right) - (1 - e^1)$$

$$= e - e^2 - \frac{1}{2}$$

TECHNOLOGY CONNECTION

Find the area under the graph on the interval indicated.

49. $f(x) = x^3 - 4x;\quad [-2, 0]$

50. $f(x) = x - x^3;\quad [0, 1]$

51. $f(x) = x^3 - 9x^2 + 27x + 50;\quad [-1.2, 6.3]$

52. $f(x) = x^4 + 4x^3 - 36x^2 - 160x + 300;\quad [-8, 1.4]$

53. $f(x) = \dfrac{4x}{x^2 + 1};\quad [0, 20]$

54. $f(x) = 1 - \sqrt{1 - x^2};\quad [-1, 1]$

55. $f(x) = \sqrt{4 - x^2};\quad [0, 2]$

56. $f(x) = x^3 - 6x^2 + 15x;\quad [0, 10]$

57. $f(x) = x(x - 5)^4;\quad [0, 5]$

58. $f(x) = x^{2/3}\left(\dfrac{5}{2} - x\right);\quad [-2, 2]$

59. $f(x) = \dfrac{x^2 - 4}{x^2 - 3};\quad [2, 4]$

60. $f(x) = \dfrac{8}{x^2 + 4};\quad [-10, 10]$

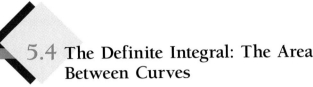

5.4 The Definite Integral: The Area Between Curves

OBJECTIVES

* Evaluate the definite integral of a continuous function.
* Find the area bounded by two graphs.
* Solve application problems involving the area between two graphs.

We have considered the definite integral for functions that are positive on an interval $[a, b]$ (the function might be 0 at one or both endpoints). Now we will consider functions that have negative values. First, let us evaluate the integral of a function without negative values:

$$\int_0^2 x^2 \, dx = \left[\frac{x^3}{3} \right]_0^2 = \frac{2^3}{3} - \frac{0^3}{3} = \frac{8}{3}.$$

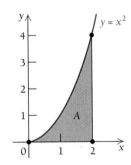

Thus the area of the shaded region is $\frac{8}{3}$.

Now let us consider the function $y = -x^2$ on the interval $[0, 2]$. Even though we have not defined the definite integral for functions with negative values, let us apply the evaluation procedure and see what we get:

$$\int_0^2 -x^2 \, dx = \left[-\frac{x^3}{3} \right]_0^2 = -\frac{2^3}{3} + \frac{0^3}{3} = -\frac{8}{3}.$$

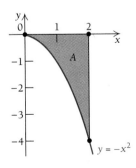

The graphs of the functions $y = x^2$ and $y = -x^2$ are reflections of each other across the x-axis. Thus the areas of the shaded regions are the same—that is, $\frac{8}{3}$. The evaluation procedure in the second case gave us $-\frac{8}{3}$. This illustrates that for negative-valued functions, the definite integral gives us the opposite, or additive inverse, of the area between the curve and the x-axis.

Now let us consider the function $x^2 - 1$ on $[-1, 2]$. It has both positive and negative values. We will apply the preceding evaluation procedure, even though function values are not all nonnegative. We will do this in two ways.

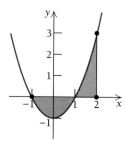

First,

$$\int_{-1}^{2} (x^2 - 1)\, dx = \int_{-1}^{1} (x^2 - 1)\, dx + \int_{1}^{2} (x^2 - 1)\, dx$$

$$= \left[\frac{x^3}{3} - x \right]_{-1}^{1} + \left[\frac{x^3}{3} - x \right]_{1}^{2}$$

$$= \left[\left(\frac{1^3}{3} - 1 \right) - \left(\frac{(-1)^3}{3} - (-1) \right) \right]$$

$$+ \left[\left(\frac{2^3}{3} - 2 \right) - \left(\frac{1^3}{3} - 1 \right) \right]$$

$$= \left[-\frac{4}{3} \right] + \left[\frac{4}{3} \right] = 0.$$

This shows that the area of the region under the x-axis is the same as the area of the region over the x-axis.

Now let us evaluate in another way:

$$\int_{-1}^{2} (x^2 - 1)\, dx = \left[\frac{x^3}{3} - x \right]_{-1}^{2}$$

$$= \left(\frac{2^3}{3} - 2 \right) - \left[\frac{(-1)^3}{3} - (-1) \right]$$

$$= \left(\frac{8}{3} - 2 \right) - \left(-\frac{1}{3} + 1 \right)$$

$$= \frac{8}{3} - 2 + \frac{1}{3} - 1 = 0.$$

This result is consistent with the first. Thus we are motivated to extend our definition of the definite integral to include any continuous function having positive, negative, or zero values.

DEFINITION

For any function with antiderivative F on an interval $[a, b]$,

$$\int_a^b f(x)\, dx = F(b) - F(a).$$

The definite integral turns out to be the area above the x-axis minus the area below.

EXAMPLE 1   Decide whether

$$\int_a^b f(x)\, dx$$

is positive, negative, or zero.

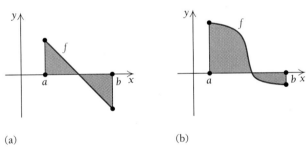

(a)   (b)   (c)

Solution

a) In this figure, there is the same area above the x-axis as below. Thus,

$$\int_a^b f(x)\, dx = 0.$$

b) In this figure, there is more area above the x-axis than below. Thus,

$$\int_a^b f(x)\, dx > 0.$$

c) In this figure, there is more area below the x-axis than above. Thus,

$$\int_a^b f(x)\, dx < 0.$$

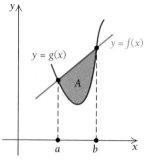

FIGURE 1

The Area of a Region Bounded by Two Graphs

Suppose we want to find the area of a region bounded by the graphs of two functions, $y = f(x)$ and $y = g(x)$, as shown in Fig. 1.

Note that the area of the desired region A in Fig. 2(a) is that of A_2 in Fig. 2(b) minus that of A_1 in Fig. 2(c).

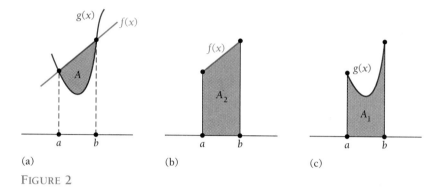

(a)   (b)   (c)

FIGURE 2

Thus,

$$A = \int_a^b f(x)\, dx - \int_a^b g(x)\, dx, \quad \text{or} \quad A = \int_a^b [f(x) - g(x)]\, dx.$$

In general, we have the following.

THEOREM 6

Let f and g be continuous functions and suppose that $f(x) \geq g(x)$ over the interval $[a, b]$. Then the area of the region between the two curves, from $x = a$ to $x = b$, is

$$\int_a^b [f(x) - g(x)]\, dx.$$

EXAMPLE 2   Find the area of the region bounded by the graphs of $y = 2x + 1$ and $y = x^2 + 1$.

Solution

a) First, make a reasonably accurate sketch, as in the following figure, to ensure that you have the right configuration. Note which is the *upper* graph. In this case, $2x + 1 \geq x^2 + 1$ over the interval $[0, 2]$.

TECHNOLOGY CONNECTION

Some graphers can approximate the area of a region bounded by two or more graphs. It is assumed that one function has outputs greater than or equal to the other. To find the area, you first graph each function and use the trace feature to determine the points of intersection. Some graphers have a shade feature to show the area and an fnint feature to approximate the area. The area bounded by the curves

$$y_1 = 3x^5 - 20x^3, \qquad y_2 = \frac{100x}{x^2 + 5},$$

$$x = 0, \quad \text{and} \quad x = 2$$

is shaded and approximated as follows:

$$\text{Area} = \int_0^2 \left(\frac{100x}{x^2 + 5} - 3x^5 + 20x^3 \right) dx$$

$$\approx 77.3893.$$

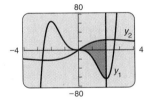

Find the area of the region in Example 2 using your grapher, if it performs such a task.

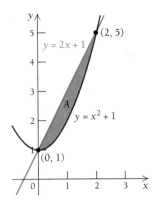

b) Second, if boundaries are not stated, determine the first coordinates of possible points of intersection. Occasionally you can do this just by looking at the graph. If not, you can solve the system of equations as follows. At the points of intersection, $y = x^2 + 1$ and $y = 2x + 1$, so

$$x^2 + 1 = 2x + 1$$

$$x^2 - 2x = 0$$

$$x(x - 2) = 0$$

$$x = 0 \quad \text{or} \quad x = 2.$$

Thus the interval with which we are concerned is $[0, 2]$.

c) Compute the area as follows:

$$\int_0^2 [(2x + 1) - (x^2 + 1)] \, dx = \int_0^2 (2x - x^2) \, dx$$

$$= \left[x^2 - \frac{x^3}{3} \right]_0^2$$

$$= \left(2^2 - \frac{2^3}{3} \right) - \left(0^2 - \frac{0^3}{3} \right)$$

$$= 4 - \frac{8}{3} = \frac{4}{3}. \qquad \blacklozenge$$

An Application

EXAMPLE 3 *Life science: Emission control.* A clever college student develops an engine that is believed to meet federal standards for emission control. The engine's rate of emission is given by

$$E(t) = 2t^2,$$

where $E(t)$ = the emissions, in billions of pollution particulates per year, at time t, in years. The emission rate of a conventional engine is given by

$$C(t) = 9 + t^2.$$

The curves for both are shown in the following graph.

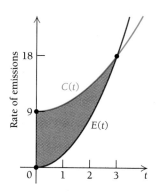

a) At what point in time will the emission rates be the same?

b) What is the reduction in emissions resulting from using the student's engine?

Solution

a) The rate of emission will be the same when $E(t) = C(t)$, or

$$2t^2 = 9 + t^2$$
$$t^2 - 9 = 0$$
$$(t - 3)(t + 3) = 0$$
$$t = 3 \quad \text{or} \quad t = -3.$$

Since negative time has no meaning in this problem, the emission rates will be the same when $t = 3$ years.

b) The reduction in emissions is represented by the area of the shaded region in the figure above. It is the area between $y = 9 + t^2$ and $y = 2t^2$, from $t = 0$ to $t = 3$, and is computed as follows:

$$\int_0^3 [(9 + t^2) - 2t^2]\, dt = \int_0^3 (9 - t^2)\, dt$$

$$= \left[9t - \frac{t^3}{3} \right]_0^3$$

$$= \left(9 \cdot 3 - \frac{3^3}{3} \right) - \left(9 \cdot 0 - \frac{0^3}{3} \right)$$

$$= 27 - 9$$

$$= 18 \text{ billion pollution particulates.}$$

5.4 Exercise Set

Find the area of the region bounded by the given graphs.

1. $y = x$, $y = x^3$, $x = 0$, $x = 1$

2. $y = x$, $y = x^4$

3. $y = x + 2$, $y = x^2$

4. $y = x^2 - 2x$, $y = x$

5. $y = 6x - x^2$, $y = x$

6. $y = x^2 - 6x$, $y = -x$

7. $y = 2x - x^2$, $y = -x$

8. $y = x^2$, $y = \sqrt{x}$

9. $y = x$, $y = \sqrt{x}$

10. $y = 3$, $y = x$, $x = 0$

11. $y = 5$, $y = \sqrt{x}$, $x = 0$

12. $y = x^2$, $y = x^3$

13. $y = 4 - x^2$, $y = 4 - 4x$

14. $y = x^2 + 1$, $y = x^2$, $x = 1$, $x = 3$

15. $y = x^2 + 3$, $y = x^2$, $x = 1$, $x = 2$

Find the area of the shaded region.

16. $f(x) = 2x + x^2 - x^3$, $g(x) = 0$

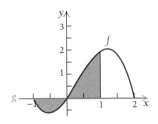

17. $f(x) = x^3 + 3x^2 - 9x - 12$, $g(x) = 4x + 3$

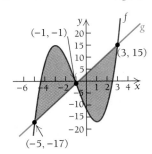

18. $f(x) = x^4 - 8x^3 + 18x^2$, $g(x) = x + 28$

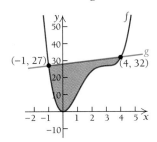

19. $f(x) = 4x - x^2$, $g(x) = x^2 - 6x + 8$

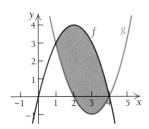

APPLICATIONS

◆ **Business and Economics**

20. *Total profit.* A company determines that its marginal revenue per day is given by

$$R'(t) = 100e^t, \quad R(0) = 0,$$

where $R(t) = $ the revenue, in dollars, on the tth day. The company's marginal cost per day is given by

$$C'(t) = 100 - 0.2t, \quad C(0) = 0,$$

where $C(t) = $ the cost, in dollars, on the tth day. Find the total profit from $t = 0$ to $t = 10$ (the first 10 days). *Note:*

$$P(T) = R(T) - C(T)$$

$$= \int_0^T [R'(t) - C'(t)] \, dt.$$

◆   **Social Sciences**

21. **Memorizing.** In a certain memory experiment, subject A is able to memorize words at the rate given by

$$m'(t) = -0.009t^2 + 0.2t \quad \text{(words per minute)}.$$

In the same memory experiment, subject B is able to memorize at the rate given by

$$M'(t) = -0.003t^2 + 0.2t \quad \text{(words per minute)}.$$

a) Which subject has the higher rate of memorization?
b) How many more words does that subject memorize from $t = 0$ to $t = 10$ (during the first 10 min)?

SYNTHESIS

Find the area of the region bounded by the given graphs.

22. $y = x^2$, $y = x^{-2}$, $x = 1$, $x = 5$

23. $y = e^x$, $y = e^{-x}$, $x = 0$, $x = 1$

24. $y = x^2$, $y = \sqrt[3]{x^2}$, $x = 1$, $x = 8$

25. $y = x^2$, $y = x^3$, $x = -1$, $x = 1$

26. $x + 2y = 2$, $y - x = 1$, $2x + y = 7$

27. Find the area of the region bounded by $y = 3x^5 - 20x^3$, the x-axis, and the first coordinates of the relative maximum and minimum values of the function.

28. Find the area of the region bounded by $y = x^3 - 3x + 2$, the x-axis, and the first coordinates of the relative maximum and minimum values of the function.

29. **Life science: Poiseuille's Law.** The flow of blood in a blood vessel is faster toward the center of the vessel and slower toward the outside. The speed of the blood is given by

$$V = \frac{p}{4Lv}(R^2 - r^2),$$

where $R = $ the radius of the blood vessel, $r = $ the distance of the blood from the center of the vessel, and p, v, and L are physical constants related to the pressure and viscosity of the blood and the length of the blood vessel. If R is constant, we can think of V as a function of r:

$$V(r) = \frac{p}{4Lv}(R^2 - r^2).$$

The *total blood flow* Q is given by

$$Q = \int_0^R 2\pi \cdot V(r) \cdot r \cdot dr.$$

Find Q.

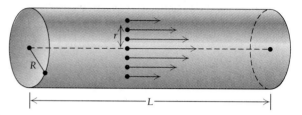

In each exercise, determine whether $\int_a^b f(x)\, dx$ is positive, negative, or zero, and express $\int_a^b f(x)\, dx$ in terms of A. Explain your result.

30. ◈ a)

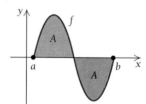

b)

31. ◈ a)

b)

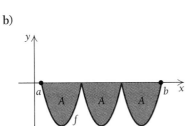

TECHNOLOGY CONNECTION

Find the area of the region bounded by the given graphs.

32. $y = x + 6$, $y = -2x$, $y = x^3$

33. $y = x^2 + 4x$, $y = \sqrt{16 - x^2}$

34. $y = x\sqrt{4 - x^2}$, $y = \dfrac{-4x}{x^2 + 1}$, $x = 0$, $x = 2$

35. $y = 2x^2 + x - 4$, $y = 1 - x + 8x^2 - 4x^4$

36. $y = \sqrt{1 - x^2}$, $y = 1 - x^2$, $x = -1$, $x = 1$

37. Consider the following functions:

$$f(x) = 3.8x^5 - 18.6x^3,$$
$$g(x) = 19x^4 - 55.8x^2.$$

a) Graph each of these functions over the interval $[-3, 3]$.
b) Estimate the first coordinates a, b, and c of the three points of intersection of the two graphs.
c) Estimate the area between the curves over the interval $[a, b]$.
d) Estimate the area between the curves over the interval $[b, c]$.

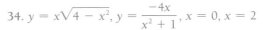

5.5 Integration Techniques: Substitution

OBJECTIVES

• Evaluate integrals using substitution.
• Solve application problems involving integration by substitution.

The following formulas provide a basis for an integration technique called *substitution*.

A. $\displaystyle\int u^r \, du = \dfrac{u^{r+1}}{r + 1} + C$,   provided $r \neq -1$

B. $\displaystyle\int e^u \, du = e^u + C$

C. $\displaystyle\int \dfrac{1}{u} \, du = \ln u + C$, $u > 0$;   or

$\displaystyle\int \dfrac{1}{u} \, du = \ln |u| + C$, $u < 0$

(We will generally consider $u > 0$.)

Recall the Leibniz notation, dy/dx, for a derivative. We gave specific definitions of the differentials dy and dx in Section 3.6. Recall that

$$\frac{dy}{dx} = f'(x)$$

and

$$dy = f'(x)\,dx.$$

We will make extensive use of this notation in this section.

EXAMPLE 1   For $y = f(x) = x^3$, find dy.

Solution   We have

$$\frac{dy}{dx} = f'(x) = 3x^2,$$

so

$$dy = f'(x)\,dx = 3x^2\,dx.$$ ◆

EXAMPLE 2   For $u = g(x) = \ln x$, find du.

Solution   We have

$$\frac{du}{dx} = g'(x) = \frac{1}{x},$$

so

$$du = g'(x)\,dx = \frac{1}{x}\,dx, \quad \text{or} \quad \frac{dx}{x}.$$ ◆

EXAMPLE 3   For $y = f(x) = e^{x^2}$, find dy.

Solution   Using the Chain Rule, we have

$$\frac{dy}{dx} = f'(x) = 2xe^{x^2},$$

so

$$dy = f'(x)\,dx = 2xe^{x^2}\,dx.$$ ◆

So far the dx in

$$\int f(x)\,dx$$

has played no role in integrating other than to indicate the variable of integration. Now it will be convenient to make use of dx. Consider the integral

$$\int 2xe^{x^2}\,dx.$$

If we set

$$u = x^2,$$

then

$$du = 2x \, dx.$$

If we substitute u for x^2 and du for $2x \, dx$, the integral takes on the form

$$\int e^u \, du.$$

Since

$$\int e^u \, du = e^u + C,$$

it follows that

$$\int 2xe^{x^2} \, dx = \int e^u \, du$$

$$= e^u + C$$

$$= e^{x^2} + C.$$

In effect, we have used the Chain Rule in reverse. We can check the result by differentiating. The procedure is referred to as *substitution,* or *change of variable.* It is a *trial-and-error* procedure that you will become more proficient with after much practice. If you try a substitution that doesn't result in an integrand that can be easily integrated, try another. There are many integrations that cannot be carried out using substitution. We do know that any integrations that fit rules A, B, or C can be done with substitution.

Let us consider some additional examples.

EXAMPLE 4   Evaluate $\displaystyle \int \frac{2x \, dx}{1 + x^2}$.

Solution

$$\int \frac{2x \, dx}{1 + x^2} = \int \frac{du}{u} \qquad \text{Substitution} \qquad \boxed{\text{Let } u = 1 + x^2; \\ \text{then } du = 2x \, dx.}$$

$$= \ln u + C$$

$$= \ln (1 + x^2) + C$$

EXAMPLE 5   Evaluate $\displaystyle \int \frac{2x \, dx}{(1 + x^2)^2}$.

Solution

$$\int \frac{2x\,dx}{(1+x^2)^2} = \int \frac{du}{u^2}$$ Substitution $\boxed{\begin{array}{l} u = 1 + x^2, \\ du = 2x\,dx \end{array}}$

$$= \int u^{-2}\,du$$

$$= -u^{-1} + C$$

$$= -\frac{1}{u} + C$$

$$= \frac{-1}{1+x^2} + C$$ ◆

EXAMPLE 6   Evaluate $\displaystyle\int \frac{\ln 3x\,dx}{x}$.

Solution

$$\int \frac{\ln 3x\,dx}{x} = \int u\,du$$ Substitution $\boxed{\begin{array}{l} u = \ln 3x, \\ du = \dfrac{1}{x}\,dx \end{array}}$

$$= \frac{u^2}{2} + C$$

$$= \frac{(\ln 3x)^2}{2} + C$$ ◆

EXAMPLE 7   Evaluate $\displaystyle\int xe^{x^2}\,dx$.

Solution   Suppose we try

$$u = x^2;$$

then we have

$$du = 2x\,dx.$$

We don't have $2x\,dx$ in $\int xe^{x^2}\,dx$. We have $x\,dx$ and need to supply a 2. We do this by multiplying by 1, using $\frac{1}{2} \cdot 2$:

$$\frac{1}{2} \cdot 2 \cdot \int xe^{x^2}\,dx = \frac{1}{2} \int 2xe^{x^2}\,dx$$

$$= \frac{1}{2} \int e^{x^2}(2x\,dx) = \frac{1}{2} \int e^u\,du$$

$$= \frac{1}{2}e^u + C = \frac{1}{2}e^{x^2} + C.$$ ◆

EXAMPLE 8   Evaluate $\displaystyle\int \frac{dx}{x + 3}$.

Solution

$$\int \frac{dx}{x + 3} = \int \frac{du}{u} \qquad \underset{\boxed{\begin{aligned} u &= x + 3, \\ du &= dx \end{aligned}}}{\text{Substitution}}$$

$$= \ln u + C$$

$$= \ln (x + 3) + C \qquad \blacklozenge$$

With practice, you will be able to make certain substitutions mentally and just write down the answer. Example 8 is a good illustration of this.

EXAMPLE 9   Evaluate $\displaystyle\int x^2(x^3 + 1)^{10}\, dx$.

Solution

$$\int x^2(x^3 + 1)^{10}\, dx = \frac{1}{3} \int (x^3 + 1)^{10}(3x^2\, dx) \qquad \underset{\boxed{\begin{aligned} u &= x^3 + 1, \\ du &= 3x^2\, dx \end{aligned}}}{\text{Substitution}}$$

$$= \frac{1}{3} \int u^{10}\, du$$

$$= \frac{1}{3} \cdot \frac{u^{11}}{11} + C$$

$$= \frac{1}{33}(x^3 + 1)^{11} + C \qquad \blacklozenge$$

EXAMPLE 10   Evaluate $\displaystyle\int_0^1 x^2(x^3 + 1)^{10}\, dx$.

Solution

a) First we find the indefinite integral (shown in Example 9).

b) Then we evaluate the definite integral on $[0, 1]$:

$$\int_0^1 x^2(x^3 + 1)^{10}\, dx = \left[\frac{1}{33}(x^3 + 1)^{11} \right]_0^1$$

$$= \frac{1}{33}[(1^3 + 1)^{11} - (0^3 + 1)^{11}]$$

$$= \frac{1}{33}(2^{11} - 1^{11})$$

$$= \frac{2^{11} - 1}{33}. \qquad \blacklozenge$$

TECHNOLOGY
CONNECTION

Use your grapher to approximate

$$\int_0^1 x^2(x^3 + 1)^{10}\, dx.$$

5.5 Exercise Set

Evaluate. (Be sure to check by differentiating!)

1. $\displaystyle\int \frac{3x^2\ dx}{7 + x^3}$

2. $\displaystyle\int \frac{3x^2\ dx}{1 + x^3}$

3. $\displaystyle\int e^{4x}\ dx$

4. $\displaystyle\int e^{3x}\ dx$

5. $\displaystyle\int e^{x/2}\ dx$

6. $\displaystyle\int e^{x/3}\ dx$

7. $\displaystyle\int x^3 e^{x^4}\ dx$

8. $\displaystyle\int x^4 e^{x^5}\ dx$

9. $\displaystyle\int t^2 e^{-t^3}\ dt$

10. $\displaystyle\int t e^{-t^2}\ dt$

11. $\displaystyle\int \frac{\ln 4x\ dx}{x}$

12. $\displaystyle\int \frac{\ln 5x\ dx}{x}$

13. $\displaystyle\int \frac{dx}{1 + x}$

14. $\displaystyle\int \frac{dx}{5 + x}$

15. $\displaystyle\int \frac{dx}{4 - x}$

16. $\displaystyle\int \frac{dx}{1 - x}$

17. $\displaystyle\int t^2 (t^3 - 1)^7\ dt$

18. $\displaystyle\int t(t^2 - 1)^5\ dt$

19. $\displaystyle\int (x^4 + x^3 + x^2)^7 (4x^3 + 3x^2 + 2x)\ dx$

20. $\displaystyle\int (x^3 - x^2 - x)^9 (3x^2 - 2x - 1)\ dx$

21. $\displaystyle\int \frac{e^x\ dx}{4 + e^x}$

22. $\displaystyle\int \frac{e^t\ dt}{3 + e^t}$

23. $\displaystyle\int \frac{\ln x^2}{x}\ dx$

24. $\displaystyle\int \frac{(\ln x)^2}{x}\ dx$

25. $\displaystyle\int \frac{dx}{x \ln x}$

26. $\displaystyle\int \frac{dx}{x \ln x^2}$

27. $\displaystyle\int \sqrt{ax + b}\ dx$

28. $\displaystyle\int x\sqrt{ax^2 + b}\ dx$

29. $\displaystyle\int b e^{ax}\ dx$

30. $\displaystyle\int P_0 e^{kt}\ dt$

31. $\displaystyle\int \frac{3x^2\ dx}{(1 + x^3)^5}$

32. $\displaystyle\int \frac{x^3\ dx}{(2 - x^4)^7}$

33. $\displaystyle\int 7x \sqrt[3]{4 - x^2}\ dx$

34. $\displaystyle\int 12x \sqrt[5]{1 + 6x^2}\ dx$

Evaluate.

35. $\displaystyle\int_0^1 2x e^{x^2}\ dx$

36. $\displaystyle\int_0^1 3x^2 e^{x^3}\ dx$

37. $\displaystyle\int_0^1 x(x^2 + 1)^5\ dx$

38. $\displaystyle\int_1^2 x(x^2 - 1)^7\ dx$

39. $\displaystyle\int_1^3 \frac{dt}{1 + t}$

40. $\displaystyle\int_1^3 e^{2x}\ dx$

41. $\displaystyle\int_1^4 \frac{2x + 1}{x^2 + x - 1}\ dx$

42. $\displaystyle\int_1^3 \frac{2x + 3}{x^2 + 3x}\ dx$

43. $\displaystyle\int_0^b e^{-x}\ dx$

44. $\displaystyle\int_0^b 2e^{-2x}\ dx$

45. $\displaystyle\int_0^b m e^{-mx}\ dx$

46. $\displaystyle\int_0^b k e^{-kx}\ dx$

47. $\displaystyle\int_0^4 (x - 6)^2\ dx$

48. $\displaystyle\int_0^3 (x - 5)^2\ dx$

49. $\displaystyle\int_0^2 \frac{3x^2\ dx}{(1 + x^3)^5}$

50. $\displaystyle\int_{-1}^0 \frac{x^3\ dx}{(2 - x^4)^7}$

51. $\displaystyle\int_0^{\sqrt{7}} 7x \sqrt[3]{1 + x^2}\ dx$

52. $\displaystyle\int_0^1 12x \sqrt[5]{1 - x^2}\ dx$

APPLICATIONS

◆ **Business and Economics**

53. *Demand from marginal demand.* A firm has the marginal-demand function

$$D'(p) = \frac{-2000p}{\sqrt{25 - p^2}}.$$

Find the demand function given that $D = 13,000$ when $p = \$3$ per unit.

54. **Profit from marginal profit.** A firm has the marginal-profit function

$$\frac{dP}{dx} = \frac{9000 - 3000x}{(x^2 - 6x + 10)^2}.$$

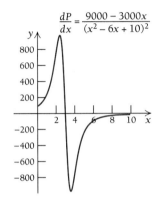

$$\frac{dP}{dx} = \frac{9000 - 3000x}{(x^2 - 6x + 10)^2}$$

Find the total-profit function given that $P = \$1500$ at $x = 3$.

55. **Value of an investment.** A company buys a new machine for $250,000. The marginal revenue from the sale of products produced by the machine is projected to be

$$R'(t) = 4000t.$$

The salvage value of the machine decreases at the rate of

$$V'(t) = 25{,}000e^{-0.1t}.$$

The total profit from the machine after T years is given by

$$P(T) = \begin{pmatrix} \text{Revenue} \\ \text{from} \\ \text{sale of} \\ \text{product} \end{pmatrix} + \begin{pmatrix} \text{Revenue} \\ \text{from} \\ \text{sale of} \\ \text{machine} \end{pmatrix} - \begin{pmatrix} \text{Cost} \\ \text{of} \\ \text{machine} \end{pmatrix}$$

$$= \int_0^T R'(t)\, dt + \int_0^T V'(t)\, dt - \$250{,}000.$$

a) Find $P(T)$.
b) Find $P(10)$.

◆ **Social Sciences**

56. **Divorce.** The divorce rate in the United States is approximated by

$$D(t) = 100{,}000e^{0.025t},$$

where $D(t)$ = the number of divorces occurring at time t and t = the number of years measured from 1900. That is, $t = 0$ corresponds to 1900, $t = 98\frac{9}{365}$ corresponds to January 9, 1998, and so on.

a) Find the total number of divorces from 1900 to 1998. Note that this is

$$\int_0^{98} D(t)\, dt.$$

b) Find the total number of divorces from 1980 to 1998. Note that this is

$$\int_{80}^{98} D(t)\, dt.$$

SYNTHESIS

Find the area of the shaded region.

57.

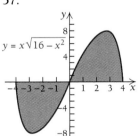

$y = x\sqrt{16 - x^2}$

58.

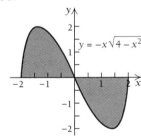

$y = -x\sqrt{4 - x^2}$

Evaluate.

59. $\displaystyle\int 5x\sqrt{1 - 4x^2}\, dx$

60. $\displaystyle\int \frac{dx}{ax + b}$

61. $\displaystyle\int \frac{x^2}{e^{x^3}}\, dx$

62. $\displaystyle\int \frac{e^{\sqrt{t}}}{\sqrt{t}}\, dt$

63. $\displaystyle\int \frac{e^{1/t}}{t^2}\, dt$

64. $\displaystyle\int \frac{(\ln x)^{99}}{x}\, dx$

65. $\displaystyle\int \frac{dx}{x(\ln x)^4}$

66. $\displaystyle\int (e^t + 2)e^t\, dt$

67. $\int x^2\sqrt{x^3 + 1}\, dx$

68. $\int \dfrac{t^2}{\sqrt[4]{2 + t^3}}\, dt$

69. $\int \dfrac{x - 3}{(x^2 - 6x)^{1/3}}\, dx$

70. $\int \dfrac{[(\ln x)^2 + 3(\ln x) + 4]}{x}\, dx$

71. $\int \dfrac{t^3 \ln (t^4 + 8)}{t^4 + 8}\, dt$

72. $\int \dfrac{t^2 + 2t}{(t + 1)^2}\, dt$

$\left(\text{Hint: } \dfrac{t^2 + 2t}{(t + 1)^2} = \dfrac{t^2 + 2t + 1 - 1}{t^2 + 2t + 1} = 1 - \dfrac{1}{(t + 1)^2}. \right)$

73. $\int \dfrac{x^2 + 6x}{(x + 3)^2}\, dx$

74. $\int \dfrac{x + 3}{x + 1}\, dx$

$\left(\text{Hint: Divide } \dfrac{x + 3}{x + 1} = 1 + \dfrac{2}{x + 1}. \right)$

75. $\int \dfrac{t - 5}{t - 4}\, dt$

76. $\int \dfrac{dx}{x(\ln x)^n}$

77. $\int \dfrac{dx}{e^x + 1}$

$\left(\text{Hint: } \dfrac{1}{e^x + 1} = \dfrac{e^{-x}}{1 + e^{-x}}. \right)$

78. $\int \dfrac{e^x - e^{-x}}{e^x + e^{-x}}\, dx$

79. $\int \dfrac{(\ln x)^n}{x}\, dx$

80. $\int \dfrac{e^{-mx}}{1 + ae^{-mx}}\, dx$

81. $\int \dfrac{dx}{x \ln x \,[\ln (\ln x)]}$

82. $\int 5x^2(2x^3 - 7)^n\, dx$

83. $\int 9x(7x^2 + 9)^n\, dx$

84. ◆ Determine whether the following is a theorem:

$$\int [f(x)]^2\, dx = 2 \int f(x)\, dx.$$

  TECHNOLOGY CONNECTION

85.–102. Use your grapher to check the results of any of Exercises 35–52 that were assigned.

5.6 Integration Techniques: Integration by Parts

OBJECTIVES

- Evaluate integrals using the formula for integration by parts.
- Solve application problems involving integration by parts.

Recall the Product Rule for derivatives:

$$\frac{d}{dx}\, uv = u\frac{dv}{dx} + v\frac{du}{dx}.$$

Integrating both sides with respect to x, we get

$$uv = \int u\frac{dv}{dx}\, dx + \int v\frac{du}{dx}\, dx$$

$$= \int u\, dv + \int v\, du.$$

Solving for $\int u\, dv$, we get the following.

THEOREM 7

The Integration-by-Parts Formula

$$\int u\,dv = uv - \int v\,du$$

This equation can be used as a formula for integrating in certain situations—that is, situations in which an integrand is a product of two functions, and one of the functions can be integrated using the techniques we have already developed. For example,

$$\int xe^x\,dx$$

can be considered as

$$\int x(e^x\,dx) = \int u\,dv,$$

where we let

$$u = x \quad \text{and} \quad dv = e^x\,dx.$$

If so, we have

$$du = dx, \quad \text{by differentiating}$$

and

$$v = e^x, \quad \text{by integrating and using the simplest antiderivative.}$$

Then the integration-by-parts formula gives us

$$\int \overset{u}{(x)}\,\overset{dv}{(e^x\,dx)} = \overset{u}{(x)}\overset{v}{(e^x)} - \int \overset{v}{(e^x)}\overset{du}{(dx)}$$

$$= xe^x - e^x + C.$$

This method of integrating is called **integration by parts.**

Note that integration by parts, like substitution, is a trial-and-error process. In the preceding example, suppose we had reversed the roles of x and e^x. We would have obtained

$$u = e^x, \qquad dv = x\,dx,$$

$$du = e^x\,dx, \qquad v = \frac{x^2}{2},$$

and

$$\int (e^x)(x\,dx) = (e^x)\left(\frac{x^2}{2}\right) - \int \left(\frac{x^2}{2}\right)(e^x\,dx).$$

Now the integrand on the right is more difficult to integrate than the one with which we began. When we integrate *both* factors of an integrand, and thus have a choice as to how to apply the integration-by-parts formula, it can happen that only one (and maybe none) of the possibilities will work.

Tips on Using Integration by Parts

1. If you have tried substitution, and have had no success, then try integration by parts.
2. Use integration by parts when an integral is of the form

$$\int f(x)\,g(x)\,dx.$$

Then match it with an integral of the form

$$\int u\,dv$$

by choosing a function to be $u = f(x)$, where $f(x)$ can be differentiated, and the remaining factor to be $dv = g(x)\,dx$, where $g(x)$ can be integrated.
3. Find du by differentiating and v by integrating.
4. If the resulting integral is more complicated than the original, make some other choice for $u = f(x)$ and $dv = g(x)\,dx$.

Let us consider some additional examples.

EXAMPLE 1   Evaluate $\int \ln x\,dx$.

Solution   Note that $\int (dx/x) = \ln x + C$, but we do not yet know how to find $\int \ln x\,dx$ since we do not know how to find an antiderivative of $\ln x$. Since we can differentiate $\ln x$, we let

$$u = \ln x \quad \text{and} \quad dv = dx.$$

Then

$$du = \frac{1}{x}\,dx \quad \text{and} \quad v = x.$$

Using the integration-by-parts formula gives

$$\int \overset{u}{(\ln\ x)} \overset{dv}{(dx)} = \overset{u}{(\ln\ x)} \overset{v}{x} - \int \overset{v}{x} \overset{du}{\left(\frac{1}{x}\ dx\right)}$$

$$= x\ \ln\ x - \int dx$$

$$= x\ \ln\ x - x + C.$$ ◆

EXAMPLE 2   Evaluate $\int x\ \ln\ x\ dx$.

Solution   Let us examine several choices, as follows.

   Choice 1:   We let

$$u = 1 \quad \text{and} \quad dv = x\ \ln\ x\ dx.$$

This will not work because we are back to our original integral, in which we do not know how to integrate $dv = x\ \ln\ x\ dx$.

   Choice 2:   We let

$$u = x\ \ln\ x \qquad\qquad \text{and} \quad dv = dx.$$

Then

$$du = \left[x\left(\frac{1}{x}\right) + 1(\ln\ x)\right] dx \quad \text{and} \quad v = x.$$
$$= (1 + \ln\ x)\ dx$$

Using the integration-by-parts formula, we have

$$\int u\ dv = uv - \int v\ du$$

$$= (x\ \ln\ x)x - \int x(1 + \ln\ x)\ dx$$

$$= x^2\ \ln\ x - \int (x + x\ \ln\ x)\ dx.$$

This integral is more complicated than the original.

   Choice 3:   We let

$$u = \ln\ x \quad \text{and} \quad dv = x\ dx.$$

Then

$$du = \frac{1}{x}\ dx \quad \text{and} \quad v = \frac{x^2}{2}.$$

Using the integration-by-parts formula, we have

$$\int u\, dv = uv - \int v\, du = (\ln x)\frac{x^2}{2} - \int \frac{x^2}{2}\cdot\frac{1}{x}\, dx$$

$$= \frac{x^2}{2}\ln x - \int \frac{x}{2}\, dx$$

$$= \frac{x^2}{2}\ln x - \frac{x^2}{4} + C.$$

This choice allows us to evaluate the integral.   ◆

EXAMPLE 3   Evaluate $\int x\sqrt{x+1}\, dx$.

Solution   We let

$$u = x \quad \text{and} \quad dv = (x+1)^{1/2}\, dx.$$

Then

$$du = dx \quad \text{and} \quad v = \tfrac{2}{3}(x+1)^{3/2}.$$

Note that we had to use substitution in order to integrate dv. We see this as follows:

$$\int (x+1)^{1/2}\, dx = \int w^{1/2}\, dw \qquad \xrightarrow{\text{Substitution}} \qquad \boxed{\begin{array}{l} w = x+1, \\ dw = dx \end{array}}$$

$$= \frac{w^{1/2+1}}{\tfrac{1}{2}+1} = \tfrac{2}{3}w^{3/2} = \tfrac{2}{3}(x+1)^{3/2}.$$

Using the integration-by-parts formula gives us

$$\int x\sqrt{x+1}\, dx = x\cdot\tfrac{2}{3}(x+1)^{3/2} - \int \tfrac{2}{3}(x+1)^{3/2}\, dx$$

$$= \tfrac{2}{3}x(x+1)^{3/2} - \tfrac{2}{3}\cdot\tfrac{2}{5}(x+1)^{5/2} + C$$

$$= \tfrac{2}{3}x(x+1)^{3/2} - \tfrac{4}{15}(x+1)^{5/2} + C.$$ ◆

EXAMPLE 4   Evaluate $\int_1^2 \ln x\, dx$.

Solution

a) First find the indefinite integral (Example 1).

b) Then evaluate the definite integral:

$$\int_1^2 \ln x\, dx = [x\ln x - x]_1^2$$

$$= (2\ln 2 - 2) - (1\cdot\ln 1 - 1)$$

$$= 2\ln 2 - 2 + 1$$

$$= 2\ln 2 - 1.$$ ◆

Repeated Integration by Parts

Sometimes we may need to apply the integration-by-parts formula more than once.

EXAMPLE 5   Evaluate $\int_0^7 x^2 e^{-x} \, dx$ to find the area of the shaded region below.

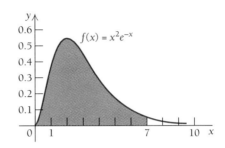

Solution

a) We let

$$u = x^2 \qquad \text{and} \quad dv = e^{-x} \, dx.$$

Then

$$du = 2x \, dx \quad \text{and} \quad v = -e^{-x}.$$

Using the integration-by-parts formula gives

$$\int u \, dv = uv - \int v \, du = x^2(-e^{-x}) - \int -e^{-x}(2x \, dx)$$

$$= -x^2 e^{-x} + \int 2x e^{-x} \, dx.$$

To evaluate the integral on the right, we can apply integration by parts again, as follows. We let

$$u = 2x \qquad \text{and} \quad dv = e^{-x} \, dx.$$

Then

$$du = 2 \, dx \quad \text{and} \quad v = -e^{-x}.$$

Using the integration-by-parts formula once again, we get

$$\int u \, dv = uv - \int v \, du$$

$$= 2x(-e^{-x}) - \int -e^{-x}(2 \, dx)$$

$$= -2x e^{-x} - 2e^{-x} + C.$$

Thus the original integral becomes

$$\int x^2 e^{-x}\, dx = -x^2 e^{-x} + \int 2x e^{-x}\, dx = -x^2 e^{-x} - 2x e^{-x} - 2e^{-x} + C$$

$$= -e^{-x}(x^2 + 2x + 2) + C.$$

b) We now evaluate the definite integral:

$$\int_0^7 x^2 e^{-x}\, dx = [-e^{-x}(x^2 + 2x + 2)]_0^7$$

$$= [-e^{-7}(7^2 + 2(7) + 2)] - [-e^{-0}(0^2 + 2(0) + 2)]$$

$$= -65e^{-7} + 2 \approx 1.94.$$

Tabular Integration by Parts

In situations like that in Example 5, we have an integral

$$\int f(x)\, g(x)\, dx,$$

where $f(x)$ can be repeatedly differentiated easily to a derivative that is eventually 0. The function $g(x)$ can also be repeatedly integrated easily. In such cases, we can use integration by parts more than once to evaluate the integral. As we saw in Example 5, the procedures can get complicated. When this happens, we can use **tabular integration**, which is illustrated in Example 6.

EXAMPLE 6   Evaluate $\int x^3 e^x\, dx$.

Solution   We let $f(x) = x^3$ and $g(x) = e^x$. Then we make a tabulation as follows.

$f(x)$ and repeated derivatives		$g(x)$ and repeated integrals
x^3	$(+)$	e^x
$3x^2$	$(-)$	e^x
$6x$	$(+)$	e^x
6	$(-)$	e^x
0		e^x

We then add products along the arrows, making the alternating sign changes to obtain

$$\int x^3 e^x\, dx = x^3 e^x - 3x^2 e^x + 6x e^x - 6e^x + C.$$

5.6 Exercise Set

Evaluate using integration by parts. Check by differentiating.

1. $\int 5xe^{5x}\, dx$ 2. $\int 2xe^{2x}\, dx$ 3. $\int x^3(3x^2\, dx)$

4. $\int x^2(2x\, dx)$ 5. $\int xe^{2x}\, dx$ 6. $\int xe^{3x}\, dx$

7. $\int xe^{-2x}\, dx$ 8. $\int xe^{-x}\, dx$

9. $\int x^2 \ln x\, dx$ 10. $\int x^3 \ln x\, dx$

11. $\int x \ln x^2\, dx$ 12. $\int x^2 \ln x^3\, dx$

13. $\int \ln (x + 3)\, dx$ 14. $\int \ln (x + 1)\, dx$

15. $\int (x + 2) \ln x\, dx$ 16. $\int (x + 1) \ln x\, dx$

17. $\int (x - 1) \ln x\, dx$ 18. $\int (x - 2) \ln x\, dx$

19. $\int x\sqrt{x + 2}\, dx$ 20. $\int x\sqrt{x + 4}\, dx$

21. $\int x^3 \ln 2x\, dx$ 22. $\int x^2 \ln 5x\, dx$

23. $\int x^2 e^x\, dx$ 24. $\int (\ln x)^2\, dx$

25. $\int x^2 e^{2x}\, dx$ 26. $\int x^{-5} \ln x\, dx$

27. $\int x^3 e^{-2x}\, dx$ 28. $\int x^5 e^{4x}\, dx$

29. $\int (x^4 + 1)e^{3x}\, dx$ 30. $\int (x^3 - x + 1)e^{-x}\, dx$

Evaluate using integration by parts.

31. $\int_1^2 x^2 \ln x\, dx$ 32. $\int_1^2 x^3 \ln x\, dx$

33. $\int_2^6 \ln (x + 3)\, dx$ 34. $\int_0^5 \ln (x + 1)\, dx$

35. $\int_0^1 xe^x\, dx$ 36. $\int_0^1 xe^{-x}\, dx$

APPLICATIONS

◆ **Business and Economics**

37. *Cost from marginal cost.* A company determines that its marginal-cost function is given by

$$C'(x) = 4x\sqrt{x + 3}.$$

Find the total cost given that $C(13) = \$1126.40$.

38. *Profit from marginal profit.* A firm determines that its marginal-profit function is given by

$$P'(x) = 1000x^2 e^{-0.2x}.$$

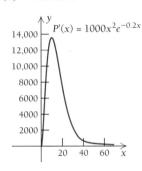

Find the total profit given that $P = -\$2000$ when $x = 0$.

◆ **Life and Physical Sciences**

39. *Electrical energy use.* The rate of electrical energy used by a family, in kilowatt hours per day, is given by

$$K(t) = 10te^{-t},$$

where $t = $ the time, in hours. That is, t is in the interval $[0, 24]$.

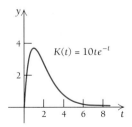

a) How many kilowatt hours does the family use in the first T hours of a day ($t = 0$ to $t = T$)?

b) How many kilowatt hours does the family use in the first 4 hours of the day?

40. **Drug dosage.** Suppose that an oral dose of a drug is taken. From that time, the drug is assimilated in the body and excreted through the urine. The total amount of the drug that has passed through the body in time T is given by

$$\int_0^T E(t)\, dt,$$

where $E =$ the rate of excretion of the drug. A typical rate-of-excretion function is

$$E(t) = te^{-kt},$$

where $k > 0$ and $t =$ the time, in hours.

a) Use integration by parts to find a formula for

$$\int_0^T E(t)\, dt.$$

b) Find

$$\int_0^{10} E(t)\, dt, \quad \text{when } k = 0.2 \text{ mg/hr.}$$

45. $\displaystyle \int \frac{\ln x}{\sqrt{x}}\, dx$

46. $\displaystyle \int x^n (\ln x)^2\, dx$

47. $\displaystyle \int \frac{13t^2 - 48}{\sqrt[5]{4t + 7}}\, dt$

48. $\displaystyle \int (27x^3 + 83x - 2)\, \sqrt[6]{3x + 8}\, dx$

49. Verify that for any positive integer n,

$$\int x^n e^x\, dx = x^n e^x - n \int x^{n-1} e^x\, dx.$$

50. Verify that for any positive integer n,

$$\int (\ln x)^n\, dx = x(\ln x)^n - n \int (\ln x)^{n-1}\, dx.$$

51. ◆ Determine whether the following is a theorem:

$$\int f(x)\, g(x)\, dx = \int f(x)\, dx \cdot \int g(x)\, dx.$$

Explain.

52. ◆ Compare the procedures of differentiation and integration. Which seems to be the most complicated or difficult and why?

Evaluate using integration by parts.

41. $\displaystyle \int \sqrt{x}\, \ln x\, dx$

42. $\displaystyle \int x^n \ln x\, dx$

43. $\displaystyle \int \frac{te^t}{(t + 1)^2}\, dt$

44. $\displaystyle \int x^2 (\ln x)^2\, dx$

 TECHNOLOGY CONNECTION

53. Graph the function

$$f(x) = x^5 \ln x$$

over the interval $[1, 10]$ and approximate the area.

5.7 Integration Techniques: Using Tables

OBJECTIVE

• Evaluate integrals using a table of integration formulas.

Tables of Integration Formulas

You have probably noticed that, generally speaking, integration is more difficult and "tricky" than differentiation. Because of this, integral formulas that are reasonable and/or important have been gathered into tables. Table 1 at the back of the book, though quite brief, is such an example. Entire books of integration formulas are available in libraries, and lengthy tables are also available in mathematics handbooks. Such tables are usually classified by the form of the integrand. The idea is to properly match the integral in question with a formula in the table. Sometimes some algebra or a technique such as integration by substitution or parts may be needed as well as a table.

EXAMPLE 1   Evaluate

$$\int \frac{dx}{x(3 - x)}.$$

Solution   This integral fits *Formula 20* in Table 1:

$$\int \frac{1}{x(ax + b)} \, dx = \frac{1}{b} \ln \left(\frac{x}{ax + b} \right) + C.$$

In our integral, $a = -1$ and $b = 3$, so we have, by the formula,

$$\int \frac{1}{x(3 - x)} \, dx = \int \frac{dx}{x(-1 \cdot x + 3)}$$

$$= \frac{1}{3} \ln \left(\frac{x}{-1 \cdot x + 3} \right) + C$$

$$= \frac{1}{3} \ln \left(\frac{x}{3 - x} \right) + C. \qquad \blacklozenge$$

EXAMPLE 2   Evaluate

$$\int \frac{5x}{7x - 8} \, dx.$$

Solution   If we first factor 5 out of the integral, then the integral fits *Formula 18* in Table 1:

$$\int \frac{x}{ax + b} \, dx = \frac{b}{a^2} + \frac{x}{a} - \frac{b}{a^2} \ln (ax + b) + C.$$

In our integral, $a = 7$ and $b = -8$, so we have, by the formula,

$$\int \frac{5x}{7x - 8} \, dx = 5 \int \frac{x}{7x - 8} \, dx$$

$$= 5 \left[\frac{-8}{7^2} + \frac{x}{7} - \frac{-8}{7^2} \ln (7x - 8) \right] + C$$

$$= 5 \left[\frac{-8}{49} + \frac{x}{7} + \frac{8}{49} \ln (7x - 8) \right] + C$$

$$= -\frac{40}{49} + \frac{5x}{7} + \frac{40}{49} \ln (7x - 8) + C. \qquad \blacklozenge$$

EXAMPLE 3   Evaluate

$$\int \sqrt{16x^2 + 3}\, dx.$$

Solution   This integral almost fits *Formula 22* in Table 1:

$$\int \sqrt{x^2 \pm a^2}\, dx = \tfrac{1}{2}\left[x\sqrt{x^2 \pm a^2} \pm a^2 \ln\left(x + \sqrt{x^2 \pm a^2}\right) \right] + C.$$

But the x^2-coefficient needs to be 1. To achieve this, we first factor out 16 as follows. Then we apply *Formula 22* in Table 1:

$$\int \sqrt{16x^2 + 3}\, dx = \int \sqrt{16\left(x^2 + \tfrac{3}{16}\right)}\, dx$$

$$= \int 4\sqrt{x^2 + \tfrac{3}{16}}\, dx$$

$$= 4\int \sqrt{x^2 + \tfrac{3}{16}}\, dx$$

$$= 4 \cdot \tfrac{1}{2}\left[x\sqrt{x^2 + \tfrac{3}{16}} + \tfrac{3}{16}\ln\left(x + \sqrt{x^2 + \tfrac{3}{16}}\right) \right] + C$$

$$= 2\left[x\sqrt{x^2 + \tfrac{3}{16}} + \tfrac{3}{16}\ln\left(x + \sqrt{x^2 + \tfrac{3}{16}}\right) \right] + C.$$

In our integral, $a^2 = 3/16$ and $a = \sqrt{3}/4$, though we did not need to use a in this form when applying the formula.   ◆

EXAMPLE 4   Evaluate

$$\int \frac{dx}{x^2 - 25}.$$

Solution   This integral fits *Formula 14* in Table 1:

$$\int \frac{1}{x^2 - a^2}\, dx = \frac{1}{2a} \ln\left(\frac{x - a}{x + a}\right) + C.$$

In our integral, $a = 5$, so we have, by the formula,

$$\int \frac{dx}{x^2 - 25} = \frac{1}{10} \ln\left(\frac{x - 5}{x + 5}\right) + C.$$ ◆

EXAMPLE 5   Evaluate $\int (\ln x)^3\, dx.$

Solution   This integral fits *Formula 9* in Table 1:

$$\int (\ln x)^n\, dx = x(\ln x)^n - n\int (\ln x)^{n-1}\, dx + C, \quad n \neq -1.$$

We must apply the formula three times:

$$\int (\ln x)^3 \, dx$$

$$= x(\ln x)^3 - 3 \int (\ln x)^2 \, dx + C \qquad \text{Formula 9}$$

$$= x(\ln x)^3 - 3\left[x(\ln x)^2 - 2 \int \ln x \, dx \right] + C \qquad \begin{array}{l}\text{Applying Formula 9} \\ \text{again}\end{array}$$

$$= x(\ln x)^3 - 3\left[x(\ln x)^2 - 2\left(x \ln x - \int dx \right) \right] + C \qquad \begin{array}{l}\text{Applying} \\ \text{Formula 9 for} \\ \text{the third time}\end{array}$$

$$= x(\ln x)^3 - 3x(\ln x)^2 + 6x \ln x - 6x + C. \qquad \blacklozenge$$

5.7 Exercise Set

Evaluate using Table 1.

1. $\displaystyle\int xe^{-3x} \, dx$

2. $\displaystyle\int xe^{4x} \, dx$

3. $\displaystyle\int 5^x \, dx$

4. $\displaystyle\int \frac{1}{\sqrt{x^2 - 9}} \, dx$

5. $\displaystyle\int \frac{1}{16 - x^2} \, dx$

6. $\displaystyle\int \frac{1}{x\sqrt{4 + x^2}} \, dx$

7. $\displaystyle\int \frac{x}{5 - x} \, dx$

8. $\displaystyle\int \frac{x}{(1 - x)^2} \, dx$

9. $\displaystyle\int \frac{1}{x(5 - x)^2} \, dx$

10. $\displaystyle\int \sqrt{x^2 + 9} \, dx$

11. $\displaystyle\int \ln 3x \, dx$

12. $\displaystyle\int \ln \frac{4}{5}x \, dx$

13. $\displaystyle\int x^4 e^{5x} \, dx$

14. $\displaystyle\int x^3 e^{-2x} \, dx$

15. $\displaystyle\int x^3 \ln x \, dx$

16. $\displaystyle\int 5x^4 \ln x \, dx$

17. $\displaystyle\int \frac{dx}{\sqrt{x^2 + 7}}$

18. $\displaystyle\int \frac{3 \, dx}{x\sqrt{1 - x^2}}$

19. $\displaystyle\int \frac{10 \, dx}{x(5 - 7x)^2}$

20. $\displaystyle\int \frac{2}{5x(7x + 2)} \, dx$

21. $\displaystyle\int \frac{-5}{4x^2 - 1} \, dx$

22. $\displaystyle\int \sqrt{9t^2 - 1} \, dt$

23. $\displaystyle\int \sqrt{4m^2 + 16} \, dm$

24. $\displaystyle\int \frac{3 \ln x}{x^2} \, dx$

25. $\displaystyle\int \frac{-5 \ln x}{x^3} \, dx$

26. $\displaystyle\int (\ln x)^4 \, dx$

27. $\displaystyle\int \frac{e^x}{x^{-3}} \, dx$

28. $\displaystyle\int \frac{3}{\sqrt{4x^2 + 100}} \, dx$

APPLICATIONS

◆ **Business and Economics**

29. **Supply from marginal supply.** A lawn machinery company introduces a new kind of lawn seeder. It finds that its marginal supply for the seeder satisfies the function

$$S'(p) = \frac{100p}{(20 - p)^2},$$

$$0 \le p \le 19,$$

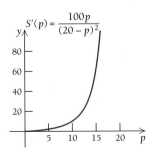

where S = the quantity purchased when the price is p thousand dollars per seeder. Find the supply function $S(p)$ given that the company will sell 2000 seeders when the price is \$19 thousand.

◆   **Social Sciences**

30. *Learning rate.* The rate of change of the probability that an employee learns a task on a new assembly line is given by

$$p'(t) = \frac{1}{t(2 + t)^2},$$

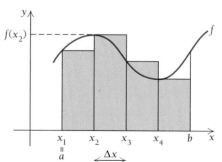

$$p'(t) = \frac{1}{t(2 + t)^2}$$

where p = the probability of learning the task after time t, in months. Find the function $p(t)$ given that $p = 0.8267$ when $t = 2$.

SYNTHESIS

Evaluate using Table 1.

31. $\displaystyle\int \frac{8}{3x^2 - 2x}\, dx$

32. $\displaystyle\int \frac{x\, dx}{4x^2 - 12x + 9}$

33. $\displaystyle\int \frac{dx}{x^3 - 4x^2 + 4x}$

34. $\displaystyle\int e^x \sqrt{e^{2x} + 1}\, dx$

35. $\displaystyle\int \frac{-e^{-2x}\, dx}{9 - 6e^{-x} + e^{-2x}}$

36. $\displaystyle\int \frac{\sqrt{(\ln x)^2 + 49}}{2x}\, dx$

5.8 The Definite Integral as a Limit of Sums

OBJECTIVES

• Approximate definite integrals by adding areas of rectangles.

• Find the average value of a function.

We now consider approximating the area of a region by dividing that region into subregions that are almost rectangles. In the following figure, $[a, b]$ has been divided into 4 subintervals, each having width Δx, or $(b - a)/4$.

The area under a curve can be approximated by a sum of rectangular areas.

The heights of the rectangles shown are

$$f(x_1), \qquad f(x_2), \qquad f(x_3), \quad \text{and} \quad f(x_4).$$

The area of the region under the curve is approximately the sum of the areas of the four rectangles:

$$f(x_1)\,\Delta x + f(x_2)\,\Delta x + f(x_3)\,\Delta x + f(x_4)\,\Delta x.$$

We can name this sum using **summation notation,** which utilizes the Greek capital letter sigma, Σ:

$$\sum_{i=1}^{4} f(x_i)\,\Delta x, \quad \text{or} \quad \sum_{i=1}^{4} f(x_i)\,\Delta x.$$

This is read "the sum of the numbers $f(x_i)\,\Delta x$ from $i = 1$ to $i = 4$." To recover the original expression, we substitute the numbers 1 through 4 successively into $f(x_i)\,\Delta x$ and write plus signs between the results.

EXAMPLE 1   Write summation notation for $2 + 4 + 6 + 8 + 10$.

Solution

$$2 + 4 + 6 + 8 + 10 = \sum_{i=1}^{5} 2i \qquad \blacklozenge$$

EXAMPLE 2   Write summation notation for

$$g(x_1)\,\Delta x + g(x_2)\,\Delta x + \cdots + g(x_{19})\,\Delta x.$$

Solution

$$g(x_1)\,\Delta x + g(x_2)\,\Delta x + \cdots + g(x_{19})\,\Delta x = \sum_{i=1}^{19} g(x_i)\,\Delta x \qquad \blacklozenge$$

EXAMPLE 3   Express $\sum_{i=1}^{4} 3^i$ without using summation notation.

Solution

$$\sum_{i=1}^{4} 3^i = 3^1 + 3^2 + 3^3 + 3^4, \quad \text{or} \quad 120 \qquad \blacklozenge$$

EXAMPLE 4   Express $\sum_{i=1}^{30} h(x_i)\,\Delta x$ without using summation notation.

Solution

$$\sum_{i=1}^{30} h(x_i)\,\Delta x = h(x_1)\,\Delta x + h(x_2)\,\Delta x + \cdots + h(x_{30})\,\Delta x \qquad \blacklozenge$$

Approximation of area by rectangles becomes more accurate as we use more rectangles and smaller subintervals, as shown in the figures on the following page.

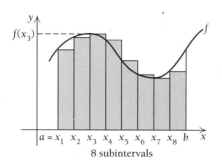

8 subintervals

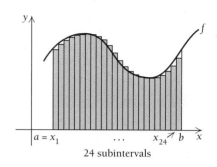

24 subintervals

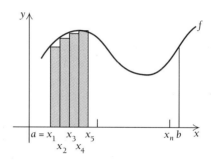

In general, the interval $[a, b]$ is divided into n equal subintervals, each of width $\Delta x = (b - a)/n$. The heights of the rectangles are

$$f(x_1), f(x_2), \ldots, f(x_n).$$

The width of each rectangle is Δx, so the first rectangle has area

$$f(x_1)\, \Delta x,$$

the second rectangle has area

$$f(x_2)\, \Delta x,$$

and so on. The area of the region under the curve is approximated by the sum of the areas of the rectangles:

$$\sum_{i=1}^{n} f(x_i)\, \Delta x.$$

We now obtain the actual area by letting the number of intervals increase indefinitely and then by taking the limit. The exact area is thus given by

$$A = \lim_{n \to \infty} \sum_{i=1}^{n} f(x_i)\, \Delta x.$$

The area is also given by a definite integral:

$$\int_a^b f(x)\, dx = \lim_{n \to \infty} \sum_{i=1}^{n} f(x_i)\, \Delta x.$$

The fact that we can express the integral of a function (positive or otherwise) as a limit of a sum or in terms of an antiderivative is so important that it has a name: *The Fundamental Theorem of Integral Calculus.*

> **The Fundamental Theorem of Integral Calculus**
>
> If a function f has an antiderivative F on $[a, b]$, then
>
> $$\lim_{n \to \infty} \sum_{i=1}^{n} f(x_i)\, \Delta x = \int_a^b f(x)\, dx = F(b) - F(a).$$

It is interesting to envision that, as we take the limit, the summation sign stretches into something reminiscent of an S (the integral sign) and Δx is defined to be dx. This is why dx appears in the integral notation.

This theorem allows us to approximate the value of a definite integral by a sum, making it as good as we please by taking n sufficiently large.

EXAMPLE 5   Consider the function

$$f(x) = 600x - x^2$$

over the interval $[0, 600]$.

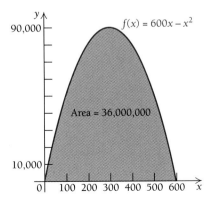

a) Approximate the area by dividing the interval into 6 subintervals.

b) Approximate the area by dividing the interval into 12 subintervals.

c) Find the exact area using integration.

Solution

a) The interval $[0, 600]$ is divided into 6 subintervals, each of length $\Delta x = (600 - 0)/6 = 100$. Now x_i is varying from $x_1 = 0$ to $x_6 = 500$ and $b = 600$. Thus we have

$$\sum_{i=1}^{6} f(x_i)\, \Delta x = f(0) \cdot 100 + f(100) \cdot 100 + f(200) \cdot 100$$
$$+ f(300) \cdot 100 + f(400) \cdot 100 + f(500) \cdot 100$$
$$= 0 \cdot 100 + 50{,}000 \cdot 100 + 80{,}000 \cdot 100$$
$$+ 90{,}000 \cdot 100 + 80{,}000 \cdot 100 + 50{,}000 \cdot 100$$
$$= 35{,}000{,}000.$$

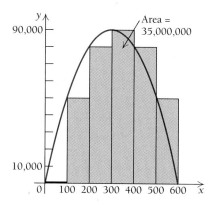

b) The interval $[0, 600]$ is divided into 12 subintervals, each of length $\Delta x = (600 - 0)/12 = 50$. Now x_i is varying from $x_1 = 0$ to $x_{12} = 550$ and $b = 600$.

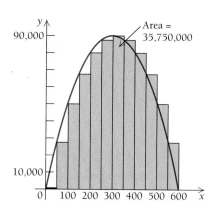

Thus we have

$$\sum_{i=1}^{12} f(x_i) \, \Delta x = f(0) \cdot 50 + f(50) \cdot 50 + f(100) \cdot 50 + f(150) \cdot 50$$
$$+ f(200) \cdot 50 + f(250) \cdot 50 + f(300) \cdot 50 + f(350) \cdot 50$$
$$+ f(400) \cdot 50 + f(450) \cdot 50 + f(500) \cdot 50 + f(550) \cdot 50$$

$$= 0 \cdot 50 + 27{,}500 \cdot 50 + 50{,}000 \cdot 50 + 67{,}500 \cdot 50$$
$$+ 80{,}000 \cdot 50 + 87{,}500 \cdot 50 + 90{,}000 \cdot 50 + 87{,}500 \cdot 50$$
$$+ 80{,}000 \cdot 50 + 67{,}500 \cdot 50 + 50{,}000 \cdot 50 + 27{,}500 \cdot 50$$

$$= 35{,}750{,}000.$$

c) We find the exact area by integrating as follows:

$$\int_0^{600} (600x - x^2) \, dx = \left[300x^2 - \frac{x^3}{3} \right]_0^{600}$$

$$= \left(300 \cdot 600^2 - \frac{600^3}{3} \right) - \left(300 \cdot 0^2 - \frac{0^3}{3} \right)$$

$$= 36{,}000{,}000. \qquad \blacklozenge$$

EXAMPLE 6   *Business: Cost from marginal cost.* Raggs, Ltd., determines that the marginal cost per suit is given by

$$C'(x) = 0.0003x^2 - 0.2x + 50.$$

a) Approximate the total cost of producing 400 suits by computing the sum $\sum_{i=1}^{4} C'(x_i) \, \Delta x$.

b) Find the exact total cost of producing 400 suits by integrating.

Solution

a) The interval $[0, 400]$ is divided into 4 subintervals, each of length $\Delta x = (400 - 0)/4 = 100$. Now x_i is varying from $x_1 = 0$ to $x_4 = 300$ and $b = 400$.

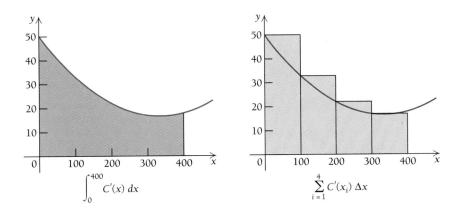

Thus we have

$$\sum_{i=1}^{4} C'(x_i) \, \Delta x = C'(0) \cdot 100 + C'(100) \cdot 100 + C'(200) \cdot 100 \\ + C'(300) \cdot 100$$

$$= 50 \cdot 100 + 33 \cdot 100 + 22 \cdot 100 + 17 \cdot 100$$

$$= \$12{,}200.$$

b) The exact total cost is

$$\int_{0}^{400} C'(x) \, dx = \$10{,}400. \qquad \text{See Example 3 of Section 5.2.}$$

Thus the approximation is not too far off, even though the number of subintervals is small. ◆

The fact that an integral can be approximated by a sum is useful when the antiderivative of a function does not have an elementary formula. For example, for the function $e^{-x^2/2}$, important in probability, there is no formula for the antiderivative. Thus tables of approximate values of its integral have been computed using summation methods.

The Average Value of a Function

Suppose that

$$T = f(t)$$

is the temperature at time t recorded at a weather station on a certain day. The station uses a 24-hr clock, so the domain of the temperature function is the interval $[0, 24]$. The function is continuous, as shown below.

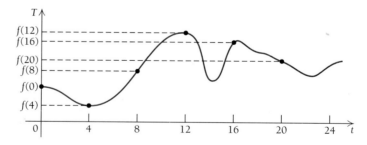

To find the average temperature for the given day, we might take six temperature readings at 4-hr intervals, starting at midnight:

$$T_0 = f(0), \qquad T_1 = f(4),$$
$$T_2 = f(8), \qquad T_3 = f(12),$$
$$T_4 = f(16), \qquad T_5 = f(20).$$

The average reading would then be the sum of these six readings divided by 6:

$$T_{av} = \frac{T_0 + T_1 + T_2 + T_3 + T_4 + T_5}{6}.$$

This computation of the average temperature may not give the most useful answer. For example, suppose it is a hot summer day, and at 2:00 in the afternoon (hour 14 on the 24-hr clock), there is a short thunderstorm that cools the air for an hour between our readings. This temporary dip would not show up in the average computed above.

What can we do? We could take 48 readings at half-hour intervals. This should give us a better result. In fact, the shorter the time between readings, the better the result should be. It seems reasonable that we might define the **average value** of T over the interval $[0, 24]$ to be the limit, as n approaches ∞, of the average of n values:

$$\text{Average value of } T = \lim_{n \to \infty} \frac{1}{n} \sum_{i=1}^{n} T_i$$

$$= \lim_{n \to \infty} \frac{1}{n} \sum_{i=1}^{n} f(t_i).$$

Note that this is not too far from our definition of an integral. All we would need is to get Δt, which is $(24 - 0)/n$, or $24/n$, into the summation:

$$\text{Average value of } T = \lim_{n \to \infty} \frac{1}{n} \sum_{i=1}^{n} f(t_i)$$

$$= \lim_{n \to \infty} \frac{1}{\Delta t} \cdot \frac{1}{n} \sum_{i=1}^{n} f(t_i) \, \Delta t$$

$$= \lim_{n \to \infty} \frac{n}{24} \cdot \frac{1}{n} \sum_{i=1}^{n} f(t_i) \, \Delta t \qquad \Delta t = \frac{24}{n}, \text{ or } \frac{1}{\Delta t} = \frac{n}{24}$$

$$= \frac{1}{24} \lim_{n \to \infty} \sum_{i=1}^{n} f(t_i) \, \Delta t$$

$$= \frac{1}{24} \int_0^{24} f(t) \, dt.$$

DEFINITION

Let f be a continuous function over a closed interval $[a, b]$. Its *average value*, y_{av}, is given by

$$y_{av} = \frac{1}{b - a} \int_a^b f(x) \, dx.$$

Let us consider average value in another way. If we multiply on both sides of

$$y_{av} = \frac{1}{b-a} \int_a^b f(x)\, dx$$

by $b-a$, we get

$$(b-a)y_{av} = \int_a^b f(x)\, dx.$$

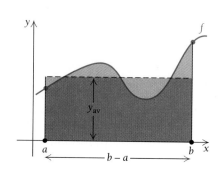

Now the expression on the left will give the area of a rectangle of length $b-a$ and height y_{av}. The area of such a rectangle is the same as the area bounded by $y = f(x)$ on the interval $[a, b]$, as shown in the figure.

EXAMPLE 7   Find the average value of $f(x) = x^2$ over the interval $[0, 2]$.

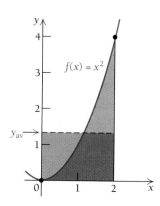

Solution   The average value is

$$\frac{1}{2-0} \int_0^2 x^2\, dx = \frac{1}{2}\left[\frac{x^3}{3}\right]_0^2$$

$$= \frac{1}{2}\left(\frac{2^3}{3} - \frac{0^3}{3}\right)$$

$$= \frac{1}{2} \cdot \frac{8}{3} = \frac{4}{3}, \quad \text{or} \quad 1\frac{1}{3}.$$

Note that although the values of $f(x)$ increase from 0 to 4, we would not expect the average value to be 2, because we see from the graph that $f(x)$ is less than 2 over more than half the interval. ◆

TECHNOLOGY CONNECTION

Graph $f(x) = x^4$. Compute the average value of the function over the interval $[0, 2]$, using the method of Example 7. Then use that value y_{av} and draw a graph of it as a horizontal line using the same set of axes. What does this line represent in comparison to the graph of $f(x) = x^4$?

EXAMPLE 8   *Life science: Engine emissions.* The emissions of an engine are given by

$$E(t) = 2t^2,$$

where $E(t)$ = the engine's rate of emission, in billions of pollution particulates, at time t, in years. Find the average emissions from $t = 1$ to $t = 5$.

Solution   The average emissions are

$$\frac{1}{5-1}\int_1^5 2t^2\,dt = \frac{1}{4}\left[\frac{2}{3}t^3\right]_1^5$$

$$= \frac{1}{4}\cdot\frac{2}{3}(5^3 - 1^3)$$

$$= \frac{1}{6}(125 - 1)$$

$$= 20\tfrac{2}{3} \text{ billion pollution particulates.} \qquad \blacklozenge$$

5.8 Exercise Set

1. a) Approximate

$$\int_1^7 \frac{dx}{x^2}$$

by computing the area of each rectangle to four decimal places and then adding.

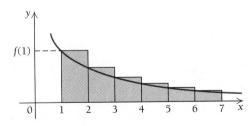

b) Evaluate

$$\int_1^7 \frac{dx}{x^2}.$$

Compare the answer to part (a).

2. a) Approximate

$$\int_0^5 (x^2 + 1)\,dx$$

by computing the area of each rectangle and then adding.

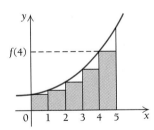

b) Evaluate

$$\int_0^5 (x^2 + 1)\,dx.$$

Compare the answer to part (a).

Find the average value over the given interval.

3. $y = 2x^3$; $[-1, 1]$ 　　**4.** $y = 4 - x^2$; $[-2, 2]$

5. $y = e^x$; $[0, 1]$ 　　**6.** $y = e^{-x}$; $[0, 1]$

7. $y = x^2 - x + 1$; $[0, 2]$

8. $y = x^2 + x - 2$; $[0, 4]$

9. $y = 3x + 1$; $[2, 6]$

10. $y = 4x + 1$; $[3, 7]$

11. $y = x^n$; $[0, 1]$

12. $y = x^n$; $[1, 2]$

APPLICATIONS

◆   Business and Economics

13. **Results of practice.** A typist's speed over a 5-min interval is given by

$$W(t) = -6t^2 + 12t + 90, \quad t \text{ in } [0, 5],$$

where $W(t) = $ the speed, in words per minute, at time t.

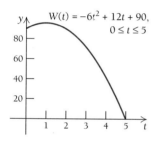

a) Find the speed at the beginning of the interval.

b) Find the maximum speed and when it occurs.

c) Find the average speed over the 5-min interval.

◆   Life and Physical Sciences

14. **Average drug dosage.** The amount of a drug in the body at time t is given by

$$A(t) = 3e^{-0.1t},$$

where A is in cubic centimeters and $t = $ the time, in hours.

a) What is the initial dosage of the drug?

b) What is the average amount in the body over the first 4 hr?

◆   Social Sciences

15. **Average population.** The population of the United States is given by

$$P(t) = 241e^{0.009t},$$

where P is in millions and $t = $ the number of years since 1986. Find the average value of the population from 1986 to 1998.

16. **Average population of a city.** The population of a city increased and then decreased over an 8-yr period according to the function

$$P(t) = -0.1t^2 + t + 3, \quad 0 \le t \le 8,$$

where P is in millions and $t = $ the time.

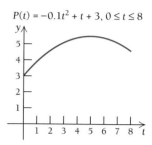

a) Find the average population.

b) Find the minimum population.

c) Find the maximum population.

17. **Results of studying.** A student's score on a test is given by the function

$$S(t) = t^2, \quad t \text{ in } [0, 10],$$

where $S(t) = $ the score after t hours of study.

a) Find the maximum score that the student can achieve and the number of hours of study required to attain it.

b) Find the average score over the 10-hr interval.

A student's test score is a function of the time spent studying.

18. **Outside temperature.** The temperature over a 10-hr period is given by

$$f(t) = -t^2 + 5t + 40, \quad 0 \le t \le 10.$$

a) Find the average temperature.
b) Find the minimum temperature.
c) Find the maximum temperature.

SYNTHESIS

The Trapezoidal Rule. Another way to approximate an integral is to replace each rectangle in the sum (see Fig. 1) by a trapezoid (see Fig. 2). The area of a trapezoid is $h(c_1 + c_2)/2$, where c_1 and c_2 are the lengths of the parallel sides. Thus, in Fig. 2,

$$\int_a^b f(x)\,dx = \int_a^m f(x)\,dx + \int_m^b f(x)\,dx$$

$$\approx \Delta x\,\frac{f(a) + f(m)}{2} + \Delta x\,\frac{f(m) + f(b)}{2}$$

$$\approx \Delta x\left[\frac{f(a)}{2} + f(m) + \frac{f(b)}{2}\right].$$

For an interval $[a, b]$ subdivided into n equal subintervals of length $\Delta x = (b - a)/n$, we get the approximation

$$\int_a^b f(x)\,dx \approx$$

$$\Delta x\left[\frac{f(a)}{2} + f(x_2) + f(x_3) + \cdots + f(x_n) + \frac{f(b)}{2}\right],$$

where $x_1 = a$ and

$$x_n = x_{n-1} + \Delta x \quad \text{or} \quad x_n = a + (n - 1)\,\Delta x.$$

This is called the **Trapezoidal Rule.**

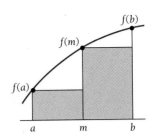

FIGURE 1

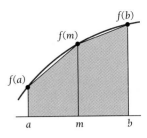

FIGURE 2

19. Use the Trapezoidal Rule and the interval subdivision of Exercise 1 to approximate

$$\int_1^7 \frac{dx}{x^2}.$$

20. Use the Trapezoidal Rule and the interval subdivision of Exercise 2 to approximate

$$\int_0^5 (x^2 + 1)\,dx.$$

TECHNOLOGY CONNECTION

Graph each function and approximate the integral using a grapher.

21. $\displaystyle\int_0^4 \sqrt{x}\,dx$

22. $\displaystyle\int_0^2 (x^4 + 1)\,dx$

23. $\displaystyle\int_2^4 \ln x\,dx$

24. $\displaystyle\int_0^3 \frac{2}{\sqrt{x + 1}}\,dx$

5 Chapter Summary and Review

TERMS TO KNOW

REVIEW EXERCISES

The review exercises are for test preparation. They can also be used as a lengthened practice test. Answers are at the back of the book. The answers also contain bracketed section references, which tell you where to restudy if your answer is incorrect.

Evaluate.

1. $\int 8x^4 \, dx$

2. $\int (3e^x + 2) \, dx$

3. $\int \left(3t^2 + 7t + \dfrac{1}{t}\right) dt$

Find the area under the curve on the interval indicated.

4. $y = 4 - x^2;$ $[-2, 1]$

5. $y = x^2 + 3x + 6;$ $[0, 2]$

In each case, give two interpretations of the shaded region.

6.

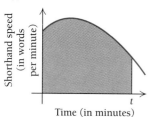

7.

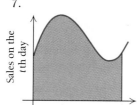

Evaluate.

8. $\int_a^b x^5 \, dx$

9. $\int_{-1}^1 (x^3 - x^4) \, dx$

10. $\int_0^1 (e^x + x) \, dx$

11. $\int_1^3 \dfrac{3}{x} \, dx$

Decide whether $\int_a^b f(x) \, dx$ is positive, negative, or zero.

12.

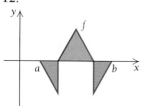

13.

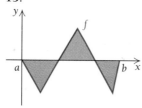

14.

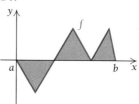

15. Find the area of the region bounded by $y = 3x^2$ and $y = 9x.$

Evaluate using substitution. Do not use Table 1.

16. $\int x^3 e^{x^4} \, dx$

17. $\int \dfrac{24t^5}{4t^6 + 3} \, dt$

18. $\int \dfrac{\ln 4x}{2x} \, dx$

19. $\int 2e^{-3x} \, dx$

Evaluate using integration by parts. Do not use Table 1.

20. $\int 3x \, e^{3x} \, dx$

21. $\int \ln x^7 \, dx$

22. $\int 3x^2 \ln x \, dx$

Evaluate using Table 1.

23. $\int \dfrac{1}{49 - x^2} \, dx$

24. $\int x^2 e^{5x} \, dx$

25. $\int \dfrac{x}{7x + 1} \, dx$

26. $\int \dfrac{dx}{\sqrt{x^2 - 36}}$

27. $\int x^6 \ln x \, dx$

28. $\int xe^{8x} \, dx$

29. Approximate $\int_1^4 (2/x) \, dx$ by computing the area of each rectangle to three decimal places and adding.

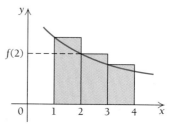

30. Find the average value of $y = e^{-x} + 5$ over $[0, 2].$

31. A particle starts out from the origin. Its velocity at any time t, $t \geq 0$, is given by $v(t) = 3t^2 + 2t$. Find the distance that the particle travels during the first 4 hr (from $t = 0$ to $t = 4$).

32. **Business: Accumulated sales.** A company estimates that its sales will grow continuously according to the function $S'(t) = 3e^{3t}$, where $S'(t) = $ the sales, in dollars, on the tth day. Find the accumulated sales for the first 4 days.

Integrate using any method.

33. $\int x^3 e^{0.1x} \, dx$

34. $\int \dfrac{12t^2}{4t^3 + 7} \, dt$

35. $\int \dfrac{x \, dx}{\sqrt{4 + 5x}}$

36. $\int 5x^4 e^{x^5} \, dx$

37. $\int \dfrac{dx}{x + 9}$

38. $\int t^7 (t^8 + 3)^{11} \, dt$

39. $\int \ln 7x \, dx$

40. $\int x \ln 8x \, dx$

SYNTHESIS

Evaluate.

41. $\int \dfrac{t^4 \ln (t^5 + 3)}{t^5 + 3} \, dt$

42. $\int \dfrac{dx}{e^x + 2}$

43. $\int \dfrac{\ln \sqrt{x}}{x} \, dx$

44. $\int x^{91} \ln x \, dx$

45. $\int \ln \left(\dfrac{x - 3}{x - 4} \right) dx$

46. $\int \dfrac{dx}{x \, (\ln x)^4}$

 TECHNOLOGY CONNECTION

47. Use your grapher to approximate the area between the following curves:

$$y = 2x^2 - 2x, \quad y = 12x^2 - 12x^3,$$
$$x = 0, \quad x = 1.$$

 5 Chapter Test

Evaluate.

1. $\int dx$

2. $\int 1000x^4 \, dx$

3. $\int \left(e^x + \dfrac{1}{x} + x^{3/8} \right) dx$

Find the area under the curve on the interval indicated.

4. $y = x - x^2$; $[0, 1]$

5. $y = \dfrac{4}{x}$; $[1, 3]$

6. Give two interpretations of the shaded area.

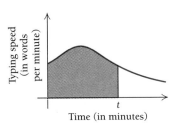

Time (in minutes)

Evaluate.

7. $\displaystyle\int_{-1}^{2} (2x + 3x^2) \, dx$

8. $\displaystyle\int_{0}^{1} e^{-2x} \, dx$

9. $\int_a^b \dfrac{dx}{x}$

10. Decide whether $\int_a^b f(x)\,dx$ is positive, negative, or zero.

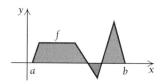

Evaluate using substitution. Do not use Table 1.

11. $\int \dfrac{dx}{x + 8}$

12. $\int e^{-0.5x}\,dx$

13. $\int t^3(t^4 + 1)^9\,dt$

Evaluate using integration by parts. Do not use Table 1.

14. $\int xe^{5x}\,dx$

15. $\int x^3 \ln x^4\,dx$

Evaluate using Table 1.

16. $\int 2^x\,dx$

17. $\int \dfrac{dx}{x(7 - x)}$

18. Find the average value of $y = 4t^3 + 2t$ over $[-1, 2]$.

19. Find the area of the region bounded by $y = x$, $y = x^5$, $x = 0$, and $x = 1$.

20. Approximate

$$\int_0^5 (25 - x^2)\,dx$$

by computing the area of each rectangle and adding.

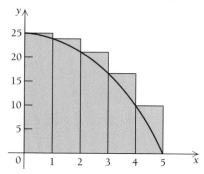

21. **Business: Cost from marginal cost.** An air conditioning company determines that the marginal cost of the xth air conditioner is given by

$$C'(x) = -0.2x + 500, \quad C(0) = 0.$$

Find the total cost of producing 100 air conditioners.

22. **Social science: Learning curve.** A typist's speed over a 4-min interval is given by

$$W(t) = -6t^2 + 12t + 90, \quad t \text{ in } [0, 4],$$

where $W(t) =$ the speed, in words per minute, at time t. How many words are typed during the second minute (from $t = 1$ to $t = 2$)?

Integrate using any method.

23. $\int \dfrac{dx}{x(10 - x)}$

24. $\int x^5 e^x\,dx$

25. $\int x^5 e^{x^6}\,dx$

26. $\int \sqrt{x}\,\ln x\,dx$

27. $\int x^3\sqrt{x^2 + 4}\,dx$

28. $\int \dfrac{dx}{64 - x^2}$

29. $\int x^4 e^{0.1x}\,dx$

30. $\int x \ln 13x\,dx$

SYNTHESIS

Evaluate using any method.

31. $\int \dfrac{[(\ln x)^3 - 4(\ln x)^2 + 5]}{x}\,dx$

32. $\int \ln\left(\dfrac{x + 3}{x + 5}\right)\,dx$

33. $\int \dfrac{8x^3 + 10}{\sqrt[3]{5x - 4}}\,dx$

TECHNOLOGY CONNECTION

34. Use your grapher to approximate the area between the following curves:

$$y = 3x - x^2, \quad y = 2x^3 - x^2 - 5x,$$
$$x = -2, \quad x = 0.$$

EXTENDED TECHNOLOGY APPLICATION

Total Sales of Sherwin-Williams, Intel, Sprint, and Alcoa

One of the many tasks of corporate leaders is the analysis of all kinds of data regarding the company they represent. The data might be total revenue, total costs, operating income, interest expense, costs of sales and marketing, and so on.

One very important use of the data can be the prediction, or forecasting, of the future. In this application, we will ask you to analyze factual data of several companies. You will be asked to do curve fitting to find models and make predictions.

EXERCISES

Sherwin-Williams Company specializes in many types of paint products and coatings. The following table lists the total sales of Sherwin-Williams Company for several years.

Year	Total sales (in millions)
0. 1988	$1950
1. 1989	2123
2. 1990	2267
3. 1991	2541
4. 1992	2748
5. 1993	2949
6. 1994	3100

cubic

1. Make a scatter plot of the data. What kind of function—linear, quadratic, cubic polynomial, quartic polynomial, exponential, or logarithmic—seems to best fit the data?

2. Use the regression feature of your grapher to fit the chosen function to the data.

3. Use the function to determine the total sales of Sherwin-Williams Company in 1997, 1998, 2001, and 2010.

4. Use integration to find the total sales of the company from 1990 through 2020.

Intel® is a company that specializes in microprocessors and other semiconductor products, such as computer chips. The growth of personal computer sales is a top priority of the company. The following table lists the total sales of Intel® for several years.

Year	Total sales (in millions)
0. 1986	$ 1,265
1. 1987	1,907
2. 1988	2,875
3. 1989	3,127
4. 1990	3,921
5. 1991	4,779
6. 1992	5,844
7. 1993	8,782
8. 1994	11,521

cubic

5. Make a scatter plot of the data. What kind of function—linear, quadratic, cubic polynomial, quartic polynomial, exponential, or logarithmic—seems to best fit the data?

6. Use the regression feature of your grapher to fit the chosen function to the data.

7. Use the function to determine the total sales of Intel® in 1997, 2000, 2005, and 2010.

8. Use integration to find the total sales of the company from 1986 through 2020.

Sprint® is a diversified telecommunications company providing global voice, data, and video conferencing services and related products. The following table lists the total net operating revenues of Sprint® for several years.

Year	Total net operating revenues (in billions)
0. 1985	$ 4.0
1. 1986	3.9
2. 1987	3.8
3. 1988	7.4
4. 1989	8.6
5. 1990	9.5
6. 1991	9.9
7. 1992	10.4
8. 1993	11.4
9. 1994	12.7

{eliminate,

9. Make a scatter plot of the data. What kind of function—linear, quadratic, cubic polynomial, quartic polynomial, exponential, or logarithmic—seems to best fit the data?

10. Use the regression feature of your grapher to fit the chosen function to the data.

11. Use the function to determine the total sales of Sprint® in 1999, 2004, 2009, and 2014.

12. Use integration to find the total sales of the company from 1985 through 2020.

Alcoa® is a company that specializes in aluminum products such as beverage cans, foil, automobile, and airplane sheeting. The following table lists the total sales of Alcoa® for several years.

Year	Total sales (in millions)
0. 1989	$10,910.0
1. 1990	10,710.2
2. 1991	9,884.1
3. 1992	9,491.5
4. 1993	9,055.9
5. 1994	9,904.3

cubic

13. Make a scatter plot of the data. What kind of function—linear, quadratic, cubic polynomial, quartic polynomial, exponential, or logarithmic—seems to best fit the data?

14. Use the regression feature of your grapher to fit the chosen function to the data.

15. Use the function to determine the total sales of Alcoa® in 1998, 2003, 2008, and 2011.

16. Use integration to find the total sales of the company from 1989 through 2020.

6 Applications of Integration

INTRODUCTION

In this chapter, we study a wide variety of applications of integration. We first consider an application in economics to finding consumer's surplus and producer's surplus. Then we study the integration of functions involved in exponential growth and decay. Here we find applications not only to business, but also to such environmental concerns as the depletion of resources and the buildup of radioactivity in the atmosphere. Integration has extensive application to probability and statistics, to finding volume, and to many situations involving the solution of differential equations.

Topics in this chapter can be chosen to fit the needs of the student and the course.

AN APPLICATION

The depletion of copper. The world reserves of copper ore are 2,300,000,000 metric tons. In 1990 ($t = 0$), the world use of copper was 8,830,000 metric tons, and the demand for copper was growing exponentially at the rate of 15% per year. Assuming that this growth rate continues and that no new reserves of copper are discovered, when will the world reserves of copper be exhausted?

THE MATHEMATICS

To find the time T at which the world reserves of copper will be exhausted, we solve the equation

$$2,300,000,000 = \int_0^T 8,830,000(e^{0.15t})\, dt.$$

This problem appears as Example 5 in Section 6.2.

6.1 Economic Application: Consumer's Surplus and Producer's Surplus

OBJECTIVE

• Given a demand and supply function, find the consumer's surplus and producer's surplus at the equilibrium point.

In our preceding functions, it was convenient to think of demand and supply as quantities that are functions of price. For purposes of this section, it will be convenient to think of them as prices that are functions of quantity: $p = D(x)$ and $p = S(x)$. Indeed, such interpretation is common in a study of economics.

The consumer's demand curve, $p = D(x)$, gives the demand price per unit that the consumer is willing to pay for x units. It is a decreasing function. The producer's supply curve, $p = S(x)$, gives the price per unit at which the seller is willing to supply x units. It is an increasing function. The equilibrium point (x_E, p_E) is the intersection of the two curves.

TECHNOLOGY CONNECTION

Graph the demand and supply functions

$$D(x) = (x - 5)^2 \quad \text{and} \quad S(x) = x^2 + x + 3$$

using the viewing window $[0, 5, 0, 30]$, with xScl $= 1$ and yScl $= 1$. Find the equilibrium point using the trace and zoom features.

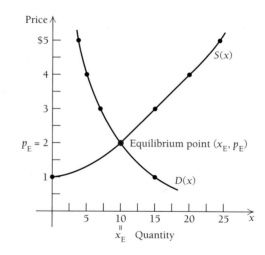

Suppose the graphs in Figs. 1 and 2 show the demand curve of college students for movies. Suppose we consider this curve for just one such student, Samantha.

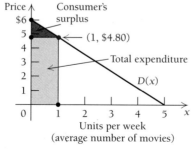

FIGURE 1

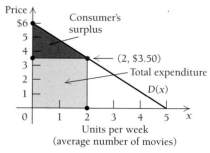

FIGURE 2

We want to examine the *utility,* or consumer satisfaction, that Samantha receives from going to the movies. Samantha goes to 0 movies per week if the price is $6. She will go to 1 movie per week if the price is $4.80. Suppose she goes to 1 movie. Then the total expenditure to her is $4.80 · 1, or $4.80, as shown in Fig. 1. The area of the blue region represents her total expenditure, $4.80. Look at the area of the orange region. It is $\frac{1}{2}(1)($1.20)$, or $0.60. The total area under the curve is a measure of the *total utility* of going to 1 movie, and is $5.40. The area of the orange region, $0.60, is a measure of the satisfaction Samantha gets but for which she does not have to pay. Economists define this amount as **consumer's surplus.** It is the utility that consumers derive from living in a society in which the price consumers are willing to pay decreases when more units are purchased. It is the extra utility that is gained from this purchase. The consumer surplus is the source of revenue that becomes available to a firm that discriminates in its prices, say, by offering discounts to college students, senior citizens, and children, or by selling discount coupons.

Suppose that Samantha goes to 2 movies per week if the price is $3.50 (Fig. 2). The total amount that Samantha actually spends is $3.50(2), or $7, and is the area of the blue region. It is her total expenditure. The total area under the curve is $9.50 and represents the total satisfaction to Samantha of going to 2 movies. Look at the area of the orange region. It (the consumer's surplus) is $\frac{1}{2}(2)($2.50)$, or $2.50. It is a measure of the utility Samantha received but for which she did not have to pay.

Suppose the graph of the demand function is a curve, as shown below.

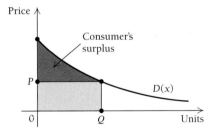

If Samantha goes to Q movies when the price is P, then Samantha's total expenditure is QP. The total area under the curve is the total utility, or the total satisfaction received, and is

$$\int_0^Q D(x)\ dx.$$

The *consumer's surplus* is the total area under the curve minus the total expenditure, quantity times price, or QP, and it is the total utility minus the total cost, which is given by

$$\int_0^Q D(x)\ dx - QP.$$

> **DEFINITION**
>
> Suppose that $p = D(x)$ describes the demand function for a commodity. Then the *consumer's surplus* is defined for point (Q, P) as
>
> $$\int_0^Q D(x)\ dx - QP.$$

EXAMPLE 1 Find the consumer's surplus for the demand function $D(x) = (x - 5)^2$ when $x = 3$.

Solution When $x = 3$, $D(3) = (3 - 5)^2 = (-2)^2 = 4$. Then

$$\text{Consumer's surplus} = \int_0^3 (x - 5)^2\ dx - 3 \cdot 4$$

$$= \int_0^3 (x^2 - 10x + 25)\ dx - 12$$

$$= \left[\frac{x^3}{3} - 5x^2 + 25x \right]_0^3 - 12$$

$$= \left(\frac{3^3}{3} - 5(3)^2 + 25(3) \right)$$

$$\quad - \left(\frac{0^3}{3} - 5(0)^2 + 25(0) \right) - 12$$

$$= (9 - 45 + 75) - 0 - 12$$

$$= \$27. \qquad \blacklozenge$$

TECHNOLOGY CONNECTION

Graph $D(x) = (x - 5)^2$ of Example 1 using the viewing window [0, 5, 0, 30], with xScl = 1 and yScl = 10. To find the consumer's surplus at $x = 3$, we first find $D(3)$. Then we graph $y = D(3)$. What is the point of intersection?

From the intersection, draw a vertical line down to the x-axis. What does the area of the resulting rectangle represent? What does the area above the horizontal line and below the curve represent?

Suppose we now look at the supply curve for the movies, as shown in Figs. 3 and 4. At a price of \$0 per movie, the producer is willing to supply 0 movies. At a price of \$1.75 per movie, the producer is willing to supply 1 movie and makes total

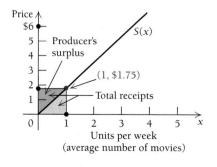

FIGURE 3

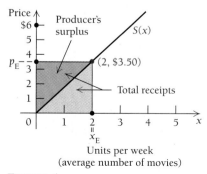

FIGURE 4

receipts of 1($1.75), or $1.75. The area of the beige triangle in Fig. 3 represents the total cost to the producer of producing 1 movie, and is $\frac{1}{2}(1)(\$1.75)$, or $0.875. The area of the green triangle is also $0.875, and represents the surplus over cost. It is a contribution to profit. Economists call this number the **producer's surplus.** It is the utility, or satisfaction, to the producer of living in a society in which a greater amount of a commodity will be supplied when the price increases. It is the extra revenue that the producer receives when the consumer cannot purchase the individual units along the producer's supply curve.

At a price of $3.50, the producer is willing to supply 2 movies and makes total receipts of 2($3.50), or $7. The area of the beige triangle in Fig. 4 represents the total cost to the producer of actually making the 2 movies, and is $\frac{1}{2}(2)(\$3.50)$, or $3.50. The area of the green triangle is also $3.50 and is the producer's surplus. It is a contribution to the profit of the producer.

Suppose the graph of the supply function is a curve, as shown at left. The producer will supply Q movies if the price is P. The total receipts are QP. The *producer's surplus* is the total receipts minus the area under the curve and is given by

$$QP - \int_0^Q S(x)\, dx.$$

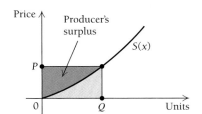

DEFINITION

Suppose that $p = S(x)$ is the supply function for a commodity. Then the *producer's surplus* is defined for the point (Q, P) as

$$QP - \int_0^Q S(x)\, dx.$$

EXAMPLE 2   Find the producer's surplus for $S(x) = x^2 + x + 3$ when $x = 3$.

Solution   When $x = 3$, $S(3) = 3^2 + 3 + 3 = 15$. Then

$$\text{Producer's surplus} = 3 \cdot 15 - \int_0^3 (x^2 + x + 3)\, dx$$

$$= 45 - \left[\frac{x^3}{3} + \frac{x^2}{2} + 3x \right]_0^3$$

$$= 45 - \left(\left[\frac{3^3}{3} + \frac{3^2}{2} + 3(3) \right] - \left[\frac{0^3}{3} + \frac{0^2}{2} + 3(0) \right] \right)$$

$$= 45 - \left(9 + \frac{9}{2} + 9 - 0 \right)$$

$$= \$22.50.$$

The **equilibrium point** (x_E, p_E) in Fig. 5 is the point at which the supply and demand curves intersect. It is that point at which the sellers and buyers come together and purchases and sales actually occur. In Fig. 6, we see the equilibrium point and the consumer's and producer's surpluses for the movie curves.

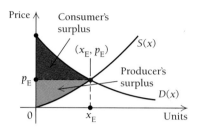

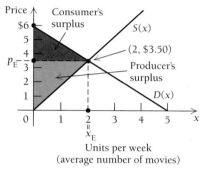

FIGURE 5   FIGURE 6

EXAMPLE 3   Given

$$D(x) = (x - 5)^2 \quad \text{and} \quad S(x) = x^2 + x + 3,$$

find each of the following.

a) The equilibrium point

b) The consumer's surplus at the equilibrium point

c) The producer's surplus at the equilibrium point

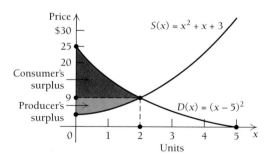

Solution

a) To find the equilibrium point, we set $D(x) = S(x)$ and solve:

$$(x - 5)^2 = x^2 + x + 3$$
$$x^2 - 10x + 25 = x^2 + x + 3$$
$$-10x + 25 = x + 3$$
$$22 = 11x$$
$$2 = x.$$

Thus, $x_E = 2$. To find p_E, we substitute x_E into either $D(x)$ or $S(x)$. If we choose $D(x)$, then

$$
\begin{aligned}
p_E = D(x_E) &= D(2) \\
&= (2 - 5)^2 = (-3)^2 \\
&= \$9 \text{ per unit.}
\end{aligned}
$$

Thus the equilibrium point is (2, $9).

b) The consumer's surplus at the equilibrium point is

$$
\int_0^{x_E} D(x)\ dx - x_E\, p_E,
$$

or

$$
\begin{aligned}
\int_0^2 (x - 5)^2\ dx - 2 \cdot 9 &= \left[\frac{(x - 5)^3}{3} \right]_0^2 - 18 \\
&= \left[\frac{(2 - 5)^3}{3} - \frac{(0 - 5)^3}{3} \right] - 18 \\
&= \frac{(-3)^3}{3} - \frac{(-5)^3}{3} - 18 \\
&= -\frac{27}{3} + \frac{125}{3} - \frac{54}{3} \\
&= \frac{44}{3} \approx \$14.67.
\end{aligned}
$$

c) The producer's surplus at the equilibrium point is

$$
x_E\, p_E - \int_0^{x_E} S(x)\ dx,
$$

or

$$
\begin{aligned}
2 \cdot 9 - \int_0^2 (x^2 + x + 3)\ dx \\
= 2 \cdot 9 - \left[\frac{x^3}{3} + \frac{x^2}{2} + 3x \right]_0^2 \\
= 18 - \left[\left(\frac{2^3}{3} + \frac{2^2}{2} + 3 \cdot 2 \right) - \left(\frac{0^3}{3} + \frac{0^2}{2} + 3 \cdot 0 \right) \right] \\
= 18 - \left(\frac{8}{3} + 2 + 6 \right) \\
= \frac{22}{3} \approx \$7.33.
\end{aligned}
$$

◆

6.1 Exercise Set

In each exercise, find (a) the equilibrium point, (b) the consumer's surplus at the equilibrium point, and (c) the producer's surplus at the equilibrium point.

1. $D(x) = -\frac{5}{6}x + 10,\quad S(x) = \frac{1}{2}x + 2$

2. $D(x) = -2x + 8,\quad S(x) = x + 2$

3. $D(x) = (x - 4)^2,\quad S(x) = x^2 + 2x + 6$

4. $D(x) = (x - 3)^2,\quad S(x) = x^2 + 2x + 1$

5. $D(x) = (x - 6)^2,\quad S(x) = x^2$

6. $D(x) = (x - 8)^2,\quad S(x) = x^2$

7. $D(x) = 1000 - 10x,\quad S(x) = 250 + 5x$

8. $D(x) = 8800 - 30x,\quad S(x) = 7000 + 15x$

9. $D(x) = 5 - x, 0 \le x \le 5;\quad S(x) = \sqrt{x + 7}$

10. $D(x) = 7 - x, 0 \le x \le 7;\quad S(x) = 2\sqrt{x + 1}$

SYNTHESIS

11. $D(x) = e^{-x+4.5},\quad S(x) = e^{x-5.5}$

12. $D(x) = \sqrt{56 - x},\quad S(x) = x$

13. ◆ Do some research on consumer's surplus in an economics book. Write a brief description.

14. ◆ Do some research on producer's surplus in an economics book. Write a brief description.

   TECHNOLOGY CONNECTION

Graph each pair of demand and supply functions. Then:

a) Find the equilibrium point using the zoom and trace features or another feature that will allow you to find this point of intersection.

b) Then determine the appropriate horizontal line, sketch it, and shade the regions of both consumer's and producer's surpluses.

c) Find the consumer's surplus.

d) Find the producer's surplus.

15. $D(x) = (x - 4)^2,\quad S(x) = x^2 + 2x + 6$

16. $D(x) = 5 - x,\quad S(x) = \sqrt{x + 7}$

6.2 Applications of the Models $\int_0^T P_0 e^{kt}\, dt$ and $\int_0^T P_0 e^{-kt}\, dt$

OBJECTIVES

- Do computations involving interest compounded continuously and continuous money flow.
- Find the total use of a natural resource.
- Find the present value of an investment.
- Find the accumulated present value of an investment.

Recall the basic model of exponential growth (Section 4.3):

$$P'(t) = k \cdot P(t), \quad \text{or} \quad \frac{dP}{dt} = kP, \quad \text{where } P = P_0 \text{ when } t = 0.$$

The function that satisfies the equation is

$$P(t) = P_0 e^{kt}. \tag{1}$$

One application of Eq. (1) is to compute the balance of a savings account after t years from an initial investment of P_0 at interest rate k, compounded continuously.

EXAMPLE 1   *Business: Growth in an investment.* Find the balance in a savings account after 3 years from an initial investment of $1000 at interest rate 8%, compounded continuously.

Solution   Using Eq. (1) with $k = 0.08$, $t = 3$, and $P_0 = \$1000$, we get

$$
\begin{aligned}
P(3) &= 1000e^{0.08(3)} \\
&= 1000e^{0.24} \\
&= 1000(1.271249) \\
&\approx \$1271.25.
\end{aligned}
$$

◆

The Integral $\int_0^T P_0 e^{kt}\, dt$

Suppose we consider the integral of $P_0 e^{kt}$ over the interval $[0, T]$:

$$
\int_0^T P_0 e^{kt}\, dt = \left[\frac{P_0}{k} \cdot e^{kt} \right]_0^T = \frac{P_0}{k}(e^{kT} - e^{k \cdot 0}) = \frac{P_0}{k}(e^{kT} - 1).
$$

We have a formula for this integral, given as

$$
\int_0^T P(t)\, dt = \int_0^T P_0 e^{kt}\, dt = \frac{P_0}{k}(e^{kT} - 1). \tag{2}
$$

We know that this formula represents the area under the graph of $P(t) = P_0 e^{kt}$ over the interval $[0, T]$.

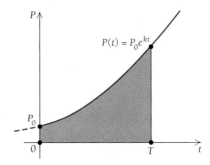

In the remainder of this section, we consider two other applications or interpretations of this integral.

Continuous Money Flow

Suppose that money is flowing continuously into a savings account at an annual rate of $1000 per year at interest rate 8%, compounded continuously. This means that over a small amount of time dt, the bank pays interest on all money accumulated in the account up to this time (since $t = 0$). The amount that is paid in

TECHNOLOGY
CONNECTION

Graph the continuous
money flow $f(x) = 1000e^{0.08x}$ using the viewing
window [0, 3, 0, 1500],
with xScl = 1 and
yScl = 100. Then evaluate
the integral

$$\int_0^3 \$1000e^{0.08x}\, dx.$$

Describe your answer.

over time dt is

$$\$1000e^{0.08t}\, dt.$$

Suppose we want to find the accumulation of all these amounts over a 5-year period.
That accumulation is given by the integral

$$\int_0^5 \$1000e^{0.08t}\, dt = \left[\frac{1000}{0.08}e^{0.08t}\right]_0^5$$
$$= 12{,}500(e^{0.08(5)} - e^{0.08(0)})$$
$$= 12{,}500(e^{0.4} - 1)$$
$$= 12{,}500(1.491825 - 1)$$
$$\approx \$6147.81.$$

Economists call \$6147.81 the *amount of a continuous money flow*. In this case,
the money is flowing according to a constant function $R(t) = \$1000$. Money could
also be flowing according to some variable function, say, $R(t) = 2t - 7$ or $R(t) = t^2$.

THEOREM 1

If the rate of flow of money into an investment is given by some constant
function $R(t)$, then the *amount of continuous money flow* at interest rate k,
compounded continuously, over time T is given by

$$\int_0^T R(t)e^{kt}\, dt.$$

TECHNOLOGY CONNECTION

Graph $f(x) = 1000e^{0.08x}$ using the viewing
window [1, 15, 0, 5000], with xScl = 2 and
yScl = 1000. Then graph the line $y = f(0)$.
Consider the rectangle formed by the lines
$y = f(0)$, $x = 15$, the y-axis, and the x-axis.
Graph it if you can. What does the area of the
rectangle represent? What does the area above
the rectangle and below the curve represent?
What does the entire area under the curve
represent?

EXAMPLE 2   *Business: Amount of a continuous money flow.*
Find the amount of a continuous money flow where \$1000 per
year is being invested at 8%, compounded continuously, for
15 years.

Solution

$$\int_0^{15} \$1000e^{0.08t}\, dt = \frac{1000}{0.08}(e^{0.08(15)} - 1) \qquad \text{Using Eq. (2)}$$
$$= 12{,}500(e^{1.2} - 1)$$
$$= 12{,}500(3.320117 - 1)$$
$$\approx \$29{,}001.46 \qquad\qquad \blacklozenge$$

Sometimes we might want to know how money should be
flowing into an investment so that we end up with a specified amount.

EXAMPLE 3 *Business: Continuous money flow.* Consider a continuous flow of money into an investment at the constant rate of P_0 dollars per year. What should P_0 be so that the amount of a continuous money flow over 20 years at interest rate 8%, compounded continuously, will be $10,000?

Solution We find P_0 such that

$$10{,}000 = \int_0^{20} P_0 e^{0.08t}\, dt.$$

We solve the following equation:

$$10{,}000 = \frac{P_0}{0.08}(e^{0.08(20)} - 1) \qquad \text{Using Eq. (2)}$$

$$800 = P_0(e^{1.6} - 1)$$

$$800 = P_0(4.953032 - 1)$$

$$800 = P_0(3.953032)$$

$$\$202.38 \approx P_0. \qquad\qquad\qquad\qquad\qquad\qquad \blacklozenge$$

Life and Physical Sciences: Depletion of Natural Resources

Another application of the integral of exponential growth concerns

$$P(t) = P_0 e^{kt}$$

as a model of the demand for natural resources. Suppose that P_0 represents the amount of a natural resource (such as coal or oil) used at time $t = 0$ and that the growth rate for the use of this resource is k. Then, assuming exponential growth (which is the case for the use of many resources), the amount to be used at time t is $P(t)$, given by

$$P(t) = P_0 e^{kt}.$$

The total amount used during an interval $[0, T]$ is given by

$$\int_0^T P(t)\, dt = \int_0^T P_0 e^{kt}\, dt$$

$$= \frac{P_0}{k}(e^{kT} - 1). \qquad\qquad\qquad\qquad (2)$$

EXAMPLE 4 *Physical science: Demand for copper.* In 1990 ($t = 0$), the world use of copper was 8,830,000 metric tons, and the demand for copper was growing exponentially at the rate of 15% per year. If the growth continues at this rate, how many tons of copper will the world use from 1990 to 2000?

Solution   Using Eq. (2), we have

$$\int_0^{10} 8{,}830{,}000 e^{0.15t}\, dt = \frac{8{,}830{,}000}{0.15}(e^{0.15(10)} - 1) \qquad \text{Using Eq. (2)}$$

$$\approx 58{,}866{,}667(e^{1.5} - 1)$$

$$= 58{,}866{,}667(4.481689 - 1)$$

$$= 58{,}866{,}667(3.481689)$$

$$\approx 204{,}955{,}400. \qquad \text{Rounded to the nearest hundred}$$

Thus from 1990 to 2000, the world will use 204,955,400 metric tons of copper. ◆

EXAMPLE 5   *Physical science: Depletion of copper ore.* The world reserves of copper ore are estimated to be 2,300,000,000 metric tons. Assuming that the growth rate in Example 4 continues and that no new reserves are discovered, when will the world reserves of copper ore be exhausted?

Solution   Using Eq. (2), we want to find T such that

$$2{,}300{,}000{,}000 = \frac{8{,}830{,}000}{0.15}(e^{0.15T} - 1).$$

Bingham Canyon mine in Utah has produced more copper than any other mine in history. The grade of ore, however, has dropped from 1.93 percent copper in 1906 to 0.6 percent today. At present, it is planned that mining will cease when the percentage of copper reaches 0.4.

We solve for T as follows:

$$2{,}300{,}000{,}000 \approx 58{,}866{,}667(e^{0.15T} - 1)$$

$$\frac{2{,}300{,}000{,}000}{58{,}866{,}667} = e^{0.15T} - 1$$

$$39.071347 \approx e^{0.15T} - 1$$

$$40.071347 = e^{0.15T}$$

$$\ln 40.071347 = \ln e^{0.15T} \qquad \text{Taking the natural logarithm on both sides}$$

$$\ln 40.071347 = 0.15T \qquad \text{Recall: } \ln e^k = k.$$

$$\frac{\ln 40.071347}{0.15} = T$$

$$\frac{3.690661}{0.15} = T$$

$$25 \approx T. \qquad \text{Rounding to the nearest one}$$

Thus 25 years from 1990 (or by 2015), the world reserves of copper ore will be exhausted. ◆

Applications of the Model $\int_0^T P e^{-kt}\, dt$

Recall our study of present value in Section 4.4. If you did not study that topic there, you should do so now. The **present value** P_0 of an amount P due t years later is found by solving the following equation for P_0:

$$P_0 e^{kt} = P$$

$$P_0 = \frac{P}{e^{kt}} = P e^{-kt}.$$

THEOREM 2

The *present value* P_0 of an amount P due t years later at interest rate k, compounded continuously, is given by

$$P_0 = P e^{-kt}.$$

Note that this can be interpreted as exponential decay from the future back to the present.

EXAMPLE 6 *Business: Present value.* Find the present value of \$200,000 due 25 years from now at 8.7%, compounded continuously.

Solution   We substitute $200,000 for P, 0.087 for k, and 25 for t in the equation for present value:

$$P_0 = 200,000e^{-0.087(25)} \approx \$22,721.63.$$

Thus the present value is $22,721.63.   ◆

Suppose we know that a continuous flow of money will go into an investment at the constant rate of P dollars per year, from now until some time T in the future. If an infinitesimal amount of time dt passes,

$$P \cdot dt$$

dollars will have accumulated. The present value of that amount is

$$(P \cdot dt)e^{-kt},$$

where k is the current interest rate, compounded continuously. The accumulation of all the present values is given by the integral

$$\int_0^T Pe^{-kt}\, dt$$

and is called the **accumulated present value.** Evaluating this integral, we get

$$\int_0^T Pe^{-kt}\, dt = \frac{P}{-k}(e^{-kT} - e^{-k \cdot 0})$$

$$= \frac{P}{k}(1 - e^{-kT}). \tag{3}$$

In the preceding case, money is flowing according to a constant function $R(t) = P$. Money could also be flowing according to some variable function, such as $R(t) = 2t + 8$ or $R(t) = t^3$.

THEOREM 3

The *accumulated present value* of a continuous money flow into an investment at a constant rate of $R(t)$ dollars per year from now until some time T in the future is given by

$$\int_0^T R(t)e^{-kt}\, dt,$$

where k is the current interest rate, compounded continuously.

EXAMPLE 7   *Business: Accumulated present value.* Find the accumulated present value of an investment over a 5-year period if there is a continuous money flow of $2400 per year and the current interest rate is 14%, compounded continuously.

Solution   The accumulated present value is

$$\int_0^5 \$2400 e^{-0.14t} \, dt = \frac{2400}{0.14}(1 - e^{-0.14 \cdot 5}) \qquad \text{Using Eq. (3)}$$

$$\approx 17{,}142.86(1 - e^{-0.7})$$

$$= 17{,}142.86(1 - 0.496585)$$

$$\approx \$8629.97. \qquad \qquad \blacklozenge$$

The preceding example is an application of the model

$$\int_0^T P e^{-kt} \, dt = \frac{P}{k}(1 - e^{-kT}).$$

This model can be applied to a calculation of the buildup of a specific amount of radioactive material released into the atmosphere annually. Some of the material decays, but more continues to be released. The amount present at time T is given by the integral above.

6.2 Exercise Set

APPLICATIONS

◆   Business and Economics

1. Find the amount in a savings account after 3 years from an initial investment of $100 at interest rate 9%, compounded continuously.

2. Find the amount in a savings account after 4 years from an initial investment of $100 at interest rate 10%, compounded continuously.

3. Find the amount of a continuous money flow in which $100 per year is being invested at 9%, compounded continuously, for 20 years.

4. Find the amount of a continuous money flow in which $100 per year is being invested at 10%, compounded continuously, for 20 years.

5. Find the amount of a continuous money flow in which $1000 per year is being invested at 8.5%, compounded continuously, for 40 years.

6. Find the amount of a continuous money flow in which $1000 per year is being invested at 7.5%, compounded continuously, for 40 years.

7. What should P_0 be so that the amount of a continuous money flow over 20 years at interest rate 8.5%, compounded continuously, will be $50,000?

8. What should P_0 be so that the amount of a continuous money flow over 20 years at interest rate 7.5%, compounded continuously, will be $50,000?

9. What should P_0 be so that the amount of a continuous money flow over 30 years at interest rate 9%, compounded continuously, will be $40,000?

10. What should P_0 be so that the amount of a continuous money flow over 30 years at interest rate 10%, compounded continuously, will be $40,000?

11. Following the birth of a child, a parent wants to make an initial investment P_0 that will grow to $50,000 by the child's 20th birthday. Interest is compounded continuously at 9%. What should the initial investment be?

12. Following the birth of a child, a parent wants to make an initial investment P_0 that will grow to $60,000 by the child's 20th birthday. Interest is compounded continuously at 10%. What should the initial investment be?

13. Find the present value of $60,000 due 8 years later at 8.8%, compounded continuously.

14. Find the present value of $50,000 due 16 years later at 7.4%, compounded continuously.

15. Find the accumulated present value of an investment over a 10-year period if there is a continuous money flow of $2700 per year and the current interest rate is 9%, compounded continuously.

16. Find the accumulated present value of an investment over a 10-year period if there is a continuous money flow of $2700 per year and the current interest rate is 10%, compounded continuously.

17. A woman with an MBA accepts a position as president of a company at the age of 35. Assuming retirement at age 65 and an annual salary of $85,000 that is paid in a continuous money flow, what is the president's accumulated present value? The current interest rate is 8%, compounded continuously.

18. A college dropout takes a job as a truck driver at the age of 25. Assuming retirement at age 65 and an annual salary of $34,000 that is paid in a continuous money flow, what is the truck driver's accumulated present value? The current interest rate is 7%, compounded continuously.

◆   Life and Physical Sciences

19. *The demand for aluminum ore (bauxite).* In 1990 ($t = 0$), the world use of aluminum ore was 101,000,000 tons, and the demand for aluminum ore was growing exponentially at the rate of 12% per year. If the demand continues to grow at this rate, how many tons of aluminum ore will the world use from 1990 to 2004?

20. *The demand for natural gas.* In 1990 ($t = 0$), the world use of natural gas was 72,137 billion cubic feet, and the demand for natural gas was growing exponentially at the rate of 4% per year. If the demand continues to grow at this rate, how many cubic feet of natural gas will the world use from 1990 to 2020?

21. *The depletion of aluminum ore.* The world reserves of aluminum ore are 75,000,000,000 tons. Assuming that the growth rate of Exercise 19 continues and that no new reserves are discovered, when will the world reserves of aluminum ore be exhausted?

22. *The depletion of natural gas.* The world reserves of natural gas are 3,926,000 billion cubic feet. Assuming

that the growth rate of Exercise 20 continues and that no new reserves are discovered, when will the world reserves of natural gas be exhausted?

23. *Radioactive buildup.* Plutonium has a decay rate of 0.003% per year. Suppose plutonium is released into the atmosphere each year for 20 years at the rate of 1 pound per year. What is the total amount of radioactive buildup?

24. *Radioactive buildup.* Cesium-137 has a decay rate of 2.3% per year. Suppose cesium-137 is released into the atmosphere each year for 20 years at the rate of 1 pound per year. What is the total amount of radioactive buildup?

25. *Demand for oil.* In 1990 ($t = 0$), the world use of oil was 6570 million barrels, and the demand for oil was growing exponentially at the rate of 10% per year. If the demand continues at this rate, how many barrels of oil will the world use from 1990 to 2000?

26. *Depletion of oil.* The world reserves of oil are 920,700 million barrels. In 1990 ($t = 0$), the world use of oil was 6570 million barrels, and the growth rate for the use of oil was 10%. Assuming that this growth rate continues and that no new reserves are discovered, when will the world reserves of oil be exhausted?

SYNTHESIS

For a nonconstant function $R(t)$, the amount of a continuous money flow is

$$\int_0^T R(t)e^{k(T-t)} \, dt.$$

Find the amount of a continuous money flow for each of the following.

27. $R(t) = 2000t + 7$, $k = 8\%$, and $T = 30$ years

28. $R(t) = t^2$, $k = 7\%$, and $T = 40$ years

For a nonconstant function $R(t)$, the accumulated present value is

$$\int_0^T R(t)e^{-k(T-t)} \, dt.$$

Find the accumulated present value for each of the following.

29. $R(t) = t$, $k = 8\%$, and $T = 20$ years

30. $R(t) = e^t$, $k = 7\%$, and $T = 10$ years

31. ◆ Look up some data on rate of usage and world reserves of a natural resource not considered in this section. Predict when the world reserves for that resource will be depleted.

32. ◆ Describe the idea of present value to a friend who is not a business major. Then describe accumulated present value.

TECHNOLOGY CONNECTION

33. Graph the area under the curve represented by the continuous money flow in Exercise 3. Then compute the area and compare the result with the exercise answer.

34. Graph the area under the curve represented by the demand for aluminum ore in Exercise 19. Then compute the area and compare the result with the exercise answer.

6.3 Improper Integrals

OBJECTIVES

• Determine whether an improper integral is convergent or divergent.
• Solve application problems involving improper integrals.

Let us try to find the area of the region under the graph of $y = 1/x^2$ on the interval $[1, \infty)$.

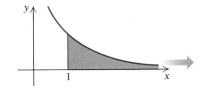
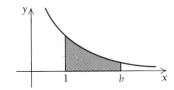

Note that this region is of infinite extent. We have not yet considered how to find the area of such a region. Let us find the area under the curve on the interval from 1 to b, and then see what happens as b gets very large. The area under the graph on $[1, b]$ is

$$\int_1^b \frac{dx}{x^2} = \left[-\frac{1}{x} \right]_1^b$$

$$= \left(-\frac{1}{b} \right) - \left(-\frac{1}{1} \right)$$

$$= -\frac{1}{b} + 1$$

$$= 1 - \frac{1}{b}.$$

Then

$$\lim_{b \to \infty} [\text{area from 1 to } b] = \lim_{b \to \infty} \left(1 - \frac{1}{b}\right).$$

Now let us investigate this limit in the following input–output table.

TECHNOLOGY
CONNECTION

Graph $y = 1 - 1/x$ and
create an input–output table.
Make an assertion about

$$\lim_{x \to \infty} \left(1 - \frac{1}{x}\right).$$

b	$1 - \dfrac{1}{b}$
2	$\frac{1}{2}$
3	$\frac{2}{3}$
10	$\frac{9}{10}$
100	$\frac{99}{100}$
1000	$\frac{999}{1000}$

Note that as $b \to \infty$, $1/b \to 0$, so $(1 - 1/b) \to 1$. Thus,

$$\lim_{b \to \infty} [\text{area from 1 to } b] = \lim_{b \to \infty} \left(1 - \frac{1}{b}\right) = 1.$$

We *define* the area from 1 to infinity to be this limit. Here we have an example of an infinitely long region with a finite area.

Such areas may not always be finite. Let us try to find the area of the region under the graph of $y = 1/x$ on the interval $[1, \infty)$.

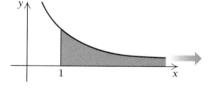
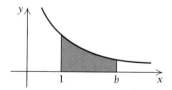

By definition, the area A from 1 to infinity is the limit as $b \to \infty$ of the area from 1 to b, so

$$A = \lim_{b \to \infty} \int_1^b \frac{dx}{x} = \lim_{b \to \infty} [\ln x]_1^b$$

$$= \lim_{b \to \infty} (\ln b - \ln 1)$$

$$= \lim_{b \to \infty} \ln b.$$

In Section 4.2, we learned that $\ln b$ increases indefinitely as b increases. Therefore, the limit does not exist.

Thus we have an infinitely long region with an infinite area. Note that the graphs of $y = 1/x^2$ and $y = 1/x$ have similar shapes, but the region under one of them has a finite area and the other does not.

An integral such as

$$\int_a^\infty f(x)\ dx,$$

with an upper limit of infinity, is called an **improper integral.** Its value is defined to be the following limit.

DEFINITION

$$\int_a^\infty f(x)\ dx = \lim_{b \to \infty} \int_a^b f(x)\ dx$$

If the limit exists, then we say that the improper integral **converges,** or is **convergent.** If the limit does not exist, then we say that the improper integral **diverges,** or is **divergent.** Thus,

$$\int_1^\infty \frac{dx}{x^2} = 1 \ converges; \quad \text{and} \quad \int_1^\infty \frac{dx}{x} \ diverges.$$

EXAMPLE 1   Determine whether the following integral is convergent or divergent, and calculate its value if it is convergent:

$$\int_0^\infty 2e^{-2x}\ dx.$$

Solution   We have

$$\int_0^\infty 2e^{-2x}\ dx = \lim_{b \to \infty} \int_0^b 2e^{-2x}\ dx$$

$$= \lim_{b \to \infty} \left[\frac{2}{-2} e^{-2x} \right]_0^b$$

$$= \lim_{b \to \infty} \left[-e^{-2x} \right]_0^b$$

$$= \lim_{b \to \infty} \left[-e^{-2b} - (-e^{-2 \cdot 0}) \right]$$

$$= \lim_{b \to \infty} (-e^{-2b} + 1)$$

$$= \lim_{b \to \infty} \left(1 - \frac{1}{e^{2b}} \right).$$

Now as $b \to \infty$, we know that $e^{2b} \to \infty$ (from Chapter 4), so

$$\frac{1}{e^{2b}} \to 0 \quad \text{and} \quad \left(1 - \frac{1}{e^{2b}}\right) \to 1.$$

Thus,

$$\int_0^\infty 2e^{-2x}\, dx = \lim_{b \to \infty}\left(1 - \frac{1}{e^{2b}}\right) = 1.$$

The integral is convergent.   ◆

Following are definitions of two other types of improper integrals.

DEFINITIONS

1. $\displaystyle \int_{-\infty}^b f(x)\, dx = \lim_{a \to -\infty} \int_a^b f(x)\, dx$

2. $\displaystyle \int_{-\infty}^\infty f(x)\, dx = \int_{-\infty}^c f(x)\, dx + \int_c^\infty f(x)\, dx$

In order for $\int_{-\infty}^\infty f(x)\, dx$ to converge, both integrals on the right in Definition 2 above must converge.

Applications

In Section 6.2, we learned that the accumulated present value of a continuous money flow of P dollars per year from now until time T in the future is given by

$$\int_0^T Pe^{-kt}\, dt = \frac{P}{k}(1 - e^{-kT}),$$

where k is the current interest rate. Suppose the money flow is to continue perpetually. Under this assumption, the accumulated present value over this infinite time period would be

$$\int_0^\infty Pe^{-kt}\, dt = \lim_{T \to \infty} \int_0^T Pe^{-kt}\, dt$$

$$= \lim_{T \to \infty} \frac{P}{k}(1 - e^{-kT})$$

$$= \lim_{T \to \infty} \frac{P}{k}\left(1 - \frac{1}{e^{kT}}\right) = \frac{P}{k}.$$

THEOREM 4

The *accumulated present value* of a continuous money flow into an investment at the rate of P dollars per year perpetually is given by

$$\frac{P}{k},$$

where k is the current interest rate, compounded continuously.

 TECHNOLOGY CONNECTION

Consider Example 2 with your grapher. To evaluate the integral $\int_0^\infty 2000e^{-0.08x}\, dx$, we would first consider

$$\int_0^t 2000e^{-0.08x}\, dx = \frac{2000}{0.08}(1 - e^{-0.08t}).$$

Let us examine what happens as t gets large. Graph

$$f(x) = \frac{2000}{0.08}(1 - e^{-0.08x})$$

using the viewing window [0, 10, 0, 30000], with xScl = 20 and yScl = 5000. Using the same set of axes, graph $y = 25,000$. Then change the viewing window to [0, 50, 0, 30000] and finally to [0, 100, 0, 30000]. What happens as x gets larger? What is the significance of 25,000?

EXAMPLE 2   *Business: Accumulated present value.* Find the accumulated present value of an investment for which there is a perpetual continuous money flow of $2000 per year. The current interest rate is 8%, compounded continuously.

Solution   The accumulated present value is 2000/0.08, or $25,000. ◆

When an amount P of radioactive material is being released into the atmosphere annually, the amount present at time T is given by

$$\int_0^T Pe^{-kt}\, dt = \frac{P}{k}(1 - e^{-kT}).$$

As $T \to \infty$ (the radioactive material is to be released forever), the buildup of radioactive material approaches a limiting value P/k. It is no wonder that scientists and environmentalists are so concerned about continued nuclear detonations. Eventually the buildup is "here to stay."

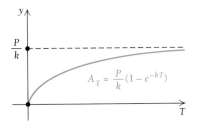

6.3 Exercise Set

Determine whether the improper integral is convergent or divergent, and calculate its value if it is convergent.

1. $\int_{3}^{\infty} \dfrac{dx}{x^2}$

2. $\int_{4}^{\infty} \dfrac{dx}{x^2}$

3. $\int_{3}^{\infty} \dfrac{dx}{x}$

4. $\int_{4}^{\infty} \dfrac{dx}{x}$

5. $\int_{0}^{\infty} 3e^{-3x}\, dx$

6. $\int_{0}^{\infty} 4e^{-4x}\, dx$

7. $\int_{1}^{\infty} \dfrac{dx}{x^3}$

8. $\int_{1}^{\infty} \dfrac{dx}{x^4}$

9. $\int_{0}^{\infty} \dfrac{dx}{1+x}$

10. $\int_{0}^{\infty} \dfrac{4\, dx}{1+x}$

11. $\int_{1}^{\infty} 5x^{-2}\, dx$

12. $\int_{1}^{\infty} 7x^{-2}\, dx$

13. $\int_{0}^{\infty} e^{x}\, dx$

14. $\int_{0}^{\infty} e^{2x}\, dx$

15. $\int_{3}^{\infty} x^{2}\, dx$

16. $\int_{5}^{\infty} x^{4}\, dx$

17. $\int_{0}^{\infty} xe^{x}\, dx$

18. $\int_{1}^{\infty} \ln x\, dx$

19. $\int_{0}^{\infty} me^{-mx}\, dx,\ m > 0$

20. $\int_{0}^{\infty} Qe^{-kt}\, dt,\ k > 0$

21. Find the area, if it exists, of the region under the graph of $y = 1/x^2$ on the interval $[2, \infty)$.

22. Find the area, if it exists, of the region under the graph of $y = 1/x$ on the interval $[2, \infty)$.

23. Find the area, if it exists, of the region bounded by $y = 2xe^{-x^2}$ and the lines $x = 0$ and $y = 0$.

24. Find the area, if it exists, of the region bounded by $y = 1/\sqrt{(3x - 2)^3}$ and the lines $x = 6$ and $y = 0$.

APPLICATIONS

◆   **Business and Economics**

25. *Accumulated present value.* Find the accumulated present value of an investment for which there is a perpetual continuous money flow of $3600 per year. The current interest rate is 8%.

26. *Accumulated present value.* Find the accumulated present value of an investment for which there is a perpetual continuous money flow of $3500 per year. The current interest rate is 7%.

27. *Total profit from marginal profit.* A firm is able to determine that its marginal profit is given by

$$P'(x) = 200e^{-0.032x}.$$

Suppose it were possible for the firm to make infinitely many units of this product. What would its total profit be?

28. *Total profit from marginal profit.* In Exercise 27, find the total profit if

$$P'(x) = 200x^{-1.032}, \quad \text{where } x \geq 1.$$

29. *Total cost from marginal cost.* A company determines that its marginal cost for the production of x units of a product is given by

$$C'(x) = 3600x^{-1.8}, \quad \text{where } x \geq 1.$$

Suppose it were possible for the firm to make infinitely many units of this product. What would the total cost be?

30. **Total production.** A firm determines that it can produce tires at a rate of

$$r(t) = 2000e^{-0.42t},$$

where t = the time, in years. Assuming the firm endures forever, how many tires can it make?

◆   **Life and Physical Sciences** $= P_o e^{-.003t}$

31. **Radioactive buildup.** Plutonium has a decay rate of 0.003% per year. Suppose plutonium is released into the atmosphere each year perpetually at the rate of 1 pound per year. What is the limiting value of the radioactive buildup?

32. **Radioactive buildup.** Cesium-137 has a decay rate of 2.3% per year. Suppose cesium-137 is released into the atmosphere each year perpetually at the rate of 1 pound per year. What is the limiting value of the radioactive buildup?

SYNTHESIS

Determine whether the improper integral is convergent or divergent, and calculate its value if it is convergent.

33. $\int_0^\infty \dfrac{dx}{x^{2/3}}$

34. $\int_1^\infty \dfrac{dx}{\sqrt{x}}$

35. $\int_0^\infty \dfrac{dx}{(x+1)^{3/2}}$

36. $\int_{-\infty}^0 e^{2x}\,dx$

37. $\int_0^\infty xe^{-x^2}\,dx$

38. $\int_{-\infty}^\infty xe^{-x^2}\,dx$

Life science: Drug dosage. Suppose an oral dose of a drug is taken. From that time, the drug is assimilated in the body and excreted through the urine. The total amount of the drug that has passed through the body in time T is given by

$$\int_0^T E(t)\,dt,$$

where E = the rate of excretion of the drug. A typical rate-of-excretion function is $E(t) = te^{-kt}$, where $k > 0$ and t = the time, in hours. (Now do Exercises 39 and 40.)

39. Find $\int_0^\infty E(t)\,dt$ and interpret the answer. That is, what does the integral represent?

40. A physician prescribes a dosage of 100 mg. Find k.

41. ◈   Consider the functions

$$y = \dfrac{1}{x^2} \quad \text{and} \quad y = \dfrac{1}{x}.$$

Suppose you went to a paint store to buy paint to cover the region under each graph on the interval $[1, \infty)$. Discuss whether you could be successful and why or why not.

42. ◈   Suppose you are the owner of a building that yields a continuous series of rental payments, and suppose that you decide to sell the building. Explain how you would use the concept of the accumulated present value of a perpetual continuous money flow to determine a fair selling price.

TECHNOLOGY CONNECTION

43. Graph the function E and shade the area under the curve for the situation in Exercises 39 and 40.

Approximate the integral.

44. $\int_1^\infty \dfrac{4}{1+x^2}\,dx$

45. $\int_1^\infty \dfrac{6}{5+e^x}\,dx$

6.4 Probability

OBJECTIVES

- Verify certain properties of probability density functions.
- Solve application problems involving probability density functions.

The definite integral plays a role in the theory of probability. Briefly, the **probability** of an event is a number from 0 to 1 that represents the chances of the event occurring. It is the "relative frequency" of occurrence—that is, the proportion or percentage of times that we might expect an event to occur in a large number of trials.

EXAMPLE 1 What is the probability of drawing an ace from a well-shuffled deck of cards?

Solution Since there are 52 possible outcomes and each card has the same chance of being drawn, and since there are 4 aces, the probability of drawing an ace is $\frac{4}{52}$ or $\frac{1}{13}$, or about 7.7%.

In practice, we may not draw an ace 7.7% of the time, but in a large number of trials, after shuffling the cards and drawing one, replacing the card, and shuffling the cards and drawing one, we would expect to draw an ace about 7.7% of the time. That is, the more draws we make, the closer we expect to get to 7.7%. ◆

A desire to calculate odds in games of chance gave rise to the theory of probability.

EXAMPLE 2 A jar contains 7 black balls, 6 yellow balls, 4 green balls, and 3 red balls. The jar is shaken well and you remove 1 ball without looking. What is the probability that the ball is red? that it is white?

Solution There are 20 balls altogether and of these 3 are red, so the probability of drawing a red ball is $\frac{3}{20}$. There are no white balls, so the probability of drawing a white one is $\frac{0}{20}$, or 0. ◆

Color	Probability
Black (B)	$\frac{7}{20}$
Yellow (Y)	$\frac{6}{20}$
Green (G)	$\frac{4}{20}$
Red (R)	$\frac{3}{20}$

Let us consider a table of probabilities from Example 2. Note that the sum of these probabilities is 1. We are certain that we will draw either a black, yellow, green, or red ball. The probability of that event is 1. Let us arrange these data from the table into what is called a *relative frequency graph,* or *histogram,* which shows the proportion of times each event occurs (the probability of each event).

If we assign a width of 1 to each rectangle, then the sum of the areas of the rectangles is 1.

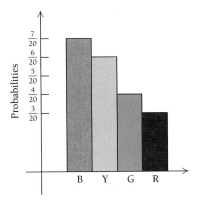

Continuous Random Variables

Suppose we throw a dart at a number line in such a way that it always lands in the interval [1, 3]. Let x be the number that the dart hits. Note that x is a quantity that can be observed (or measured) repeatedly and whose possible values consist of an entire interval of real numbers. Such a variable is called a **continuous random variable.**

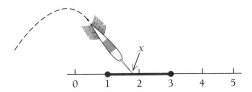

Suppose we throw the dart a large number of times and it lands 43% of the time in the subinterval [1.6, 2.8] of the main interval [1, 3]. The probability, then, that the dart lands in the interval [1.6, 2.8] is 0.43.

Let us consider some other examples of continuous random variables.

EXAMPLE 3 Suppose that x is the arrival time of buses at a bus stop in a 3-hour period from 2:00 P.M. to 5:00 P.M. The interval is [2, 5]. Then x is a continuous random variable distributed over the interval [2, 5].

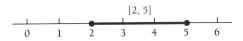

EXAMPLE 4 Suppose that x is the corn acreage of any farm in the United States and Canada. The interval is [0, a], where a is the highest acreage. (Not knowing what the highest acreage might be, we could say that the interval is [0, ∞) to allow

for all possibilities. *Note:* It might be argued that there is a value in $[0, a]$ or $[0, \infty)$ for which no farm has that acreage, but for practical convenience, all values are included in our consideration.)

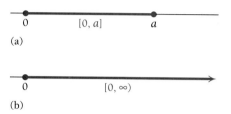

(a)

(b)

Then x is a continuous random variable distributed over the interval $[0, a]$ or $[0, \infty)$. ◆

Considering Example 3 on the arrival times of buses, suppose that we want to know the probability that a bus will arrive between 4:00 P.M. and 5:00 P.M., as represented by

$$P([4, 5]), \quad \text{or} \quad P(4 \le x \le 5).$$

It is possible to find functions over $[2, 5]$ such that the areas under the graphs over subintervals give the probabilities that a bus will arrive during these subintervals. For example, suppose we have a constant function $f(x) = \frac{1}{3}$ that will give us these probabilities. Look at its graph.

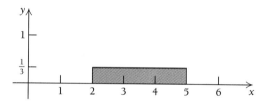

The area under the curve is $3 \cdot \frac{1}{3}$, or 1. The probability that a bus will arrive between 4:00 P.M. and 5:00 P.M. is that fraction of the large area that lies over the interval $[4, 5]$. That is,

$$P([4, 5]) = \frac{1}{3} = 33\frac{1}{3}\%.$$

This is the area of the rectangle over $[4, 5]$.

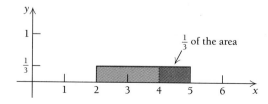

The probability that a bus will arrive between 2:00 P.M. and 4:30 P.M. is $\frac{5}{6}$, or $83\frac{1}{3}$%. This is the area of the rectangle over $[2, 4.5]$.

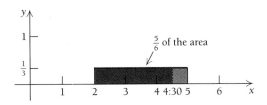

Note that in the above example, where $f(x) = \frac{1}{3}$, any interval between 2:00 P.M. and 5:00 P.M. of length 1 has probability $\frac{1}{3}$. This might not always happen. Suppose instead that

$$f(x) = \tfrac{3}{117}x^2.$$

As shown in the graph below, the area under the graph of $f(x)$ from 4 to 5 is given by the definite integral over the interval $[4, 5]$ and would yield the probability that a bus will arrive between 4:00 P.M. and 5:00 P.M.

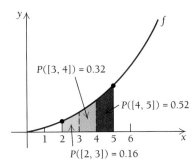

Thus,

$$
\begin{aligned}
P([4, 5]) &= \int_4^5 f(x)\, dx \\
&= \int_4^5 \frac{3}{117}x^2\, dx \\
&= \frac{3}{117}\left[\frac{x^3}{3}\right]_4^5 \\
&= \frac{1}{117}\left[x^3\right]_4^5 \\
&= \frac{1}{117}(5^3 - 4^3) = \frac{61}{117} \approx 0.52.
\end{aligned}
$$

TECHNOLOGY CONNECTION

Graph the function

$$f(x) = \tfrac{3}{117}x^2$$

using a viewing window of $[0, 5, 0, 1]$, with $xScl = 1$ and $yScl = 1$. Then successively evaluate each of the following integrals, shading the appropriate area if possible:

$$\int_2^3 \tfrac{3}{117}x^2\, dx, \qquad \int_3^4 \tfrac{3}{117}x^2\, dx,$$

and

$$\int_4^5 \tfrac{3}{117}x^2\, dx.$$

Then add your results and explain the meaning of the answer.

Thus there is a probability of 0.52 that at least one bus will arrive between 4:00 P.M. and 5:00 P.M. The function f is called a **probability density function.** Its integral over *any* subinterval gives the probability that x "lands" in that subinterval.

Similar calculations are shown in the following table.

Time interval	Probability that a bus arrives during interval
2:00 P.M. and 3:00 P.M.	$\int_{2}^{3} \frac{3}{117}x^2\,dx = 0.16 = P([2, 3])$
3:00 P.M. and 4:00 P.M.	$\int_{3}^{4} \frac{3}{117}x^2\,dx = 0.32 = P([3, 4])$
4:00 P.M. and 5:00 P.M.	$\int_{4}^{5} \frac{3}{117}x^2\,dx = 0.52 = P([4, 5])$
2:00 P.M. and 5:00 P.M.	$\int_{2}^{5} \frac{3}{117}x^2\,dx = 1.00 = P([2, 5])$

The results in the table lead us to the following definition of a probability density function.

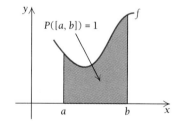

FIGURE 1

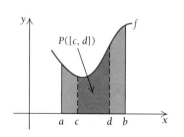

FIGURE 2

DEFINITION

Let x be a continuous random variable distributed over some interval $[a, b]$. A function f is said to be a *probability density function* for x if:

1. f is nonnegative over $[a, b]$; that is, $f(x) \geq 0$ for all x in $[a, b]$.

2. The probability that x lands in $[a, b]$ is 1 (see Figure 1):

$$\int_{a}^{b} f(x)\,dx = 1;$$

   that is, we are "certain" that x is in the interval $[a, b]$.

3. For any subinterval $[c, d]$ of $[a, b]$ (see Figure 2), the probability $P([c, d])$, or $P(c \leq x \leq d)$, that x lands in that subinterval is given by

$$P([c, d]) = \int_{c}^{d} f(x)\,dx.$$

EXAMPLE 5 Verify Property 2 of the definition above for

$$f(x) = \frac{3}{117}x^2, \quad \text{for } 2 \le x \le 5.$$

Solution

$$\int_2^5 \frac{3}{117}x^2\, dx = \frac{3}{117}\left[\frac{1}{3}x^3\right]_2^5 = \frac{1}{117}\left[x^3\right]_2^5 = \frac{1}{117}(5^3 - 2^3) = \frac{117}{117} = 1 \qquad \blacklozenge$$

EXAMPLE 6 *Business: Life of a product.* A company that produces transistors determines that the life t of a transistor is from 3 to 6 yr and that the probability density function for t is given by

$$f(t) = \frac{24}{t^3}, \quad \text{for } 3 \le t \le 6.$$

a) Verify Property 2.

b) Find the probability that a transistor will last no more than 4 yr.

c) Find the probability that a transistor will last at least 4 yr and at most 5 yr.

Solution

a) We want to show that $\int_3^6 f(t)\, dt = 1$. Now

$$\int_3^6 \frac{24}{t^3}\, dt = 24\int_3^6 t^{-3}\, dt = 24\left[\frac{t^{-2}}{-2}\right]_3^6$$

$$= -12\left[\frac{1}{t^2}\right]_3^6 = -12\left(\frac{1}{6^2} - \frac{1}{3^2}\right)$$

$$= -12\left(\frac{1}{36} - \frac{1}{9}\right) = -12\left(-\frac{3}{36}\right) = 1.$$

b) The probability that a transistor will last no more than 4 yr is

$$P(3 \le t \le 4) = \int_3^4 \frac{24}{t^3}\, dt = 24\int_3^4 t^{-3}\, dt$$

$$= 24\left[\frac{t^{-2}}{-2}\right]_3^4 = -12\left[\frac{1}{t^2}\right]_3^4$$

$$= -12\left(\frac{1}{4^2} - \frac{1}{3^2}\right) = -12\left(\frac{1}{16} - \frac{1}{9}\right)$$

$$= -12\left(-\frac{7}{144}\right) = \frac{7}{12} \approx 0.58.$$

c) The probability that a transistor will last at least 4 yr and at most 5 yr is

$$P(4 \le t \le 5) = \int_4^5 \frac{24}{t^3} \, dt = 24 \int_4^5 t^{-3} \, dt$$

$$= 24 \left[\frac{t^{-2}}{-2} \right]_4^5 = -12 \left[\frac{1}{t^2} \right]_4^5$$

$$= -12 \left(\frac{1}{5^2} - \frac{1}{4^2} \right) = -12 \left(\frac{1}{25} - \frac{1}{16} \right)$$

$$= -12 \left(-\frac{9}{400} \right) = \frac{27}{100} = 0.27.$$ ◆

Constructing Probability Density Functions

Suppose that you have an arbitrary nonnegative function $f(x)$ whose definite integral over some interval $[a, b]$ is K. Then

$$\int_a^b f(x) \, dx = K.$$

Multiplying on both sides by $1/K$ gives us

$$\frac{1}{K} \int_a^b f(x) \, dx = \frac{1}{K} \cdot K = 1, \quad \text{or} \quad \int_a^b \frac{1}{K} \cdot f(x) \, dx = 1.$$

Thus when we multiply the function $f(x)$ by $1/K$, we have a function whose area over the given interval is 1.

EXAMPLE 7   Find k such that

$$f(x) = kx^2$$

is a probability density function over the interval $[2, 5]$. Then write the probability density function.

Solution   We have

$$\int_2^5 x^2 \, dx = \left[\frac{x^3}{3} \right]_2^5 = \frac{5^3}{3} - \frac{2^3}{3} = \frac{125}{3} - \frac{8}{3} = \frac{117}{3}.$$

Thus,

$$k = \frac{1}{\frac{117}{3}} = \frac{3}{117}$$

and the probability density function is

$$f(x) = \frac{3}{117} x^2.$$ ◆

Uniform Distributions

Suppose that the probability density function of a continuous random variable is constant. How is it described? Consider the graph shown below.

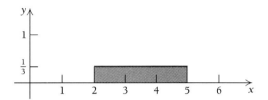

The length of the shaded rectangle is the length of the interval $[2, 5]$, which is 3. In order for the shaded area to be 1, the height of the rectangle must be $\frac{1}{3}$. Thus, $f(x) = \frac{1}{3}$.

The length of the shaded rectangle is the length of the interval $[a, b]$, which is $b - a$. In order for the shaded area to be 1, the height of the rectangle must be $1/(b - a)$. Thus, $f(x) = 1/(b - a)$.

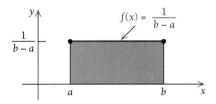

DEFINITION

A continuous random variable x is said to be *uniformly distributed* over an interval $[a, b]$ if it has a probability density function f given by

$$f(x) = \frac{1}{b - a}, \quad \text{for } a \leq x \leq b.$$

EXAMPLE 8 A number x is selected at random from the interval $[40, 50]$. The probability density function for x is given by

$$f(x) = \frac{1}{10}, \quad \text{for } 40 \leq x \leq 50.$$

Find the probability that a number selected is in the subinterval $[42, 48]$.

Solution The probability is

$$P(42 \leq x \leq 48) = \int_{42}^{48} \tfrac{1}{10} \, dx = \tfrac{1}{10}[x]_{42}^{48}$$

$$= \tfrac{1}{10}(48 - 42) = \tfrac{6}{10} = 0.6. \qquad \blacklozenge$$

EXAMPLE 9 *Business: Quality control.* A company produces sirens for tornado warnings. The maximum loudness L of the sirens ranges from 70 to 100 decibels. The probability density function for L is

$$f(L) = \tfrac{1}{30}, \quad \text{for } 70 \le L \le 100.$$

A siren is selected at random off the assembly line. Find the probability that its maximum loudness is from 70 to 92 decibels.

Solution   The probability is

$$P(70 \le L \le 92) = \int_{70}^{92} \tfrac{1}{30}\, dL = \tfrac{1}{30}[L]_{70}^{92}$$

$$= \tfrac{1}{30}(92 - 70) = \tfrac{22}{30} = \tfrac{11}{15} \approx 0.73. \qquad \blacklozenge$$

Exponential Distributions

The duration of a phone call, the distance between successive cars on a highway, and the amount of time required to learn a task are all examples of exponentially distributed random variables. That is, their probability density functions are exponential.

DEFINITION

A continuous random variable is *exponentially distributed* if it has a probability density function given by

$$f(x) = ke^{-kx}, \quad \text{over the interval } [0, \infty).$$

The function $f(x) = 2e^{-2x}$ is such a probability density function. That

$$\int_0^\infty 2e^{-2x}\, dx = 1$$

is shown in Section 6.3. The general case

$$\int_0^\infty ke^{-kx}\, dx = 1$$

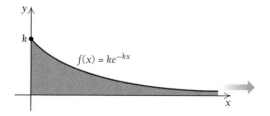

$f(x) = ke^{-kx}$

can be verified in a similar way.

Why is it reasonable to assume that the distance between cars is exponentially distributed? Part of the reason is that there are many more cases in which distances are small, though we can find other distributions that are "skewed" in this manner. The same argument holds for the duration of a phone call. That is, there are more short calls than long ones. The rest of the reason might lie in an analysis of the data involving such distances or phone calls.

EXAMPLE 10   *Business: Transportation planning.* The distance x, in feet, between successive cars on a certain stretch of highway has a probability density function

$$f(x) = ke^{-kx}, \quad \text{for } 0 \le x < \infty,$$

where $k = 1/a$ and $a =$ the average distance between successive cars over some period of time.

A transportation planner determines that the average distance between cars on a certain stretch of highway is 166 ft. What is the probability that if we choose two successive cars at random, the distance between those cars is 50 ft or less?

Solution   We first determine k:

$$k = \frac{1}{166}$$
$$\approx 0.006024.$$

The probability density function for x is

$$f(x) = 0.006024e^{-0.006024x}, \quad \text{for } 0 \le x < \infty.$$

The probability that the distance between the cars is 50 ft or less is

$$P(0 \le x \le 50) = \int_0^{50} 0.006024e^{-0.006024x}\, dx$$

$$= \left[\frac{0.006024}{-0.006024}e^{-0.006024x}\right]_0^{50}$$

$$= \left[-e^{-0.006024x}\right]_0^{50}$$

$$= \left(-e^{-0.006024 \cdot 50}\right) - \left(-e^{-0.006024 \cdot 0}\right)$$

$$= -e^{-0.301200} + 1$$

$$= 1 - e^{-0.301200}$$

$$= 1 - 0.739930$$

$$\approx 0.260.$$

A transportation planner can determine probabilities that cars are certain distances apart.

6.4 Exercise Set

Verify Property 2 of the definition of a probability density function over the given interval.

1. $f(x) = 2x$, $[0, 1]$

2. $f(x) = \frac{1}{4}x$, $[1, 3]$

3. $f(x) = \frac{1}{3}$, $[4, 7]$

4. $f(x) = \frac{1}{4}$, $[9, 13]$

5. $f(x) = \frac{3}{26}x^2$, $[1, 3]$

6. $f(x) = \frac{3}{64}x^2$, $[0, 4]$

7. $f(x) = \frac{1}{x}$, $[1, e]$

8. $f(x) = \frac{1}{e-1}e^x$, $[0, 1]$

9. $f(x) = \frac{3}{2}x^2$, $[-1, 1]$

10. $f(x) = \frac{1}{3}x^2$, $[-2, 1]$

11. $f(x) = 3e^{-3x}$, $[0, \infty)$

12. $f(x) = 4e^{-4x}$, $[0, \infty)$

Find k such that the function is a probability density function over the given interval. Then write the probability density function.

13. $f(x) = kx$, $[1, 3]$

14. $f(x) = kx$, $[1, 4]$

15. $f(x) = kx^2$, $[-1, 1]$

16. $f(x) = kx^2$, $[-2, 2]$

17. $f(x) = k$, $[2, 7]$ 18. $f(x) = k$, $[3, 9]$

19. $f(x) = k(2 - x)$, $[0, 2]$

20. $f(x) = k(4 - x)$, $[0, 4]$

21. $f(x) = \dfrac{k}{x}$, $[1, 3]$ 22. $f(x) = \dfrac{k}{x}$, $[1, 2]$

23. $f(x) = ke^x$, $[0, 3]$ 24. $f(x) = ke^x$, $[0, 2]$

25. A dart is thrown at a number line in such a way that it always lands in the interval $[0, 10]$. Let $x =$ the number that the dart hits. Suppose the probability density function for x is given by

$$f(x) = \tfrac{1}{50}x, \quad \text{for } 0 \le x \le 10.$$

Find $P(2 \le x \le 6)$, the probability that it lands in $[2, 6]$.

26. In Exercise 25, suppose that the dart always lands in the interval $[0, 5]$, and that the probability density function for x is given by

$$f(x) = \tfrac{3}{125}x^2, \quad \text{for } 0 \le x \le 5.$$

Find $P(1 \le x \le 4)$, the probability that it lands in $[1, 4]$.

27. A number x is selected at random from the interval $[4, 20]$. The probability density function for x is given by

$$f(x) = \tfrac{1}{16}, \quad \text{for } 4 \le x \le 20.$$

Find the probability that a number selected is in the subinterval $[9, 17]$.

28. A number x is selected at random from the interval $[5, 29]$. The probability density function for x is given by

$$f(x) = \tfrac{1}{24}, \quad \text{for } 5 \le x \le 29.$$

Find the probability that a number selected is in the subinterval $[13, 29]$.

APPLICATIONS

◆ **Business and Economics**

29. *Transportation planning.* A transportation planner determines that the average distance between cars on a certain highway is 100 ft. What is the probability that if we chose two successive cars at random, the distance between those cars is 40 ft or less?

30. *Transportation planning.* A transportation planner determines that the average distance between cars on a certain highway is 200 ft. What is the probability that if we chose two successive cars at random, the distance between those cars is 10 ft or less?

31. *Duration of a phone call.* A telephone company determines that the duration t of a phone call is an exponentially distributed random variable with probability density function

$$f(t) = 2e^{-2t}, \quad 0 \le t < \infty.$$

Find the probability that a phone call will last no more than 5 min.

32. *Duration of a phone call.* Referring to Exercise 31, find the probability that a phone call will last no more than 2 min.

33. *Time to failure.* The *time to failure* t, in hours, of a certain machine can often be assumed to be exponentially distributed with probability density function

$$f(t) = ke^{-kt}, \quad 0 \le t < \infty,$$

where $k = 1/a$ and $a =$ the average time that will pass before a failure occurs. Suppose the average time that will pass before a failure occurs is 100 hr. What is the probability that a failure will occur in 50 hr or less?

34. *Reliability of a machine.* The *reliability* of the machine (the probability that it will work) in Exercise 33 is defined as

$$R(T) = 1 - \int_0^T 0.01e^{-0.01t}\, dt,$$

where $R(T)$ is the reliability at time T. Find $R(T)$.

◆ **Social Sciences**

35. *Time in a maze.* In a psychology experiment, the time t, in seconds, that it takes a rat to learn its way through a maze is an exponentially distributed random variable with probability density function

$$f(t) = 0.02e^{-0.02t}, \quad 0 \le t < \infty.$$

Find the probability that a rat will learn its way through a maze in 150 sec or less.

36. *Time in a maze.* Using the situation and the equation in Exercise 35, find the probability that a rat will learn its way through the maze in 50 sec or less.

The time it takes a rat to learn its way through a maze is an exponentially distributed random variable.

SYNTHESIS

37. The function $f(x) = x^3$ is a probability density function on $[0, b]$. What is b?

38. The function $f(x) = 12x^2$ is a probability density function on $[-a, a]$. What is a?

39. ◆ Explain the idea of a probability density function.

40. ◆ Give as many examples as you can of the use of probability in daily life.

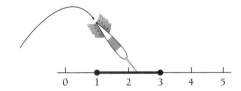

TECHNOLOGY CONNECTION

41.–52. Verify Property 2 of the definition of a probability density function for the functions in Exercises 1–12.

6.5 Probability: Expected Value; The Normal Distribution

OBJECTIVES

• Find $E(x)$, $E(x^2)$, the mean, the variance, and the standard deviation.

• Evaluate normal distribution probabilities using a table.

Expected Value

Let us again consider throwing a dart at a number line in such a way that it always lands in the interval $[1, 3]$.

Suppose we throw the dart at the line 100 times and keep track of the numbers it hits. Then suppose we calculate the arithmetic mean (or average) $\bar{x}$ of all these numbers:

$$\bar{x} = \frac{x_1 + x_2 + x_3 + \cdots + x_{100}}{100} = \frac{\sum\limits_{i=1}^{100} x_i}{100} = \sum\limits_{i=1}^{100} x_1 \cdot \frac{1}{100}.$$

The expression

$$\sum_{i=1}^{n} x_i \cdot \frac{1}{n}, \quad \text{or} \quad \sum_{i=1}^{n} x_i \cdot \frac{1}{2} \cdot \frac{2}{n}$$

is analogous to the integral

$$\int_{1}^{3} x \cdot f(x) \, dx,$$

where f is the probability density function for x and where, in this case, $f(x)$ is the constant function $\frac{1}{2}$. The probability density function gives a "weight" to x. Also, $2/n$ can be considered to be Δx. We add all the

$$x_i \cdot \frac{1}{2} \cdot \left(\frac{2}{n}\right)$$

values when we find $\sum_{i=1}^{n} x_i \cdot \frac{1}{2} \cdot (2/n)$; and, similarly, we add all the

$$x \cdot f(x) \cdot \Delta x$$

values when we find $\int_{1}^{3} x \cdot f(x) \, dx$.

Suppose we have the probability density function $f(x) = \frac{1}{4}x$ over the interval $[1, 3]$. As we can see from the graph below, this function gives more "weight" to the right side of the interval than to the left. Perhaps more points are awarded when the dart lands on the right. Then

$$\int_{1}^{3} x \cdot f(x) \, dx = \int_{1}^{3} x \cdot \frac{1}{4}x \, dx$$

$$= \frac{1}{4} \int_{1}^{3} x^2 \, dx = \frac{1}{4} \left[\frac{x^3}{3} \right]_{1}^{3}$$

$$= \frac{1}{12}(3^3 - 1^3) = \frac{26}{12} \approx 2.17.$$

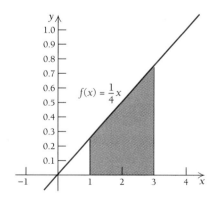

Suppose we continue to throw the dart and compute averages. The more times we throw the dart, the closer we expect the averages to come to 2.17.

Let x be a continuous random variable over the interval $[a, b]$ with probability density function f.

DEFINITION

The *expected value* of x is defined by

$$E(x) = \int_a^b x \cdot f(x) \, dx,$$

where f is a probability density function for x.

The concept of expected value of a random variable generalizes to other functions of a random variable. Suppose $y = g(x)$ is a function of the random variable x. Then we have the following.

DEFINITION

The *expected value* of $g(x)$ is defined by

$$E(g(x)) = \int_a^b g(x) \cdot f(x) \, dx,$$

where f is a probability density function for x.

For example,

$$E(x) = \int_a^b x f(x) \, dx,$$

$$E(x^2) = \int_a^b x^2 f(x) \, dx,$$

$$E(e^x) = \int_a^b e^x f(x) \, dx,$$

and

$$E(2x + 3) = \int_a^b (2x + 3) f(x) \, dx.$$

EXAMPLE 1   Given the probability density function

$$f(x) = \tfrac{1}{2}x \quad \text{over } [0, 2],$$

find $E(x)$ and $E(x^2)$.

Solution

$$E(x) = \int_0^2 x \cdot \frac{1}{2}x \, dx = \int_0^2 \frac{1}{2}x^2 \, dx$$

$$= \frac{1}{2}\left[\frac{x^3}{3}\right]_0^2 = \frac{1}{6}\left[x^3\right]_0^2$$

$$= \frac{1}{6}(2^3 - 0^3)$$

$$= \frac{1}{6} \cdot 8 = \frac{4}{3};$$

$$E(x^2) = \int_0^2 x^2 \cdot \frac{1}{2}x \, dx = \int_0^2 \frac{1}{2}x^3 \, dx = \frac{1}{2}\left[\frac{x^4}{4}\right]_0^2$$

$$= \frac{1}{8}\left[x^4\right]_0^2 = \frac{1}{8}(2^4 - 0^4)$$

$$= \frac{1}{8} \cdot 16 = 2$$

◆

DEFINITION

The *mean* μ of a continuous random variable is defined to be $E(x)$. That is,

$$\mu = E(x) = \int_a^b xf(x) \, dx,$$

where f is a probability density function for x. (The symbol μ is the lower-case Greek letter "mu.")

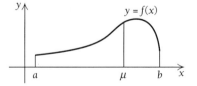
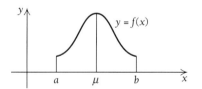

We can get a physical idea of a mean of a random variable by pasting the graph of the probability density function on cardboard and cutting out the area under the curve over the interval $[a, b]$. Then we try to find a fulcrum, or balance point, on the x-axis. That balance point is the mean.

DEFINITION

The *variance* σ^2 of a continuous random variable is defined as

$$\sigma^2 = E(x^2) - \mu^2$$
$$= E(x^2) - [E(x)]^2$$
$$= \int_a^b x^2 f(x)\, dx - \left[\int_a^b x f(x)\, dx\right]^2.$$

The *standard deviation* σ of a continuous random variable is defined as

$$\sigma = \sqrt{\text{variance}}.$$

(The symbol σ is the lower-case Greek letter "sigma.")

EXAMPLE 2   Given the probability density function

$$f(x) = \tfrac{1}{2}x \quad \text{over } [0, 2],$$

find the mean, the variance, and the standard deviation.

Solution   From Example 1, we have

$$E(x) = \tfrac{4}{3} \quad \text{and} \quad E(x^2) = 2.$$

Then

$$\text{the mean} = \mu = E(x) = \tfrac{4}{3};$$
$$\text{the variance} = \sigma^2 = E(x^2) - [E(x)]^2$$
$$= 2 - \left(\tfrac{4}{3}\right)^2 = 2 - \tfrac{16}{9}$$
$$= \tfrac{18}{9} - \tfrac{16}{9} = \tfrac{2}{9};$$
$$\text{the standard deviation} = \sigma = \sqrt{\tfrac{2}{9}}$$
$$= \tfrac{1}{3}\sqrt{2} \approx 0.47.$$

Loosely speaking, we say that the standard deviation is a measure of how close the graph of f is to the mean. Note the examples that follow.

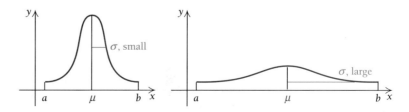

The Normal Distribution

Suppose the average on a test is 70. Usually there are about as many scores above the average as there are below the average; and the further away from the average a particular score is, the fewer people there are who get that score. In this example, it is probable that more people would score in the 80s than in the 90s, and more people would score in the 60s than in the 50s. Test scores, heights of human beings, and weights of human beings are all examples of random variables that are often *normally* distributed.

Consider the function

$$g(x) = e^{-x^2/2} \quad \text{over the interval } (-\infty, \infty).$$

This function has the entire set of real numbers as its domain. Its graph is the bell-shaped curve shown below. We can find function values by using a calculator:

$$y = e^{-x^2/2}.$$

x	y
0	1
1	0.6
2	0.1
3	0.01
−1	0.6
−2	0.1
−3	0.01

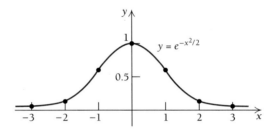

This function has an antiderivative, but that antiderivative has no elementary formula. Nevertheless, it has been shown that its improper integral converges over the integral $(-\infty, \infty)$ to a number given by

$$\int_{-\infty}^{\infty} e^{-x^2/2}\, dx = \sqrt{2\pi}.$$

That is, although an elementary expression for the antiderivative cannot be found, there is a numerical value for the improper integral evaluated over the set of real numbers. Note that since the area is not 1, the function g is not a probability density function, but the following is:

$$\frac{1}{\sqrt{2\pi}} e^{-x^2/2}.$$

DEFINITION

A continuous random variable x has a *standard normal distribution* if its probability density function is

$$f(x) = \frac{1}{\sqrt{2\pi}} e^{-x^2/2} \quad \text{over } (-\infty, \infty).$$

TECHNOLOGY CONNECTION

Use your grapher to approximate

$$\int_{-b}^{b} \frac{1}{\sqrt{2\pi}} e^{-x^2/2} \, dx$$

for $b = 10$, 100, and 1000. What does this suggest about

$$\int_{-\infty}^{\infty} \frac{1}{\sqrt{2\pi}} e^{-x^2/2} \, dx?$$

This is a way to verify part of the assertion that

$$f(x) = \frac{1}{\sqrt{2\pi}} e^{-x^2/2}$$

is a probability density function. Then use a similar approximation procedure to show that the mean is 0 and the standard deviation is 1.

This distribution has a mean of 0 and a standard deviation of 1. Its graph follows.

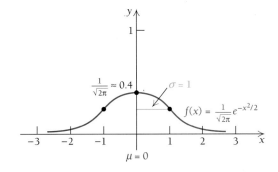

The general case is defined as follows.

DEFINITION

A continuous random variable x is *normally distributed* with mean μ and standard deviation σ if its probability density function is given by

$$f(x) = \frac{1}{\sigma\sqrt{2\pi}} e^{-(1/2)[(x-\mu)/\sigma]^2} \quad \text{over } (-\infty, \infty).$$

The graph is a transformation of the graph of the standard density function. This can be shown by translating the graph along the x-axis and changing the way in which the graph is clustered about the mean. Some examples follow.

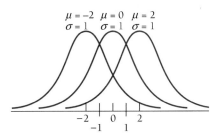

Normal distributions with same standard deviations but different means

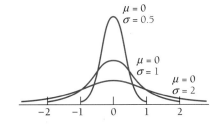

Normal distributions with same means but different standard deviations

The normal distribution is extremely important in statistics; it underlies much of the research in the behavioral and social sciences. Because of this, tables of approximate values of the definite integral of the standard density functions have been prepared using numerical approximation methods like the Trapezoidal Rule given in Exercise Set 5.8. Table 2 at the back of the book is such a table. It contains values of

$$P(0 \leq x \leq t) = \int_0^t \frac{1}{\sqrt{2\pi}} e^{-x^2/2} \, dx.$$

The symmetry of the graph about the mean allows many types of probabilities to be computed from the table.

EXAMPLE 3 Let x be a continuous random variable with standard normal density. Using Table 2, find each of the following.

a) $P(0 \leq x \leq 1.68)$ b) $P(-0.97 \leq x \leq 0)$

c) $P(-2.43 \leq x \leq 1.01)$ d) $P(1.90 \leq x \leq 2.74)$

e) $P(-2.98 \leq x \leq -0.42)$ f) $P(x \geq 0.61)$

Solution

a) $P(0 \leq x \leq 1.68)$ is the area bounded by the standard normal curve and the lines $x = 0$ and $x = 1.68$. We look this up in Table 2 by going down the left column to 1.6, then moving to the right to the column headed 0.08. There we read 0.4535. Thus,

$$P(0 \leq x \leq 1.68) = 0.4535.$$

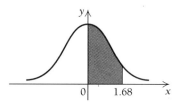

b) Because of the symmetry of the graph,

$$P(-0.97 \leq x \leq 0)$$
$$= P(0 \leq x \leq 0.97)$$
$$= 0.3340.$$

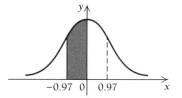

c) $P(-2.43 \leq x \leq 1.01)$
   $= P(-2.43 \leq x \leq 0) + P(0 \leq x \leq 1.01)$
   $= P(0 \leq x \leq 2.43) + P(0 \leq x \leq 1.01)$
   $= 0.4925 + 0.3438$
   $= 0.8363$

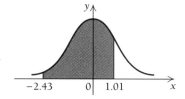

d) $P(1.90 \le x \le 2.74)$
$= P(0 \le x \le 2.74) - P(0 \le x \le 1.90)$
$= 0.4969 - 0.4713$
$= 0.0256$

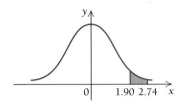

e) $P(-2.98 \le x \le -0.42)$
$= P(0.42 \le x \le 2.98)$
$= P(0 \le x \le 2.98) - P(0 \le x \le 0.42)$
$= 0.4986 - 0.1628$
$= 0.3358$

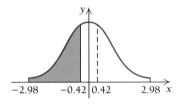

f) $P(x \ge 0.61)$
$= P(x \ge 0) - P(0 \le x \le 0.61)$
$= 0.5000 - 0.2291$
$= 0.2709$

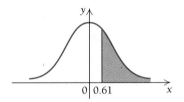

(Because of the symmetry about the line $x = 0$, half the area is on each side of the line, and since the entire area is 1, $P(x \ge 0) = 0.5000$.)   ◆

In many applications, a normal distribution is not standard. It would be a hopeless task to make tables for all values of the mean μ and the standard deviation σ. In such cases, the transformation

$$X = \frac{x - \mu}{\sigma}$$

standardizes the distribution, permitting the use of Table 2 at the back of the book. That is,

$$P(a \le x \le b) = P\left(\frac{a - \mu}{\sigma} \le X \le \frac{b - \mu}{\sigma}\right),$$

and the probability on the right can be found using Table 2. To see this, consider

$$P(a \le x \le b) = \int_a^b \frac{1}{\sigma\sqrt{2\pi}} e^{-(1/2)[(x-\mu)/\sigma]^2} \, dx,$$

and make the substitution

$$X = \frac{x - \mu}{\sigma} = \frac{x}{\sigma} - \frac{\mu}{\sigma}.$$

Then

$$dX = \frac{1}{\sigma} dx.$$

When $x = a$, $X = (a - \mu)/\sigma$; and when $x = b$, $X = (b - \mu)/\sigma$. Then

$$P(a \le x \le b) = \int_a^b \frac{1}{\sigma\sqrt{2\pi}} e^{-(1/2)[(x-\mu)/\sigma]^2} dx$$

and

$$P(a \le x \le b) = \int_{(a-\mu)/\sigma}^{(b-\mu)/\sigma} \frac{1}{\sqrt{2\pi}} e^{-(1/2)X^2} dX \qquad \text{The integrand is now in the form of the standard density.}$$

$$= P\left(\frac{a - \mu}{\sigma} \le X \le \frac{b - \mu}{\sigma}\right). \qquad \text{We can look this up in Table 2.}$$

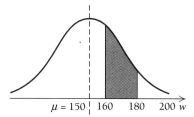

EXAMPLE 4   The weights w of the students in a calculus class are normally distributed with mean 150 lb and standard deviation 25 lb. Find the probability that a student's weight is from 160 lb to 180 lb.

Solution   We first standardize the weights:

$$180 \text{ is standardized to } \frac{b - \mu}{\sigma} = \frac{180 - 150}{25} = 1.2;$$

$$160 \text{ is standardized to } \frac{a - \mu}{\sigma} = \frac{160 - 150}{25} = 0.4.$$

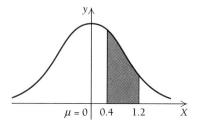

Then

$$
\begin{aligned}
P(160 \le w \le 180) &= P(0.4 \le X \le 1.2) \qquad \text{Now we can use Table 2.} \\
&= P(0 \le X \le 1.2) - P(0 \le X \le 0.4) \\
&= 0.3849 - 0.1554 \\
&= 0.2295.
\end{aligned}
$$

Thus the probability that a student's weight is from 160 lb to 180 lb is 0.2295. That is, about 23% of the students have weights from 160 lb to 180 lb.   ◆

6.5 Exercise Set

For each probability density function, over the given interval, find $E(x)$, $E(x^2)$, the mean, the variance, and the standard deviation.

1. $f(x) = \frac{1}{3}$, $[2, 5]$

2. $f(x) = \frac{1}{4}$, $[3, 7]$

3. $f(x) = \frac{2}{9}x$, $[0, 3]$

4. $f(x) = \frac{1}{8}x$, $[0, 4]$

5. $f(x) = \frac{2}{3}x$, $[1, 2]$

6. $f(x) = \frac{1}{4}x$, $[1, 3]$

7. $f(x) = \frac{1}{3}x^2$, $[-2, 1]$

8. $f(x) = \frac{3}{2}x^2$, $[-1, 1]$

9. $f(x) = \dfrac{1}{\ln 3} \cdot \dfrac{1}{x}$, $[1, 3]$

10. $f(x) = \dfrac{1}{\ln 2} \cdot \dfrac{1}{x}$, $[1, 2]$

Let x be a continuous random variable with standard normal density. Using Table 2, find each of the following.

11. $P(0 \le x \le 2.69)$ 12. $P(0 \le x \le 0.04)$

13. $P(-1.11 \le x \le 0)$ 14. $P(-2.61 \le x \le 0)$

15. $P(-1.89 \le x \le 0.45)$ 16. $P(-2.94 \le x \le 2.00)$

17. $P(1.76 \le x \le 1.86)$ 18. $P(0.76 \le x \le 1.45)$

19. $P(-1.45 \le x \le -0.69)$

20. $P(-2.45 \le x \le -1.69)$

21. $P(x \ge 3.01)$ 22. $P(x \ge 1.01)$

23. a) $P(-1 \le x \le 1)$
    b) What percentage of the area is from -1 to 1?

24. a) $P(-2 \le x \le 2)$
    b) What percentage of the area is from -2 to 2?

Let x be a continuous random variable that is normally distributed with mean $\mu = 22$ and standard deviation $\sigma = 5$. Using Table 2, find each of the following.

25. $P(24 \le x \le 30)$ 26. $P(22 \le x \le 27)$

27. $P(19 \le x \le 25)$ 28. $P(18 \le x \le 26)$

APPLICATIONS

◆ **Business and Economics**

29. *Mail orders.* The number of daily orders N received by a mail-order firm is normally distributed with mean 250 and standard deviation 20. The company has to hire extra help or pay overtime on those days when the number of orders received is 300 or higher. What percentage of the days will the company have to hire extra help or pay overtime?

30. *Stereo production.* The daily production N of stereos by a recording company is normally distributed with mean 1000 and standard deviation 50. The company promises to pay bonuses to its employees on those days when the production of stereos is 1100 or more. What percentage of the days will the company have to pay a bonus?

◆ **Social Sciences**

31. *Test score distribution.* The scores S on a psychology test are normally distributed with mean 65 and stan-

dard deviation 20. A score of 80 to 89 is a B. What is the probability of getting a B?

◆ **General Interest**

32. *Bowling scores.* At the time of this writing, the bowling scores S of the author of this text were normally distributed with mean 196 and standard deviation 23.

a) Find the probability that a score is from 190 to 213.
b) Find the probability that a score is from 160 to 175.
c) Find the probability that a score is greater than 200.

SYNTHESIS

For each probability density function, over the given interval, find $E(x)$, $E(x^2)$, the mean, the variance, and the standard deviation.

33. The uniform probability density

$$f(x) = \frac{1}{b - a} \quad \text{over } [a, b]$$

34. The probability density

$$f(x) = \frac{3a^3}{x^4} \quad \text{over } [a, \infty)$$

Median. Let x be a continuous random variable over $[a, b]$ with probability density function f. Then the *median* of x is that number m for which

$$\int_a^m f(x)\, dx = \tfrac{1}{2}.$$

Find the median.

35. $f(x) = \frac{1}{2}x$, $[0, 2]$

36. $f(x) = \frac{3}{2}x^2$, $[-1, 1]$

37. $f(x) = ke^{-kx}$, $[0, \infty)$

38. **Business: Sugar production.** Suppose the amount of sugar going in a sack has a mean μ, which can be adjusted on the filling machine. Suppose the amount dispensed is normally distributed with $\sigma = 0.1$ oz. What should be the setting of μ to ensure that only one bag in 20 will have less than 60 oz?

39. **Business: Does thy cup overflow?** Suppose the mean amount of coffee μ dispensed by a coffee machine can be set. If a cup holds 6.5 oz and the amount of coffee dispensed is normally distributed with $\sigma = 0.3$ oz,

what should be the setting of μ be to ensure that the cup will overflow only one time in a hundred?

40. ◆ Explain the uses of integration in the study of probability.

41. ◆ You are told that "The antiderivative of the function $f(x) = e^{-x^2/2}$ has no elementary formula." Make some guesses of functions that might seem reasonable to you as antiderivatives and show why they are not.

TECHNOLOGY CONNECTION

42. Approximate the integral

$$\int_{-\infty}^{\infty} e^{-x^2}\, dx.$$

6.6 Volume

OBJECTIVE

• Find the volume of a solid of revolution.

Consider the graph of $y = f(x)$. If the upper half-plane is rotated about the x-axis, then each point on the graph has a circular path, and the whole graph sweeps out a certain surface, called a *surface of revolution*.

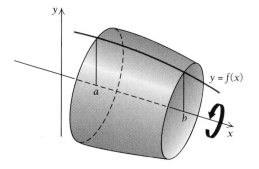

The plane region bounded by the graph, the x-axis, and the interval $[a, b]$ sweeps out a *solid of revolution*. To calculate the volume of this solid, we first approximate it by a finite sum of thin right circular cylinders (Fig. 1). We divide the interval $[a, b]$ into equal subintervals, each of length Δx. Thus the height of each cylinder is Δx (Fig. 2). The radius of each cylinder is $f(x_i)$, where x_i is the righthand endpoint of the subinterval that determines that cylinder.

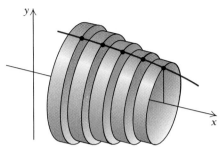

FIGURE 1   FIGURE 2

Since the volume of a right circular cylinder is given by

$$V = \pi r^2 h,$$

each of the approximating cylinders has volume

$$\pi[f(x_i)]^2 \, \Delta x.$$

The volume of the solid of revolution is approximated by the sum of the volumes of all the cylinders:

$$V \approx \sum_{i=1}^{n} \pi[f(x_i)]^2 \, \Delta x.$$

The actual volume is the limit as the thickness of the cylinders approaches zero or the number of them approaches infinity:

$$V = \lim_{n \to \infty} \sum_{i=1}^{n} \pi[f(x_i)]^2 \, \Delta x.$$

This is just the definite integral of the function $y = \pi[f(x)]^2$.

THEOREM 5

$$V = \int_a^b \pi[f(x)]^2 \, dx$$ *Volume of a solid of revolution rotated about the x-axis*

EXAMPLE 1   Find the volume of the solid of revolution generated by rotating the region under the graph of

$$y = \sqrt{x}$$

from $x = 0$ to $x = 1$ about the x-axis.

Solution

$$V = \int_0^1 \pi[f(x)]^2 \, dx$$

$$= \int_0^1 \pi[\sqrt{x}]^2 \, dx$$

$$= \int_0^1 \pi x \, dx$$

$$= \pi\left[\frac{x^2}{2}\right]_0^1$$

$$= \frac{\pi}{2}\left[x^2\right]_0^1$$

$$= \frac{\pi}{2}(1^2 - 0^2) = \frac{\pi}{2}$$

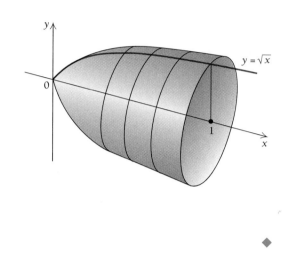

EXAMPLE 2   Find the volume of the solid of revolution, "horn," generated by rotating the region under the graph of

$$y = e^x$$

from $x = -1$ to $x = 2$ about the x-axis.

Solution

$$V = \int_{-1}^2 \pi[f(x)]^2 \, dx$$

$$= \int_{-1}^2 \pi[e^x]^2 \, dx$$

$$= \int_{-1}^2 \pi e^{2x} \, dx$$

$$= \left[\frac{\pi}{2} e^{2x}\right]_{-1}^2$$

$$= \frac{\pi}{2}\left[e^{2x}\right]_{-1}^2$$

$$= \frac{\pi}{2}(e^{2 \cdot 2} - e^{2(-1)})$$

$$= \frac{\pi}{2}(e^4 - e^{-2})$$

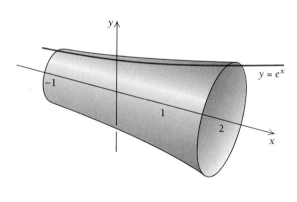

Explain how this could be interpreted as a solid of revolution.

6.6 Exercise Set

Find the volume generated by revolving about the *x*-axis the regions bounded by the following graphs.

1. $y = x, x = 0, x = 1$

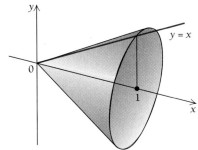

2. $y = \sqrt{x}, x = 0, x = 2$ **3.** $y = x, x = 1, x = 2$

4. $y = x, x = 1, x = 3$ **5.** $y = e^x, x = -2, x = 5$

6. $y = e^x, x = -3, x = 2$ **7.** $y = \dfrac{1}{x}, x = 1, x = 3$

8. $y = \dfrac{1}{x}, x = 1, x = 4$ **9.** $y = \dfrac{1}{\sqrt{x}}, x = 1, x = 3$

10. $y = \dfrac{1}{\sqrt{x}}, x = 1, x = 4$ **11.** $y = 4, x = 1, x = 3$

12. $y = 5, x = 1, x = 3$

13. $y = x^2, x = 0, x = 2$

14. $y = x + 1, x = -1, x = 2$

15. $y = \sqrt{1 + x}, x = 2, x = 10$

16. $y = 2\sqrt{x}, x = 1, x = 2$

17. $y = \sqrt{4 - x^2}, x = -2, x = 2$

18. $y = \sqrt{r^2 - x^2}, x = -r, x = r$

Make a drawing for Exercise 18. Here you will derive a general formula for the volume of a sphere.

SYNTHESIS

Find the volume generated by revolving about the *x*-axis the regions bounded by the following graphs.

19. $y = \sqrt{\ln x}, x = e, x = e^3$

20. $y = \sqrt{xe^{-x}}, x = 1, x = 2$

21. Consider the function $y = 1/x$ over the interval $[1, \infty)$. We showed in Section 6.3 that the area under the curve does not exist; that is,

$$\int_1^\infty \frac{1}{x}\,dx$$

diverges.

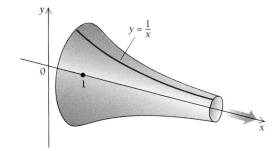

Find the volume of the solid of revolution; that is, find

$$\int_1^\infty \pi\left[\frac{1}{x}\right]^2 dx.$$

This solid is sometimes referred to as *Gabriel's horn*.

TECHNOLOGY CONNECTION

22. *Paradox of Gabriel's horn or the infinite paint can.* Though we cannot prove it here, the surface area of Gabriel's horn is given by

$$S = \int_1^\infty \frac{2\pi}{x}\sqrt{1 + \frac{1}{x^4}}\,dx.$$

Show that the surface area of Gabriel's horn does not exist. The paradox occurs because the volume of the horn exists, but the surface area does not. This is like a can of paint that has a finite volume, but not enough paint to paint the outside of the can.

6.7 Differential Equations

OBJECTIVES

- Solve differential equations.
- Verify that a given function is a solution of a differential equation.
- Solve differential equations using separation of variables.

A **differential equation** is an equation that involves derivatives or differentials. In Chapter 4, we studied one very important differential equation,

$$\frac{dP}{dt} = kP,$$

where P, or $P(t)$, is the population at time t. This equation is a model of uninhibited population growth. Its solution is the function

$$P = P_0 e^{kt},$$

where the constant P_0 is the size of the initial population, that is, at $t = 0$. As this one equation illustrates, differential equations are rich in application.

Solving Certain Differential Equations

In this chapter, we will frequently use the notation y' for a derivative—mainly because it is simple. Thus, if $y = f(x)$, then

$$y' = \frac{dy}{dx} = f'(x).$$

We have already found solutions of certain differential equations when we found antiderivatives or indefinite integrals. The differential equation

$$\frac{dy}{dx} = g(x), \quad \text{or} \quad y' = g(x),$$

has the solution

$$y = \int g(x)\, dx.$$

EXAMPLE 1   Solve $y' = 2x$.

Solution

$$y = \int 2x\, dx = x^2 + C \qquad \blacklozenge$$

Waves can be represented by differential equations.

Look again at the solution to Example 1. Note the constant of integration. This solution is called a *general solution* because taking all values of C gives *all* the solutions. Taking specific values of C gives *particular solutions*. For example, the following are particular solutions to $y' = 2x$:

$$y = x^2 + 3,$$
$$y = x^2,$$
$$y = x^2 - 3.$$

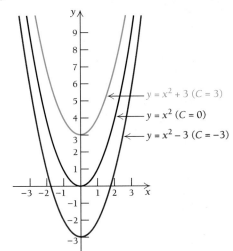

$y = x^2 + 3$ $(C = 3)$

$y = x^2$ $(C = 0)$

$y = x^2 - 3$ $(C = -3)$

TECHNOLOGY
CONNECTION

For $y' = 3x^2$, write the general solution. Then graph the particular solutions for $C = -2$, $C = 0$, and $C = 1$.

The graph shows the curves of a few particular solutions. The general solution can be regarded as the set of all particular solutions, a *family* of curves.

Knowing the value of a function at a particular point may allow us to select a particular solution from the general solution.

EXAMPLE 2   Solve

$$f'(x) = e^x + 5x - x^{1/2},$$

given that $f(0) = 8$.

Solution

a) We first find the general solution:

$$f(x) = \int f'(x)\, dx$$
$$= e^x + \tfrac{5}{2}x^2 - \tfrac{2}{3}x^{3/2} + C.$$

b) Since $f(0) = 8$, we substitute to find C:

$$8 = e^0 + \tfrac{5}{2} \cdot 0^2 - \tfrac{2}{3} \cdot 0^{3/2} + C$$
$$8 = 1 + C$$
$$7 = C.$$

Thus the particular solution is

$$f(x) = e^x + \tfrac{5}{2}x^2 - \tfrac{2}{3}x^{3/2} + 7.$$

Verifying Solutions

To verify that a function is a solution to a differential equation, we find the necessary derivatives and substitute.

EXAMPLE 3   Show that $y = 4e^x + 5e^{3x}$ is a solution to

$$y'' - 4y' + 3y = 0.$$

Solution

a) We first find y' and y'':

$$y' = 4e^x + 15e^{3x};$$
$$y'' = 4e^x + 45e^{3x}.$$

b) Then we substitute in the differential equation, as follows:

$y'' - 4y' + 3y$	$=$	0
$(4e^x + 45e^{3x}) - 4(4e^x + 15e^{3x}) + 3(4e^x + 5e^{3x})$		0
$4e^x + 45e^{3x} - 16e^x - 60e^{3x} + 12e^x + 15e^{3x}$		
0		

◆

Separation of Variables

Consider the differential equation

$$\frac{dy}{dx} = 2xy. \tag{1}$$

We treat dy/dx as a quotient, as we did in Chapter 5. We multiply Eq. (1) by dx and then by $1/y$ to get

$$\frac{dy}{y} = 2x \, dx, \quad y \neq 0. \tag{2}$$

We say that we have **separated the variables,** meaning that all the expressions involving y are on one side and all those involving x are on the other. We then integrate both sides of Eq. (2):

$$\int \frac{dy}{y} = \int 2x \, dx$$

$$\ln y = x^2 + C, \quad y > 0.$$

We use only one constant because any two antiderivatives differ by a constant. Recall that the definition of logarithms says that if $\log_a b = t$, then $b = a^t$. Now, $\ln y = \log_e y = x^2 + C$, so by the definition of logarithms, we have

$$y = e^{x^2+C} = e^{x^2} \cdot e^C.$$

Thus the solution to differential equation (1) is

$$y = C_1 e^{x^2}, \quad \text{where } C_1 = e^C.$$

In fact, C_1 is still an arbitrary constant.

EXAMPLE 4   Solve:

$$3y^2 \frac{dy}{dx} + x = 0, \quad \text{where } y = 5 \text{ when } x = 0.$$

Solution

a) We first separate the variables as follows:

$$3y^2 \frac{dy}{dx} = -x$$

$$3y^2 \, dy = -x \, dx.$$

We then integrate both sides:

$$\int 3y^2 \, dy = \int -x \, dx$$

$$y^3 = -\frac{x^2}{2} + C$$

$$= C - \frac{x^2}{2}$$

$$y = \sqrt[3]{C - \frac{x^2}{2}}. \qquad \text{Taking the cube root}$$

b) Since $y = 5$ when $x = 0$, we substitute to find C:

$$5 = \sqrt[3]{C - \frac{0^2}{2}} \qquad \text{Substituting 5 for } y \text{ and 0 for } x$$

$$5 = \sqrt[3]{C}$$

$$125 = C. \qquad \text{Cubing both sides}$$

The particular solution is

$$y = \sqrt[3]{125 - \frac{x^2}{2}}.$$

*At the 1968 Olympic Games in Mexico City, Bob Beamon made a miraculous long jump of 29 ft, $2\frac{1}{2}$ in. Many believed this was due to the altitude, which was 7400 ft. Using differential equations for analysis, M. N. Bearley refuted the altitude theory in "The Long Jump Miracle of Mexico City," (Mathematics Magazine, **45**, November 1972, pp. 241–246). Bearley argues that the world record jump was a result of Beamon's exceptional speed (9.5 sec in the 100-yd dash) and the fact that he hit the take-off board in perfect position.*

EXAMPLE 5   Solve:

$$\frac{dy}{dx} = \frac{x}{y}.$$

Solution   We first separate the variables:

$$y \frac{dy}{dx} = x$$

$$y \, dy = x \, dx.$$

We then integrate both sides:

$$\int y \, dy = \int x \, dx$$

$$\frac{y^2}{2} = \frac{x^2}{2} + C$$

$$y^2 = x^2 + 2C$$

$$y^2 = x^2 + C_1,$$

where $C_1 = 2C$. We make this substitution in order to simplify the equation. We then obtain the solutions

$$y = \sqrt{x^2 + C_1}$$

and

$$y = -\sqrt{x^2 + C_1}.$$

◆

EXAMPLE 6   Solve: $y' = x - xy$.

Solution   Before we separate the variables, we replace y' by dy/dx:

$$\frac{dy}{dx} = x - xy.$$

Then we separate the variables:

$$dy = (x - xy) \, dx$$

$$dy = x(1 - y) \, dx$$

$$\frac{dy}{1 - y} = x \, dx.$$

Next we integrate both sides:

$$\int \frac{dy}{1 - y} = \int x \, dx$$

$$-\ln (1 - y) = \frac{x^2}{2} + C \qquad 1 - y > 0$$

$$\ln (1 - y) = -\frac{x^2}{2} - C.$$

TECHNOLOGY
CONNECTION

Solve $y' = 2x + xy$. Graph
the particular solutions for
$C_1 = -2$, $C_1 = 0$, and
$C_1 = 1$.

Then

$$1 - y = e^{-x^2/2-C}$$
$$-y = e^{-x^2/2-C} - 1$$
$$y = -e^{-x^2/2-C} + 1$$
$$= -e^{-x^2/2} \cdot e^{-C} + 1.$$

Thus,

$$y = 1 + C_1 e^{-x^2/2}, \quad \text{where } C_1 = -e^{-C}. \qquad \blacklozenge$$

An Economic Application: Elasticity

EXAMPLE 7   Suppose for a certain product that the elasticity of demand is 1 for all
$p > 0$. That is, $E(p) = 1$ for all $p > 0$. Find the demand function $x = D(p)$. (See
Section 4.6.)

Solution   Since $E(p) = 1$ for all $p > 0$,

$$1 = E(p) = -\frac{p\, D'(p)}{D(p)}$$
$$= -\frac{p}{x} \cdot \frac{dx}{dp}.$$

Then

$$\frac{dx}{dp} = -\frac{x}{p}. \tag{3}$$

Separating the variables, we get

$$\frac{dp}{p} = -\frac{dx}{x}.$$

Now we integrate both sides:

$$\int \frac{dp}{p} = -\int \frac{dx}{x}$$
$$\ln p = -\ln x + C. \qquad \begin{array}{l} p > 0 \text{ and } dx/dp < 0, \text{ since} \\ \text{demand functions are decreasing.} \\ \text{So } x > 0. \text{ See Eq. (3).} \end{array}$$

We can express $C = \ln C_1$, since any real number C is the natural logarithm of
some number C_1. Then

$$\ln p = \ln C_1 - \ln x$$
$$= \ln \frac{C_1}{x},$$

so

$$p = \frac{C_1}{x} \quad \text{and} \quad x = \frac{C_1}{p}.$$

This characterizes those demand functions for which the elasticity is always 1. ◆

A Psychological Application: Reaction to a Stimulus

The Weber–Fechner Law

In psychology, one model of stimulus–response asserts that the rate of change dR/dS of the reaction R with respect to a stimulus S is inversely proportional to the stimulus; that is,

$$\frac{dR}{dS} = \frac{k}{S},$$

where k is some positive constant.

To solve this equation, we first separate the variables:

$$dR = k \cdot \frac{dS}{S}.$$

We then integrate both sides:

$$\int dR = \int k \cdot \frac{dS}{S}$$
$$R = k \ln S + C. \tag{4}$$

Now suppose we let S_0 be the lowest level of the stimulus that can be detected consistently. This is the *threshold value,* or the *detection threshold.* For example, the lowest level of sound that can be consistently detected is the tick of a watch at 20 feet, under very quiet conditions. If S_0 is the lowest level of sound that can be detected, it seems reasonable that the reaction to it would be 0. That is, $R(S_0) = 0$. Substituting this condition into Eq. (4), we get

$$0 = k \ln S_0 + C,$$

or

$$-k \ln S_0 = C.$$

Replacing C in Eq. (4) by $-k \ln S_0$ gives us

$$R = k \cdot \ln S - k \cdot \ln S_0$$
$$R = k(\ln S - \ln S_0).$$

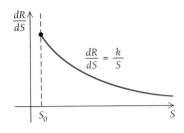

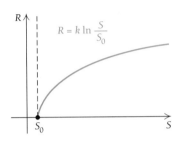

Using a property of logarithms, we have

$$R = k \cdot \ln \frac{S}{S_0}.$$

Look at the graphs of dR/dS and R shown at left. Note that as the stimulus gets larger, the rate of change decreases; that is, the change in reaction becomes smaller as the stimulation received becomes stronger. For example, suppose you are in a room with one lamp and that lamp has a 50-watt bulb in it. If the bulb were suddenly changed to 100 watts, you would probably be very aware of the difference. That is, your reaction would be strong. If the bulb were then changed to 150 watts, your reaction would not be as great as it was to the change from 50 to 100 watts. A change from a 150- to a 200-watt bulb would cause even less reaction, and so on.

For your interest, here are some other detection thresholds.

Stimulus	Detection threshold
Light	The flame of a candle 30 miles away on a dark night
Taste	Water diluted with sugar in the ratio of one teaspoon to two gallons
Smell	One drop of perfume diffused into the volume of three average-size rooms
Touch	The wing of a bee dropped on your cheek at a distance of 1 centimeter $\left(\text{about } \frac{3}{8} \text{ of an inch}\right)$

6.7 Exercise Set

Find the general solution and three particular solutions.

1. $y' = 4x^3$

2. $y' = 6x^5$

3. $y' = e^{2x} + x$

4. $y' = e^{3x} - x$

5. $y' = \dfrac{3}{x} - x^2 + x^5$.

6. $y' = \dfrac{5}{x} + x^2 - x^4$

Find the particular solution determined by the given condition.

7. $y' = x^2 + 2x - 3; \quad y = 4$ when $x = 0$

8. $y' = 3x^2 - x + 5; \quad y = 6$ when $x = 0$

9. $f'(x) = x^{2/3} - x; \quad f(1) = -6$

10. $f'(x) = x^{2/5} + x; \quad f(1) = -7$

11. Show that $y = x \ln x + 3x - 2$ is a solution to

$$y'' - \frac{1}{x} = 0.$$

12. Show that $y = x \ln x - 5x + 7$ is a solution to

$$y'' - \frac{1}{x} = 0.$$

13. Show that $y = e^x + 3xe^x$ is a solution to

$$y'' - 2y' + y = 0.$$

14. Show that $y = -2e^x + xe^x$ is a solution to

$$y'' - 2y' + y = 0.$$

Solve.

15. $\dfrac{dy}{dx} = 4x^3y$ 16. $\dfrac{dy}{dx} = 5x^4y$

17. $3y^2\dfrac{dy}{dx} = 5x$ 18. $3y^2\dfrac{dy}{dx} = 7x$

19. $\dfrac{dy}{dx} = \dfrac{2x}{y}$ 20. $\dfrac{dy}{dx} = \dfrac{x}{2y}$

21. $\dfrac{dy}{dx} = \dfrac{3}{y}$ 22. $\dfrac{dy}{dx} = \dfrac{4}{y}$

23. $y' = 3x + xy$; $y = 5$ when $x = 0$

24. $y' = 2x - xy$; $y = 9$ when $x = 0$

25. $y' = 5y^{-2}$; $y = 3$ when $x = 2$

26. $y' = 7y^{-2}$; $y = 3$ when $x = 1$

27. $\dfrac{dy}{dx} = 3y$ 28. $\dfrac{dy}{dx} = 4y$

29. $\dfrac{dP}{dt} = 2P$ 30. $\dfrac{dP}{dt} = 4P$

31. Solve

$$f'(x) = \frac{1}{x} - 4x + \sqrt{x},$$

given that $f(1) = \frac{23}{3}$.

APPLICATIONS

◆ **Business and Economics**

32. *Total revenue from marginal revenue.* The marginal revenue for a certain product is given by $R'(x) = 300 - 2x$. Find the total-revenue function $R(x)$, assuming that $R(0) = 0$.

33. *Total cost from marginal cost.* The marginal cost for a certain product is given by $C'(x) = 2.6 - 0.02x$. Find the total-cost function $C(x)$ and the average cost $A(x)$, assuming that fixed costs are \$120; that is, $C(0) = \$120$.

34. *Domar's capital expansion model* is

$$\frac{dI}{dt} = hkI,$$

where I = the investment, h = the investment productivity (constant), k = the marginal productivity to the consumer (constant), and t = the time.

a) Use separation of variables to solve the differential equation.

b) Rewrite the solution in terms of the condition $I_0 = I(0)$.

35. *Total profit from marginal profit.* A firm's marginal profit P as a function of its total cost C is given by

$$\frac{dP}{dC} = \frac{-200}{(C + 3)^{3/2}}.$$

a) Find the profit function $P(C)$ if $P = \$10$ when $C = \$61$.

b) At what cost will the firm break even $(P = 0)$?

36. *Stock growth.* The *growth rate of a certain stock* is modeled by

$$\frac{dV}{dt} = k(L - V), \quad V = \$20 \text{ when } t = 0,$$

where V = the value of the stock, per share, after time t (in months), $L = \$24.81$, the *limiting value* of the stock, and k = a constant. Find the solution to the differential equation in terms of t and k.

37. *Utility.* The reaction R in pleasure units by a consumer receiving S units of a product can be modeled by the differential equation

$$\frac{dR}{dS} = \frac{k}{S + 1},$$

where k is a positive constant.

a) Use separation of variables to solve the differential equation.

b) Rewrite the solution in terms of the initial condition $R(0) = 0$.

c) Explain why the condition $R(0) = 0$ is reasonable.

Elasticity. Find the demand function $p = D(x)$ given the following elasticity conditions.

38. $E(p) = \dfrac{4}{p}$; $x = e$ when $p = 4$

39. $E(p) = \dfrac{p}{200 - p}$; $x = 190$ when $p = 10$

40. $E(p) = 2$ for all $p > 0$

41. $E(p) = n$ for some constant n and all $p > 0$

◆ **Life and Physical Sciences**

42. a) Use separation of variables to solve the differential-equation model of uninhibited growth,

$$\frac{dP}{dt} = kP.$$

b) Rewrite the solution in terms of the condition $P_0 = P(0)$.

◆ **Social Sciences**

43. *The Brentano–Stevens Law.* The Weber–Fechner Law has been the subject of great debate among psychologists as to its validity. The model

$$\frac{dR}{dS} = k \cdot \frac{R}{S},$$

where k is a positive constant, has also been conjectured and experimented with. Find the general solution to this equation. (This has also been referred to as the *Power Law of Stimulus–Response.*)

SYNTHESIS

44. ◈ Discuss as many applications as you can of the use of integration in this chapter.

TECHNOLOGY CONNECTION

45. Solve $dy/dx = 5/y$. Graph the particular solutions for $C_1 = 5$, $C_1 = -200$, and $C_1 = 100$.

6 Chapter Summary and Review

TERMS TO KNOW

REVIEW EXERCISES

The review exercises are for test preparation. They can also be used as a lengthened practice test. Answers are at the back of the book. The answers also contain bracketed section references, which tell you where to restudy if your answer is incorrect.

Given $p = D(x) = (x - 6)^2$ and $p = S(x) = x^2 + 12$, find each of the following.

1. The equilibrium point

2. The consumer's surplus at the equilibrium point

3. The producer's surplus at the equilibrium point

4. **Business: Amount of a continuous money flow.** Find the amount of a continuous money flow in which $2400 per year is being invested at 10%, compounded continuously, for 15 yr.

5. **Business: Continuous money flow.** Consider a continuous money flow into an investment at the constant rate of P_0 dollars per year. What should P_0 be so that the amount of a continuous money flow over 25 yr at 12%, compounded continuously, will be $40,000?

6. **Physical science: Demand for potash.** In 1990 ($t = 0$), the world use of potash was 31,270 thousand tons, and the demand for potash was growing at the rate of 3% per year. If the demand continues to grow at this rate, how many tons of potash will the world use from 1990 to 1998?

7. **Physical science: Depletion of potash.** The world reserves of potash is 250,000,000 thousand tons. Assuming that the growth rate in Exercise 6 continues and that no new reserves are discovered, when will the world reserves of potash be exhausted?

8. **Business: Present value.** Find the present value of $100,000 due 50 yr later at 12%, compounded continuously.

9. **Business: Accumulated present value.** Find the accumulated present value of an investment over a 20-yr period if there is a continuous money flow of $4800 per year and the current interest rate is 9%.

10. **Business: Accumulated present value.** Find the accumulated present value of an investment for which the continuous money flow in Exercise 9 is perpetual.

Determine whether the improper integral is convergent or divergent, and calculate its value if it is convergent.

11. $\int_{1}^{\infty} \frac{1}{x^2}\, dx$

12. $\int_{1}^{\infty} e^{4x}\, dx$

13. $\int_{0}^{\infty} e^{-2x}\, dx$

14. Find k such that $f(x) = k/x^3$ is a probability density function over the interval $[1, 2]$. Then write the probability density function.

15. **Business: Waiting time.** A person randomly arrives at a bus stop where the waiting time t for a bus is no more than 25 min. The probability density function

for t is $f(t) = \frac{1}{25}$, for $0 \le t \le 25$. Find the probability that a person will have to wait no more than 15 min for a bus.

Given the probability density function $f(x) = 3x^2$ over $[0, 1]$, find each of the following.

16. $E(x^2)$

17. $E(x)$

18. The mean

19. The variance

20. The standard deviation

Let x be a continuous random variable with standard normal density. Using Table 2, find each of the following.

21. $P(0 \le x \le 1.85)$

22. $P(-1.74 \le x \le 1.43)$

23. $P(-2.08 \le x \le -1.18)$

24. $P(x \ge 0)$

25. **Business: Pizza sales.** The number of pizzas sold daily at Benito's Pizzeria is normally distributed with mean 400 and standard deviation 60. What is the probability that the number sold during one day is 480 or more?

Find the volume generated by revolving about the x-axis the region bounded by each of the following graphs.

26. $y = x^3$, $x = 1$, $x = 2$

27. $y = \dfrac{1}{x + 2}$, $x = 0$, $x = 1$

Solve the differential equation.

28. $\dfrac{dy}{dx} = 11x^{10}y$

29. $\dfrac{dy}{dx} = \dfrac{2}{y}$

30. $\dfrac{dy}{dx} = 4y$; $y = 5$ when $x = 0$

31. $\dfrac{dv}{dt} = 5v^{-2}$; $v = 4$ when $t = 3$

32. $y' = \dfrac{3x}{y}$

33. $y' = 8x - xy$

34. **Economics: Elasticity.** Find the demand function given the elasticity condition

$$E(p) = \frac{p}{p - 100}, \quad x = 70 \text{ when } p = 30.$$

35. **Business: Stock growth.** The growth rate of a stock is modeled by

$$\frac{dV}{dt} = k(L - V), \quad V = \$30 \text{ when } t = 0,$$

where V = the value of the stock, per share, after time t (in months), L = \$36.37, the limiting value of the stock, and k = a constant. Find the solution to the differential equation in terms of t and k.

SYNTHESIS

36. The function $f(x) = x^8$ is a probability density function on the interval $[-c, c]$. Find c.

Determine whether the improper integral is convergent or divergent, and calculate its value if it is convergent.

37. $\displaystyle\int_{-\infty}^{0} x^4 e^{-x^5}\, dx$

38. $\displaystyle\int_{0}^{\infty} \frac{dx}{(x+1)^{4/3}}$

 TECHNOLOGY CONNECTION

39. Approximate the integral

$$\int_{1}^{\infty} \frac{\ln x}{x^2}\, dx.$$

6 Chapter Test

Given the demand and supply functions

$$p = D(x) = (x-7)^2 \text{ and } p = S(x) = x^2 + x + 4,$$

find each of the following.

1. The equilibrium point

2. The consumer's surplus at the equilibrium point

3. The producer's surplus at the equilibrium point

4. **Business: Amount of a continuous money flow.** Find the amount of a continuous money flow if \$1200 per year is being invested at 6%, compounded continuously, for 15 yr.

5. **Business: Continuous money flow.** Consider a continuous money flow into an investment at the rate of P_0 dollars per year. What should P_0 be so that the amount of a continuous money flow over 25 yr at 6%, compounded continuously, will be \$20,000?

6. **Physical science: Demand for iron ore.** In 1990 ($t = 0$), the world use of iron ore was 960,000 thousand tons, and the demand for iron ore was growing exponentially at the rate of 6% per year. If the demand continues to grow at this rate, how many tons of iron ore will the world use from 1990 to 2010?

7. **Physical science: Depletion of iron ore.** The world reserves of iron ore are 147,400,000 thousand tons. Assuming that the growth rate of 6% per year continues and that no new reserves are discovered, when will the world reserves of iron ore be exhausted?

8. **Business: Present value.** Following the birth of a child, a parent wants to make an initial investment P_0 that will grow to \$10,000 by the child's 20th birthday. Interest is compounded continuously at 7%. What should the initial investment be?

9. **Business: Accumulated present value.** Find the accumulated present value of an investment over a 20-yr period if there is a continuous money flow of \$3800 per year and the current interest rate is 11%.

10. **Business: Accumulated present value.** Find the accumulated present value of an investment for which the continuous money flow in Question 9 is perpetual.

Determine whether the improper integral is convergent or divergent, and calculate its value if it is convergent.

11. $\displaystyle\int_{1}^{\infty} \frac{dx}{x^5}$

12. $\displaystyle\int_{0}^{\infty} \frac{3}{1+x}\, dx$

13. Find k such that $f(x) = kx^3$ is a probability density function over the interval $[0, 2]$. Then find the probability density function.

14. **Business: Times of telephone calls.** A telephone company determines that the length of time t of a phone call is an exponentially distributed random variable with probability density function

$$f(t) = 2e^{-2t}, \quad 0 \le t < \infty.$$

Find the probability that a phone call will last no more than 1 min.

Given the probability density function $f(x) = \frac{1}{4}x$ over $[1, 3]$, find each of the following.

15. $E(x)$

16. $E(x^2)$

17. The mean

18. The variance

19. The standard deviation

Let x be a continuous random variable with standard normal density. Using Table 2, find each of the following.

20. $P(0 \le x \le 1.5)$

21. $P(0.12 \le x \le 2.32)$

22. $P(-1.61 \le x \le 1.76)$

23. The price per pound p of a T-bone steak at various stores in a certain city is normally distributed with mean \$4.75 and standard deviation \$0.25. What is the probability that the price per pound is \$4.80 or more?

Find the volume generated by revolving about the x-axis the regions bounded by the following graphs.

24. $y = \dfrac{1}{\sqrt{x}}, x = 1, x = 5$

25. $y = \sqrt{2 + x}, x = 0, x = 1$

Solve the differential equation.

26. $\dfrac{dy}{dx} = 8x^7 y$

27. $\dfrac{dy}{dx} = \dfrac{9}{y}$

28. $\dfrac{dy}{dt} = 6y; \quad y = 11$ when $t = 0$

29. $y' = 5x^2 - x^2 y$

30. $\dfrac{dv}{dt} = 2v^{-3}$

31. $y' = 4y + xy$

32. **Economics: Elasticity.** Find the demand function given the elasticity condition

$$E(p) = 4 \quad \text{for all } p > 0.$$

33. **Business: Stock growth.** The growth rate of stock for Glamour Industries is modeled by

$$\frac{dV}{dt} = k(L - V),$$

where $V =$ the value of the stock per share, after time t (in months), $L = \$36$, the *limiting value* of the stock, $k = $ a constant, and $V(0) = \$0$.

a) Write the solution $V(t)$ in terms of L and k.
b) If $V(6) = \$18$, determine k to the nearest hundredth.
c) Rewrite $V(t)$ in terms of t and k using the value of k found in part (b).
d) Use the equation in part (c) to find $V(12)$, the value of the stock after 12 months.
e) In how many months will the value be \$30?

SYNTHESIS

34. The function $f(x) = x^5$ is a probability density on $[0, b]$. What is b?

35. Determine whether the following improper integral is convergent or divergent, and calculate its value if it is convergent:

$$\int_{-\infty}^{0} x^3 e^{-x^4} \, dx.$$

 TECHNOLOGY CONNECTION

36. Approximate the integral

$$\int_{-\infty}^{\infty} \frac{1}{1 + x^2} \, dx.$$

EXTENDED TECHNOLOGY APPLICATION

Curve Fitting and the Volume of a Bottle of Soda

Consider the urn or vase below. How could we estimate the volume? One way would be to simply pour some fluid into the container and then pour it into a measuring device.

Another way, using calculus and the curve-fitting or regression features of your grapher, would be to turn the urn on its side, as shown below, take vertical measurements from the center to the top, do a curve-fitting or regression procedure with the grapher, and then estimate the volume using the integration technique of the grapher.

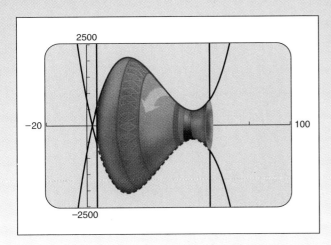

Suppose the following is a table of values for the red curve.

x	y
5	547.5
10	1440.0
15	1932.5
20	2100.0
25	2017.5
30	1760.0
35	1402.5
40	1020.0
45	687.5
50	480.0
55	472.5
60	740.0
61	833.1

EXERCISES

1. Use the regression feature on your grapher to fit a cubic polynomial function to the data.

2. Use the function found in Exercise 1 and the integration feature to estimate the volume of the urn. Integrate over the interval [5, 61].

Now consider the soda bottle pictured to the right. To find its volume in a similar manner, we can turn the bottle on its side and use a measuring device to take vertical measurements and proceed as we did with the urn.

The table of measurements is as follows.

x	y
1	0.938
2	1.375
3	1.156
4	1.250
5	1.313
6	1.313
7	1.281
8	0.813
9	0.500
9.875	0.625

EXERCISES

3. Use the regression feature on your grapher to fit a quartic polynomial function to the data.

4. Use the function found in Exercise 3 and the integration feature to estimate the volume of the bottle. Your answer will be in cubic inches. Convert it to fluid ounces using the fact that $1 \text{ in}^3 = 0.55424$ fluid ounce.

5. The bottle in question has a capacity of 20 oz. How good was our curve-fitting procedure for making this estimate?

6. Try to find a curve that gives a better estimate of the volume. What is the curve and what is the estimated volume?

7 Functions of Several Variables

INTRODUCTION

Functions that have more than one input are called *functions of several variables*. We introduce these functions in this chapter and learn to differentiate them to find what are called *partial derivatives*. Then we use these functions and their partial derivatives to solve maximum–minimum problems. Finally, we consider the integration of such functions.

AN APPLICATION

According to a recent census, the population of San Francisco was 713,000, and the population of Oakland was 352,000. The distance between the cities is 8 miles. What is the average number of telephone calls in a day between these two cities?

THE MATHEMATICS

The average number of telephone calls in a day between any two cities is given by

$$N(d, P_1, P_2) = \frac{2.8 P_1 P_2}{d^{2.4}}.$$

This is a *function of several variables*.

To find the number of telephone calls between San Francisco and Oakland, we substitute 713,000 for P_1, 352,000 for P_2, and 8 for d.

This problem appears as both Example 4 and a Technology Connection in Section 7.1.

7.1 Functions of Several Variables

OBJECTIVE

- Find a function value for a function of several variables.

Suppose that a one-product firm produces x items of its product at a profit of $4 per item. Then its total profit $P(x)$ is given by

$$P(x) = 4x.$$

This is a function of one variable.

Suppose that a two-product firm produces x items of one product at a profit of $4 per item and y items of a second at a profit of $6 per item. Then its total profit P is a function of the *two* variables x and y, and is given by

$$P(x, y) = 4x + 6y.$$

This function assigns to the input pair (x, y) a unique output number $4x + 6y$.

> **DEFINITION**
>
> A *function of two variables* assigns to each input pair (x, y) exactly one output number $f(x, y)$.

We can also think of a function of two variables as a machine that has two inputs. The domain of a function of two variables is a set of pairs (x, y) in the plane. Unless otherwise restricted, when such a function is given by a formula, the domain consists of all ordered pairs (x, y) that are meaningful replacements in the formula.

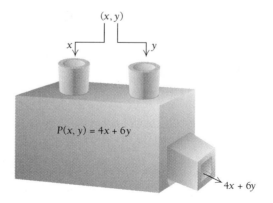

EXAMPLE 1   For $P(x, y) = 4x + 6y$, find $P(25, 10)$.

Solution $P(25, 10)$ is defined to be the value of the function found by substituting 25 for x and 10 for y:

$$P(25, 10) = 4 \cdot 25 + 6 \cdot 10$$
$$= 100 + 60$$
$$= \$160.$$

This means that by selling 25 items of the first product and 10 of the second, the two-product firm will make a profit of $160.   ◆

The following are examples of **functions of several variables,** that is, functions of two or more variables. If there are n variables, then there are n inputs for such a function.

EXAMPLE 2   *Business: Total cost.* The total cost of a company, in thousands of dollars, is given by

$$C(x, y, z, w) = 4x^2 + 5y + z - \ln (w + 1),$$

where x dollars is spent for labor, y dollars for raw materials, z dollars for advertising, and w dollars for machinery. This is a function of four variables (all in thousands of dollars). Find $C(3, 2, 0, 10)$.

Solution   We substitute 3 for x, 2 for y, 0 for z, and 10 for w:

$$C(3, 2, 0, 10) = 4 \cdot 3^2 + 5 \cdot 2 + 0 - \ln (10 + 1)$$
$$= 4 \cdot 9 + 10 + 0 - 2.397895$$
$$\approx \$43.6 \text{ thousand.}$$ ◆

EXAMPLE 3   *Business: Cost of storage equipment.* A business purchases a piece of storage equipment that costs C_1 dollars and has capacity V_1. Later it wishes to replace the original with a new piece of equipment that costs C_2 dollars and has capacity V_2. The ratio of the new capacity to the original is

$$\frac{V_2}{V_1} = k,$$

so $V_2 = kV_1.$

It has been found in industrial economics that in this case, the cost of the new piece of equipment can be estimated by the function of three variables

$$C_2 = \left(\frac{V_2}{V_1} \right)^{0.6} C_1 = k^{0.6} C_1. \tag{1}$$

For $45,000, a beverage company buys a manufacturing tank that has a capacity of 10,000 gallons. Later it decides to buy a tank with double the capacity of the original. Estimate the cost of the new tank.

Solution We substitute 20,000 for V_2, 10,000 for V_1, and 45,000 for C_1 in Eq. (1):

$$C_2 = \left(\frac{20{,}000}{10{,}000}\right)^{0.6}(45{,}000)$$

$$= 2^{0.6}(45{,}000)$$

$$= (1.515717)(45{,}000)$$

$$\approx \$68{,}207.25.$$

Note that a 100% increase in capacity was achieved by about a 52% increase in cost. This is independent of any increase in the costs of labor, management, or other equipment resulting from the purchase of the tank. ◆

TECHNOLOGY CONNECTION

According to a recent census, the population of San Francisco was 713,000, and the population of Oakland was 352,000. The distance between the cities is 8 miles. What is the average number of telephone calls in a day between these two cities?

EXAMPLE 4 *Social science: The gravity model.* The average number of telephone calls in a day between two cities is given by

$$N(d, P_1, P_2) = \frac{2.8 P_1 P_2}{d^{2.4}},$$

where d is the distance, in miles, between the cities and P_1 and P_2 are their populations. ◆

This NASA photograph shows an overhead view of San Francisco and Oakland. Sociologists say that as two cities merge, the communication between them increases.

A constant can also be thought of as a function of several variables.

EXAMPLE 5   The constant function f is given by

$$f(x, y) = -3 \quad \text{for all inputs } x \text{ and } y.$$

Find $f(5, 7)$ and $f(-2, 0)$.

Solution   Since this is a constant function, it has the value -3 for any x and y. Thus,

$$f(5, 7) = -3 \quad \text{and} \quad f(-2, 0) = -3. \qquad \blacklozenge$$

Geometric Interpretations

Consider a function of two variables

$$z = f(x, y).$$

Recall the mapping interpretation of function that we considered in Chapter 1. As a mapping, a function of two variables can be thought of as mapping a point (x_1, y_1) in an xy-plane onto a point z_1 on a number line.

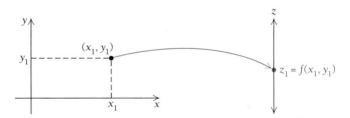

To graph a function of two variables, we need a three-dimensional coordinate system. The axes are generally placed as shown below. The line z, called the z-axis, is placed perpendicular to the xy-plane at the origin.

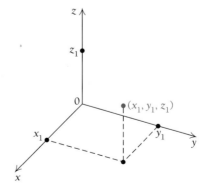

To help visualize this, think of looking into the corner of a room, where the floor is the xy-plane and the z-axis is the intersection of the two walls. To plot a point (x_1, y_1, z_1), we locate the point (x_1, y_1) in the xy-plane and move up or down in space according to the value of z_1.

EXAMPLE 6   Plot these points:

$$P_1(2, 3, 5), \quad P_2(2, -2, -4), \quad P_3(0, 5, 2), \quad \text{and} \quad P_4(2, 3, 0).$$

Solution   The solution is shown below.

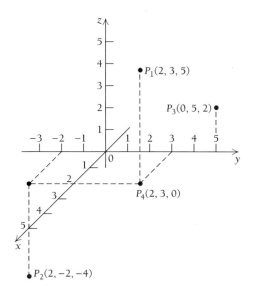

The *graph* of a function of two variables

$$z = f(x, y)$$

consists of ordered triples (x_1, y_1, z_1), where $z_1 = f(x_1, y_1)$. The domain of f is a region D in the xy-plane, and the graph of f is a surface S.

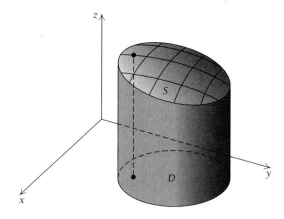

TECHNOLOGY
CONNECTION

There are many 3D graphers capable of generating graphs of functions of two variables. Use one to verify the graphs shown here. What problems do you have with viewing windows?

The following graphs have been generated from their equations by a computer graphics program called *Mathematica*. There are many excellent graphics programs for generating graphs of functions of two variables. The graphs can be generated with very few keystrokes. Seeing one of these graphs often reveals more insight to the behavior of the function than does its formula.

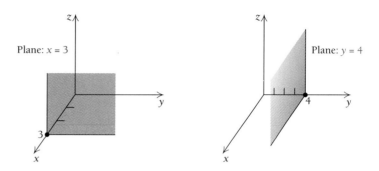

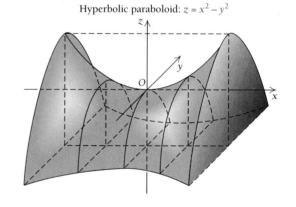

Elliptic paraboloid: $z = x^2 + y^2$

Hyperbolic paraboloid: $z = x^2 - y^2$

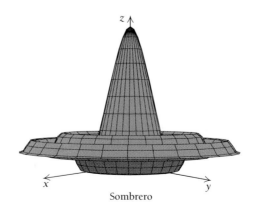

Sombrero

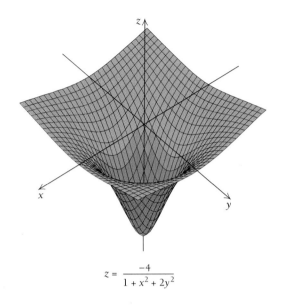

$$z = \frac{-4}{1 + x^2 + 2y^2}$$

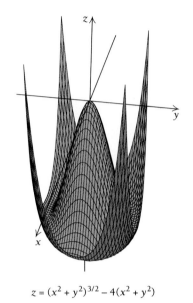

$$z = (x^2 + y^2)^{3/2} - 4(x^2 + y^2)$$

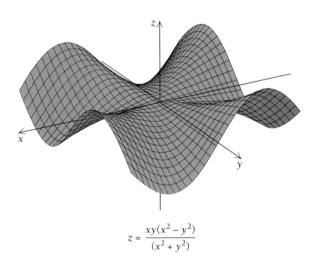

$$z = \frac{xy(x^2 - y^2)}{(x^2 + y^2)}$$

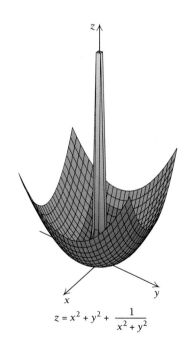

$$z = x^2 + y^2 + \frac{1}{x^2 + y^2}$$

7.1 Exercise Set

1. For $f(x, y) = x^2 - 2xy$, find $f(0, -2)$, $f(2, 3)$, and $f(10, -5)$.

2. For $f(x, y) = (y^2 + 3xy)^3$, find $f(-2, 0)$, $f(3, 2)$, and $f(-5, 10)$.

3. For $f(x, y) = 3^x + 7xy$, find $f(0, -2)$, $f(-2, 1)$, and $f(2, 1)$.

4. For $f(x, y) = \log_{10} x - 5y^2$, find $f(10, 2)$, $f(1, -3)$, and $f(100, 4)$.

5. For $f(x, y) = \ln x + y^3$, find $f(e, 2)$, $f(e^2, 4)$, and $f(e^3, 5)$.

6. For $f(x, y) = 2^x - 3^y$, find $f(0, 0)$, $f(1, 1)$, and $f(2, 2)$.

7. For $f(x, y, z) = x^2 - y^2 + z^2$, find $f(-1, 2, 3)$ and $f(2, -1, 3)$.

8. For $f(x, y, z) = 2^x + 5zy - x$, find $f(0, 1, -3)$ and $f(1, 0, -3)$.

APPLICATIONS

◆   **Business and Economics**

9. *Cost of storage equipment.* Consider the storage model in Example 3. For $100,000, a company buys a storage tank that has a capacity of 80,000 gallons. Later it replaces the tank with a new tank with double the capacity of the original. Estimate the cost of the new tank.

10. *Price–earnings ratio.* The *price–earnings ratio* of a stock is given by

$$R(P, E) = \frac{P}{E},$$

where P = the price per share of the stock and E = the earnings per share. Recently, the price per share of IBM stock was $287\frac{3}{8}$ and the earnings per share were $23.30. Find the price–earnings ratio. Give decimal notation to the nearest tenth.

11. *Yield of a stock.* The *yield* of a stock is given by

$$Y(D, P) = \frac{D}{P},$$

where D = the dividends per share of a stock and P = the price per share. Recently, the price per share of Goodyear stock was $16\frac{7}{8}$ and the dividends per share were $1.30. Find the yield. Give percent notation to the nearest tenth of a percent.

◆   **Life and Physical Sciences**

12. *Poiseuille's Law.* The speed of blood in a vessel is given by

$$V(L, p, R, r, v) = \frac{p}{4Lv}(R^2 - r^2),$$

where R = the radius of the vessel, r = the distance of the blood from the center of the vessel, L = the length of the blood vessel, p = the pressure, and v = the viscosity. Find $V(1, 100, 0.0075, 0.0025, 0.05)$.

13. *Wind speed of a tornado.* Under certain conditions, the *wind speed S*, in miles per hour, of a tornado at a distance d from its center can be approximated by the function

$$S = \frac{aV}{0.51d^2},$$

where a = an atmospheric constant that depends on certain atmospheric conditions and V = the approximate volume of the tornado, in cubic feet. Approximate the wind speed 100 ft from the center of a tornado when its volume is 1,600,000 ft^3 and $a = 0.78$.

◆   **Social Sciences**

14. *Intelligence quotient.* The *intelligence quotient* in psychology is given by

$$Q(m, c) = 100 \cdot \frac{m}{c},$$

where m = a person's mental age and c = his or her chronological, or actual, age. Find $Q(21, 20)$ and $Q(19, 20)$.

SYNTHESIS

15. ◆ Explain the difference between a function of two variables and a function of one variable.

16. ◆ Find some examples of functions of several variables not considered in the text, even some that may not have formulas.

TECHNOLOGY CONNECTION

◆   **General Interest**

Wind chill temperature. Because wind speed enhances the loss of heat from the skin, we feel colder when there is wind than when there is not. The *wind chill temperature* is what the temperature would have to be with no wind in order to give the same chilling effect. The wind chill temperature W is given by

$$W(v, T) = 91.4 - \frac{(10.45 + 6.68\sqrt{v} - 0.447v)(457 - 5T)}{110},$$

where T = the actual temperature as given by a thermometer, in degrees Fahrenheit, and v = the speed of the wind, in miles per hour. Find the wind chill temperature in each case. Round to the nearest one degree.

17. $T = 30°F$, $v = 25$ mph

18. $T = 20°F$, $v = 20$ mph

19. $T = 20°F$, $v = 40$ mph

20. $T = -10°F$, $v = 30$ mph

Use a 3D grapher to generate the graph of the function.

21. $f(x, y) = y^2$

22. $f(x, y) = x^2 + y^2$

23. $f(x, y) = (x^4 - 16x^2)e^{-y^2}$

24. $f(x, y) = 4(x^2 + y^2) - (x^2 + y^2)^2$

25. $f(x, y) = x^3 - 3xy^2$

26. $f(x, y) = \dfrac{1}{x^2 + 4y^2}$

7.2 Partial Derivatives

OBJECTIVES

• Find the partial derivatives of a given function.

• Evaluate partial derivatives.

Finding Partial Derivatives

Consider the function f given by

$$z = f(x, y) = x^2y^3 + xy + 4y^2.$$

Suppose for the moment that we fix y at 3. Then

$$f(x, 3) = x^2(3^3) + x(3) + 4(3^2) = 27x^2 + 3x + 36.$$

Note that we now have a function of only one variable. Taking the first derivative with respect to x, we have

$$54x + 3.$$

Now, without replacing y by a specific number, let us consider y fixed. Then f becomes a function of x alone and we can calculate its derivative with respect to x. This derivative is called the *partial derivative of f with respect to x*. Notation for this partial derivative is

$$\frac{\partial f}{\partial x} \quad \text{or} \quad \frac{\partial z}{\partial x}.$$

Thus let us again consider the function

$$z = f(x, y) = x^2y^3 + xy + 4y^2.$$

The color blue indicates the variable x when we fix y and treat it as a constant. The expressions y^3, y, and y^2 are then constants. We have

$$\frac{\partial f}{\partial x} = \frac{\partial z}{\partial x} = 2xy^3 + y.$$

Similarly, we find $\partial f/\partial y$ or $\partial z/\partial y$ by fixing x (treating it as a constant) and calculating the derivative with respect to y. From

$$z = f(x, y) = x^2y^3 + xy + 4y^2, \qquad \text{The color blue indicates the variable.}$$

we get

$$\frac{\partial f}{\partial y} = \frac{\partial z}{\partial y} = 3x^2y^2 + x + 8y.$$

A definition of partial derivatives is as follows.

DEFINITION

For $z = f(x, y)$,

$$\frac{\partial z}{\partial x} = \lim_{h \to 0} \frac{f(x + h, y) - f(x, y)}{h},$$

$$\frac{\partial z}{\partial y} = \lim_{k \to 0} \frac{f(x, y + k) - f(x, y)}{k}.$$

We can find partial derivatives of functions of any number of variables.

EXAMPLE 1   For $w = x^2 - xy + y^2 + 2yz + 2z^2 + z$, find

$$\frac{\partial w}{\partial x}, \quad \frac{\partial w}{\partial y}, \quad \text{and} \quad \frac{\partial w}{\partial z}.$$

Solution   In order to find $\partial w/\partial x$, we consider x the variable and the other letters the constants. From

$$w = x^2 - xy + y^2 + 2yz + 2z^2 + z,$$

we get

$$\frac{\partial w}{\partial x} = 2x - y.$$

From

$$w = x^2 - xy + y^2 + 2yz + 2z^2 + z,$$

we get

$$\frac{\partial w}{\partial y} = -x + 2y + 2z;$$

and from

$$w = x^2 - xy + y^2 + 2yz + 2z^2 + z,$$

we get

$$\frac{\partial w}{\partial z} = 2y + 4z + 1. \qquad \blacklozenge$$

We will often make use of a simpler notation f_x for the partial derivative of f with respect to x and f_y for the partial derivative of f with respect to y.

EXAMPLE 2   For $f(x, y) = 3x^2y + xy$, find f_x and f_y.

Solution   We have

$$f_x = 6xy + y,$$
$$f_y = 3x^2 + x. \qquad \blacklozenge$$

For the function in the preceding example, let us evaluate f_x at $(2, -3)$:

$$f_x(2, -3) = 6 \cdot 2 \cdot (-3) + (-3) = -39.$$

If we use the notation $\partial f/\partial x = 6xy + y$, where $f = 3x^2y + xy$, the value of the partial derivative at $(2, -3)$ is given by

$$\frac{\partial f}{\partial x}\bigg|_{(2, -3)} = 6 \cdot 2 \cdot (-3) + (-3) = -39,$$

but this notation is not as convenient as $f_x(2, -3)$.

Consider finding values of a partial derivative of $f(x, y) = 3x^3y + 2xy$ using a grapher that finds derivatives of functions of one variable. How can you find $f_x(-4, 1)$? Then how can you find $f_y(2, 6)$?

EXAMPLE 3 For $f(x, y) = e^{xy} + y \ln x$, find f_x and f_y.

Solution

$$f_x = y \cdot e^{xy} + y \cdot \frac{1}{x} = ye^{xy} + \frac{y}{x},$$

$$f_y = x \cdot e^{xy} + 1 \cdot \ln x = xe^{xy} + \ln x \qquad \blacklozenge$$

The Geometric Interpretation of Partial Derivatives

The *graph* of a function of two variables $z = f(x, y)$ is a surface S, as follows.

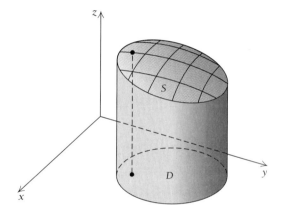

Now suppose we hold x fixed, say, at the value a. The set of all points for which $x = a$ is a plane parallel to the yz-plane, so when x is fixed at a, y and z vary along the plane, as shown below. The plane in the figure cuts the surface in some curve C_1. The partial derivative f_y gives the slope of tangent lines to this curve.

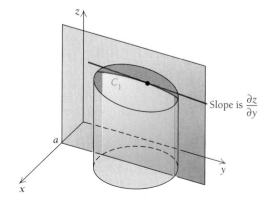

Similarly, if we hold y fixed, say, at the value b, we obtain a curve C_2, as follows. The partial derivative f_x gives the slope of tangent lines to this curve.

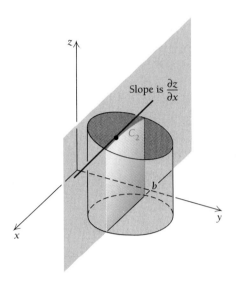

An Economic Application: The Cobb–Douglas Production Function

One model of production that is frequently considered in business and economics is the *Cobb–Douglas production function*:

$$p(x, y) = Ax^a y^{1-a}, \qquad A > 0 \text{ and } 0 < a < 1,$$

where p = the number of units produced with x units of labor and y units of capital. (Capital is the cost of machinery, buildings, tools, and other supplies.) The partial derivatives

$$\frac{\partial p}{\partial x} \quad \text{and} \quad \frac{\partial p}{\partial y}$$

are called, respectively, the *marginal productivity of labor* and the *marginal productivity of capital*.

EXAMPLE 4 A company has the following production function for a certain product:

$$p(x, y) = 50x^{2/3} y^{1/3}.$$

a) Find the production from 125 units of labor and 64 units of capital.

b) Find the marginal productivities.

c) Evaluate the marginal productivities at $x = 125$ and $y = 64$.

Solution

a) $p(125, 64) = 50(125)^{2/3}(64)^{1/3} = 50(25)(4) = 5000$ units

b) $\dfrac{\partial p}{\partial x} = 50\left(\dfrac{2}{3}\right)x^{-1/3}y^{1/3} = \dfrac{100y^{1/3}}{3x^{1/3}}$, $\quad$ or $\quad \dfrac{100}{3}\left(\dfrac{y}{x}\right)^{1/3}$;

$\dfrac{\partial p}{\partial y} = 50\left(\dfrac{1}{3}\right)x^{2/3}y^{-2/3} = \dfrac{50x^{2/3}}{3y^{2/3}}$, $\quad$ or $\quad \dfrac{50}{3}\left(\dfrac{x}{y}\right)^{2/3}$

c) $\left.\dfrac{\partial p}{\partial x}\right|_{(125,\,64)} = \dfrac{100(64)^{1/3}}{3(125)^{1/3}} = \dfrac{100(4)}{3(5)} = 26\dfrac{2}{3}$;

$\left.\dfrac{\partial p}{\partial y}\right|_{(125,\,64)} = \dfrac{50(125)^{2/3}}{3(64)^{2/3}} = \dfrac{50(25)}{3(16)} = 26\dfrac{1}{24}$ ◆

How can we interpret these marginal productivities? Suppose the amount spent on capital is fixed at 64. Then if the amount of labor changes by one unit from 125, production will change by about 27 units. Suppose the amount of labor is held fixed at 125. Then if the amount of capital spent changes by one unit from 64, production will change by about 26 units.

A Cobb–Douglas production function is consistent with the law of diminishing returns. That is, if one input (of either labor or capital) is held fixed while the other increases infinitely, then production will eventually increase at a decreasing rate. With such functions, it also turns out that if a certain maximum production is possible, then the expense of more labor, for example, will not prevent that maximum output from still being attainable.

7.2 Exercise Set

Find $\dfrac{\partial z}{\partial x}$, $\dfrac{\partial z}{\partial y}$, $\left.\dfrac{\partial z}{\partial x}\right|_{(-2,\,-3)}$ and $\left.\dfrac{\partial z}{\partial y}\right|_{(0,\,-5)}$

1. $z = 2x - 3xy$ $\qquad\qquad$ **2.** $z = (x - y)^3$

3. $z = 3x^2 - 2xy + y$ $\qquad$ **4.** $z = 2x^3 + 3xy - x$

Find f_x, f_y, $f_x(-2, 4)$, and $f_y(4, -3)$.

5. $f(x, y) = 2x - 3y$ $\qquad\qquad$ **6.** $f(x, y) = 5x + 7y$

Find f_x, f_y, $f_x(-2, 1)$, and $f_y(-3, -2)$.

7. $f(x, y) = \sqrt{x^2 + y^2}$ $\qquad$ **8.** $f(x, y) = \sqrt{x^2 - y^2}$

Find f_x and f_y.

9. $f(x, y) = e^{2x+3y}$ $\qquad\qquad$ **10.** $f(x, y) = e^{3x-2y}$

11. $f(x, y) = e^{xy}$ $\qquad\qquad$ **12.** $f(x, y) = e^{2xy}$

13. $f(x, y) = y \ln (x + y)$ $\qquad$ **14.** $f(x, y) = x \ln (x + y)$

15. $f(x, y) = x \ln (xy)$ $\qquad\quad$ **16.** $f(x, y) = y \ln (xy)$

17. $f(x, y) = \dfrac{x}{y} - \dfrac{y}{x}$ $\qquad\quad$ **18.** $f(x, y) = \dfrac{x}{y} + \dfrac{y}{x}$

19. $f(x, y) = 3(2x + y - 5)^2$

20. $f(x, y) = 4(3x + y - 8)^2$

Find $\dfrac{\partial f}{\partial b}$ and $\dfrac{\partial f}{\partial m}$.

21. $f(b, m) = (m + b - 4)^2 + (2m + b - 5)^2 +$ $(3m + b - 6)^2$

22. $f(b, m) = (m + b - 6)^2 + (2m + b - 8)^2 + (3m + b - 9)^2$

Find f_x, f_y, and f_λ.

23. $f(x, y, \lambda) = 3xy - \lambda(2x + y - 8)$

24. $f(x, y, \lambda) = 4xy - \lambda(3x - y + 7)$

25. $f(x, y, \lambda) = x^2 + y^2 - \lambda(10x + 2y - 4)$

26. $f(x, y, \lambda) = x^2 - y^2 - \lambda(4x - 7y - 10)$

APPLICATIONS

◆ **Business and Economics**

27. *The Cobb–Douglas model.* A company has the following production function for a certain product:

$$p(x, y) = 1800x^{0.621}y^{0.379},$$

where p = the number of units produced with x units of labor and y units of capital.

a) Find the production from 2500 units of labor and 1700 units of capital.

b) Find the marginal productivities.

c) Evaluate the marginal productivities at $x = 2500$ and $y = 1700$.

28. *The Cobb–Douglas model.* A company has the following production function for a certain product:

$$p(x, y) = 2400x^{2/5}y^{3/5},$$

where p = the number of units produced with x units of labor and y units of capital.

a) Find the production from 32 units of labor and 1024 units of capital.

b) Find the marginal productivities.

c) Evaluate the marginal productivities at $x = 32$ and $y = 1024$.

◆ **Life and Physical Sciences**

Temperature–humidity heat index. In the summer, humidity affects the actual temperature, making a person feel hotter due to a reduced heat loss from the skin caused by higher humidity. The *temperature–humidity index, T_h,* is what the temperature would have to be with no humidity in order to give the same heat effect. One index often used is given by

$$T_h = 1.98T - 1.09(1 - H)(T - 58) - 56.9,$$

where T = the air temperature, in degrees Fahrenheit, and H = the relative humidity, which is the ratio of the amount of water vapor in the air to the maximum amount of water vapor possible in the air at that temperature. H is usually expressed as a percentage. Find the temperature–humidity index in each case. Round to the nearest tenth of a degree.

29. $T = 85°$ and $H = 60\%$

30. $T = 90°$ and $H = 90\%$

31. $T = 90°$ and $H = 100\%$

32. $T = 78°$ and $H = 100\%$

33. Find $\dfrac{\partial T_h}{\partial H}$.

34. Find $\dfrac{\partial T_h}{\partial T}$.

◆ **Social Sciences**

Reading ease. The following formula is used by psychologists and educators to predict the *reading ease E* of a passage of words:

$$E = 206.835 - 0.846w - 1.015s,$$

where w = the number of syllables in a 100-word section and s = the average number of words per sentence. Find the reading ease in each case.

35. $w = 146$ and $s = 5$

36. $w = 180$ and $s = 6$

37. Find $\dfrac{\partial E}{\partial w}$.

38. Find $\dfrac{\partial E}{\partial s}$.

41. $f(x, t) = \dfrac{2\sqrt{x} - 2\sqrt{t}}{1 + 2\sqrt{t}}$ **42.** $f(x, t) = \sqrt[4]{x^3 t^5}$

43. $f(x, t) = 6x^{2/3} - 8x^{1/4}t^{1/2} - 12x^{-1/2}t^{3/2}$

44. $f(x, t) = \left(\dfrac{x^2 + t^2}{x^2 - t^2}\right)^5$

45. ◈ Do some library research on the Cobb–Douglas production function. Can you find how it was developed?

46. ◈ Explain the meaning of the first partial derivatives of a function of two variables in terms of slopes of tangent lines.

SYNTHESIS

Find f_x and f_t.

39. $f(x, t) = \dfrac{x^2 + t^2}{x^2 - t^2}$ **40.** $f(x, t) = \dfrac{x^2 - t}{x^3 + t}$

7.3 Higher-Order Partial Derivatives

OBJECTIVE

- Find the four second-order partial derivatives of a function.

Consider

$$z = f(x, y) = 3xy^2 + 2xy + x^2.$$

Then

$$\frac{\partial z}{\partial x} = \frac{\partial f}{\partial x} = 3y^2 + 2y + 2x.$$

Suppose we continue and find the first partial derivative of $\partial z/\partial x$ with respect to y. This will be a **second-order partial derivative** of the original function z. Its notation is as follows:

$$\frac{\partial}{\partial y}\left(\frac{\partial z}{\partial x}\right) = \frac{\partial}{\partial y}\left(\frac{\partial f}{\partial x}\right)$$

$$= \frac{\partial^2 z}{\partial y\, \partial x}$$

$$= \frac{\partial^2 f}{\partial y\, \partial x}$$

$$= 6y + 2.$$

We could also denote the preceding partial derivative using the notation f_{xy}. Then

$$f_{xy} = 6y + 2.$$

Note that in the notation f_{xy}, x and y are in the order (left to right) in which the differentiation is done, but in

$$\frac{\partial^2 f}{\partial y\, \partial x},$$

the order of x and y is reversed. In both notations, the differentiation with respect to x is done first, followed by differentiation with respect to y.

Notation for the four second-order partial derivatives is as follows.

DEFINITION

1. $\dfrac{\partial^2 z}{\partial x\, \partial x} = \dfrac{\partial^2 f}{\partial x\, \partial x} = \dfrac{\partial^2 z}{\partial x^2} = \dfrac{\partial^2 f}{\partial x^2} = f_{xx}$

Take the partial with respect to x, and then with respect to x again.

2. $\dfrac{\partial^2 z}{\partial y\, \partial x} = \dfrac{\partial^2 f}{\partial y\, \partial x} = f_{xy}$

Take the partial with respect to x, and then with respect to y.

3. $\dfrac{\partial^2 z}{\partial x\, \partial y} = \dfrac{\partial^2 f}{\partial x\, \partial y} = f_{yx}$

Take the partial with respect to y, and then with respect to x.

4. $\dfrac{\partial^2 z}{\partial y\, \partial y} = \dfrac{\partial^2 f}{\partial y\, \partial y} = \dfrac{\partial^2 z}{\partial y^2} = \dfrac{\partial^2 f}{\partial y^2} = f_{yy}$

Take the partial with respect to y, and then with respect to y again.

EXAMPLE 1 For

$$z = f(x, y) = x^2 y^3 + x^4 y + xe^y,$$

find the four second-order partial derivatives.

Solution

a) $\dfrac{\partial^2 f}{\partial x^2} = f_{xx} = \dfrac{\partial}{\partial x}(2xy^3 + 4x^3 y + e^y)$

           $= 2y^3 + 12x^2 y$

Differentiate twice with respect to x.

b) $\dfrac{\partial^2 f}{\partial y\, \partial x} = f_{xy} = \dfrac{\partial}{\partial y}(2xy^3 + 4x^3 y + e^y)$

           $= 6xy^2 + 4x^3 + e^y$

Differentiate with respect to x, and then with respect to y.

c) $\dfrac{\partial^2 f}{\partial x\, \partial y} = f_{yx} = \dfrac{\partial}{\partial x}(3x^2 y^2 + x^4 + xe^y)$

           $= 6xy^2 + 4x^3 + e^y$

Differentiate with respect to y, and then with respect to x.

d) $\dfrac{\partial^2 f}{\partial y^2} = f_{yy} = \dfrac{\partial}{\partial y}(3x^2 y^2 + x^4 + xe^y)$ Differentiate twice with respect to y.

$\qquad = 6x^2 y + xe^y$

We see by comparing parts (b) and (c) in Example 1 that

$$\frac{\partial^2 f}{\partial y\,\partial x} = \frac{\partial^2 f}{\partial x\,\partial y} \quad \text{and} \quad f_{xy} = f_{yx}.$$

Although this will be true for virtually all functions that we consider in this text, it is *not* true for all functions. Such a function is given in Exercise 19 of Exercise Set 7.3.

7.3 Exercise Set

Find the four second-order partial derivatives.

1. $f(x, y) = 3x^2 - xy + y$
2. $f(x, y) = 5x^2 + xy - x$
3. $f(x, y) = 3xy$
4. $f(x, y) = 4xy$
5. $f(x, y) = x^5 y^4 + x^3 y^2$
6. $f(x, y) = x^4 y^3 - x^2 y^3$

Find $f_{xx}, f_{xy}, f_{yx},$ and f_{yy}. (Remember, f_{yx} means to differentiate with respect to y, and then to x.)

7. $f(x, y) = 2x - 3y$
8. $f(x, y) = 3x + 5y$
9. $f(x, y) = e^{2xy}$
10. $f(x, y) = e^{xy}$
11. $f(x, y) = x + e^y$
12. $f(x, y) = y - e^x$
13. $f(x, y) = y \ln x$
14. $f(x, y) = x \ln y$

SYNTHESIS

Find $f_{xx}, f_{xy}, f_{yx},$ and f_{yy}.

15. $f(x, y) = \dfrac{x}{y^2} - \dfrac{y}{x^2}$

16. $f(x, y) = \dfrac{xy}{x - y}$

17. Consider $f(x, y) = \ln(x^2 + y^2)$. Show that f is a solution to the partial differential equation

$$\frac{\partial^2 f}{\partial x^2} + \frac{\partial^2 f}{\partial y^2} = 0.$$

18. Consider $f(x, y) = x^3 - 5xy^2$. Show that f is a solution to the partial differential equation

$$xf_{xy} - f_y = 0.$$

19. Consider the function f defined as follows:

$$f(x, y) = \begin{cases} \dfrac{xy(x^2 - y^2)}{x^2 + y^2}, & \text{for } (x, y) \neq (0, 0), \\ 0, & \text{for } (x, y) = (0, 0). \end{cases}$$

a) Find $f_x(0, y)$ by evaluating the limit

$$\lim_{h \to 0} \frac{f(h, y) - f(0, y)}{h}.$$

b) Find $f_y(x, 0)$ by evaluating the limit

$$\lim_{h \to 0} \frac{f(x, h) - f(x, 0)}{h}.$$

c) Now find and compare $f_{yx}(0, 0)$ and $f_{xy}(0, 0)$.

20. For $f = [\ln(x^3 + e^y)]^5$, find $f_{xx}, f_{xy}, f_{yx},$ and f_{yy}.

7.4 Maximum–Minimum Problems

We will now find maximum and minimum values of functions of two variables.

DEFINITION

A function f of two variables:

1. has a *relative maximum* at (a, b) if

$$f(x, y) \leq f(a, b)$$

   for all points in a rectangular region containing (a, b);

2. has a *relative minimum* at (a, b) if

$$f(x, y) \geq f(a, b)$$

   for all points in a rectangular region containing (a, b).

This definition is illustrated in Figs. 1 and 2. A relative maximum (minimum) may not be an "absolute" maximum (minimum), as illustrated in Fig. 3.

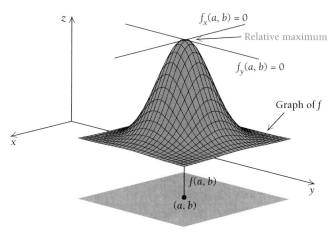

FIGURE 1

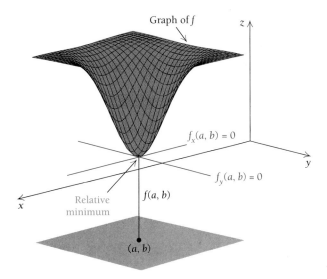

$f_x(a, b) = 0$

$f_y(a, b) = 0$

$f(a, b)$

Relative
minimum

(a, b)

Graph of f

FIGURE 2

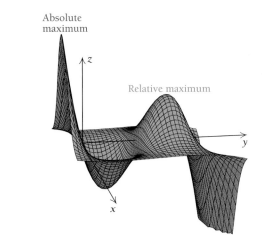

Absolute
maximum

Relative maximum

FIGURE 3

Determining Maximum and Minimum Values

Suppose a function f assumes a relative maximum (or minimum) value at some point (a, b) inside its domain. We assume that f and its partial derivatives exist and are "continuous" inside its domain, though we will not take the space to define continuity. If we hold y fixed at the value b, then $f(x, b)$ is a function of one variable x

Where is the saddle point?

having its relative maximum value at $x = a$. Thus its derivative must be 0 there—that is, $f_x = 0$ at the point (a, b). Similarly, $f_y = 0$ at (a, b). The equations

$$f_x = 0 \quad \text{and} \quad f_y = 0 \tag{1}$$

are thus satisfied by the point (a, b) at which the relative maximum occurs. We call a point (a, b) at which both partial derivatives are 0 a *critical point*. This is comparable to the earlier definition for functions of one variable. Thus one strategy for finding relative maximum or minimum values is to solve the system of equations (1) above to find critical points. Just as for functions of one variable, this strategy does *not* guarantee that we will have a relative maximum or minimum value. We have argued only that *if* f has a maximum or minimum value at (a, b), *then* both its partial derivatives must be 0 at that point. Look at Figs. 1 and 2. Then note Fig. 4, which illustrates a case in which the partial derivatives are 0 but the function does not have a relative maximum or minimum value at (a, b).

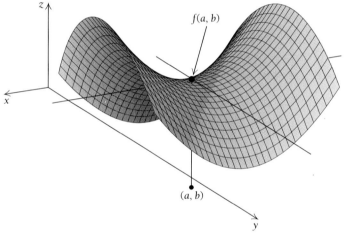

FIGURE 4

In Fig. 4, suppose we fix y at a value b. Then $f(x, b)$, considered as a function of one variable, has a minimum at a, but f does not. Similarly, if we fix x at a, then $f(a, y)$, considered as a function of one variable, has a maximum at b, but f does not. The point $f(a, b)$ is called a **saddle point.** In other words, $f_x(a, b) = 0$ and $f_y(a, b) = 0$ [the point (a, b) is a critical point], but f does not attain a relative maximum or minimum value at (a, b). A saddle point for a function of two variables is comparable to a point of inflection (which is simultaneously a critical point) for a function of one variable.

A test for finding relative maximum and minimum values that involves the use of first- and second-order partial derivatives is stated below. We will not prove this theorem.

> ## THEOREM 1
>
> The *D*-test
>
> To find the relative maximum and minimum values of f:
>
> 1. Find f_x, f_y, f_{xx}, f_{yy}, and f_{xy}.
> 2. Solve the system of equations $f_x = 0$, $f_y = 0$. Let (a, b) represent a solution.
> 3. Evaluate D, where $D = f_{xx}(a, b) \cdot f_{yy}(a, b) - [f_{xy}(a, b)]^2$.
> 4. Then:
>
>   a) f has a maximum at (a, b) if $D > 0$ and $f_{xx}(a, b) < 0$.
>
>   b) f has a minimum at (a, b) if $D > 0$ and $f_{xx}(a, b) > 0$.
>
>   c) f has neither a maximum nor a minimum at (a, b) if $D < 0$. The function has a *saddle point* at (a, b). See Fig. 4.
>
>   d) This test is not applicable if $D = 0$.

A relative maximum or minimum *may not be an absolute maximum or minimum value*. Tests for absolute maximum or minimum values are rather complicated. We will restrict our attention to finding *relative* maximum or minimum values. Fortunately, in most of our applications, relative maximum or minimum values turn out to be absolute as well.

The shape of a perfect tent. To give a tent roof the maximum strength possible, designers draw the fabric into a series of three-dimensional shapes that, viewed in profile, resemble a horse's saddle and that mathematicians call an anticlastic curve. Two people with a stretchy piece of fabric such as Lycra Spandex can duplicate the shape, as shown above. One person pulls up and out on two diagonal corners; the other person pulls down and out on the other two corners. The opposing tensions draw each point of the fabric's surface into rigid equilibrium. The more pronounced the curve, the stiffer the surface.

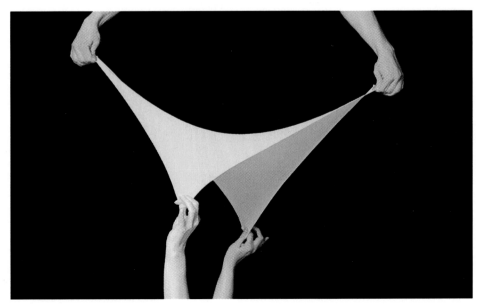

EXAMPLE 1 Find the relative maximum and minimum values of

$$f(x, y) = x^2 + xy + y^2 - 3x.$$

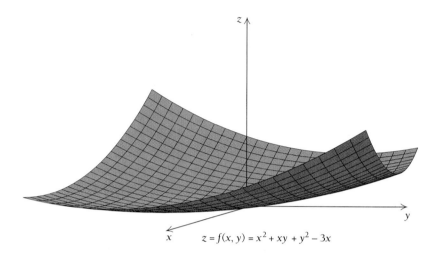

$z = f(x, y) = x^2 + xy + y^2 - 3x$

Solution

1. Find $f_x, f_y, f_{xx}, f_{yy},$ and f_{xy}:

$$f_x = 2x + y - 3, \qquad f_y = x + 2y,$$
$$f_{xx} = 2; \qquad\qquad f_{yy} = 2;$$
$$f_{xy} = 1.$$

2. Solve the system of equations $f_x = 0, f_y = 0$:

$$2x + y - 3 = 0, \tag{1}$$
$$x + 2y = 0. \tag{2}$$

Solving Eq. (2) for x, we get $x = -2y$. Substituting $-2y$ for x in Eq. (1) and solving, we get

$$2(-2y) + y - 3 = 0$$
$$-4y + y - 3 = 0$$
$$-3y = 3$$
$$y = -1.$$

To find x when $y = -1$, we substitute -1 for y in either Eq. (1) or Eq. (2). We choose Eq. (2):

$$x + 2(-1) = 0$$
$$x = 2.$$

Thus, $f(2, -1)$ is our candidate for a maximum or minimum value.

3. We must check to see whether $f(2, -1)$ is a maximum or minimum value:

$$D = f_{xx}(2, -1) \cdot f_{yy}(2, -1) - [f_{xy}(2, -1)]^2$$
$$= 2 \cdot 2 - [1]^2$$
$$= 3.$$

4. Thus, $D = 3$ and $f_{xx}(2, -1) = 2$. Since $D > 0$ and $f_{xx}(2, -1) > 0$, it follows that f has a relative minimum at $(2, -1)$ and that the minimum value is found as follows:

$$f(2, -1) = 2^2 + 2(-1) + (-1)^2 - 3 \cdot 2$$
$$= 4 - 2 + 1 - 6$$
$$= -3. \qquad \blacklozenge$$

EXAMPLE 2   Find the relative maximum and minimum values of

$$f(x, y) = xy - x^3 - y^2.$$

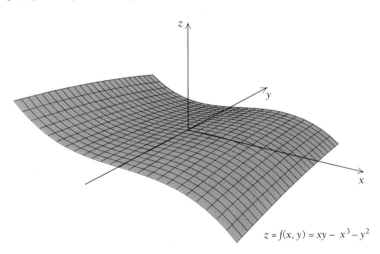

$$z = f(x, y) = xy - x^3 - y^2$$

Solution

1. Find f_x, f_y, f_{xx}, f_{yy}, and f_{xy}:

$$f_x = y - 3x^2, \qquad f_y = x - 2y,$$
$$f_{xx} = -6x; \qquad\quad f_{yy} = -2;$$
$$f_{xy} = 1.$$

2. Solve the system of equations $f_x = 0$, $f_y = 0$:

$$y - 3x^2 = 0, \qquad\qquad\qquad\qquad (1)$$
$$x - 2y = 0. \qquad\qquad\qquad\qquad (2)$$

Solving Eq. (1) for y, we get $y = 3x^2$. Substituting $3x^2$ for y in Eq. (2) and solving, we get

$$x - 2(3x^2) = 0$$
$$x - 6x^2 = 0$$
$$x(1 - 6x) = 0. \qquad \textbf{Factoring}$$

Setting each factor equal to 0 and solving, we have

$$x = 0 \quad \text{or} \quad 1 - 6x = 0$$
$$x = 0 \quad \text{or} \qquad x = \tfrac{1}{6}.$$

To find y when $x = 0$, we substitute 0 for x in either Eq. (1) or Eq. (2). We choose Eq. (2):

$$0 - 2y = 0$$
$$-2y = 0$$
$$y = 0.$$

Thus, $f(0, 0)$ is one candidate for a maximum or minimum value. To find the other, we substitute $\tfrac{1}{6}$ for x in either Eq. (1) or Eq. (2). We choose Eq. (2):

$$\tfrac{1}{6} - 2y = 0$$
$$-2y = -\tfrac{1}{6}$$
$$y = \tfrac{1}{12}.$$

Thus, $f\left(\tfrac{1}{6}, \tfrac{1}{12}\right)$ is another candidate for a maximum or minimum value.

3. We must check both $(0, 0)$ and $\left(\tfrac{1}{6}, \tfrac{1}{12}\right)$ to see whether they yield maximum or minimum values:

$$\text{For } (0, 0): \quad D = f_{xx}(0, 0) \cdot f_{yy}(0, 0) - [\, f_{xy}(0, 0)]^2$$
$$= (-6 \cdot 0) \cdot (-2) - [1]^2$$
$$= -1.$$

Since $D < 0$, it follows that $f(0, 0)$ is neither a maximum nor a minimum value, but a saddle point.

$$\text{For } \left(\tfrac{1}{6}, \tfrac{1}{12}\right): \quad D = f_{xx}\left(\tfrac{1}{6}, \tfrac{1}{12}\right) \cdot f_{yy}\left(\tfrac{1}{6}, \tfrac{1}{12}\right) - \left[f_{xy}\left(\tfrac{1}{6}, \tfrac{1}{12}\right)\right]^2$$
$$= \left(-6 \cdot \tfrac{1}{6}\right) \cdot (-2) - [1]^2$$
$$= -1(-2) - 1$$
$$= 1.$$

4. Thus, $D = 1$ and $f_{xx}\left(\tfrac{1}{6}, \tfrac{1}{12}\right) = -1$. Since $D > 0$ and $f_{xx}\left(\tfrac{1}{6}, \tfrac{1}{12}\right) < 0$, it follows that f has a relative maximum at $\left(\tfrac{1}{6}, \tfrac{1}{12}\right)$ and that maximum value is found as follows:

$$f\left(\tfrac{1}{6}, \tfrac{1}{12}\right) = \tfrac{1}{6} \cdot \tfrac{1}{12} - \left(\tfrac{1}{6}\right)^3 - \left(\tfrac{1}{12}\right)^2$$
$$= \tfrac{1}{72} - \tfrac{1}{216} - \tfrac{1}{144} = \tfrac{1}{432}. \qquad \blacklozenge$$

EXAMPLE 3   *Business: Maximizing profit.* A firm produces two kinds of golf ball, one that sells for \$3 each and the other for \$2 each. The total revenue, in thousands of dollars, from the sale of x thousand balls at \$3 each and y thousand at \$2 each is given by

$$R(x, y) = 3x + 2y.$$

The company determines that the total cost, in thousands of dollars, of producing x thousand of the \$3 ball and y thousand of the \$2 ball is given by

$$C(x, y) = 2x^2 - 2xy + y^2 - 9x + 6y + 7.$$

Find the amount of each type of ball that must be produced and sold in order to maximize profit.

Solution   The total profit $P(x, y)$ is given by

$$
\begin{aligned}
P(x, y) &= R(x, y) - C(x, y) \\
&= 3x + 2y - (2x^2 - 2xy + y^2 - 9x + 6y + 7) \\
P(x, y) &= -2x^2 + 2xy - y^2 + 12x - 4y - 7.
\end{aligned}
$$

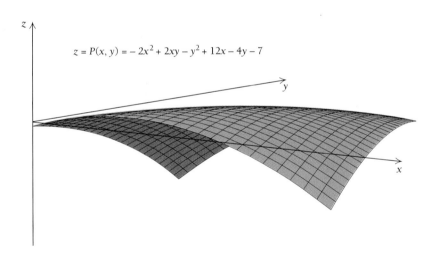

$$z = P(x, y) = -2x^2 + 2xy - y^2 + 12x - 4y - 7$$

1. Find P_x, P_y, P_{xx}, P_{yy}, and P_{xy}:

$$P_x = -4x + 2y + 12, \qquad P_y = 2x - 2y - 4,$$
$$P_{xx} = -4; \qquad\qquad\qquad P_{yy} = -2;$$
$$P_{xy} = 2.$$

2. Solve the system of equations $P_x = 0$, $P_y = 0$:

$$-4x + 2y + 12 = 0, \tag{1}$$
$$2x - 2y - 4 = 0. \tag{2}$$

Adding these equations, we get

$$-2x + 8 = 0.$$

Then

$$-2x = -8$$
$$x = 4.$$

To find y when $x = 4$, we substitute 4 for x in either Eq. (1) or Eq. (2). We choose Eq. (2):

$$2 \cdot 4 - 2y - 4 = 0$$
$$-2y + 4 = 0$$
$$-2y = -4$$
$$y = 2.$$

Thus, $P(4, 2)$ is a candidate for a maximum or minimum value.

3. We must check to see whether $P(4, 2)$ is a maximum or minimum value:

$$D = P_{xx}(4, 2) \cdot P_{yy}(4, 2) - [P_{xy}(4, 2)]^2$$
$$= (-4)(-2) - 2^2$$
$$= 4.$$

4. Thus, $D = 4$ and $P_{xx}(4, 2) = -4$. Since $D > 0$ and $P_{xx}(4, 2) < 0$, it follows that P has a relative maximum at $(4, 2)$. Thus in order to maximize profit, the company must produce and sell 4 thousand of the \$3 golf balls and 2 thousand of the \$2 golf balls. ◆

7.4 Exercise Set

Find the relative maximum and minimum values.

1. $f(x, y) = x^2 + xy + y^2 - y$

2. $f(x, y) = x^2 + xy + y^2 - 5y$

3. $f(x, y) = 2xy - x^3 - y^2$

4. $f(x, y) = 4xy - x^3 - y^2$

5. $f(x, y) = x^3 + y^3 - 3xy$

6. $f(x, y) = x^3 + y^3 - 6xy$

7. $f(x, y) = x^2 + y^2 - 2x + 4y - 2$

8. $f(x, y) = x^2 + 2xy + 2y^2 - 6y + 2$

9. $f(x, y) = x^2 + y^2 + 2x - 4y$

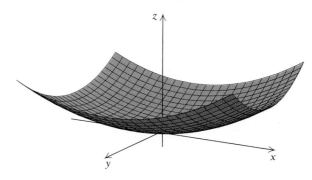

10. $f(x, y) = 4y + 6x - x^2 - y^2$

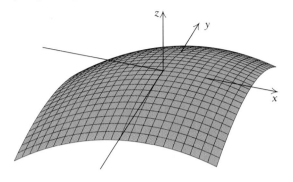

11. $f(x, y) = 4x^2 - y^2$

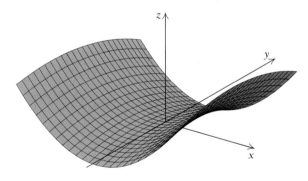

12. $f(x, y) = x^2 - y^2$

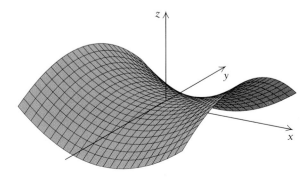

APPLICATIONS _____

◆ **Business and Economics**

In these problems, assume that relative maximum and minimum values are absolute maximum and minimum values.

13. *Maximizing profit.* A firm produces two kinds of radio, one that sells for $17 each and the other for $21 each. The total revenue from the sale of x thousand radios at $17 each and y thousand at $21 each is given by

$$R(x, y) = 17x + 21y.$$

The company determines that the total cost, in thousands of dollars, of producing x thousand of the $17 radio and y thousand of the $21 radio is given by

$$C(x, y) = 4x^2 - 4xy + 2y^2 - 11x + 25y - 3.$$

Find the amount of each type of radio that must be produced and sold in order to maximize profit.

14. *Maximizing profit.* A firm produces two kinds of baseball glove, one that sells for $18 each and the other for $25 each. The total revenue from the sale of x thousand gloves at $18 each and y thousand at $25 each is given by

$$R(x, y) = 18x + 25y.$$

The company determines that the total cost, in thousands of dollars, of producing x thousand of the $18 glove and y thousand of the $25 glove is given by

$$C(x, y) = 4x^2 - 6xy + 3y^2 + 20x + 19y - 12.$$

Find the amount of each type of glove that must be produced and sold in order to maximize profit.

15. *Maximizing profit.* A one-product company finds that its profit, in millions of dollars, is a function P given by

$$P(a, p) = 2ap + 80p - 15p^2 - \tfrac{1}{10}a^2p - 100,$$

where a = the amount spent on advertising, in millions of dollars, and p = the price charged per item of the product, in dollars. Find the maximum value of P and the values of a and p at which it is attained.

16. *Maximizing profit.* A one-product company finds that its profit, in millions of dollars, is a function P given

by

$$P(a, n) = -5a^2 - 3n^2 + 48a - 4n + 2an + 300,$$

where $a = $ the amount spent on advertising, in millions of dollars, and $n = $ the number of items sold, in thousands. Find the maximum value of P and the values of a and n at which it is attained.

17. Minimizing the cost of a container. A trash company is designing an open-top, rectangular container that will have a volume of 320 ft³. The cost of making the bottom of the container is $5 per square foot, and the cost of the sides is $4 per square foot. Find the dimensions of the container that will minimize total cost. (*Hint:* Make a substitution using the formula for volume.)

18. Two-variable revenue maximization. Boxowitz, Inc., a computer firm, markets two kinds of electronic calculator that compete with one another. Their demand functions are expressed by the following relationships:

$$q_1 = 78 - 6p_1 - 3p_2. \tag{1}$$
$$q_2 = 66 - 3p_1 - 6p_2, \tag{2}$$

where p_1 and $p_2 = $ the price of each calculator, in multiples of $10, and q_1 and $q_2 = $ the quantity of each calculator demanded, in hundreds of units.

a) Find a formula for the total-revenue function R in terms of the variables p_1 and p_2. [*Hint:* $R = p_1q_1 + p_2q_2$; then substitute expressions from Eqs. (1) and (2) to find $R(p_1, p_2)$.]
b) What prices p_1 and p_2 should be charged for each product in order to maximize total revenue?
c) How many units will be demanded?
d) What is the maximum total revenue?

19. Two-variable revenue maximization. Repeat Exercise 18, where

$$q_1 = 64 - 4p_1 - 2p_2, \quad \text{and}$$
$$q_2 = 56 - 2p_1 - 4p_2.$$

20. Temperature. A flat metal plate is located on a coordinate plane. The temperature of the plate, in degrees Fahrenheit, at point (x, y) is given by

$$T(x, y) = x^2 + 2y^2 - 8x + 4y.$$

Find the minimum temperature and where it occurs. Is there a maximum temperature?

SYNTHESIS

Find the relative maximum and minimum values and the saddle points.

21. $f(x, y) = e^x + e^y - e^{x+y}$

22. $f(x, y) = xy + \dfrac{2}{x} + \dfrac{4}{y}$

23. $f(x, y) = 2y^2 + x^2 - x^2y$

24. $S(b, m) = (m + b - 72)^2 + (2m + b - 73)^2 + (3m + b - 75)^2$

25. ◆ Describe the D-test and how it is used.

26. ◆ Explain the difference between a relative minimum and an absolute minimum of a function of two variables.

 TECHNOLOGY CONNECTION

Use a 3D grapher to graph each of the following functions. Then estimate any relative extrema.

27. $f(x, y) = \dfrac{-5}{x^2 + 2y^2 + 1}$

28. $f(x, y) = x^3 + y^3 + 3xy$

29. $f(x, y) = \dfrac{3xy(x^2 - y^2)}{(x^2 + y^2)}$

30. $f(x, y) = \dfrac{y + x^2y^2 - 8x}{xy}$

7.5 An Application: The Least-Squares Technique

OBJECTIVES

- Find a regression line.
- Solve application problems involving regression lines.

The problem of fitting an equation to a set of data occurs frequently. We considered one procedure for doing this in Section 1.6. Such an equation provides a model of the phenomena from which predictions can be made. For example, in business, one might want to predict future sales on the basis of past data. In ecology, one might want to predict future demands for natural gas on the basis of past need. Suppose we are trying to find a linear equation

$$y = mx + b$$

to fit the data. To determine this equation is to determine the values of m and b. But how? Let us consider some factual data.

The graph shown in Fig. 1 appeared in a newspaper advertisement for the Indianapolis Life Insurance Company. It pertains to the total amount of life insurance in force in various years. The same data are compiled in the table following the graph.

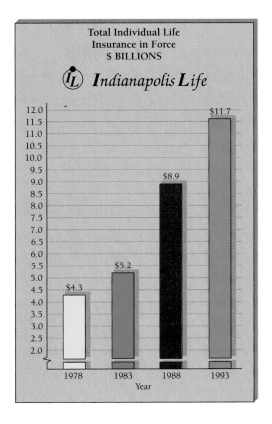

FIGURE 1

Year, x	1. 1978	2. 1983	3. 1988	4. 1993	5. 1998
Total amount of individual life insurance in force (in billions), y	$4.3	$5.2	$8.9	$11.7	?

Suppose we plot these points and try to draw a line through them that fits. Note that there are several ways in which this might be done (see Figs. 2 and 3). Each would give a different estimate of the total amount of insurance in force in 1998.

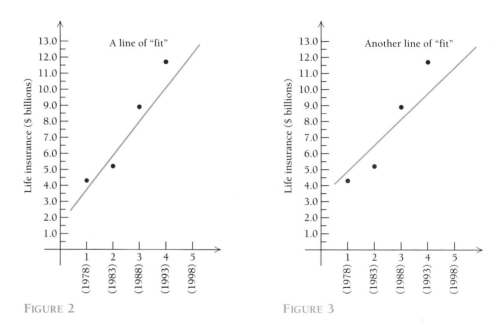

FIGURE 2   FIGURE 3

Note that the time is given in increments of five years, making computations easier. Consider the data points (1, 4.3), (2, 5.2), (3, 8.9), and (4, 11.7), as plotted in Fig. 4.

We will try to fit these data with a line

$$y = mx + b$$

by determining the values of m and b. Note the y-errors, or y-deviations, $y_1 - 4.3$, $y_2 - 5.2$, $y_3 - 8.9$, and $y_4 - 11.7$ between the observed points (1, 4.3), (2, 5.2), (3, 8.9), and (4, 11.7) and the points (1, y_1), (2, y_2), (3, y_3), and (4, y_4) on the line. We would like, somehow, to minimize these deviations in order to have a good fit. One way of minimizing the deviations is based on the *least-squares assumption*.

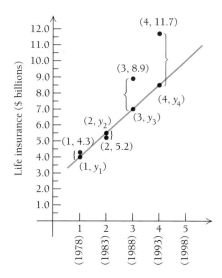

FIGURE 4

The Least-Squares Assumption

The line of best fit is the line for which the sum of the squares of the
y-deviations is a minimum. This is called the *regression line*.

Using the least-squares assumption for the life insurance data, we would
minimize

$$(y_1 - 4.3)^2 + (y_2 - 5.2)^2 + (y_3 - 8.9)^2 + (y_4 - 11.7)^2, \tag{1}$$

and since the points $(1, y_1)$, $(2, y_2)$, $(3, y_3)$, and $(4, y_4)$ must be solutions of
$y = mx + b$, it follows that

$$y_1 = m(1) + b = m + b,$$
$$y_2 = m(2) + b = 2m + b,$$
$$y_3 = m(3) + b = 3m + b,$$
$$y_4 = m(4) + b = 4m + b.$$

Substituting $m + b$ for y_1, $2m + b$ for y_2, $3m + b$ for y_3, and $4m + b$ for y_4 in
Eq. (1), we have

$$(m + b - 4.3)^2 + (2m + b - 5.2)^2 + (3m + b - 8.9)^2$$
$$+ (4m + b - 11.7)^2. \tag{2}$$

Thus, to find the regression line for the given set of data, we must find the values of
m and b that minimize the function S given by the sum in Eq. (2).

To apply the D-test, we first find the partial derivatives $\partial S/\partial b$ and $\partial S/\partial m$:

$$\frac{\partial S}{\partial b} = 2(m + b - 4.3) + 2(2m + b - 5.2) + 2(3m + b - 8.9)$$
$$+ 2(4m + b - 11.7)$$
$$= 20m + 8b - 60.2;$$

$$\frac{\partial S}{\partial m} = 2(m + b - 4.3) + 2(2m + b - 5.2)2 + 2(3m + b - 8.9)3$$
$$+ 2(4m + b - 11.7)4$$
$$= 60m + 20b - 176.4.$$

We set these derivatives equal to 0 and solve the resulting system:

$$20m + 8b - 60.2 = 0, \qquad 5m + 2b = 15.05,$$
$$\text{or}$$
$$60m + 20b - 176.4 = 0; \qquad 15m + 5b = 44.1.$$

The solution of this system is

$$b = 1.05, \qquad m = 2.59.$$

We leave it to the reader to complete the D-test to verify that $(1.05, 2.59)$ does, in fact, yield the minimum of S. We need not bother to compute $S(1.05, 2.59)$.

The values of m and b are all we need to determine $y = mx + b$. The regression line is

$$y = 2.59x + 1.05.$$

We can extrapolate from the data to predict the total amount of life insurance in force in 1998:

$$y = 2.59(5) + 1.05 = \$14.0.$$

Thus the total amount of life insurance in force in 1998 will be about \$14.0 billion.

The method of least squares is a statistical process illustrated here with only four data points in order to simplify the explanation. Most statistical researchers would warn that many more than four data points should be used to get a "good" regression line. Furthermore, making predictions too far in the future from any linear model may not be valid. It can be done, but the further into the future the prediction is made, the more dubious one should be about the prediction.

*The Regression Line for an Arbitrary Collection of Data Points $(c_1, d_1), (c_2, d_2), \ldots, (c_n, d_n)$

Look again at the regression line

$$y = 2.59x + 1.05$$

---

*This part is considered optional and can be omitted without loss of continuity.

for the data points (1, 4.3), (2, 5.2), (3, 8.9), and (4, 11.7). Let us consider the arithmetic averages, or means, of the x-coordinates, denoted $\bar{x}$, and the y-coordinates, denoted $\bar{y}$:

$$\bar{x} = \frac{1 + 2 + 3 + 4}{4} = 2.5,$$

$$\bar{y} = \frac{4.3 + 5.2 + 8.9 + 11.7}{4} = 7.525.$$

It turns out that the point $(\bar{x}, \bar{y})$, or (2.5, 7.525), is on the regression line since

$$7.525 = 2.59(2.5) + 1.05.$$

Thus the regression line is

$$y - \bar{y} = m(x - \bar{x}),$$

or

$$y - 7.525 = m(x - 2.5).$$

All that remains, in general, is to determine m.

Suppose we want to find the regression line for an arbitrary number of points $(c_1, d_1), (c_2, d_2), \ldots, (c_n, d_n)$. To do so, we find the values m and b that minimize the function S given by

$$S(b, m) = (y_1 - d_1)^2 + (y_2 - d_2)^2 + \cdots + (y_n - d_n)^2$$
$$= \sum_{i=1}^{n} (y_i - d_i)^2,$$

where $y_i = mc_i + b$.

Using a procedure like the one we used earlier to minimize S, we can show that $y = mx + b$ takes the form

$$y - \bar{y} = m(x - \bar{x}),$$

where

$$\bar{x} = \frac{\sum_{i=1}^{n} c_i}{n}, \qquad \bar{y} = \frac{\sum_{i=1}^{n} d_i}{n},$$

and

$$m = \frac{\sum_{i=1}^{n} (c_i - \bar{x})(d_i - \bar{y})}{\sum_{i=1}^{n} (c_i - \bar{x})^2}.$$

Let us see how this works out for the individual life-insurance example used previously.

c_i	d_i	$c_i - \bar{x}$	$(c_i - \bar{x})^2$	$(d_i - \bar{y})$	$(c_i - \bar{x})(d_i - \bar{y})$
1	4.3	-1.5	2.25	-3.225	4.8375
2	5.2	-0.5	0.25	-2.325	1.1625
3	8.9	0.5	0.25	1.375	0.6875
4	11.7	1.5	2.25	4.175	6.2625

$$\sum_{i=1}^{4} c_i = 10 \qquad \sum_{i=1}^{4} d_i = 30.1 \qquad \sum_{i=1}^{4}(c_i - \bar{x})^2 = 5 \qquad \sum_{i=1}^{4}(c_i - \bar{x})(d_i - \bar{y}) = 12.95$$

$$\bar{x} = 2.5 \qquad \bar{y} = 7.525 \qquad\qquad\qquad m = \frac{12.95}{5} = 2.59$$

Thus the regression line is

$$y - 7.525 = 2.59(x - 2.5),$$

which simplifies to

$$y = 2.59x + 1.05.$$

TECHNOLOGY CONNECTION

Some graphers can perform linear regression. Use such a grapher to fit a linear equation to the life-insurance data. Compare your answer to what was found here in the text.

With some graphers, you will also obtain a number r, called the **coefficient of correlation**. Although we cannot develop that concept in detail in this text, keep in mind that r is used to describe the strength of the linear relationship between x and y. The closer r is to 1, the better the correlation.

For the life-insurance data, $r \approx 0.977$, which indicates a fairly good linear relationship. Keep in mind that a high linear correlation does not necessarily indicate a "cause-and-effect" connection between the variables.

*Nonlinear Regression

It can happen that data do not seem to fit a linear equation, but when logarithms of either the x-values or the y-values (or both) are taken, a linear relationship will exist. Indeed, when considering the graph in Fig. 1, it is not unreasonable to expect these data to fit an exponential function.

EXAMPLE 1   Use logarithms and regression to find an exponential function

$$y = Be^{kx}$$

that fits the data. Then estimate the total amount of life insurance in force in 1998.

---

*This part is considered optional, and can be omitted without loss of continuity.

Year, x	1. 1978	2. 1983	3. 1988	4. 1993
Total individual life insurance in force (in billions), y	$4.3	$5.2	$8.9	$11.7

Solution   If we take the natural logarithm of both sides of

$$y = Be^{kx},$$

we get

$$\ln y = \ln B + kx, \quad \text{or} \quad kx + \ln B.$$

Note that $\ln B$ and k are constants. Thus, if we replace $\ln y$ by a new variable Y, the equation takes the form of a linear function

$$Y = mx + b,$$

where $m = k$ and $b = \ln B$.

We are going to find this regression line, but before beginning, we need to find the logarithms of the y-values.

x	1	2	3	4
$Y = \ln y$	1.4586	1.6487	2.1861	2.4596

To find the regression line, we use the abbreviated procedure described in the preceding part of this section.

c_i	d_i	$c_i - \bar{x}$	$(c_i - \bar{x})^2$	$(d_i - \bar{Y})$	$(c_i - \bar{x})(d_i - \bar{Y})$
1	1.4586	−1.5	2.25	−0.47965	0.719475
2	1.6487	−0.5	0.25	−0.28955	0.144775
3	2.1861	0.5	0.25	0.24785	0.123925
4	2.4596	1.5	2.25	0.52135	0.782025

$$\sum_{i=1}^{4} c_i = 10 \qquad \sum_{i=1}^{4} d_i \qquad \sum_{i=1}^{4} (c_i - \bar{x})^2 \qquad \sum_{i=1}^{4} (c_1 - \bar{x})(d_i - \bar{Y})$$

$$\bar{x} = \frac{10}{4} = 2.5 \qquad = 7.753 \qquad = 5 \qquad\qquad = 1.7702$$

$$\bar{Y} = \frac{7.753}{4} \qquad\qquad m = \frac{1.7702}{5}$$

$$= 1.93825 \qquad\qquad = 0.35404$$

Thus the regression line is

$$Y - 1.93825 = 0.35404(x - 2.5),$$

which simplifies to

$$Y = 0.35404x + 1.05315.$$

Recall that we were to find k and B. From this equation, we know that

$$m = k = 0.35404$$

and $b = \ln B = 1.05315.$

To find B, we use the definition of logarithms (or take the antilog) and get

$$B = e^{1.05315} = 2.866667.$$

Then the desired equation is

$$y = 2.866667e^{0.35404x}.$$

Using this equation, we find that the total amount of life insurance in force in 1998 will be

$$y = 2.866667e^{0.35404(5)} \approx \$16.8 \text{ billion},$$

which seems like a reasonable prediction.   ◆

In conclusion, there are other kinds of nonlinear regression besides exponential. For example, a set of data might fit a quadratic equation

$$y = ax^2 + bx + c.$$

In such a case, one can still use regression to find the numbers a, b, and c that minimize the sums of squares of deviations. Many software–grapher packages allow curve fitting to polynomials up to the fourth degree.

TECHNOLOGY
CONNECTION

Use a grapher to fit an exponential equation to the life-insurance data. Compare your answer to what was found here in the text.

7.5 Exercise Set

APPLICATIONS _____

◆   Business and Economics

1. *Predicting the sales of Compaq Computer Corporation.* Consider the factual data in the following table showing the second-quarter sales, in millions of dollars, of Compaq Computer Corporation during three recent years of operation.

Year, x	Second-quarter sales (in millions), y
1. 1992	$0.83
2. 1993	1.63
3. 1994	2.50

a) Find the regression line $y = mx + b$.
b) Use the regression line to predict sales in 2000; in 2020.

2. *Predicting the stock value of Eli Lilly and Co.* Consider the factual data below showing the book value of the common stock of Eli Lilly and Co., a pharmaceutical firm, during four recent years of operation.

Year, x	Book value of stock, y
1. 1990	$11.22
2. 1991	15.52
3. 1992	15.14
4. 1993	15.61

a) Find the regression line $y = mx + b$.
b) Use the regression line to predict the value of the stock in 1998; in 2010.

◆ Life and Physical Sciences

3. *Life expectancy of women.* Consider the factual data below showing the life expectancy of women in various years.

Year, x	Life expectancy of women (in years), y
1. 1950	70.9
2. 1960	73.2
3. 1970	74.8
4. 1980	77.5
5. 1990	78.6

a) Find the regression line $y = mx + b$.
b) Use the regression line to predict the life expectancy of women in 2000; in 2010.

4. *Life expectancy of men.* Consider the factual data below showing the life expectancy of men in various years.

Year, x	Life expectancy of men (in years), y
1. 1950	65.3
2. 1960	66.6
3. 1970	67.1
4. 1980	69.9
5. 1990	71.8

a) Find the regression line $y = mx + b$.
b) Use the regression line to predict the life expectancy of men in 2000; in 2010.

◆ General Interest

5. *Grade predictions.* A professor wants to predict students' final examination scores on the basis of their midterm test scores. An equation was determined on the basis of data on the scores of three students who took the same course with the same instructor the previous semester (see the following table).

Midterm score (%), x	Final exam score (%), y
70	75
60	62
85	89

a) Find the regression line $y = mx + b$. (*Hint:* The y-deviations are $70m + b - 75$, $60m + b - 62$, and so on.)
b) The midterm score of a student was 81. Use the regression line to predict the student's final exam score.

6. *Predicting the world record in the high jump.* On July 29, 1989, Javier Sotomayer of Cuba set an astounding world record of 8 ft in the high jump. It has been established that most world records in track and field can be modeled by a linear function. The following table shows world records for various years.

Year, x (Use the actual year for x.)	World record in high jump (in inches), y
1912	78.0
1956	84.5
1973	90.5
1989	96.0

a) Find the regression line $y = mx + b$.
b) Use the regression line to predict the world record in the high jump in 2000; in 2050.

7. ◆ Explain the concept of linear regression to a friend.

8. ◆ Discuss the idea of linear regression with an expert in your major. Write a brief report.

TECHNOLOGY CONNECTION

Many graphers now have STAT packages with the capability to use the least-squares technique to fit equations of various types to sets of data. The following exercises are meant to be done with such a grapher.

9. *General interest: Predicting the world record in the one-mile run.*

   a) Find the regression line $y = mx + b$ that fits the set of data in the following table. (*Hint:* Convert each time to decimal notation; for example, $4{:}24.5 = 4\frac{24.5}{60} = 4.4083$.)

Year, x (Use the actual year for x.)	World record in mile (min:sec), y
1875 (Walter Slade)	4:24.5
1894 (Fred Bacon)	4:18.2
1923 (Paavo Nurmi)	4:10.4
1937 (Sidney Wooderson)	4:06.4
1942 (Gunder Haegg)	4:06.2
1945 (Gunder Haegg)	4:01.4
1954 (Roger Bannister)	3:59.4
1964 (Peter Snell)	3:54.4
1967 (Jim Ryun)	3:51.1
1975 (John Walker)	3:49.4
1979 (Sebastian Coe)	3:49.0
1980 (Steve Ovett)	3:48.8
1985 (Steve Cram)	3:46.31

b) Use the regression line to predict the world record in the mile in 1998; in 2004.

c) In August 1993, Noureddine Morceli set a new world record of 3:44.39 for the mile. How does this compare with what can be predicted by the regression?

10. *Life sciences: Test-tube babies.* Consider the factual data below showing the number of babies born in various years by a process called in vitro fertilization (IVF).

Year (from 1989), x	Test-tube babies born, y
0	500
1	1250
2	3180
3	4160

a) Use regression to find an exponential function $y = Be^{kx}$ that fits the data.

b) Use the regression equation to predict the number of test-tube babies born in 1999.

11. *Business: Predicting the sales of Compaq Computer Corporation*

a) Although the data in Exercise 1 do not support the assumption that an exponential function can be used as a model, use regression to find an exponential function

$$y = Be^{kx}$$

that fits the data in Exercise 1.

b) Use the exponential regression equation to predict second-quarter sales in 2000; in 2020.

c) Compare your answers with those in Exercise 1.

12. *Social sciences: Predicting the death rate from number of hours of sleep*

a) Use regression to find a quadratic function

$$y = ax^2 + bx + c$$

that fits the following set of data.*

_____

*The set of data in Exercise 12 comes from a study by Dr. Harold J. Morowitz.

Average number of hours of sleep, x	Death rate in one year (per 100,000 males), y
5	1121
6	805
7	626
8	813
9	967

b) Find the death rate of those who average 4 hr of sleep; 10 hr of sleep; 7.5 hr of sleep.

13. *Life sciences: Life expectancy of women.* Use the data in Exercise 3.

a) Use regression to find a cubic polynomial function

$$y = ax^3 + bx^2 + cx + d$$

that fits the data.

b) Use the regression equation to predict the life expectancy of women in 2000; in 2010.

c) Compare your answers with those in Exercise 3.

7.6 Constrained Maximum and Minimum Values: Lagrange Multipliers

OBJECTIVES

- Find maximum and minimum values using Lagrange multipliers.
- Solve application problems involving Lagrange multipliers.

Before we proceed in detail, let us return to a problem we considered in Chapter 3.

PROBLEM A hobby store has 20 feet of fencing to fence off a rectangular electric-train area in one corner of its display room. The two sides up against the wall require no fence. What dimensions of the rectangle will maximize the area?

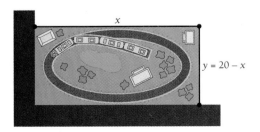

We maximize the function

$$A = xy$$

subject to the condition, or *constraint,* $x + y = 20$. Note that A is a function of two variables.

When we solved this earlier, we first solved the constraint for y:

$$y = 20 - x.$$

We then substituted $20 - x$ for y to obtain

$$A(x, y) = x(20 - x)$$
$$= 20x - x^2,$$

which is a function of one variable. Next, we found a maximum value using Maximum–Minimum Principle 1 (see Section 3.5). By itself, the function of two variables

$$A(x, y) = xy$$

has no maximum value. This can be checked using the D-test. With the constraint $x + y = 20$, however, the function does have a maximum. We see this in the following graph.

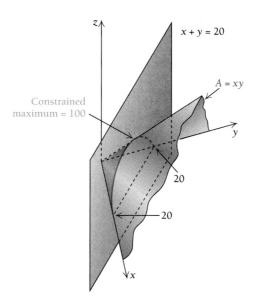

It may be quite difficult to solve a constraint for one variable. The procedure outlined below allows us to proceed without doing so.

> ### THEOREM 2
>
> The Method of Lagrange Multipliers
>
> To find a maximum or minimum value of a function $f(x, y)$ subject to the constraint $g(x, y) = 0$:
>
> 1. Form a new function:
>
> $$F(x, y, \lambda) = f(x, y) - \lambda g(x, y).$$
>
> 2. Find the first partial derivatives F_x, F_y, and F_λ.
> 3. Solve the system
>
> $$F_x = 0, \quad F_y = 0, \quad \text{and} \quad F_\lambda = 0.$$
>
> Let (a, b, λ) represent a solution of this system. We still must determine whether (a, b, λ) yields a maximum or minimum of the function f, but we will assume such a maximum or minimum for each of the problems considered here.

The variable λ (lambda) is called a **Lagrange multiplier.** We first illustrate the method of Lagrange multipliers by resolving the hobby-store problem.

EXAMPLE 1   Find the maximum value of

$$A(x, y) = xy$$

subject to the constraint $x + y = 20$.

Solution

1. We form the new function F given by

   $$F(x, y, \lambda) = xy - \lambda \cdot (x + y - 20).$$

   Note that we first had to express $x + y = 20$ as $x + y - 20 = 0$.
2. We find the first partial derivatives:

   $$F_x = y - \lambda,$$
   $$F_y = x - \lambda,$$
   $$F_\lambda = -(x + y - 20).$$

3. We set these derivatives equal to 0 and solve the resulting system:

$$y - \lambda = 0, \tag{1}$$
$$x - \lambda = 0, \tag{2}$$
$$-(x + y - 20) = 0, \quad \text{or} \quad x + y - 20 = 0. \tag{3}$$

From Eqs. (1) and (2), it follows that

$$x = y = \lambda.$$

Substituting λ for x and y in Eq. (3), we get

$$\lambda + \lambda - 20 = 0$$
$$2\lambda = 20$$
$$\lambda = 10.$$

Thus, $x = \lambda = 10$ and $y = \lambda = 10$. The maximum value of A subject to the constraint occurs at $(10, 10)$ and is

$$A(10, 10) = 10 \cdot 10$$
$$= 100.$$ ◆

EXAMPLE 2 Find the maximum value of

$$f(x, y) = 3xy$$

subject to the constraint

$$2x + y = 8.$$

(*Note:* f can be interpreted as a production function with budget constraint $2x + y = 8$.)

Solution

1. We form the new function F given by

$$F(x, y, \lambda) = 3xy - \lambda(2x + y - 8).$$

   Note that we first had to express $2x + y = 8$ as $2x + y - 8 = 0$.

2. We find the first partial derivatives:

$$F_x = 3y - 2\lambda,$$
$$F_y = 3x - \lambda,$$
$$F_\lambda = -(2x + y - 8).$$

3. We set these derivatives equal to 0 and solve the resulting system:

$$3y - 2\lambda = 0, \tag{1}$$
$$3x - \lambda = 0, \tag{2}$$
$$-(2x + y - 8) = 0, \quad \text{or} \quad 2x + y - 8 = 0. \tag{3}$$

   Solving Eq. (1) for y, we get

$$y = \frac{2}{3}\lambda.$$

Solving Eq. (2) for x, we get

$$x = \frac{\lambda}{3}.$$

Substituting $(2/3)\lambda$ for y and $(\lambda/3)$ for x in Eq. (3), we get

$$2\left(\frac{\lambda}{3}\right) + \left(\frac{2}{3}\lambda\right) - 8 = 0$$

$$\frac{4}{3}\lambda = 8$$

$$\lambda = \frac{3}{4} \cdot 8 = 6.$$

Then

$$x = \frac{\lambda}{3} = \frac{6}{3} = 2 \quad \text{and} \quad y = \frac{2}{3}\lambda = \frac{2}{3} \cdot 6 = 4.$$

The maximum value of f subject to the constraint occurs at $(2, 4)$ and is

$$f(2, 4) = 3 \cdot 2 \cdot 4 = 24. \qquad \blacklozenge$$

EXAMPLE 3   *Business: The beverage-can problem (12 oz).* The standard beverage can has a volume of 12 oz, or 21.66 in^3. What dimensions yield the minimum surface area? Find the minimum surface area.

Solution   We want to minimize the function s given by

$$s(h, r) = 2\pi rh + 2\pi r^2$$

subject to the volume constraint

$$\pi r^2 h = 21.66, \quad \text{or} \quad \pi r^2 h - 21.66 = 0.$$

Note that s does not have a minimum without the constraint.

1. We form the new function S given by

$$S(h, r, \lambda) = 2\pi rh + 2\pi r^2 - \lambda(\pi r^2 h - 21.66).$$

2. We find the first partial derivatives:

$$\frac{\partial S}{\partial h} = 2\pi r - \lambda \pi r^2,$$

$$\frac{\partial S}{\partial r} = 2\pi h + 4\pi r - 2\lambda \pi rh,$$

$$\frac{\partial S}{\partial \lambda} = -(\pi r^2 h - 21.66).$$

3. We set these derivatives equal to 0 and solve the resulting system:

$$2\pi r - \lambda\pi r^2 = 0, \tag{1}$$

$$2\pi h + 4\pi r - 2\lambda\pi rh = 0, \tag{2}$$

$$-(\pi r^2 h - 21.66) = 0, \quad \text{or} \quad \pi r^2 h - 21.66 = 0. \tag{3}$$

Note that we can solve Eq. (1) for r:

$$\pi r(2 - \lambda r) = 0$$

$$\pi r = 0 \quad \text{or} \quad 2 - \lambda r = 0$$

$$r = 0 \quad \text{or} \quad r = \frac{2}{\lambda}.$$

Since $r = 0$ cannot be a solution to the original problem, we continue by substituting $2/\lambda$ for r in Eq. (2):

$$2\pi h + 4\pi \cdot \frac{2}{\lambda} - 2\lambda\pi \cdot \frac{2}{\lambda} \cdot h = 0$$

$$2\pi h + \frac{8\pi}{\lambda} - 4\pi h = 0$$

$$\frac{8\pi}{\lambda} - 2\pi h = 0$$

$$-2\pi h = -\frac{8\pi}{\lambda},$$

so

$$h = \frac{4}{\lambda}.$$

Since $h = 4/\lambda$ and $r = 2/\lambda$, it follows that $h = 2r$. Substituting $2r$ for h in Eq. (3) yields

$$\pi r^2(2r) - 21.66 = 0$$

$$2\pi r^3 - 21.66 = 0$$

$$2\pi r^3 = 21.66$$

$$\pi r^3 = 10.83$$

$$r^3 = \frac{10.83}{\pi}$$

$$r = \sqrt[3]{\frac{10.83}{\pi}} \approx 1.51 \text{ in.}$$

Thus when $r = 1.51$ in., $h = 3.02$ in. The surface area is a minimum and is approximately

$$2\pi(1.51)(3.02) + 2\pi(1.51)^2, \quad \text{or about} \quad 42.98 \text{ in}^2.$$

The actual dimensions of a standard-sized 12-oz beverage can are $r = 1.25$ in. and $h = 4.875$ in. A natural question after studying Example 3 is, "Why don't beverage companies make cans using the dimensions found in that example?" To do this at this time would mean an enormous cost in retooling. New can-making machines would have to be purchased at a cost of millions. New beverage-filling machines would have to be purchased. Vending machines would no longer be the correct size. A partial response to the desire to save aluminum has been found in recycling and in manufacturing cans with a rippled effect at the top. These cans require less aluminum. As a result of many engineering ideas, the amount of aluminum required to make 1000 cans has been reduced from 36.5 lb to 28.1 lb. The consumer is actually a very important factor in the shape of the can. Market research has shown that a can with the dimensions found in Example 3 is not as comfortable to hold and might not be accepted by consumers.

7.6 Exercise Set

Find the maximum value of f subject to the given constraint.

1. $f(x, y) = xy; \quad 2x + y = 8$
2. $f(x, y) = 2xy; \quad 4x + y = 16$
3. $f(x, y) = 4 - x^2 - y^2; \quad x + 2y = 10$
4. $f(x, y) = 3 - x^2 - y^2; \quad x + 6y = 37$

Find the minimum value of f subject to the given constraint.

5. $f(x, y) = x^2 + y^2; \quad 2x + y = 10$
6. $f(x, y) = x^2 + y^2; \quad x + 4y = 17$
7. $f(x, y) = 2y^2 - 6x^2; \quad 2x + y = 4$
8. $f(x, y) = 2x^2 + y^2 - xy; \quad x + y = 8$
9. $f(x, y, z) = x^2 + y^2 + z^2; \quad y + 2x - z = 3$
10. $f(x, y, z) = x^2 + y^2 + z^2; \quad x + y + z = 1$

Use the method of Lagrange multipliers to solve each of the following.

11. Of all numbers whose sum is 70, find the two that have the maximum product.

12. Of all numbers whose sum is 50, find the two that have the maximum product.

13. Of all numbers whose difference is 6, find the two that have the minimum product.

14. Of all numbers whose difference is 4, find the two that have the minimum product.

APPLICATIONS

◆ **Business and Economics**

15. *Maximizing typing area.* A standard piece of typing paper has a perimeter of 39 in. Find the dimensions of the paper that will give the most typing area, subject to the perimeter constraint of 39 in. What is its area? Does the standard $8\frac{1}{2}$-in. × 11-in. paper have maximum area?

16. *Maximizing room area.* A carpenter is building a rectangular room with a fixed perimeter of 80 ft. What are the dimensions of the largest room that can be built? What is its area?

17. Minimizing surface area. An oil drum of standard size has a volume of 200 gal, or 27 ft³. What dimensions yield the minimum surface area? Find the minimum surface area.

Do these drums appear to be made in such a way as to minimize surface area?

18. Juice-can problem. A standard-sized juice can has a volume of 99 in³. What dimensions yield the minimum surface area? Find the minimum surface area.

19. Maximizing total sales. The total sales S of a one-product firm are given by

$$S(L, M) = ML - L^2,$$

where $M =$ the cost of materials and $L =$ the cost of labor. Find the maximum value of this function subject to the budget constraint

$$M + L = 80.$$

20. Maximizing total sales. The total sales S of a one-product firm are given by

$$S(L, M) = 2ML - L^2,$$

where $M =$ the cost of materials and $L =$ the cost of labor. Find the maximum value of this function subject to the budget constraint

$$M + L = 60.$$

21. Minimizing construction costs. A company is planning to construct a warehouse whose cubic footage is to be 252,000 ft³. Construction costs per square foot are estimated to be as follows:

Walls:	\$3.00
Floor:	\$4.00
Ceiling:	\$3.00

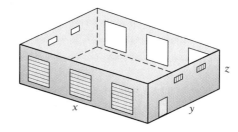

a) The total cost of the building is a function $C(x, y, z)$, where $x =$ the length, $y =$ the width, and $z =$ the height. Find a formula for $C(x, y, z)$.
b) What dimensions of the building will minimize the total cost? What is the minimum cost?

22. Minimizing the costs of container construction. A container company is going to construct a shipping container of volume 12 ft³ with a square bottom and top. The cost of the top and the sides is \$2 per square foot and for the bottom is \$3 per square foot. What dimensions will minimize the cost of the container?

23. Minimizing total cost. A product can be made entirely on either machine A or machine B, or both. The nature of the machines makes their cost functions differ:

$$\text{Machine A:} \quad C(x) = 10 + \frac{x^2}{6},$$

$$\text{Machine B:} \quad C(y) = 200 + \frac{y^3}{9}.$$

Total cost is given by $C(x, y) = C(x) + C(y)$. How many units should be made on each machine in order to minimize total costs if $x + y = 10{,}100$ units are required?

SYNTHESIS

Find the indicated maximum or minimum values of f subject to the given constraint.

24. Minimum: $f(x, y) = xy$; $x^2 + y^2 = 4$

25. Minimum: $f(x, y) = 2x^2 + y^2 + 2xy + 3x + 2y$; $y^2 = x + 1$

26. Maximum: $f(x, y, z) = x + y + z$; $x^2 + y^2 + z^2 = 1$

27. Maximum: $f(x, y, z) = x^2 y^2 z^2$; $x^2 + y^2 + z^2 = 1$

28. Maximum: $f(x, y, z) = x + 2y - 2z$; $x^2 + y^2 + z^2 = 4$

29. Maximum: $f(x, y, z, t) = x + y + z + t$; $x^2 + y^2 + z^2 + t^2 = 1$

30. Minimum: $f(x, y, z) = x^2 + y^2 + z^2$; $x - 2y + 5z = 1$

31. *Economics: The law of Equimarginal Productivity.* Suppose that $p(x, y)$ represents the production of a two-product firm. We give no formula for p. The company produces x items of the first product at a cost of c_1 each and y items of the second product at a cost of c_2 each. The budget constraint B is a constant given by

$$B = c_1 x + c_2 y.$$

Find the value of λ using the Lagrange-multiplier method in terms of p_x, p_y, c_1, and c_2. The resulting equation is called the *Law of Equimarginal Productivity.*

32. *Business: Maximizing production.* A company has the following Cobb–Douglas production function for a certain product:

$$p(x, y) = 800x^{3/4}y^{1/4},$$

where $x = $ the labor, measured in dollars, and $y = $ the capital, measured in dollars. Suppose that a company can make a total investment in labor and capital of $1,000,000. How should it allocate the investment between labor and capital in order to maximize production?

33. ◆ Write a brief report on the life and work of the mathematician Joseph Louis Lagrange (1736–1813).

34. ◆ Discuss the difference between solving a problem using Lagrange multipliers and using the method of Section 7.4. Describe the difference graphically.

TECHNOLOGY CONNECTION

35.–42. Use a 3D grapher to graph both equations in each of Exercises 1–8. Then visually check the results that you found analytically.

7.7 Multiple Integration

OBJECTIVE

• Evaluate a multiple integral.

The following is an example of a *double integral*:

$$\int_3^6 \int_{-1}^2 10xy^2 \, dx \, dy, \quad \text{or} \quad \int_3^6 \left(\int_{-1}^2 10xy^2 \, dx \right) dy.$$

Evaluating a double integral is somewhat similar to "undoing" a second partial derivative. We first evaluate the inside x-integral, treating y as a constant:

$$\int_{-1}^2 10xy^2 \, dx = 10y^2 \left[\frac{x^2}{2} \right]_{-1}^2 = 5y^2 [x^2]_{-1}^2 = 5y^2 [2^2 - (-1)^2] = 15y^2.$$

Color indicates the variable. All else is constant.

Then we evaluate the outside y-integral:

$$
\begin{aligned}
\int_3^6 15y^2 \, dy &= 15\left[\frac{y^3}{3}\right]_3^6 \\
&= 5[y^3]_3^6 \\
&= 5(6^3 - 3^3) \\
&= 945.
\end{aligned}
$$

More precisely, the above is called a **double iterated integral.** The word "iterate" means "to do again."

If the dx and dy and the limits of integration are interchanged, as follows,

$$
\int_{-1}^2 \int_3^6 10xy^2 \, dy \, dx,
$$

we first evaluate the inside y-integral, treating x as a constant:

$$
\begin{aligned}
\int_3^6 10xy^2 \, dy &= 10x\left[\frac{y^3}{3}\right]_3^6 \\
&= \frac{10x}{3}\left[y^3\right]_3^6 \\
&= \frac{10}{3}x(6^3 - 3^3) \\
&= 630x.
\end{aligned}
$$

Then we evaluate the outside x-integral:

$$
\begin{aligned}
\int_{-1}^2 630x \, dx &= 630\left[\frac{x^2}{2}\right]_{-1}^2 \\
&= 315[x^2]_{-1}^2 \\
&= 315[2^2 - (-1)^2] \\
&= 945.
\end{aligned}
$$

Note that we get the same result. This is not always true, but will be for the types of function that we consider.

Sometimes variables occur as limits of integration.

EXAMPLE 1 Evaluate

$$
\int_0^1 \int_{x^2}^x xy^2 \, dy \, dx.
$$

Solution   We first evaluate the y-integral, treating x as a constant:

$$\int_{x^2}^{x} xy^2 \, dy = x\left[\frac{y^3}{3}\right]_{x^2}^{x}$$

$$= \frac{1}{3}x[x^3 - (x^2)^3]$$

$$= \frac{1}{3}(x^4 - x^7).$$

Then we evaluate the outside integral:

$$\frac{1}{3}\int_{0}^{1}(x^4 - x^7) \, dx = \frac{1}{3}\left[\frac{x^5}{5} - \frac{x^8}{8}\right]_{0}^{1}$$

$$= \frac{1}{3}\left[\left(\frac{1^5}{5} - \frac{1^8}{8}\right) - \left(\frac{0^5}{5} - \frac{0^8}{8}\right)\right] = \frac{1}{40}.$$

Thus,

$$\int_{0}^{1}\int_{x^2}^{x} xy^2 \, dy \, dx = \frac{1}{40}.$$ ◆

The Geometric Interpretation of Multiple Integrals

Suppose the region G in the xy-plane is bounded by the functions $y_1 = g(x)$ and $y_2 = h(x)$ and the lines $x_1 = a$ and $x_2 = b$. We want the volume V of the solid above G and under the surface $z = f(x, y)$. We can think of the solid as composed of many vertical columns, one of which is shown in Fig. 1. The volume of one such column can be thought of as $l \cdot w \cdot h$, or $z \cdot \Delta y \cdot \Delta x$. Integrating such columns in

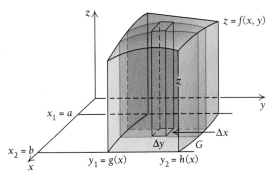

FIGURE 1

the y-direction, we obtain

$$\left[\int_{y_1}^{y_2} z \, dy\right] dx,$$

which can be pictured as a "slab," or slice. Then integrating such slices in the x-direction, we obtain the entire volume:

$$V = \int_a^b \left[\int_{y_1}^{y_2} z \, dy\right] dx,$$

or

$$V = \int_a^b \int_{g(x)}^{h(x)} z \, dy \, dx,$$

where $z = f(x, y)$.

    In Example 1, the region of integration G is the plane region between the graphs of $y = x^2$ and $y = x$, as shown in Figs. 2 and 3.

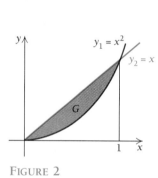

FIGURE 2   FIGURE 3

    When we evaluated the double integral in Example 1, we found the volume of the solid based on G and capped by the surface $z = xy^2$, as shown in Fig. 2.

An Application to Probability

Suppose we throw a dart at a region R in a plane. We assume that the dart lands on a point (x, y) in R (Fig. 4). We can think of (x, y) as a continuous random variable whose coordinates assume all values in some region R. A function f is said to be a **joint probability density function** if

$$f(x, y) \geq 0 \quad \text{for all } (x, y) \text{ in } R$$

and

$$\int\int_R f(x, y) \, dx \, dy = 1,$$

where $\int\int_R$ refers to the double integral evaluated over the region R.

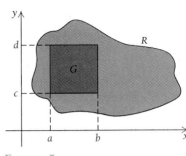

FIGURE 4   FIGURE 5

Suppose we want to know the probability that the dart hits a point (x, y) in a rectangular subregion G of R, where G is the set of points for which $a \le x \le b$ and $c \le y \le d$ (Fig. 5). This would be given by

$$\int \int_G f(x, y) \, dx \, dy = \int_c^d \int_a^b f(x, y) \, dx \, dy.$$

7.7 Exercise Set

Evaluate.

1. $\displaystyle\int_0^1 \int_0^1 2y \, dx \, dy$

2. $\displaystyle\int_0^1 \int_0^1 2x \, dx \, dy$

3. $\displaystyle\int_{-1}^1 \int_x^1 xy \, dy \, dx$

4. $\displaystyle\int_{-1}^1 \int_x^2 (x + y) \, dy \, dx$

5. $\displaystyle\int_0^1 \int_{-1}^3 (x + y) \, dy \, dx$

6. $\displaystyle\int_0^1 \int_{-1}^1 (x + y) \, dy \, dx$

7. $\displaystyle\int_0^1 \int_{x^2}^x (x + y) \, dy \, dx$

8. $\displaystyle\int_0^1 \int_{-1}^x (x^2 + y^2) \, dy \, dx$

9. $\displaystyle\int_0^2 \int_0^x (x + y^2) \, dy \, dx$

10. $\displaystyle\int_1^3 \int_0^x 2e^{x^2} \, dy \, dx$

11. Find the volume of the solid capped by the surface $z = 1 - y - x^2$ over the region bounded above and below by $y = 1 - x^2$ and $y = 0$ and left and right by $x = 0$ and $x = 1$, by evaluating the integral

$$\int_0^1 \int_0^{1-x^2} (1 - y - x^2) \, dy \, dx.$$

12. Find the volume of the solid capped by the surface $z = x + y$ over the region bounded above and below by $y = 1 - x$ and $y = 0$ and left and right by $x = 0$ and $x = 1$, by evaluating the integral

$$\int_0^1 \int_0^{1-x} (x + y) \, dy \, dx.$$

Suppose a continuous random variable has a joint probability density function given by

$$f(x, y) = x^2 + \tfrac{1}{3}xy,$$
$$0 \le x \le 1, \quad 0 \le y \le 2.$$

13. Find

$$\int_0^2 \int_0^1 f(x, y) \, dx \, dy.$$

14. Find the probability that a point (x, y) is in the region bounded by $0 \le x \le \tfrac{1}{2}$, $1 \le y \le 2$, by evaluating the integral

$$\int_1^2 \int_0^{1/2} f(x, y) \, dx \, dy.$$

SYNTHESIS

A *triple iterated integral* such as

$$\int_{r}^{s} \int_{c}^{d} \int_{a}^{b} f(x, y, z)\, dx\, dy\, dz$$

is evaluated in much the same way as the double iterated integral. We first evaluate the inside x-integral, treating y and z as constants. Then we evaluate the middle y-integral, treating z as a constant. Finally, we evaluate the outside z-integral. Evaluate these triple integrals.

15. $\displaystyle\int_{0}^{1} \int_{1}^{3} \int_{-1}^{2} (2x + 3y - z)\, dx\, dy\, dz$

16. $\displaystyle\int_{0}^{2} \int_{1}^{4} \int_{-1}^{2} (8x - 2y + z)\, dx\, dy\, dz$

17. $\displaystyle\int_{0}^{1} \int_{0}^{1-x} \int_{0}^{2-x} xyz\, dz\, dy\, dx$

18. $\displaystyle\int_{0}^{2} \int_{2-y}^{6-2y} \int_{0}^{\sqrt{4-y^2}} z\, dz\, dx\, dy$

19. ◈   Describe the geometric meaning of the multiple integral of a function of two variables.

 TECHNOLOGY CONNECTION

20. Use a grapher that does multiple integration to evaluate some double integrals found in this exercise set.

7 Chapter Summary and Review

TERMS TO KNOW

Function of two variables, p. 480
Function of several variables, p. 481
Partial derivative, p. 488
Second-order partial derivative, p. 495

Relative maximum, p. 498
Relative minimum, p. 498
Saddle point, p. 500
D-test, p. 501

Least-squares technique, p. 511
Method of Lagrange multipliers, p. 521
Multiple integration, p. 527
Double iterated integral, p. 528

REVIEW EXERCISES

These review exercises are for test preparation. They can also be used as a lengthened practice test. Answers are at the back of the book. The answers also contain bracketed section references, which tell you where to restudy if your answer is incorrect.

Given $f(x, y) = e^{y} + 3xy^{3} + 2y$, find each of the following.

1. $f(2, 0)$

2. f_x

3. f_y

4. f_{xy}

5. f_{yx}

6. f_{xx}

7. f_{yy}

Given $z = x^{2} \ln y + y^{4}$, find each of the following.

8. $\dfrac{\partial z}{\partial x}$

9. $\dfrac{\partial z}{\partial y}$

10. $\dfrac{\partial^{2} z}{\partial y\, \partial x}$

11. $\dfrac{\partial^{2} z}{\partial x\, \partial y}$

12. $\dfrac{\partial^{2} z}{\partial x^{2}}$

13. $\dfrac{\partial^{2} z}{\partial y^{2}}$

Find the relative maximum and minimum values.

14. $f(x, y) = x^3 - 6xy + y^2 + 6x + 3y - \frac{1}{5}$

15. $f(x, y) = x^2 - xy + y^2 - 2x + 4y$

16. $f(x, y) = 3x - 6y - x^2 - y^2$

17. $f(x, y) = x^4 + y^4 + 4x - 32y + 80$

18. Consider the data in the following table regarding the average salary of an NFL player over a recent six-year period.

Year, x	Average salary of an NFL player (in millions), y
1. 1991	$0.43
2. 1992	0.52
3. 1993	0.65
4. 1994	0.80
5. 1995	0.90
6. 1996	1.00

a) Find the regression line $y = mx + b$.

b) Use the regression line to predict the average NFL player salary in 1997; in 2000.

19. Consider the data in the following table regarding enrollment in colleges and universities during a recent three-year period.

Year, x	Enrollment (in millions), y
1	7.2
2	8.0
3	8.4

a) Find the regression line $y = mx + b$.

b) Use the regression line to predict enrollment in the fourth year.

20. Find the minimum value of

$$f(x, y) = x^2 - 2xy + 2y^2 + 20$$

subject to the constraint $2x - 6y = 15$.

21. Find the maximum value of

$$f(x, y) = 6xy$$

subject to the constraint $2x + y = 20$.

Evaluate.

22. $\int_0^1 \int_{x^2}^{3x} (x^3 + 2y) \, dy \, dx$

23. $\int_0^1 \int_{x^2}^{x} (x - y) \, dy \, dx$

SYNTHESIS

24. Evaluate

$$\int_0^2 \int_{1-2x}^{1-x} \int_0^{\sqrt{2-x^2}} z \, dz \, dy \, dx.$$

25. *Business: Minimizing surface area.* Suppose beverages could be packaged in either a cylindrical container or a rectangular container with a square top and bottom. Each container is designed to be of minimum surface area for its shape. If we assume a volume of 26 in³, which container would have the smaller surface area?

 TECHNOLOGY CONNECTION

26. Use a 3D grapher to graph

$$f(x, y) = x^2 + 4y^2.$$

7 Chapter Test

Given $f(x, y) = e^x + 2x^3y + y$, find each of the following.

1. $f(-1, 2)$

2. $\dfrac{\partial f}{\partial x}$

3. $\dfrac{\partial f}{\partial y}$

4. $\dfrac{\partial^2 f}{\partial x^2}$

5. $\dfrac{\partial^2 f}{\partial x \, \partial y}$

6. $\dfrac{\partial^2 f}{\partial y \, \partial x}$

7. $\dfrac{\partial^2 f}{\partial y^2}$

Find the relative maximum and minimum values.

8. $f(x, y) = x^2 - xy + y^3 - x$

9. $f(x, y) = y^2 - x^2$

10. **Business: Predicting total sales.** Consider the data in the following table regarding the total sales of a company during the first three years of operation.

Year, x	Sales (in millions), y
1	$10
2	15
3	19

a) Find the regression line $y = mx + b$.
b) Use the regression line to predict sales in the fourth year.

11. Find the maximum value of
$$f(x, y) = 6xy - 4x^2 - 3y^2$$
subject to the constraint $x + 3y = 19$.

12. Evaluate
$$\int_0^2 \int_1^x (x^2 - y) \, dy \, dx.$$

SYNTHESIS

13. **Business: Maximizing production.** A company has the following Cobb–Douglas production function for a certain product,
$$p(x, y) = 50x^{2/3}y^{1/3},$$
where $x =$ labor, measured in dollars, and $y =$ capital, measured in dollars. Suppose a company can make a total investment in labor and capital of $600,000. How should it allocate the investment between labor and capital in order to maximize production?

14. Find f_x and f_t:
$$f(x, t) = \frac{x^2 - 2t}{x^3 + 2t}.$$

    TECHNOLOGY CONNECTION

15. Use a 3D grapher to graph
$$f(x, y) = x - \tfrac{1}{2}y^2 - \tfrac{1}{3}x^3.$$

EXTENDED TECHNOLOGY APPLICATION

Minimizing Employees' Travel Time in a Building

For a business in which employees spend considerable time moving between offices, designing a building to minimize travel time can reap enormous financial savings.

Let us assume that each floor has a square grid of hallways, as shown in the figure below. Suppose you are standing at the most remote point P in the top northeast corner of such a building with 12 floors. How long will it take to reach the southwest corner on the first floor—that is, point Q?

Let us call the time t. We find a formula for t in two steps:

1. You are to go from the twelfth floor to the first floor. This is a move in a vertical direction.

2. You need to go across the first floor. This is a move in a horizontal direction.

In multilevel building design, one consideration is travel time between the most remote points in a rectangular building with a square base. We will make use of LaGrange multipliers to design such a building.

The vertical time is h, the height of the top floor from the ground, divided by a, the speed at which you can travel in a vertical direction (elevator speed). Thus vertical time is given by h/a.

The horizontal time is the time it takes to go across the first level, by way of the square grid of hallways (from R to Q above). If each floor is a square with side of length k, then the distance from R to Q is $2k$. If the walking speed is b, then the horizontal time is given by $2k/b$. Thus the time it will take to go from P to Q is a function of two variables, h and k, given by

$$t(h, k) = \text{Vertical time} + \text{Horizontal time}$$

$$= \frac{h}{a} + \frac{2k}{b}.$$

What happens if we have to choose between two (or more) building plans with the same floor area, but with different dimensions? Will the travel time be the same? Or will it be different for the two buildings? First of all, what is the total floor area of a given building? Suppose the building has n floors, each a square of side k. Then the total floor area is given by

$$A = nk^2.$$

Note that the area of the roof is not included.

If h is the height of the top of the building to the ground and c is the height of each floor, then $n = h/c$. (For simplicity, we are ignoring the thickness of each floor.) Thus,

$$A = \frac{h}{c}k^2.$$

Let us return to the problem of two buildings with the same total floor area, but with different dimensions, and see what happens to $t(h, k)$.

EXERCISES

1. Use the table-making feature on your grapher, if you have one, or a spreadsheet software program. For each case below, let the elevator speed $a = 10$ ft/sec, the walking speed $b = 4$ ft/sec, and the height of each floor $c = 15$ ft. Each case in the table covers two situations, though the floor area stays the same for a particular case. Complete the following table.

Case	Building	n	k	A	h	$t(h, k)$
1	B1	2	40	3200	30	23
	B2	3	32.66	3200	45	20.83
2	B1	2	60	7200		
	B2	3	48.99			
3	B1	4	40			
	B2	5	35.777			
4	B1	5	60			
	B2	10	42.426			
5	B1	5	150			
	B2	10	106.066			
6	B1	10	40			
	B2	17	30.679			
7	B1	10	80			
	B2	17	61.357			
8	B1	17	40			
	B2	26	32.344			
9	B1	17	50			
	B2	26	40.43			
10	B1	26	77			
	B2	50	55.525			

2. Do different dimensions, with a fixed floor area, yield different travel times?

3. Do you think there are values of h and k for a building with a given floor area that will minimize travel time? Try fixing the floor area at 40,000 ft^2 and explore the results of different dimensions. What do we need to use to know for sure?

Find the dimensions of a rectangular building with a square base that will minimize travel time t between the most remote points in the building described as follows. Each floor has a square grid of hallways. The height of the top floor from the ground is h, and the length of a side of each floor is k. The elevator speed is 10 ft/sec and the average speed of a person walking is 4 ft/sec. The total floor area of the building is 40,000 ft^2. The height of each floor is 8 ft.

4. Use the information given to find a formula for the function $t(h, k)$.

5. Find a formula for the constraint.

6. Use the method of LaGrange multipliers to find the dimensions of the building that will minimize travel time t between the most remote points in the building. Use a 3D grapher to graph both equations in Exercises 4 and 5. Then visually check the results you found analytically.

7. Find a general solution in terms of a, b, c, and A.

8 Trigonometric Functions

INTRODUCTION

Functions for which the graphs repeat themselves are called *periodic*. Of special importance in mathematics are the periodic *trigonometric functions*. These functions originated as a means of indirect measurement. For example, a surveyor might use trigonometry to measure the distance across a river without actually crossing the river. In this chapter, we will learn to differentiate and integrate these functions and to solve related problems.

AN APPLICATION

A satellite circles the earth in such a way that it is y miles from the equator (north or south, height not considered) t minutes after launch, where

$$y = 5000 \left[\cos \frac{\pi}{45}(t - 10) \right].$$

Find dy/dt.

THE MATHEMATICS

The derivative is given by

$$\frac{dy}{dt} = -\frac{1000\pi}{9} \left[\underbrace{\sin \frac{\pi}{45}(t - 10)}_{\uparrow} \right].$$

This is a *trigonometric function*.

This problem appears as Exercise 45 in Exercise Set 8.2.

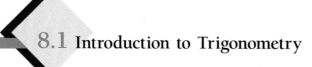

8.1 Introduction to Trigonometry

Angles and Rotations

We now introduce the trigonometric functions together with their derivatives and integrals.

We first consider a rotating ray, with its endpoint at the origin in an xy-plane. The ray starts in position along the positive half of the x-axis. A counterclockwise rotation is called *positive,* and a clockwise rotation is called *negative.*

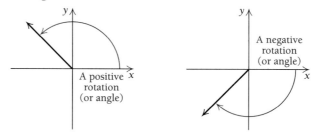

Note that the rotating ray and the positive half of the x-axis form an **angle.** Thus we often speak of "rotations" and "angles" interchangeably. The rotating ray is often called the *terminal side* of the angle, and the positive half of the x-axis is called the *initial side.*

Measures of Rotations or Angles

The size, or *measure,* of an angle, or rotation, may be given in **degrees.** A complete revolution has a measure of 360°, half a revolution has a measure of 180°, and so on. We can also speak of an *angle* of 90° or 720° or −240°.

An angle between 0° and 90° has its terminal side in the first quadrant. An angle between 90° and 180° has its terminal side in the second quadrant. An angle between 180° and 270° has its terminal side in the third quadrant. An angle between 0° and −90° has its terminal side in the fourth quadrant.

Note that angles with measures 0°, 360°, and 720° have the same terminal side, as do those with measures 270° and −90°.

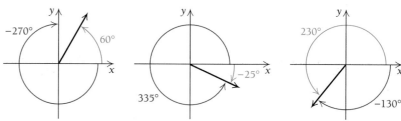

Measurements in Radians

A unit of angle or rotation measure other than the degree is very useful for many purposes. This unit is called the **radian.** Consider a circle with radius of length 1, centered at the origin. The measure of an angle, or rotation, in radians is the distance around this circle from the initial side of the angle to the terminal side.

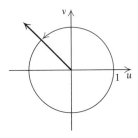

Since the circumference of the circle is $2\pi \cdot 1$, or 2π, a complete revolution (360°) has a measure of 2π radians. Half of this (180°) is π radians, and a fourth of this (90°) is $\pi/2$ radians. In general, we can convert from one measure to the other using this proportion.

THEOREM 1

$$\frac{\text{Radian measure}}{\pi} = \frac{\text{Degree measure}}{180}$$

EXAMPLE 1   Convert 270° to radians.

Solution   We use the proportion in Theorem 1, multiplying by π on both sides:

$$\frac{\text{Radian measure}}{\pi} = \frac{270}{180}$$

$$\text{Radian measure} = \frac{270}{180} \cdot \pi, \quad \text{or} \quad \frac{3}{2}\pi. \qquad \blacklozenge$$

When no unit is specified for an angle measure, it is understood to be given in radians.

EXAMPLE 2   Convert $\pi/4$ radians to degrees.

Solution

$$\frac{\pi/4}{\pi} = \frac{\text{Degree measure}}{180}$$

$$\text{Degree measure} = 180 \cdot \frac{\pi/4}{\pi}, \quad \text{or} \quad 45° \qquad \blacklozenge$$

TECHNOLOGY
CONNECTION

Graphers and other scientific calculators can make conversions between radian measure and degree measure. Convert 135° and −427° to radian measure. Convert $\pi/3$ and -300π to degree measure.

Trigonometric Functions

The concept of rotation or angle is important to functions called **trigonometric,** or **circular,** functions.

Consider an angle t, measured in radians, shown below on a circle with radius 1. The unit circle has equation $u^2 + v^2 = 1$.

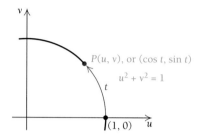

The terminal side of the angle intersects this circle at point $P(u, v)$. The arc length *around the circle* from $(1, 0)$ to P is t. We define $\cos t$ (cosine t) and $\sin t$ (sine t) as the first and second coordinates of P, respectively.

> **DEFINITION**
>
> $\cos t = $ the first coordinate of $P = u$,
>
> $\sin t = $ the second coordinate of $P = v$

Certain values of these functions are easy to determine. When $t = \pi$, for example, the terminal side of the angle is on the horizontal axis and P is one unit to the left; hence the first coordinate is -1 and the second coordinate is 0 (Fig. 1a). Thus,

$$\cos \pi = u = -1 \quad \text{and} \quad \sin \pi = v = 0,$$

which checks because $(-1)^2 + 0^2 = 1$.

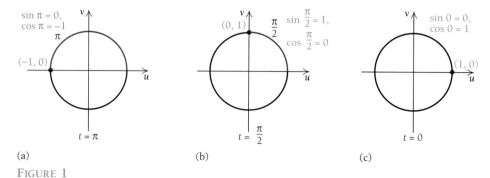

(a)   (b)   (c)

FIGURE 1

Similarly, when $t = \pi/2$, the point P is one unit up on the vertical axis; hence the first coordinate is 0 and the second coordinate is 1 (Fig. 1b). Thus,

$$\cos \frac{\pi}{2} = u = 0 \quad \text{and} \quad \sin \frac{\pi}{2} = v = 1,$$

which checks because $0^2 + 1^2 = 1$.

When $t = 0$, the terminal side is on the horizontal axis and P is one unit to the right; hence the first coordinate is 1 and the second coordinate is 0 (Fig. 1c). Thus,

$$\cos 0 = u = 1 \quad \text{and} \quad \sin 0 = v = 0.$$

We can also define the trigonometric functions by using right triangles. Suppose t is an angle of a right triangle, measured in degrees. Then we have the following.

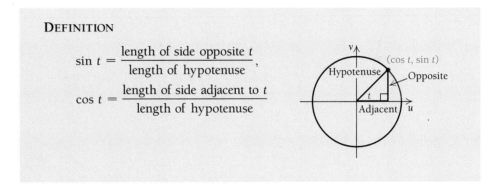

DEFINITION

$$\sin t = \frac{\text{length of side opposite } t}{\text{length of hypotenuse}},$$

$$\cos t = \frac{\text{length of side adjacent to } t}{\text{length of hypotenuse}}$$

Function Values for Special Angles

Using properties of right triangles, we can develop values of the sine and cosine functions for the special angles 45°, 30°, and 60°. First, recall the Pythagorean theorem. It says that in any right triangle, $a^2 + b^2 = c^2$, where c is the length of the hypotenuse and a and b are the lengths of the legs.

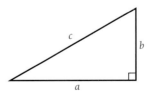

Consider $t = \pi/4 = 45°$. A right triangle with a 45° acute angle actually has two 45° angles. Thus the triangle is isosceles, and the legs are the same length. Let us consider such a triangle whose legs have length 1. Then its hypotenuse has length c:

$$1^2 + 1^2 = c^2, \quad \text{or} \quad c^2 = 2, \quad \text{or} \quad c = \sqrt{2}.$$

Such a triangle is shown below. From this diagram, we can easily determine the trigonometric function values for 45°.

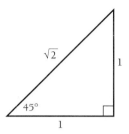

Thus,

$$\cos \frac{\pi}{4} = \cos 45° = \frac{\sqrt{2}}{2} \approx 0.707$$

and

$$\sin \frac{\pi}{4} = \sin 45° = \frac{\sqrt{2}}{2} \approx 0.707.$$

In a similar way, we can determine function values for 30° and 60°. A right triangle with 30° and 60° acute angles is half of an equilateral triangle, as shown:

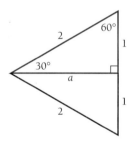

Thus if we choose an equilateral triangle whose sides have length 2 and take half of it, we obtain a right triangle that has a hypotenuse of length 2 and a leg of length 1. The other leg has length a, which can be found using the Pythagorean theorem, as follows:

$$a^2 + 1^2 = 2^2$$
$$a^2 = 3$$
$$a = \sqrt{3}.$$

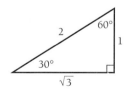

We can now determine function values for 60° and 30°.
For $t = \pi/3 = 60°$,

$$\cos \frac{\pi}{3} = \cos 60° = \frac{1}{2} = 0.5$$

and

$$\sin \frac{\pi}{3} = \sin 60° = \frac{\sqrt{3}}{2} \approx 0.866.$$

For $t = \pi/6 = 30°$,

$$\cos \frac{\pi}{6} = \cos 30° = \frac{\sqrt{3}}{2} \approx 0.866$$

and

$$\sin \frac{\pi}{6} = \sin 30° = \frac{1}{2} = 0.5.$$

The following table summarizes these important values of the sine and cosine functions. It should be memorized.

t	t	$\sin t$	$\cos t$
0°	0	0	1
30°	$\frac{\pi}{6}$	$\frac{1}{2}$	$\frac{\sqrt{3}}{2}$
45°	$\frac{\pi}{4}$	$\frac{\sqrt{2}}{2}$	$\frac{\sqrt{2}}{2}$
60°	$\frac{\pi}{3}$	$\frac{\sqrt{3}}{2}$	$\frac{1}{2}$

Other function values follow from certain symmetries on the unit circle. Some are shown in Fig. 2.

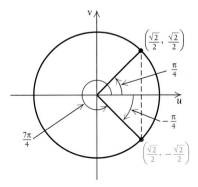

(a)

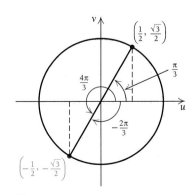

(b)

Figure 2

From Fig. 2(a), we get

$$\cos(-45°) = \cos\left(-\frac{\pi}{4}\right) = \frac{\sqrt{2}}{2} \approx 0.707,$$

$$\sin(-45°) = \sin\left(-\frac{\pi}{4}\right) = -\frac{\sqrt{2}}{2} \approx -0.707,$$

and

$$\cos 315° = \cos\frac{7\pi}{4} = \frac{\sqrt{2}}{2} \approx 0.707,$$

$$\sin 315° = \sin\frac{7\pi}{4} = -\frac{\sqrt{2}}{2} \approx -0.707.$$

From Fig. 2(b), we get

$$\cos(-120°) = \cos\left(-\frac{2\pi}{3}\right) = -\frac{1}{2} = -0.5,$$

$$\sin(-120°) = \sin\left(-\frac{2\pi}{3}\right) = -\frac{\sqrt{3}}{2} \approx -0.866,$$

and

$$\cos 240° = \cos\frac{4\pi}{3} = -\frac{1}{2} = -0.5,$$

$$\sin 240° = \sin\frac{4\pi}{3} = -\frac{\sqrt{3}}{2} \approx -0.866.$$

Approximations of the trigonometric values can be found on your calculator. Be sure to check whether it uses degrees or radians. Usually you can adapt to consider either.

Graphs of cos t and sin t

Note that $\cos t$ and $\sin t$ are functions of t defined for all real numbers t. For very large $|t|$, we may "wrap around" the circle several times before coming to the terminal point P. Nevertheless, P still has one first coordinate, $\cos t$, and one second coordinate, $\sin t$. For example,

$$\cos 3\pi = \cos \pi = -1$$

and

$$\sin 3\pi = \sin \pi = 0.$$

Also,

$$\cos\left(\frac{15\pi}{4}\right) = \cos\left(\frac{7\pi}{4}\right) = \frac{\sqrt{2}}{2}$$

and

$$\sin\left(\frac{15\pi}{4}\right) = \sin\left(\frac{7\pi}{4}\right) = -\frac{\sqrt{2}}{2}.$$

Plotting points previously obtained, and finding others with a calculator, we graph the cosine and sine functions, as shown in the following figure.

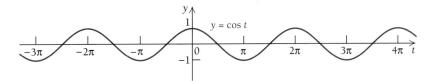

(a) The cosine function

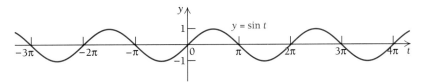

(b) The sine function

At the origin, t is, of course, 0. Thus the point P is on the horizontal axis, so $\cos 0 = 1$ and $\sin 0 = 0$. Moving to the right on the graphs corresponds to rotating the terminal side of the angle counterclockwise. Moving to the left corresponds to rotating the terminal side clockwise. Note in particular that $\sin t = 0$ has solutions $0, \pm\pi, \pm 2\pi, \ldots$, and $\cos t = 0$ has solutions $\pm(\pi/2), \pm(3\pi/2), \pm(5\pi/2), \ldots$. Note that each curve repeats itself as the terminal side makes successive revolutions. From 0 to 2π is one complete revolution, or *cycle*. The cycle repeats itself from there on. We say that the *period* of each function is 2π. Algebraically, this means that for any t,

$$\cos(t + 2\pi) = \cos t \quad \text{and} \quad \sin(t + 2\pi) = \sin t.$$

DEFINITION

A function f is *periodic* if there exists a positive number p such that

$$f(t + p) = f(t)$$

for every t in the domain of f. This means that adding p to an input does not change the output. The smallest such number p is called the *period*.

A printout from an electrocardiogram may form a periodic function.

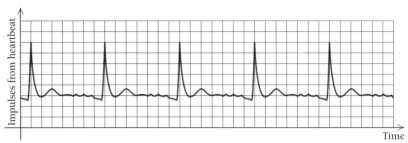

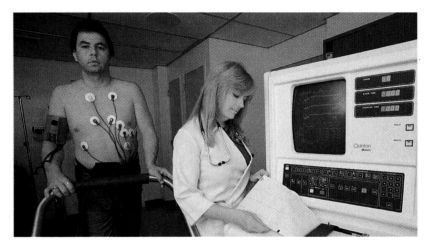

Other Trigonometric Functions

The functions sin t and cos t are the basic trigonometric functions, but there are four others—the tangent, cotangent, secant, and cosecant functions—defined as follows.

DEFINITION

$$\tan t = \frac{\sin t}{\cos t},$$

$$\cot t = \frac{\cos t}{\sin t} = \frac{1}{\tan t},$$

$$\sec t = \frac{1}{\cos t},$$

$$\csc t = \frac{1}{\sin t},$$

provided the denominators are not equal to 0.

Let us find some values of the tangent function.

EXAMPLE 3   Find tan $\pi/6$ and tan $\pi/2$.

Solution

$$\tan \frac{\pi}{6} = \frac{\sin (\pi/6)}{\cos (\pi/6)} = \frac{\frac{1}{2}}{\sqrt{3}/2}$$

$$= \frac{1}{\sqrt{3}} = \frac{1}{\sqrt{3}} \cdot \frac{\sqrt{3}}{\sqrt{3}} = \frac{\sqrt{3}}{3} \approx 0.577;$$

$$\tan \frac{\pi}{2} = \frac{\sin (\pi/2)}{\cos (\pi/2)} = \frac{1}{0} = \text{undefined} \qquad \blacklozenge$$

The graph of tan t follows. The vertical dashed lines are asymptotes. They occur at values t for which cos $t = 0$. These are $t = \pi/2 + k\pi$, where k is any integer.

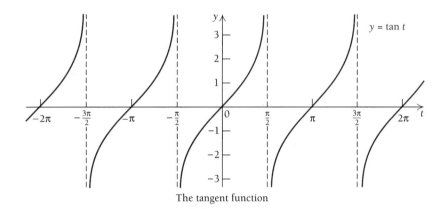

The tangent function

Identities

An **identity** is an equation that holds for all meaningful replacements of its variables by real numbers. An example is $t + 3 = 3 + t$, which is true for all real numbers. Another example is $(t^2 - 4)/(t - 2) = t + 2$. It is true for all real numbers except 2, which is *not* a meaningful replacement in the expression on the left.

The properties

$$\cos(t + 2\pi) = \cos t \quad \text{and} \quad \sin(t + 2\pi) = \sin t$$

hold for all real numbers t. They are examples of **trigonometric identities.** Another identity that holds for all real numbers is the following:

THEOREM 2

$$\sin^2 t + \cos^2 t = 1, \tag{1}$$

---

where $\sin^2 t$ means $(\sin t)^2$ and $\cos^2 t$ means $(\cos t)^2$. To see why this holds, note the right triangle inside the unit circle.

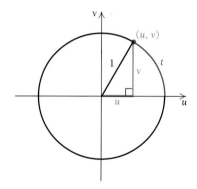

Since the length of the radius is 1, we know from the Pythagorean theorem that

$$u^2 + v^2 = 1^2, \quad \text{or} \quad u^2 + v^2 = 1;$$

and, since for any point (u, v) on the unit circle, $\cos t = u$ and $\sin t = v$, the identity follows.

If we multiply Identity (1) by $1/(\cos^2 t)$, we get another identity:

$$\frac{\sin^2 t}{\cos^2 t} + \frac{\cos^2 t}{\cos^2 t} = \frac{1}{\cos^2 t},$$

or $\tan^2 t + 1 = \sec^2 t,$

where, as before, $\tan^2 t = (\tan t)^2$ and $\sec^2 t = (\sec t)^2$. Another identity can be proved in a similar way by multiplying on both sides of Identity (1) by $1/\sin^2 t$.

The new identities are as follows.

THEOREM 3

$$\tan^2 t + 1 = \sec^2 t, \tag{2}$$
$$\cot^2 t + 1 = \csc^2 t \tag{3}$$

The following are called **sum–difference identities.**

THEOREM 4

$$\cos(u + v) = \cos u \cos v - \sin u \sin v, \tag{4}$$
$$\cos(u - v) = \cos u \cos v + \sin u \sin v, \tag{5}$$
$$\sin(u + v) = \sin u \cos v + \cos u \sin v, \tag{6}$$
$$\sin(u - v) = \sin u \cos v - \cos u \sin v \tag{7}$$

EXAMPLE 4   Use Identity (5) to find $\cos 15°$.

Solution   If we think of $15°$ as $45° - 30°$, then

$$\cos 15° = \cos(45° - 30°)$$
$$= \cos 45° \cos 30° + \sin 45° \sin 30°$$
$$= \frac{\sqrt{2}}{2} \cdot \frac{\sqrt{3}}{2} + \frac{\sqrt{2}}{2} \cdot \frac{1}{2}$$
$$= \frac{\sqrt{6} + \sqrt{2}}{4}. \qquad \blacklozenge$$

If we let $u = v$ in Identity (4), we obtain a **double-angle identity:**

$$\cos 2u = \cos(u + u)$$
$$= \cos u \cos u - \sin u \sin u$$
$$= \cos^2 u - \sin^2 u.$$

A similar identity for $\sin 2u$ can be found using Identity (6).

THEOREM 5

$$\cos 2u = \cos^2 u - \sin^2 u,$$
$$\sin 2u = 2 \sin u \cos u$$

TECHNOLOGY CONNECTION

A grapher can be used for a partial verification of an identity. Consider $\tan^2 t + 1 = \sec^2 t$. Graph each of $y_1 = \tan^2 x + 1$ and $y_2 = \sec^2 x$ and see if the graphs agree. The graphs agree as shown below. This is considered a partial verification because we can never see the complete graphs, and theoretically they might vary somewhere "way out." Similarly, we see that $3 \sin x + 2 \cos x = \sin 3x - \cos x$ is not an identity because the graphs do not match.

$\tan^2 t + 1 = \sec^2 t$ is an identity.

$y_1 = \tan^2 x + 1$ and $y_2 = \sec^2 x$

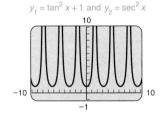

$3 \sin x + 2 \cos x = \sin 3x - \cos x$ is not an identity.

$y_1 = 3 \sin x + 2 \cos x$ $y_2 = \sin 3x - \cos x$

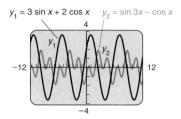

Determine whether each of the following is an identity:

$$\cot^2 t + 1 = \csc^2 t,$$
$$\cos 2u = \cos^2 u - \sin^2 u,$$
$$\sin 3x = \sin 2x + \sin x.$$

8.1 Exercise Set

In what quadrant does the terminal side of the angle lie?

1. $34°$
2. $320°$
3. $-120°$
4. $-205°$

Convert to radian measure. Leave answers in terms of π.

5. $30°$
6. $15°$
7. $60°$
8. $200°$
9. $75°$
10. $300°$

Convert to degree measure.

11. $\dfrac{3}{2}\pi$
12. $\dfrac{5}{4}\pi$
13. $-\dfrac{\pi}{4}$

14. $-\dfrac{\pi}{6}$
15. 8π
16. -12π

17. 1 radian
18. 2 radians

Find each of the following.

19. $\sin \dfrac{\pi}{3}$
20. $\cos \dfrac{\pi}{4}$
21. $\cos \dfrac{3\pi}{2}$

22. $\sin \dfrac{5\pi}{4}$
23. $\sin \dfrac{\pi}{6}$
24. $\cos \pi$

25. $\tan \dfrac{\pi}{2}$
26. $\tan \dfrac{\pi}{6}$
27. $\cot \dfrac{\pi}{3}$

28. $\cot \dfrac{\pi}{6}$
29. $\sec \pi$
30. $\csc \dfrac{\pi}{4}$

Verify each of the following identities.

31. $\tan (u + v) = \dfrac{\tan u + \tan v}{1 - \tan u \tan v}$

32. $\tan 2u = \dfrac{2 \tan u}{1 - \tan^2 u}$

Find answers to the following exercises to four decimal places, using a calculator. Check to see if it is using degree or radian measure.

33. $\sin 31.4°$

34. $\cos 1.07°$

35. $\tan 139.2°$

36. $\cot 153.5°$

37. $\cos (-1.91)$

38. $\sin (-11.2\pi)$

39. $\cot 49\pi$

40. $\tan (-17.4)$

APPLICATIONS

◆ **Business and Economics**

41. *Sales.* Sales of certain products fluctuate in cycles, as shown in this graph. A company in a northern climate has total sales of skis as given by

$$S(t) = 7\left(1 - \cos \frac{\pi}{6}t\right),$$

where S = sales, in thousands of dollars, during the tth month. Find $S(0)$, $S(1)$, $S(2)$, $S(3)$, $S(6)$, $S(12)$, and $S(15)$. Round to the nearest tenth.

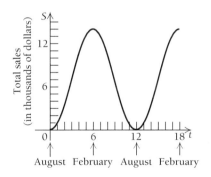

◆ **Life and Physical Sciences**

42. *Temperature during an illness.* The temperature of a patient during a 12-day illness is given by

$$T(t) = 101.6° + 3 \sin \left(\frac{\pi}{8}t\right).$$

The graph is shown here. Find $T(0)$, $T(1)$, $T(2)$, $T(4)$, and $T(12)$. Round to the nearest tenth of a degree.

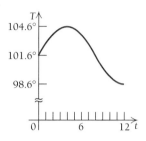

43. ◈ Compare the concepts of radian and degree.

  TECHNOLOGY CONNECTION

44. ◈ A grapher has many advantages, but it also has its disadvantages. Explain a disadvantage when graphing a function like

$$f(x) = \frac{\sin x}{x}.$$

Determine whether each of the following is an identity.

45. $\dfrac{\cos x - \sin x}{\sin x + \cos x} = \dfrac{\cos 2x}{1 + \sin 2x}$

46. $\tan^2 x - \sin^2 x = \sin^2 x \tan x$

47. $2 \sin x + \sin 2x = 2 \sin x - \cos 3x$

48. $\cos^4 x - \sin^4 x = \cos 2x$

Find each of the following limits, using the zoom and trace features if necessary. Use your grapher to create graphs and input–output tables.

49. $\lim\limits_{x \to 0} \dfrac{1 - \cos x}{x^2}$

50. $\lim\limits_{x \to 0} \dfrac{x}{\sin 2x}$

51. $\lim\limits_{x \to 0} \dfrac{3x - \sin x}{x}$

52. $\lim\limits_{x \to 0} \dfrac{x - \sin x}{x^3}$

8.2 Derivatives of the Trigonometric Functions

OBJECTIVES

• Differentiate the trigonometric functions.
• Solve application problems involving trigonometric derivatives.

The development of the derivatives of $\sin x$ and $\cos x$ is comparable to the development of the derivative of e^x. We first find the value of the derivative at 0, and then extend this to the general formula. The derivatives at 0 are given by the following limits:

$$\sin'(0) = \lim_{h \to 0} \frac{\sin(0 + h) - \sin 0}{h}$$

$$= \lim_{h \to 0} \frac{\sin h}{h};$$

$$\cos'(0) = \lim_{h \to 0} \frac{\cos(0 + h) - \cos 0}{h}$$

$$= \lim_{h \to 0} \frac{\cos h - 1}{h}.$$

We find the limits using the following input–output table as well as the graphs of each function shown below.

h (in radians) with decimal approximation	$\sin h$	$\cos h$	$\dfrac{\sin h}{h}$	$\dfrac{\cos h - 1}{h}$
$\dfrac{\pi}{2} \ (\approx 1.5708)$	1	0	0.6366	−0.6366
$\dfrac{\pi}{3} \ (\approx 1.0472)$	$\dfrac{\sqrt{3}}{2} \ (\approx 0.8660)$	$\dfrac{1}{2} \ (= 0.5000)$	0.8270	−0.4775
$\dfrac{\pi}{4} \ (\approx 0.7854)$	$\dfrac{\sqrt{2}}{2} \ (\approx 0.7071)$	$\dfrac{\sqrt{2}}{2} \ (\approx 0.7071)$	0.9003	−0.3729
$\dfrac{\pi}{6} \ (\approx 0.5236)$	$\dfrac{1}{2} \ (= 0.5000)$	$\dfrac{\sqrt{3}}{2} \ (\approx 0.8660)$	0.9549	−0.2559
$\dfrac{\pi}{20} \ (\approx 0.1571)$	0.1564	0.9877	0.9959	−0.0784
$\dfrac{\pi}{50} \ (\approx 0.0628)$	0.0628	0.9980	0.9993	−0.0314
$\dfrac{\pi}{4000} \ (\approx 0.0008)$	0.0008	0.9999997	1.0000	−0.00039

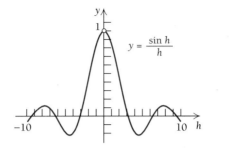

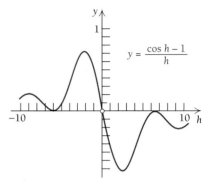

Values approaching 0 from the left can also be checked. The following values of the limits can be conjectured from the table and the graphs:

$$\sin' (0) = \lim_{h \to 0} \frac{\sin h}{h} = 1 \quad \text{and} \quad \cos' (0) = \lim_{h \to 0} \frac{\cos h - 1}{h} = 0.$$

The results can be proved more formally, but we will not do so here.

Now let us consider the general derivatives:

$$\frac{d}{dx} \sin x = \lim_{h \to 0} \frac{\sin (x + h) - \sin x}{h}.$$

Using Identity (6), we get

$$\frac{\sin (x + h) - \sin x}{h} = \frac{\sin x \cos h + \cos x \sin h - \sin x}{h}$$

$$= \frac{\sin x \cos h - \sin x}{h} + \frac{\cos x \sin h}{h}$$

$$= \sin x \left(\frac{\cos h - 1}{h} \right) + \cos x \left(\frac{\sin h}{h} \right).$$

Then using the limits just developed, we have

$$\frac{d}{dx} \sin x = \lim_{h \to 0} \frac{\sin (x + h) - \sin x}{h}$$

$$= \lim_{h \to 0} \left[\sin x \left(\frac{\cos h - 1}{h} \right) + \cos x \left(\frac{\sin h}{h} \right) \right]$$

$$= (\sin x) \cdot 0 + (\cos x) \cdot 1$$

$$= \cos x.$$

A development for the derivative of $\cos x$ is similar but uses Identity (4). The result is $-\sin x$.

TECHNOLOGY
CONNECTION

Graph $y = \sin x$ and
$y = \cos x$ using the same
set of axes. How can you
use these graphs to show
that the derivative of $\sin x$
is $\cos x$? What should the
derivative be doing as the
sine function is rising? as it
is falling? What does the
derivative do when the
curve reaches its highest
point? its lowest point? Answer the same questions for
the graphs of $y = \cos x$ and
$y = -\sin x$.

In summary, we have the following.

THEOREM 6

$$\frac{d}{dx} \sin x = \cos x \quad \text{and} \quad \frac{d}{dx} \cos x = -\sin x$$

The derivatives of the remaining trigonometric functions are computed from their definitions in terms of $\sin x$ and $\cos x$, together with the Quotient Rule and/or the Extended Power Rule. The remaining formulas are as follows.

THEOREM 7

$$\frac{d}{dx} \tan x = \sec^2 x, \qquad \frac{d}{dx} \cot x = -\csc^2 x,$$

$$\frac{d}{dx} \sec x = \tan x \sec x, \qquad \frac{d}{dx} \csc x = -\cot x \csc x$$

EXAMPLE 1 Prove that

$$\frac{d}{dx} \tan x = \sec^2 x.$$

Solution   Recall that $(\sec x)^2 = \sec^2 x$, $(\sin x)^2 = \sin^2 x$, and $(\cos x)^2 = \cos^2 x$. By definition,

$$\tan x = \frac{\sin x}{\cos x}.$$

TECHNOLOGY
CONNECTION

Graph $f(x) = \sin x$. Then
graph tangent lines to the
curve at $x = -\pi$, $x =$
$-\pi/2$, $x = 0$, $x = \pi/4$,
$x = 3\pi/4$, and $x = \pi$. Next,
using the same set of axes,
graph the derivative f'. Now
graph $f(x) = \tan x$. Then
using the same set of axes,
graph its derivative.

Thus we can find its derivative using the Quotient Rule:

$$\frac{d}{dx} \tan x = \frac{d}{dx} \left[\frac{\sin x}{\cos x} \right]$$

$$= \frac{\cos x (\cos x) - (-\sin x)(\sin x)}{\cos^2 x}$$

$$= \frac{\cos^2 x + \sin^2 x}{\cos^2 x}$$

$$= \frac{1}{\cos^2 x} = \sec^2 x. \qquad \blacklozenge$$

EXAMPLE 2 Find the derivative of $y = \sec^3 x$.

Solution   We use the Chain Rule:

$$\frac{dy}{dx} = 3 \sec^2 x \cdot \left(\frac{d}{dx} \sec x \right)$$

$$= 3 \sec^2 x \cdot \tan x \cdot \sec x. \qquad \text{By Theorem 7}$$

Replacing the factors by the definitions in terms of $\sin x$ and $\cos x$, we can simplify this as follows:

$$\frac{dy}{dx} = 3 \sec^2 x \cdot \tan x \cdot \sec x$$

$$= 3 \cdot \frac{1}{\cos^2 x} \cdot \frac{\sin x}{\cos x} \cdot \frac{1}{\cos x}$$

$$= \frac{3 \sin x}{\cos^4 x}. \qquad \blacklozenge$$

Using the Chain Rule, we can find other derivatives.

EXAMPLE 3   Differentiate $f(x) = \sin (x^3 - 5x)$.

Solution

$$f'(x) = [\cos (x^3 - 5x)] \cdot (3x^2 - 5)$$

$$= (3x^2 - 5) \cos (x^3 - 5x) \qquad \blacklozenge$$

EXAMPLE 4   Differentiate $y = \cos (e^{4x}) \sin x^2$.

Solution   We use the Product Rule and the Chain Rule:

$$\frac{dy}{dx} = \cos (e^{4x}) \cdot 2x \cdot \cos x^2 - \sin (e^{4x}) \cdot (4e^{4x}) \cdot \sin x^2$$

$$= 2x \cos (e^{4x}) \cos x^2 - 4e^{4x} \sin (e^{4x}) \sin x^2. \qquad \blacklozenge$$

TECHNOLOGY
CONNECTION

Use a grapher that graphs
both f and f'. For each of
the functions in Exam-
ples 1–4, graph the function
and its derivative. Use the
graphs to verify the results.

Applications

Equations of the type

$$y = A \sin (Bx - C) + D$$

have many applications. We can also express this equation as

$$y = A \sin \left[B \left(x - \frac{C}{B} \right) \right] + D.$$

The numbers A, B, C, and D play an important role in graphing such an equation. The number A corresponds to a vertical stretching or shrinking of the graph of $y = \sin x$, B corresponds to a horizontal stretching or shrinking, C/B corresponds to a shift of the entire graph to the right or left, and D corresponds to a vertical translation. These particular expressions are given the following names:

$$Amplitude = |A|,$$

$$Period = \frac{2\pi}{B},$$

$$Phase\ shift = \frac{C}{B}.$$

EXAMPLE 5   *Physical science: Spring oscillation.* A weight is attached to the end of a spring. When the weight is disturbed, it bobs up and down with a definite frequency. If the motion were to occur in a perfect vacuum and if the spring were perfectly elastic, then the oscillatory motion would continue undiminished forever. Suppose that a spring oscillates in such a way that its vertical position at time t, from its position at rest, is given by

$$y = 3 \sin \left(2t + \frac{\pi}{2} \right) = 3 \sin \left[2 \left(t - \left(-\frac{\pi}{4} \right) \right) \right].$$

a) Find the amplitude, the period, and the phase shift.

b) Graph the equation.

c) Find dy/dt.

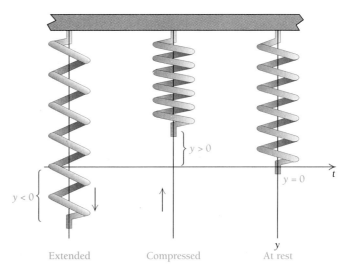

Solution

a) The amplitude $= |3| = 3$. The period is $2\pi/2 = \pi$. The phase shift $= -(\pi/4)$.

b) We plot the various equations needed to get the final graph. The graph of $y = 3 \sin t$ is a vertical stretching, by a factor of 3, of the graph of $y = \sin t$. The period of $y = 3 \sin t$ is still 2π.

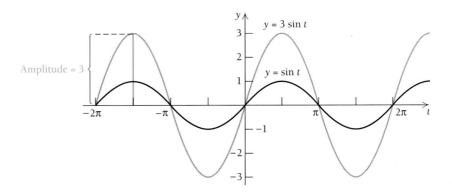

The graph of $y = 3 \sin 2t$ is a horizontal shrinking, by a factor of $\frac{1}{2}$, of the graph of $y = 3 \sin t$. The period of $y = 3 \sin 2t$ is π.

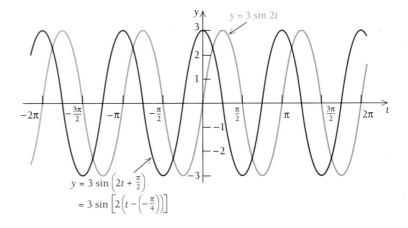

The graph of

$$y = 3 \sin \left[2 \left(t - \left(-\frac{\pi}{4} \right) \right) \right]$$

is shifted $\pi/4$ units to the left of the graph of $y = 3 \sin 2t$. If the phase shift were positive, the shift would be to the right.

c) $\dfrac{dy}{dt} = \left[3 \cos \left(2t + \dfrac{\pi}{2} \right) \right] 2 = 6 \cos \left(2t + \dfrac{\pi}{2} \right)$ ◆

The motion of the spring in the preceding example is called *simple harmonic motion*. Other types of simple harmonic motion are sound waves and light waves. For sound waves, the amplitude is the loudness. For light waves, the amplitude is the brightness. The *frequency,* which is the reciprocal of the period, is the pitch of a sound wave and the color of a light wave.

A Biological Application: Biorhythms

Some people conjecture that a person's life has cycles of good and bad days that begin at birth. There are supposedly three such cycles, or *biorhythms.*

1. *Physical.* A cycle represented by a sine function with a period of 23 days:

$$y = \sin \frac{2\pi}{23}t.$$

   This cycle is related to physical qualities such as vitality, strength, and energy.

2. *Emotional (sensitivity).* A cycle represented by a sine function with a period of 28 days:

$$y = \sin \frac{2\pi}{28}t.$$

   This cycle is related to creativity, moodiness, intuition, and cheerfulness.

3. *Mental (intellectual).* A cycle represented by a sine function with a period of 33 days:

$$y = \sin \frac{2\pi}{33}t.$$

   This cycle is related to the ability to study, think, react, and remember. (The positive part of *this* cycle would be a good time to study calculus.)

Because these cycles have periods of different lengths, there are varying combinations of highs and lows throughout one's life. This is illustrated in the following figure.*

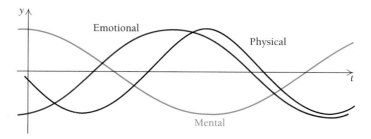

TECHNOLOGY
CONNECTION

Suppose a woman born on April 13, 1947, gave birth to a child on December 20, 1975. Investigate the physical, emotional, and mental biorhythms of the mother on the day she gave birth. Do so by sketching appropriate biorhythm graphs.

EXAMPLE 6   *Biology: Minimizing the surface area of a bee's cell.* Did you know that a honey bee constructs the cells in its comb in such a way that the minimum amount of wax is used?† One cell of a honeycomb is shown below. It is a solid whose upper base is a regular hexagon. The bottom comes together at point *A*. The surface

---

*For more on these biorhythms, write to Biorhythm Computers, Inc., 298 Fifth Ave, New York, NY 10001.
†This discovery was published in a study by Sir D'Arcy Wentworth Thompson, *On Growth and Form* (Cambridge University Press, 1917).

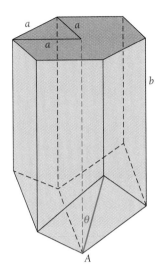

area is given by

$$S(\theta) = 6ab + \frac{3}{2}a^2\left(\frac{\sqrt{3} - \cos\theta}{\sin\theta}\right),$$

where θ is the measure of what is called the angle of inclination. To maintain the prism shape of the cell, we can vary the values of θ between $0°$ and $90°$. Thus we want to minimize S on the open interval $(0, 90)$.

Solution   We first find $dS/d\theta$:

$$\frac{dS}{d\theta} = \frac{3}{2}a^2\left[\frac{\sin\theta\,(\sin\theta) - \cos\theta\,(\sqrt{3} - \cos\theta)}{\sin^2\theta}\right]$$

$$= \frac{3}{2}a^2\left[\frac{\sin^2\theta - \sqrt{3}\cos\theta + \cos^2\theta}{\sin^2\theta}\right]$$

$$= \frac{3}{2}a^2\left[\frac{1 - \sqrt{3}\cos\theta}{\sin^2\theta}\right].$$

Critical points occur where $\sin\theta$ is 0. But if we check such values, we see that at such values the bottom of the cell would be flat, which is not the way it is to be constructed. Thus we set $dS/d\theta = 0$ and solve for θ with the assumption that $\sin\theta \neq 0$:

$$\frac{dS}{d\theta} = \frac{3}{2}a^2\left[\frac{1 - \sqrt{3}\cos\theta}{\sin^2\theta}\right] = 0$$

$$\frac{1 - \sqrt{3}\cos\theta}{\sin^2\theta} = 0$$

$$1 - \sqrt{3}\cos\theta = 0 \qquad \text{Multiplying by } \sin^2\theta$$

$$\cos\theta = \frac{1}{\sqrt{3}}$$

$$= \frac{\sqrt{3}}{3} \approx 0.5774.$$

A honeycomb.

We need to find θ such that $\cos\theta \approx 0.5774$. We can work backward on a calculator using a $\boxed{\cos^{-1}}$ key. It follows that

$$\theta \approx 54.7°.$$

The study showed that bees tend to use this angle. That they do may be explained by a genetic theory that those who survive to pass on their genetic traits are those who are more "successful" at living. Perhaps bees that waste less time and wax by making cells with the minimum amount of wax were, at one time, such successful genetic survivors.   ◆

8.2 Exercise Set

Prove the following derivative formulas.

1. $\dfrac{d}{dx} \sec x = \tan x \sec x$

2. $\dfrac{d}{dx} \csc x = -\cot x \csc x$

Differentiate.

3. $y = x \sin x$

4. $y = x \cos x$

5. $f(x) = e^x \sin x$

6. $f(x) = e^x \cos x$

7. $y = \dfrac{\sin x}{x}$

8. $y = \dfrac{\cos x}{x}$

9. $f(x) = \sin^2 x$

10. $f(x) = \cos^2 x$

11. $y = \sin x \cos x$

12. $y = \cos^2 x + \sin^2 x$

13. $f(x) = \dfrac{\sin x}{1 + \cos x}$

14. $f(x) = \dfrac{1 - \cos x}{\sin x}$

15. $y = \tan^2 x$

16. $y = \sec^2 x$

17. $f(x) = \sqrt{1 + \cos x}$

18. $f(x) = \sqrt{1 - \sin x}$

19. $y = x^2 \cos x - 2x \sin x - 2 \cos x$

20. $y = x^2 \sin x - 2x \cos x + 2 \sin x$

21. $y = e^{\sin x}$

22. $y = e^{\cos x}$

23. Find d^2y/dx^2 if $y = \sin x$.

24. Find d^2y/dx^2 if $y = \cos x$.

25. $y = \sin (x^2 + x^3)$

26. $y = \sin (x^5 - x^4)$

27. $f(x) = \cos (x^5 - x^4)$

28. $f(x) = \cos (x^2 + x^3)$

29. $f(x) = \cos \sqrt{x}$

30. $f(x) = \sin \sqrt{x}$

31. $y = \sin (\cos x)$

32. $y = \cos (\sin x)$

33. $y = \sqrt{\cos 4x}$

34. $y = \sin^2 5x$

35. $f(x) = \cot \sqrt[3]{5 - 2x}$

36. $f(x) = \sec (\tan 7x)$

37. $y = \tan^4 3x - \sec^4 3x$

38. $y = \sqrt[5]{\cot 5x - \cos 5x}$

Differentiate. Use the formula $\dfrac{d}{du} \ln |u| = \dfrac{1}{u} du$.

39. $y = \ln |\sin x|$

40. $f(x) = \ln |x - \cos x|$

APPLICATIONS

◆ **Business and Economics**

41. **Total sales.** A company determines that sales during the tth month are given by

$$S(t) = 40{,}000(\sin t + \cos t).$$

The sales are seasonal and fluctuate. Find $S'(t)$.

42. **Total sales.** A company in a northern climate has total sales of skis as given by

$$S(t) = 7\left(1 - \cos \frac{\pi}{6} t\right).$$

Find $S'(t)$.

◆ **Life and Physical Sciences**

43. **Temperature during an illness.** The temperature of a patient during a 12-day illness is given by

$$T(t) = 101.6° + 3 \sin \frac{\pi}{8} t.$$

Find $T'(t)$.

44. **Rollercoaster layout.** A rollercoaster is constructed in such a way that it is y meters above ground x meters from the starting point, where

$$y = 15 + 15 \sin \frac{\pi}{50} x.$$

Find dy/dx.

45. Satellite location. A satellite circles the earth in such a manner that it is y miles from the equator (north or south, height not considered) t minutes after its launch, where

$$y = 5000\left[\cos\frac{\pi}{45}(t - 10)\right].$$

Find dy/dt.

46. Electric current. The current i at time t of a wire passing through a magnetic field is given by

$$i = I \sin(\omega t + a).$$

a) Find the amplitude, the period, and the phase shift.
b) Find di/dt.

47. Spring oscillation. A spring oscillates in such a way that its vertical position at time t, from its position at rest, is given by

$$y = 5 \sin(4t + \pi).$$

a) Find the amplitude, the period, and the phase shift.
b) Find dy/dt.

48. Piston movement. A piston connected to a crankshaft (see the figure) moves up and down in such a way that its second coordinate after time t is given by

$$y(t) = \sin t + \sqrt{25 - \cos^2 t}.$$

Find the rate of change $y'(t)$.

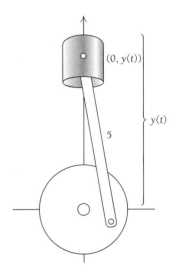

49. Corridor width. A corridor of width a meets a corridor of width b at right angles (see the figure). Workers wish to push a heavy beam of length L on dollies around the corner. Before starting, however, they want to be sure they can make the turn. How long a beam will go around the corner? (Disregard the width of the beam.)

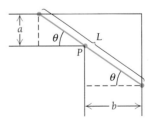

The length L of the beam is given by

$$L = \frac{a}{\sin\theta} + \frac{b}{\cos\theta},$$

where θ is the angle shown above.

a) Find L in terms of a and b such that L is of maximum length.
b) Use the formula in part (a) to find how long a beam will go around the corner when $a = 8$ and $b = 8$.

50. *Corridor width.* Referring to Exercise 49, determine how long a beam will go around the corner when $a = 3\sqrt{3}$ and $b = 5\sqrt{5}$.

51. *Sound waves.* A sound wave is given by

$$y = 0.03 \sin 1.198 \, \pi x.$$

Find dy/dx.

52. *Light waves.* A light wave is given by

$$L = A \sin \left(2\pi f T - \frac{d}{\omega} \right).$$

Find dL/dT.

♦ **General Interest**

53. A V-shaped water trough is to be made with sides that are 10 in. wide and 500 in. long. At what angle θ will the trough be able to carry the greatest volume of water?

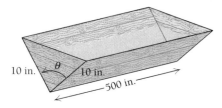

54. The function f given by $f(x) = x + \sin x$ achieves an absolute maximum on the interval $[0, \pi]$. Find the maximum value and where it occurs.

SYNTHESIS

Find the partial derivatives f_x, f_y, f_{xx}, f_{yx}, f_{xy}, and f_{yy}.
55. $f(x, y) = \sin 2y - ye^x$
56. $f(x, y) = e^{-x} \cos y$
57. $f(x, y) = \cos (2x + 3y)$
58. $f(x, y) = y \ln (\sin x)$
59. $f(x, y) = x^3 \tan (5xy)$ 60. $f(x, y) = \sin x \cos y$

Differentiate implicitly to find y'.
61. $y = x \cos y$ 62. $xy = e^x \sin y$

63.   Explain to a fellow student the proof of the formula for the derivative of the sine function.

64.   How many maximum values are there of the cosine function? Where do they occur?

TECHNOLOGY CONNECTION

Graph each of the following functions.
65. $f(x) = x \cos x$
66. $f(x) = x \sin x$
67. $f(x) = 2 \sin x + \sin 2x$
68. $f(x) = e^{-x/2} \sin x$
69. $f(x) = \ln |\sec x + \tan x|$
70. $f(x) = \sin x + \ln x$

Use a 3D grapher to graph each of these surfaces.
71. $z = 1 + \cos (x^2 + y^2)$
72. $z = e^{-y} \cos x$

73. Find the zeros of $f(x) = \sin x - \cos x$ on the interval $[-2\pi, 2\pi]$.

8.3 Integration of the Trigonometric Functions

OBJECTIVE
• Integrate trigonometric functions.

Each of the previously developed differentiation formulas yields an integration formula. For example, we have the following.

THEOREM 8

$$\int \sin x \, dx = -\cos x + C, \quad \int \cos x \, dx = \sin x + C,$$

$$\int \sec^2 x \, dx = \tan x + C, \quad \int \csc^2 x \, dx = -\cot x + C$$

EXAMPLE 1   Find the area under the graph of $y = \cos x$ on the interval $[0, \pi/2]$.

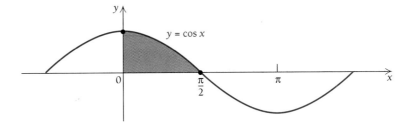

Solution

$$\int_0^{\pi/2} \cos x \, dx = [\sin x]_0^{\pi/2} = \sin \frac{\pi}{2} - \sin 0 = 1 - 0 = 1 \qquad \blacklozenge$$

We use substitution in the following examples.

EXAMPLE 2   Evaluate $\int \sin^2 x \cos x \, dx$.

Solution

$$\int \sin^2 x \cos x \, dx = \int u^2 \, du \qquad \underline{\text{Substitution}} \quad \boxed{\begin{array}{l} u = \sin x, \\ du = \cos x \, dx \end{array}}$$

$$= \frac{u^3}{3} + C$$

$$= \frac{\sin^3 x}{3} + C \qquad \blacklozenge$$

EXAMPLE 3   Evaluate $\int \sin 3x \, dx$.

Solution

$$\int \sin 3x \, dx = \tfrac{1}{3} \int \sin u \, du \qquad \underline{\text{Substitution}} \quad \boxed{\begin{array}{l} u = 3x, \\ du = 3 \, dx \end{array}}$$

$$= -\tfrac{1}{3} \cos u + C$$

$$= -\tfrac{1}{3} \cos 3x + C \qquad \blacklozenge$$

In an expression where natural logarithms of trigonometric functions are involved, absolute-value signs are usually necessary because the trigonometric functions alternate periodically between positive and negative values. Accordingly, we use the integration formula

$$\int \frac{du}{u} = \ln |u| + C.$$

EXAMPLE 4 Evaluate $\int \dfrac{1 + \cos x}{x + \sin x} \, dx$.

Solution

$$\int \frac{1 + \cos x}{x + \sin x} \, dx = \int \frac{du}{u} \qquad \underline{\text{Substitution}} \quad \boxed{\begin{array}{l} u = x + \sin x, \\ du = (1 + \cos x) \, dx \end{array}}$$

$$= \ln |u| + C$$

$$= \ln |x + \sin x| + C \qquad\qquad \blacklozenge$$

We use integration by parts in the following example.

EXAMPLE 5 Integrate $\int x \cos x \, dx$.

Solution Let

$$u = x \quad \text{and} \quad dv = \cos x \, dx.$$

Then

$$du = dx \quad \text{and} \quad v = \sin x.$$

Using the integration-by-parts formula, we get

$$\int \overset{u}{(x)} \overset{dv}{(\cos x \, dx)} = \overset{u}{(x)} \overset{v}{(\sin x)} - \int \overset{v}{(\sin x)} \overset{du}{(dx)}$$

$$= x \sin x - (-\cos x) + C$$

$$= x \sin x + \cos x + C. \qquad\qquad \blacklozenge$$

To find an integral such as $\int \sec x \, dx$, we must first use some algebra to rename the expression. Then we use substitution.

EXAMPLE 6 Integrate $\int \sec x \, dx$.

Solution

$$\int \sec x \, dx = \int (\sec x) \cdot 1 \, dx \qquad\qquad \text{Multiplying by 1}$$

$$= \int \sec x \cdot \frac{\sec x + \tan x}{\sec x + \tan x} \, dx \qquad \text{Substituting } \frac{\sec x + \tan x}{\sec x + \tan x} \text{ for 1}$$

Then

$$= \int \frac{\sec x \, (\sec x + \tan x)}{\sec x + \tan x} \, dx$$

$$= \int \frac{\sec x \tan x + \sec^2 x}{\sec x + \tan x} \, dx$$

$$= \int \frac{du}{u} \qquad \underline{\text{Substitution}} \quad \boxed{\begin{array}{l} u = \sec x + \tan x, \\ du = (\sec x \tan x + \sec^2 x) \, dx \end{array}}$$

$$= \ln |u| + C$$

$$= \ln |\sec x + \tan x| + C$$

8.3 Exercise Set

1. Find the area under the graph of $y = \sin x$ on the interval $[0, \pi/3]$.

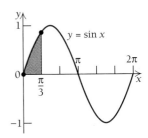

2. Find the area under the graph of $y = \cos x$ on the interval $[-\pi/2, \pi/2]$.

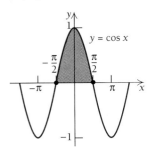

Evaluate.

3. $\displaystyle\int_{-\pi}^{\pi} \cos x \, dx$

4. $\displaystyle\int_{-\pi/2}^{\pi/2} \sin x \, dx$

5. $\displaystyle\int_{-\pi/4}^{2\pi} \sec^2 x \, dx$

6. $\displaystyle\int_{\pi/4}^{\pi/3} \csc^2 x \, dx$

Integrate using substitution.

7. $\displaystyle\int \sin^4 x \cos x \, dx$

8. $\displaystyle\int \sin^5 x \cos x \, dx$

9. $\displaystyle\int -\cos^2 x \sin x \, dx$

10. $\displaystyle\int (\cos^3 x)(-\sin x) \, dx$

11. $\displaystyle\int \cos (x + 3) \, dx$

12. $\displaystyle\int \sin (x + 4) \, dx$

13. $\displaystyle\int \sin 2x \, dx$

14. $\displaystyle\int \cos 3x \, dx$

15. $\displaystyle\int x \cos x^2 \, dx$

16. $\displaystyle\int x \sin x^2 \, dx$

17. $\displaystyle\int e^x \sin (e^x) \, dx$

18. $\displaystyle\int e^x \cos (e^x) \, dx$

19. $\displaystyle\int \tan x \, dx \left(\text{Hint: } \tan x = \frac{\sin x}{\cos x}. \right)$

20. $\displaystyle\int \cot x \, dx \left(\text{Hint: } \cot x = \frac{\cos x}{\sin x}. \right)$

Integrate by parts.

21. $\displaystyle\int x \cos 4x \, dx$

22. $\displaystyle\int x \sin 3x \, dx$

23. $\displaystyle\int 3x \cos x \, dx$

24. $\displaystyle\int 2x \sin x \, dx$

25. $\int x^2 \sin x \, dx$ (*Hint:* Let $u = x$ and $dv = x \sin x \, dx$ or use tables of integration.)

26. $\int x^2 \cos x \, dx$ (*Hint:* Let $u = x$ and $dv = x \cos x \, dx$ and use the result of Example 5, or use tables of integration.)

27. $\int \tan^2 x \, dx$ **28.** $\int \cot^2 x \, dx$

APPLICATIONS

◆ **Business and Economics**

29. *Total sales.* A company in a northern climate has sales of skis as given by

$$S(t) = 7\left(1 - \cos \frac{\pi}{6}t\right),$$

where S = sales, in thousands of dollars, during the tth month. The total sales for the first year are given by

$$\int_0^{12} S(t) \, dt.$$

Find the total sales.

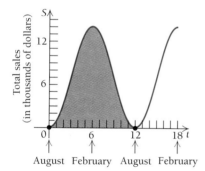

◆ **General Interest**

30. *Total area under a rollercoaster.* A rollercoaster is made in such a way that it is y meters above the ground x meters from the starting point, where

$$y = 15 + 15 \sin \frac{\pi}{50}x.$$

The area under the rollercoaster, from the starting point to a point 100 meters away, is given by

$$\int_0^{100} y \, dx.$$

Find this area.

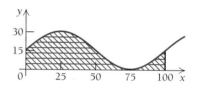

SYNTHESIS

Integrate by parts.

31. $\int \sin (\ln x) \, dx$ **32.** $\int \cos (\ln x) \, dx$

33. $\int e^x \cos x \, dx$ **34.** $\int e^x \sin x \, dx$

35. $\int x^3 \sin x \, dx$ **36.** $\int x^4 \cos x \, dx$

Integrate.

37. $\int \sec^2 7x \, dx$

38. $\int e^x \csc^2 (e^x) \, dx$

39. $\int \sec x \, (\sec x + \tan x) \, dx$

40. $\int \sec u \tan u \, du$ **41.** $\int \csc u \cot u \, du$

42. $\int e^{2x} \sec (e^{2x}) \, dx$ **43.** $\int \frac{\cos^2 x}{\sin x} \, dx$

44. $\int \cot 7x \sin 7x \, dx$ **45.** $\int (1 + \sec x)^2 \, dx$

46. $\int (1 - \csc x)^2 \, dx$

47. $\int_{\pi/4}^{\pi/2} \int_0^{\pi/2} \sin x \cos y \, dy \, dx$

48. $\int_0^{\pi} \int_0^x \frac{\sin x}{x} \, dy \, dx$

49. $\int_0^{\pi} \int_0^x x \cos y \, dy \, dx$

50. $\int_{\pi}^{2\pi} \int_0^{\pi} (\sin y + \cos x) \, dx \, dy$

51. ◈ Suppose you know that

$$\int_a^b \sin x \, dx = 0.$$

Explain the meaning of this result in terms of area.

52. ◈ For $y' = 5 \sin x$, find y. Then list and explain some of the particular solutions. (This presumes that you have studied Section 6.7.)

TECHNOLOGY CONNECTION

Graph each function over the interval indicated by the integral. Shade the area under the graph. Then approximate the integral.

53. $\int_{-\pi}^{\pi} \cos x \, dx$ **54.** $\int_{-\pi/2}^{\pi/2} \sin x \, dx$

55. $\int_{3\pi/4}^{\pi} \sec^2 x \, dx$ · **56.** $\int_{\pi/4}^{\pi/3} \csc^2 x \, dx$

Approximate the double integral.

57. $\int_0^{\pi} \int_0^x \frac{\sin x}{x} \, dy \, dx$

58. $\int_0^{\pi} \int_0^x x \cos y \, dy \, dx$

8.4 Inverse Trigonometric Functions

OBJECTIVES

• Find values of inverse trigonometric functions.

• Integrate functions whose antiderivatives involve inverse trigonometric functions.

Look back at the graph of $y = \sin x$. Note that outputs ranged from -1 to 1. Suppose we wanted to work backward from an output to an input. More specifically, suppose x is some number such that $-1 \le x \le 1$ and that we wanted to find a number y such that

$$-\frac{\pi}{2} \le y \le \frac{\pi}{2} \quad \text{and} \quad \sin y = x.$$

This determines a function, called the *inverse sine function*, given by

$$y = \sin^{-1} x, \quad \text{or} \quad \arcsin x.$$

Caution!   The -1 is *not* an exponent in this context and $\sin^{-1} x$ is *not* the reciprocal of $\sin x$. The function $\sin^{-1} x$ is an **inverse trigonometric function.**

Let us find a function value.

EXAMPLE 1   Find $\sin^{-1} (\sqrt{3}/2)$.

Solution   *Think:* What number between $-(\pi/2)$ and $\pi/2$ is such that its sine is $\sqrt{3}/2$? That number is $\pi/3$. Thus,

$$\sin^{-1}\left(\frac{\sqrt{3}}{2}\right) = \frac{\pi}{3}, \quad \text{which means that} \quad \sin\frac{\pi}{3} = \frac{\sqrt{3}}{2}. \qquad \blacklozenge$$

TECHNOLOGY
CONNECTION

Find each of the following
using your grapher:

$\sin^{-1}\left(\frac{1}{2}\right)$,

$\sin^{-1}\left(-\frac{\sqrt{3}}{2}\right)$,

$\tan^{-1}\sqrt{3}$,

$\cos^{-1}\left(\frac{\sqrt{3}}{2}\right)$,

$\cos^{-1}\left(-\frac{1}{2}\right)$.

The *inverse cosine function* is given by

$$y = \cos^{-1} x, \quad \text{or} \quad y = \arccos x,$$

where

$$x = \cos y, \quad 0 \leq y \leq \pi, \quad \text{and} \quad -1 \leq x \leq 1.$$

The *inverse tangent function* is given by

$$y = \tan^{-1} x, \quad \text{or} \quad y = \arctan x,$$

where

$$x = \tan y, \quad -\frac{\pi}{2} < y < \frac{\pi}{2}, \quad \text{and} \quad x \text{ is any real number.}$$

Graphs of these functions follow.

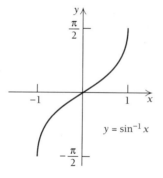

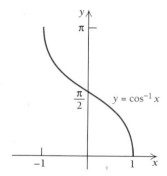

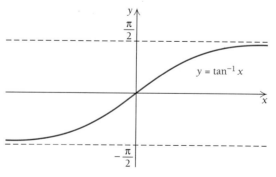

TECHNOLOGY
CONNECTION

Graph the inverse trigonometric functions. Do the
graphs agree with those in
the text? Find where the
graphs of $y = \sin^{-1} x$ and
$y = \cos^{-1} x$ intersect.

Inverse trigonometric functions are integrals of certain functions. You will often see them in tables of integrals.

EXAMPLE 2   Prove that

$$\int \frac{1}{1 + x^2} \, dx = \tan^{-1} x + C.$$

Solution   We use a method of integration known as *trigonometric substitution*. We simplify the integral using the identity $1 + \tan^2 u = \sec^2 u$:

$$\int \frac{1}{1 + x^2} \, dx = \int \frac{1}{1 + \tan^2 u} \sec^2 u \, du \quad \underline{\text{Substitution}} \quad \boxed{\begin{array}{l} u = \tan^{-1} x, \\ x = \tan u, \\ dx = \sec^2 u \, du \end{array}}$$

$$= \int \frac{1}{\sec^2 u} \sec^2 u \, du \quad 1 + \tan^2 u = \sec^2 u$$

$$= \int du = u + C = \tan^{-1} x + C. \qquad \blacklozenge$$

EXAMPLE 3   Prove that

$$\int -\frac{1}{\sqrt{1 - x^2}} \, dx = \cos^{-1} x + C.$$

Solution   Again we use substitution:

$$\int -\frac{1}{\sqrt{1 - x^2}} \, dx = \int \frac{1}{\sqrt{1 - \cos^2 u}} \sin u \, du. \quad \underline{\text{Substitution}} \quad \boxed{\begin{array}{l} u = \cos^{-1} x, \\ x = \cos u, \\ dx = -\sin u \, du \end{array}}$$

By Identity (1), $\sin^2 u + \cos^2 u = 1$, so

$$1 - \cos^2 u = \sin^2 u.$$

Then, since u is such that $0 \le u \le \pi$, $\sin u \ge 0$, so $\sqrt{\sin^2 u} = \sin u$ and the integral becomes

$$\int \frac{1}{\sqrt{\sin^2 u}} \cdot \sin u \, du \quad \text{By Identity (1)}$$

$$= \int \frac{1}{\sin u} \cdot \sin u \, du = \int du = u + C = \cos^{-1} x + C. \qquad \blacklozenge$$

8.4 Exercise Set

Find each of the following.

1. $\sin^{-1} \left(\dfrac{\sqrt{2}}{2} \right)$

2. $\sin^{-1} \left(-\dfrac{\sqrt{2}}{2} \right)$

5. $\tan^{-1} (1)$

6. $\tan^{-1} (-1)$

3. $\cos^{-1} (0)$

4. $\cos^{-1} \left(\dfrac{\sqrt{2}}{2} \right)$

7. $\sin^{-1} \left(-\dfrac{1}{2} \right)$

8. $\cos^{-1} \left(-\dfrac{\sqrt{3}}{2} \right)$

Integrate using substitution.

9. $\displaystyle\int \frac{e^t}{1 + e^{2t}}\, dt$

10. $\displaystyle\int \frac{-1}{\sqrt{1 - 25x^2}}\, dx$

11. $\displaystyle\int \frac{1}{1 + 25x^2}\, dx$

12. $\displaystyle\int -\frac{1}{\sqrt{1 - 4x^2}}\, dx$

13. Show that

$$\int \frac{1}{\sqrt{1 - x^2}}\, dx = \sin^{-1} x + C.$$

14. Integrate using the formula in Exercise 13:

$$\int \frac{1}{\sqrt{1 - 49x^2}}\, dx.$$

Find each of the following.

15. $\sin^{-1}(0.9874)$

16. $\cos^{-1}(-0.3487)$

17. $\cos^{-1}(0.9988)$

18. $\tan^{-1}(2000)$

SYNTHESIS

19. Since

$$\int \frac{1}{1 + x^2}\, dx = \tan^{-1} x + C,$$

it follows that if $u = \tan^{-1} x$, then

$$\frac{du}{dx} = \frac{1}{1 + x^2}.$$

Use this result and integration by parts to find

$$\int \tan^{-1} x\, dx.$$

8 Chapter Summary and Review

TERMS TO KNOW

Angle, p. 538
Degrees, p. 538
Radians, p. 539
Trigonometric function, p. 540
Circular function, p. 540
Sine function, pp. 540, 541

Cosine function, pp. 540, 541
Periodic function, p. 545
Tangent function, p. 546
Cotangent function, p. 546
Secant function, p. 546
Cosecant function, p. 546

Trigonometric identity, p. 548
Sum–difference identities, p. 549
Double-angle identity, p. 549
Inverse trigonometric function, p. 567

REVIEW EXERCISES

These review exercises are for test preparation. They can also be used as a lengthened practice test. Answers are at the back of the book. The answers also contain bracketed section references, which tell you where to restudy if your answer is incorrect.

Convert to radian measure.

1. $240°$

2. $315°$

Convert to degree measure.

3. $\dfrac{5\pi}{6}$

4. $\dfrac{7\pi}{3}$

Find each of the following.

5. $\sin \dfrac{2\pi}{3}$

6. $\cos\left(-\dfrac{\pi}{2}\right)$

7. $\tan \dfrac{\pi}{4}$

8. $\csc \pi$

Differentiate.

9. $y = \tan(3x^2 - x)$ **10.** $y = \sin^6 x$

11. $y = \ln(\tan x)$ **12.** $f(x) = e^{\sin 3x}$

13. $f(x) = \dfrac{1 + \cos 2x}{\sin 2x}$

14. $f(x) = \cos \sqrt{t} + \sqrt{\cos t}$

15. $y = \sqrt{\tan x}$ **16.** $y = e^x \sin x$

17. $f(x) = \sin 3x \cos 3x$ **18.** $y = \sin x - x \cos x$

19. *Physical science: Pendulum oscillation.* A pendulum oscillates in such a way that its horizontal position at time t, from its position at rest, is given by

$$y = 3 \cos(3t - 2\pi).$$

Find dy/dt.

20. *Life science: Temperature during an illness.* The temperature of a patient during an 8-day illness is given by

$$T(t) = 99.7° + 5 \sin \frac{\pi}{3} t.$$

Find $T'(t)$.

21. Find the area under the graph of $y = \cos x$ on the interval $[0, \pi/6]$.

Integrate using substitution.

22. $\displaystyle \int e^x \cos e^x \, dx$ **23.** $\displaystyle \int \sin(x - 1) \, dx$

24. $\displaystyle \int \cos 8x \, dx$

Integrate by parts.

25. $\displaystyle \int 3x \cos 2x \, dx$ **26.** $\displaystyle \int x \sin 3x \, dx$

Integrate.

27. $\displaystyle \int \frac{1}{\sqrt{1 - 64t^2}} \, dt$ **28.** $\displaystyle \int \frac{5 - \cos x}{5x - \sin x} \, dx$

SYNTHESIS _____

29. Find $\dfrac{d^3y}{dx^3}$: $y = e^{\sin x}$.

30. Find f_x and f_{xy}: $f(x, y) = \sin y \cos x$.

 TECHNOLOGY CONNECTION

Determine whether each of the following is an identity.

31. $\cos x + \sin 2x = 2 \cos x + \sin x$

32. $\dfrac{\tan x + \sin x}{1 + \sec x} = \sin x$

8 Chapter Test

1. Convert $120°$ to radian measure. (Leave the answer in terms of π.)

2. Convert $5\pi/3$ to degree measure.

Find each of the following.

3. $\sin \dfrac{\pi}{6}$ **4.** $\cos \pi$

Differentiate.

5. $y = \cos t$ **6.** $y = \sin(3x^2 - 5x)$

7. $f(x) = \dfrac{x}{\sin x}$ **8.** $f(t) = \tan^2 t$

9. $f(x) = \sqrt{\sin x + \cos x}$

10. $y = \dfrac{\sin x + \cos x}{\sin x - \cos x}$

11. **Business: Seasonal sales.** A company has seasonal sales as given by

$$S(t) = 20\left(1 + \sin \frac{\pi}{8}t\right).$$

Find $S'(t)$.

12. Find the area under $y = \sin x$ on the interval $[0, \pi/6]$.

Integrate using substitution.

13. $\displaystyle\int \sin^9 x \cos x \, dx$

14. $\displaystyle\int \cos 5t \, dt$

15. Integrate by parts:

$$\int 5x \sin 5x \, dx.$$

16. Integrate using substitution:

$$\int \frac{1}{1 + 16t^2} \, dt.$$

Integrate.

17. $\displaystyle\int \frac{2 - \cos x}{2x - \sin x} \, dx$

18. $\displaystyle\int_0^{\pi/4} \sec^2 x \, dx$

SYNTHESIS

19. Find

$$\frac{dy}{dx} \quad \text{and} \quad \frac{d^2y}{dx^2},$$

if $y = 2e^{\cos x}$.

20. Find f_x and f_{xy}: $f(x, y) = \dfrac{\cos x}{\cos y}$.

21. Use an input–output table or a graph to find the following limit:

$$\lim_{x \to 0} \ln |x + \sin x|.$$

TECHNOLOGY CONNECTION

Graph each of the following.

22. $f(x) = |3 \sin x|$

23. $f(x) = 3 \sin x - \cos 2x$

EXTENDED TECHNOLOGY APPLICATION

The Cyclical Business of New Home Sales

Housing is a large part of the business of a community, serving as an indicator of its growth and potential.

If we were to draw a bar graph to examine the data, we would have the following.

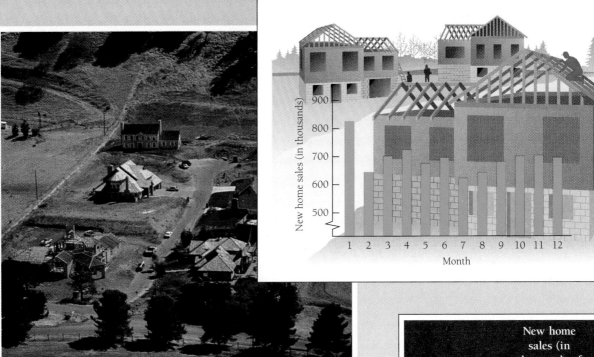

EXERCISE

1. Argue whether or not you think the number of new homes built in an area each month might follow that of a periodic or trigonometric function.

Let's explore some recent data (shown in the table at right) regarding new home sales as compiled by the U.S. Department of Commerce.

Month	New home sales (in thousands of units)
1. December	825
2. January	640
3. February	700
4. March	725
5. April	675
6. May	695
7. June	625
8. July	640
9. August	680
10. September	700
11. October	705
12. November	695

Examining the data, we see a somewhat up-and-down, cyclical nature to the graph. This suggests that new home sales may indeed be periodic or trigonometric. Unfortunately, most graphers are not equipped to fit trigonometric curves to data. Nevertheless, we can use our graphers to examine some possible curve-fitting functions given as follows.

EXERCISES

Consider each of the following functions in relation to the data above.

$$y_1 = 1982 \sin\left(\frac{\pi x}{24}\right) - 850,$$

$$y_2 = 1982 \sin\left(\frac{\pi x}{24}\right) + 244 \sin\left(\frac{\pi x}{12}\right) + 669 \sin\left(\frac{\pi x}{8}\right)$$
$$- 850,$$

$$y_3 = 1982 \sin\left(\frac{\pi x}{24}\right) + 244 \sin\left(\frac{\pi x}{12}\right) + 669 \sin\left(\frac{\pi x}{8}\right)$$
$$+ 502 \sin\left(\frac{\pi x}{6}\right) + 420 \sin\left(\frac{\pi x}{4}\right) - 950,$$

$$y_4 = 1982 \sin\left(\frac{\pi x}{24}\right) + 244 \sin\left(\frac{\pi x}{12}\right) + 335 \sin\left(\frac{\pi x}{8}\right)$$
$$+ 250 \sin\left(\frac{\pi x}{6}\right) + 210 \sin\left(\frac{\pi x}{4}\right)$$
$$+ 68 \sin\left(\frac{3\pi x}{2}\right) - 950$$

Use your grapher to do Exercises 2–6.

2. Graph the data points using the STAT PLOTS feature if available.

3. Graph each function using the viewing window [0, 24, −500, 2000].

4. Find the derivative of each function.

5. Use each function to predict new sales in the 13th, 18th, 20th, and 24th months.

6. Which function, if any, seems best to you as a predictor of new home sales? Explain.

7. Assume that the function y_3 is a good predictor of new home sales. On what intervals is the function increasing? On what intervals is it decreasing? Find the relative extrema.

Consider the function

$$y_5 = 703 - \frac{203}{x} + \frac{320}{x^2}.$$

Use your grapher to do Exercises 8–12.

8. Graph the function y_5 using the viewing window [0, 24, −500, 2000].

9. Find the derivative of the function.

10. Use the function to predict new sales in the 13th, 18th, 20th, and 24th months.

11. On what intervals is the function increasing? On what intervals is it decreasing? Find the relative extrema.

12. Analyze the function y_5 as a predictor of new home sales when compared to functions y_1, y_2, y_3, and y_4.

Cumulative Review

1. Write an equation of the line with slope -4 and containing the point $(-7, 1)$.

2. For $f(x) = x^2 - 5$, find $f(x + h)$.

3. a) Graph:

$$f(x) = \begin{cases} 5 - x, & \text{for } x \neq 2, \\ -3, & \text{for } x = 2. \end{cases}$$

   b) Find $\lim_{x \to 2} f(x)$.
   c) Find $f(2)$.
   d) Is f continuous at 2?

Find the limit, if it exists.

4. $\lim_{x \to -4} \dfrac{x^2 - 16}{x + 4}$

5. $\lim_{x \to 1} \sqrt{x^3 + 8}$

6. $\lim_{x \to 3} \dfrac{4}{x - 3}$

7. $\lim_{x \to \infty} \dfrac{12x - 7}{3x + 2}$

8. $\lim_{h \to 0} \dfrac{3x^2 + h}{x - 2h}$

9. $\lim_{x \to \infty} \dfrac{8x^3 + 5x^2 - 7}{16x^5 - 4x^3 + 9}$

Differentiate.

10. $y = -9x + 3$

11. $y = x^2 - 7x + 3$

12. $y = x^{1/4}$

13. $f(x) = x^{-6}$

14. $f(x) = (x - 3)(x + 1)^5$

15. $f(x) = \dfrac{x^3 - 1}{x^5}$

16. $y = \dfrac{e^x + x}{e^x}$

17. $y = \ln (x^2 + 5)$

18. $y = e^{\ln x}$

19. $y = e^{3x} + x^2$

20. $y = e^{-x} \sin (x^2 + 3)$

21. $y = \ln |\cos x|$

22. For $y = \tan x$, find d^2y/dx^2.

23. Differentiate implicitly to find dy/dx if $x^3 + x/y = 7$.

24. Find an equation of the tangent line to the graph of $y = e^x - x^2 - 3$ at the point $(0, -2)$.

Find the relative extrema of the function. List your answer in terms of an ordered pair. Then sketch a graph of the function.

25. $f(x) = x^3 - 3x + 1$

26. $f(x) = 2x^2 - x^4 - 3$

27. $f(x) = \dfrac{8x}{x^2 + 1}$

28. $f(x) = \dfrac{8}{x^2 - 4}$

Find the absolute maximum and minimum values, if they exist, over the indicated interval. If no interval is indicated, use the real line.

29. $f(x) = 3x^2 - 6x - 4$

30. $f(x) = -5x + 1$

31. $f(x) = \frac{1}{3}x^3 - x^2 - 3x + 5; \quad [-2, 0]$

32. **Business: Maximizing profit.** For a certain product, the total-revenue and total-cost functions are given by

$$R(x) = 4x^2 + 11x + 110,$$
$$C(x) = 4.2x^2 + 5x + 10.$$

Find the number of units that must be produced and sold in order to maximize profit.

33. **Business: Minimizing inventory costs.** An appliance store sells 450 pocket radios each year. It costs $4 to store one radio for one year. To order radios, there is a fixed cost of $1 plus $0.75 for each radio. How many times per year should the store reorder radios, and in what lot size, in order to minimize inventory costs?

34. For $f(x) = 3x^2 - 7$, $x = 5$, and $\Delta x = 0.1$, find Δy and $f'(x) \, \Delta x$.

35. **Business: Exponential growth.** A national frozen yogurt firm is experiencing a growth pattern of 10% per year in the number N of franchises that it owns; that

is,

$$\frac{dN}{dt} = 0.1N,$$

where N = the number of franchises and t = the time, in years, from 1996.

a) Given that there were 8000 franchises in 1996, find the solution of the equation, assuming $N_0 = 8000$ and $k = 0.1$.
b) How many franchises will there be in the year 2004?
c) What is the doubling time of the number of franchises?

36. **Economics: Elasticity of demand.** Consider the demand function

$$x = D(p) = 240 - 20p.$$

a) Find the elasticity.
b) Find the elasticity at $p = \$2$, stating whether the demand is elastic or inelastic.
c) Find the elasticity at $p = \$9$, stating whether the demand is elastic or inelastic.
d) At a price of $2, will a small increase in price cause total revenue to increase or decrease?
e) Find the value of p for which the total revenue is a maximum.

Integrate.

37. $\int 3x^5 \, dx$

38. $\int_{-1}^{0} (2e^x + 1) \, dx$

39. $\int \frac{x}{(7 - 3x)^2} \, dx$ (Use Table 1.)

40. $\int x^3 e^{x^4} \, dx$ (Use substitution. Do not use Table 1.)

41. $\int (x + 3) \ln x \, dx$

42. $\int 2x \sin (x^2 + 3) \, dx$

43. $\int e^x \cos (e^x) \, dx$

44. Find the area under the graph of $y = x^2 + 3x$ on the interval $[1, 5]$.

45. **Business: Present value.** Find the present value of $250,000 due in 30 years at 8%, compounded continuously.

46. **Business: Accumulated present value.** Find the accumulated present value of an investment over a 50-year period in which there is a continuous money flow of

$4500 and the current interest rate is 7.2%, compounded continuously.

47. Determine whether the following improper integral is convergent or divergent, and calculate its value if convergent:

$$\int_{3}^{\infty} \frac{1}{x^7} \, dx.$$

48. Given the probability density function

$$f(x) = \frac{3}{2x^2} \quad \text{over } [1, 3],$$

find $E(x)$.

49. Let x be a continuous random variable that is normally distributed with mean $\mu = 3$ and standard deviation $\sigma = 5$. Using Table 2, find $P(-2 \le x \le 8)$.

50. **Economics: Supply and demand.** Given the demand and supply functions

$$p = D(x) = (x - 20)^2$$

and $$p = S(x) = x^2 + 10x + 50,$$

find the equilibrium point and the consumer's surplus.

51. Find the volume of the solid of revolution generated by rotating the region under the graph of

$$y = e^{-x} \quad \text{from } x = 0 \text{ to } x = 5$$

about the x-axis.

52. Solve the differential equation $dy/dx = xy$.

53. **Business: Effect of advertising.** In an advertising experiment, it was determined that

$$\frac{dP}{dt} = k(L - P),$$

where P = the percentage of people in the city who bought the product after the ad was run t times, and $L = 100\%$, or 1.

a) Express $P(t)$ in terms of $L = 1$.
b) It was determined that $P(10) = 70\%$. Use this to determine k. Round your answer to the nearest hundredth.
c) Rewrite $P(t)$ in terms of k using the value of k found in part (b).
d) Use the equation in part (c) to find $P(20)$.

Given $f(x, y) = e^y + 4x^2y^3 + 3x$, find each of the following.

54. f_x **55.** f_{yy}

56. Find the relative maximum and minimum values of $f(x, y) = 8x^2 - y^2$.

57. Maximize $f(x, y) = 4x + 2y - x^2 - y^2 + 4$, subject to the constraint $x + 2y = 9$.

58. Evaluate

$$\int_0^3 \int_{-1}^2 e^x \, dy \, dx.$$

Find each of the following.

59. $\sin \dfrac{\pi}{4}$ **60.** $\cos \dfrac{\pi}{3}$ **61.** $\tan \pi$

Integrate.

62. $\displaystyle\int \sec 8x \, dx$ **63.** $\displaystyle\int \dfrac{dx}{1 + 9x^2}$

Differentiate.

64. $y = \cot (\sin x^3)$ **65.** $f(x) = \sin^{-1} x$

Tables

TABLE 1

Integration Formulas

(Whenever ln X is used, it is assumed that $X > 0$.)

1. $\displaystyle\int x^n \, dx = \frac{x^{n+1}}{n+1} + C, n \neq -1$

2. $\displaystyle\int \frac{dx}{x} = \ln x + C$

3. $\displaystyle\int u \, dv = uv - \int v \, du$

4. $\displaystyle\int e^x \, dx = e^x + C$

5. $\displaystyle\int e^{ax} \, dx = \frac{1}{a} \cdot e^{ax} + C$

6. $\displaystyle\int xe^{ax} \, dx = \frac{1}{a^2} \cdot e^{ax}(ax - 1) + C$

7. $\displaystyle\int x^n e^{ax} \, dx = \frac{x^n e^{ax}}{a} - \frac{n}{a} \int x^{n-1} e^{ax} \, dx + C$

8. $\displaystyle\int \ln x \, dx = x \ln x - x + C$

9. $\displaystyle\int (\ln x)^n \, dx = x(\ln x)^n - n \int (\ln x)^{n-1} \, dx + C, n \neq -1$

10. $\displaystyle\int x^n \ln x \, dx = x^{n+1}\left[\frac{\ln x}{n+1} - \frac{1}{(n+1)^2}\right] + C, n \neq -1$

11. $\displaystyle\int a^x \, dx = \frac{a^x}{\ln a} + C, a > 0, a \neq 1$

12. $\displaystyle\int \frac{1}{\sqrt{x^2 + a^2}} \, dx = \ln (x + \sqrt{x^2 + a^2}) + C$

13. $\displaystyle\int \frac{1}{\sqrt{x^2 - a^2}} \, dx = \ln (x + \sqrt{x^2 - a^2}) + C$

14. $\displaystyle\int \frac{1}{x^2 - a^2} \, dx = \frac{1}{2a} \ln\left(\frac{x - a}{x + a}\right) + C$

15. $\displaystyle\int \frac{1}{a^2 - x^2} \, dx = \frac{1}{2a} \ln\left(\frac{a + x}{a - x}\right) + C$

16. $\displaystyle\int \frac{1}{x\sqrt{a^2 + x^2}} \, dx = -\frac{1}{a} \ln\left(\frac{a + \sqrt{a^2 + x^2}}{x}\right) + C$

17. $\displaystyle\int \frac{1}{x\sqrt{a^2 - x^2}} \, dx = -\frac{1}{a} \ln\left(\frac{a + \sqrt{a^2 - x^2}}{x}\right) + C, 0 < x < a$

18. $\displaystyle\int \frac{x}{ax + b} \, dx = \frac{b}{a^2} + \frac{x}{a} - \frac{b}{a^2} \ln (ax + b) + C$

19. $\displaystyle\int \frac{x}{(ax + b)^2} \, dx = \frac{b}{a^2(ax + b)} + \frac{1}{a^2} \ln (ax + b) + C$

20. $\int \dfrac{1}{x(ax + b)}\, dx = \dfrac{1}{b}\ln\!\left(\dfrac{x}{ax + b}\right) + C$

21. $\int \dfrac{1}{x(ax + b)^2}\, dx = \dfrac{1}{b(ax + b)} + \dfrac{1}{b^2}\ln\!\left(\dfrac{x}{ax + b}\right) + C$

22. $\int \sqrt{x^2 \pm a^2}\, dx = \tfrac{1}{2}[x\sqrt{x^2 \pm a^2} \pm a^2 \ln\,(x + \sqrt{x^2 \pm a^2})] + C$

23. $\int \sin x\, dx = -\cos x + C,\ \int \cos x\, dx = \sin x + C$

24. $\int \sec^2 x\, dx = \tan x + C,\ \int \csc^2 x\, dx = -\cot x + C$

25. $\int \tan x \sec x\, dx = \sec x + C,\ \int \cot x \csc x = -\csc x + C$

26. $\int \sec x\, dx = \ln\,|\sec x + \tan x| + C,\ \int \csc x\, dx = -\ln\,|\csc x + \cot x| + C$

27. $\int \dfrac{1}{1 + x^2}\, dx = \tan^{-1} x + C$

28. $\int \dfrac{1}{\sqrt{1 - x^2}}\, dx = \sin^{-1} x + C,\ \int \dfrac{-1}{\sqrt{1 - x^2}}\, dx = \cos^{-1} x + C$

29. $\int x\sqrt{a + bx}\, dx = \dfrac{2}{15b^2}(3bx - 2a)(a + bx)^{3/2} + C$

30. $\int x^2\sqrt{a + bx}\, dx = \dfrac{2}{105b^3}(15b^2x^2 - 12abx + 8a^2)(a + bx)^{3/2} + C$

31. $\int \dfrac{x\, dx}{\sqrt{a + bx}} = \dfrac{2}{3b^2}(bx - 2a)\sqrt{a + bx} + C$

32. $\int \dfrac{x^2\, dx}{\sqrt{a + bx}} = \dfrac{2}{15b^3}(3b^2x^2 - 4abx + 8a^2)\sqrt{a + bx} + C$

33. $\int \sin^n x\, dx = -\dfrac{1}{n}\sin^{n-1} x \cos x + \dfrac{n - 1}{n}\int \sin^{n-2} x\, dx + C$

34. $\int \cos^n x\, dx = \dfrac{1}{n}\cos^{n-1} x \sin x + \dfrac{n - 1}{n}\int \cos^{n-2} x\, dx + C$

35. $\int e^{ax} \sin bx\, dx = \dfrac{e^{ax}}{a^2 + b^2}(a \sin bx - b \cos bx) + C$

36. $\int e^{ax} \cos bx\, dx = \dfrac{e^{ax}}{a^2 + b^2}(a \cos bx + b \sin bx) + C$

TABLE 2

| | Areas for a Standard Normal Distribution | | | | | | | | | |

Entries in the table represent area under the curve between $t = 0$ and a positive value of t. Because of the symmetry of the curve, area under the curve between $t = 0$ and a negative value of t would be found in a similar manner.

Area = Probability
$$= P(0 \le x \le t)$$
$$= \int_0^t \frac{1}{\sqrt{2\pi}} e^{-x^2/2}\, dx$$

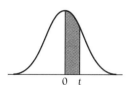

t	0.00	0.01	0.02	0.03	0.04	0.05	0.06	0.07	0.08	0.09
0.0	.0000	.0040	.0080	.0120	.0160	.0199	.0239	.0279	.0319	.0359
0.1	.0398	.0438	.0478	.0517	.0557	.0596	.0636	.0675	.0714	.0753
0.2	.0793	.0832	.0871	.0910	.0948	.0987	.1026	.1064	.1103	.1141
0.3	.1179	.1217	.1255	.1293	.1331	.1368	.1406	.1443	.1480	.1517
0.4	.1554	.1591	.1628	.1664	.1700	.1736	.1772	.1808	.1844	.1879
0.5	.1915	.1950	.1985	.2019	.2054	.2088	.2123	.2157	.2190	.2224
0.6	.2257	.2291	.2324	.2357	.2389	.2422	.2454	.2486	.2517	.2549
0.7	.2580	.2611	.2642	.2673	.2704	.2734	.2764	.2794	.2823	.2852
0.8	.2881	.2910	.2939	.2967	.2995	.3023	.3051	.3078	.3106	.3133
0.9	.3159	.3186	.3212	.3238	.3264	.3289	.3315	.3340	.3365	.3389
1.0	.3413	.3438	.3461	.3485	.3508	.3531	.3554	.3577	.3599	.3621
1.1	.3643	.3665	.3686	.3708	.3729	.3749	.3770	.3790	.3810	.3830
1.2	.3849	.3869	.3888	.3907	.3925	.3944	.3962	.3980	.3997	.4015
1.3	.4032	.4049	.4066	.4082	.4099	.4115	.4131	.4147	.4162	.4177
1.4	.4192	.4207	.4222	.4236	.4251	.4265	.4279	.4292	.4306	.4319
1.5	.4332	.4345	.4357	.4370	.4382	.4394	.4406	.4418	.4429	.4441
1.6	.4452	.4463	.4474	.4484	.4495	.4505	.4515	.4525	.4535	.4545
1.7	.4554	.4564	.4573	.4582	.4591	.4599	.4608	.4616	.4625	.4633
1.8	.4641	.4649	.4656	.4664	.4671	.4678	.4686	.4693	.4699	.4706
1.9	.4713	.4719	.4726	.4732	.4738	.4744	.4750	.4756	.4761	.4767
2.0	.4772	.4778	.4783	.4788	.4793	.4798	.4803	.4808	.4812	.4817
2.1	.4821	.4826	.4830	.4834	.4838	.4842	.4846	.4850	.4854	.4857
2.2	.4861	.4864	.4868	.4871	.4875	.4878	.4881	.4884	.4887	.4890
2.3	.4893	.4896	.4898	.4901	.4904	.4906	.4909	.4911	.4913	.4916
2.4	.4918	.4920	.4922	.4925	.4927	.4929	.4931	.4932	.4934	.4936
2.5	.4938	.4940	.4941	.4943	.4945	.4946	.4948	.4949	.4951	.4952
2.6	.4953	.4955	.4956	.4957	.4959	.4960	.4961	.4962	.4963	.4964
2.7	.4965	.4966	.4967	.4968	.4969	.4970	.4971	.4972	.4973	.4974
2.8	.4974	.4975	.4976	.4977	.4977	.4978	.4979	.4979	.4980	.4981
2.9	.4981	.4982	.4982	.4983	.4984	.4984	.4985	.4985	.4986	.4986
3.0	.4987	.4987	.4987	.4988	.4988	.4989	.4989	.4989	.4990	.4990

Photo Credits

Answers

Note: All nonlinear curves in the answer section have been computer-generated utilizing Mathematica®, a software-based symbolic math processor. The curves are actually composed of tightly packed laser exposures based on Mathematica's polygonal data output. Ordered pairs are sampled at a rate optimized to produce smooth, plotter-like curves on high-resolution imagesetters.

CHAPTER 1

Exercise Set 1.1, p. 9

1. $5 \cdot 5 \cdot 5$, or 125   **3.** $(-7)(-7)$, or 49   **5.** 1.0201

7. $\dfrac{1}{16}$ **9.** 1   **11.** t **13.** 1   **15.** $\dfrac{1}{3^2}$, or $\dfrac{1}{9}$ **17.** 8

19. 0.1   **21.** $\dfrac{1}{e^b}$ **23.** $\dfrac{1}{b}$ **25.** x^5 **27.** x^{-6}, or $\dfrac{1}{x^6}$

29. $35x^5$ **31.** x^4 **33.** 1   **35.** x^3 **37.** x^{-3}, or $\dfrac{1}{x^3}$

39. 1   **41.** e^{t-4} **43.** t^{14} **45.** t^2 **47.** a^6b^5 **49.** t^{-6},

or $\dfrac{1}{t^6}$ **51.** e^{4x} **53.** $8x^6y^{12}$ **55.** $\dfrac{1}{81}x^8y^{20}z^{-16}$, or $\dfrac{x^8y^{20}}{81z^{16}}$

57. $9x^{-16}y^{14}z^4$, or $\dfrac{9y^{14}z^4}{x^{16}}$ **59.** $5x - 35$ **61.** $x - xt$

63. $x^2 - 7x + 10$ **65.** $a^3 - b^3$ **67.** $2x^2 + 3x - 5$
69. $a^2 - 4$ **71.** $25x^2 - 4$ **73.** $a^2 - 2ah + h^2$
75. $25x^2 + 10xt + t^2$ **77.** $5x^5 + 30x^3 + 45x$
79. $a^3 + 3a^2b + 3ab^2 + b^3$ **81.** $x^3 - 15x^2 + 75x - 125$
83. $x(1 - t)$ **85.** $(x + 3y)^2$ **87.** $(x - 5)(x + 3)$
89. $(x - 5)(x + 4)$ **91.** $(7x - t)(7x + t)$

93. $4(3t - 2m)(3t + 2m)$ **95.** $ab(a + 4b)(a - 4b)$
97. $(a^4 + b^4)(a^2 + b^2)(a + b)(a - b)$
99. $10x(a + 2b)(a - 2b)$
101. $2(1 + 4x^2)(1 + 2x)(1 - 2x)$ **103.** $(9x - 1)(x + 2)$
105. $(x + 2)(x^2 - 2x + 4)$
107. $(y - 4t)(y^2 + 4yt + 16t^2)$ **109. (a)** 0.81;
(b) 0.0801;   **(c)** 0.008001   **111. (a)** \$1060;   **(b)** \$1060.90;
(c) \$1061.36;   **(d)** \$1061.83, assuming 365 days in a year;
(e) \$1061.84   **113.** \$353.52   **115.** ◈

Exercise Set 1.2, p. 16

1. $\frac{7}{4}$ **3.** -8 **5.** 120   **7.** 200   **9.** $0, -3, \frac{4}{5}$ **11.** 0, 2
13. 0, 3   **15.** 0, 7   **17.** $0, \frac{1}{3}, -\frac{1}{3}$ **19.** 1   **21.** $x \geq -\frac{4}{5}$
23. $x > -\frac{1}{12}$ **25.** $x > -\frac{4}{7}$ **27.** $x \leq -3$ **29.** $x > \frac{2}{3}$
31. $x < -\frac{2}{5}$ **33.** $2 < x < 4$ **35.** $\frac{3}{2} \leq x \leq \frac{11}{2}$
37. $-1 \leq x \leq \frac{14}{5}$ **39.** $(0, 5)$ **41.** $[-9, -4)$
43. $[x, x + h]$ **45.** (p, ∞) **47.** $[-3, 3]$
49. $[-14, -11)$ **51.** $(-\infty, -4]$ **53.** \$650   **55.** More
than 7000 units   **57.** 480 lb   **59.** 810,000
61. $60\% \leq x < 100\%$ **63.** ◈

Exercise Set 1.3, p. 31

1. (a)

Inputs	Outputs
4.1	11.2
4.01	11.02
4.001	11.002
4	11

(b) $f(5) = 13, f(-1) = 1, f(k) = 2k + 3, f(1 + t) = 2t + 5,$
$f(x + h) = 2x + 2h + 3$ **3.** $g(-1) = -2, g(0) = -3,$
$g(1) = -2, g(5) = 22, g(u) = u^2 - 3,$
$g(a + h) = a^2 + 2ah + h^2 - 3,$
$g(1 - h) = h^2 - 2h - 2$ **5. (a)** $f(4) = 1, f(-2) = 25,$
$f(0) = 9, f(a) = a^2 - 6a + 9, f(t + 1) = t^2 - 4t + 4,$
$f(t + 3) = t^2, f(x + h) = x^2 + 2xh + h^2 - 6x - 6h + 9;$
(b) Take an input, square it, subtract 6 times the input,
add 9.

7.

9.
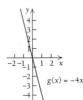

11.
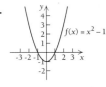

13.
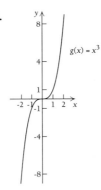

15. Yes   **17.** Yes   **19.** No   **21.** No   **23.** Yes   **25.** Yes

27. (a)

(b) no

29. $f(x + h) = x^2 + 2xh + h^2 - 3x - 3h$

31.
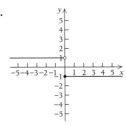

33.
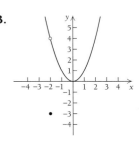

35. $R(10) = \$2050, R(100) = \$20{,}050$ **37. (a)** Yes;
(b) 205 million, 1.2 million;   **(c)** 1987;   **(d)** 30 million;
495.4 million;   **(e)** 1991;   **(f)** 1988;   **(g)** 1993   **39.** $y = 5$;
a function   **41.** $y = \pm\sqrt{x}$; not a function   **43.** ◈

45.
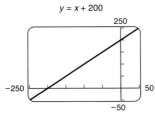

47.
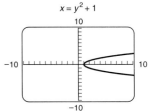

49.
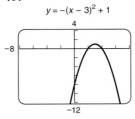

51.
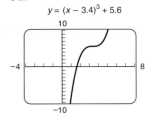

53.
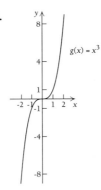

Exercise Set 1.4, p. 44

1.

3.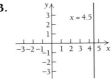

5. $m = -3$,
y-intercept: $(0, 0)$

7. $m = 0.5$,
y-intercept: $(0, 0)$

9. $m = -2$,
y-intercept: $(0, 3)$

11. $m = -1$,
y-intercept: $(0, -2)$

13. $m = -2$, y-intercept: $(0, 2)$ **15.** $m = -1$, y-intercept:
$\left(0, -\frac{5}{2}\right)$ **17.** $y + 5 = -5(x - 1)$, or $y = -5x$
19. $y - 3 = -2(x - 2)$, or $y = -2x + 7$
21. $y = \frac{1}{2}x - 6$ **23.** $y = 3$ **25.** $\frac{3}{2}$ **27.** $\frac{1}{2}$ **29.** No
slope   **31.** 0   **33.** 3   **35.** 2   **37.** $y - 1 = \frac{3}{2}(x + 2)$,
or $y + 2 = \frac{3}{2}(x + 4)$, or $y = \frac{3}{2}x + 4$
39. $y + 4 = \frac{1}{2}(x - 2)$, or $y + 3 = \frac{1}{2}(x - 4)$, or $y = \frac{1}{2}x - 5$
41. $x = 3$ **43.** $y = 3$ **45.** $y = 3x$ **47.** $y = 2x + 3$
49. (a) $A = P + 8\%P = P + 0.08P = 1.08P$;   **(b)** \$108;
(c) \$240   **51. (a)** $C(x) = 20x + 100{,}000$;
(b) $R(x) = 45x$;   **(c)** $P(x) = R(x) - C(x) = 25x - 100{,}000$;
(d) \$3,650,000, a profit;   **(e)** 4000
53. (a) $C(x) = x + 250$;   **(b)** $R(x) = 10x$. The student
charges \$10 per lawn.   **(c)** 28   **55. (a)** $R = 4.17T$;
(b) $R = 25.02$ **57. (a)** $B = 0.025W$;   **(b)** $B = 2.5\%W$.
The weight of the brain is 2.5% of the body weight.
(c) 3 lb   **59. (a)** $D(0°) = 115$ ft, $D(-20°) = 75$ ft,
$D(10°) = 135$ ft, $D(32°) = 179$ ft;
(b) Temperatures below $-57.5°$ would yield a negative
stopping distance, which has no meaning here. For
temperatures above $32°$, there would be no ice.

61. (a) 145.78 cm;   **(b)** 142.98 cm   **63. (a)** $A(0) = 19.7$,
$A(1) = 19.78$, $A(10) = 20.5$, $A(30) = 22.1$,
$A(50) = 23.7$ **(b)** 23.54;   **(c)**

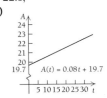

65. ◈   **67.** Answers may vary.

Exercise Set 1.5, p. 62

1.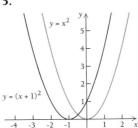

3.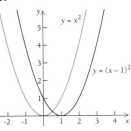

5.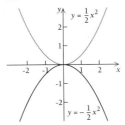

7.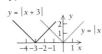

9.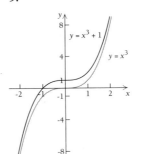

11.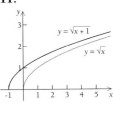

13.

15.

87. **(a)** 78,409; 84,210; 114,595;   **(b)**
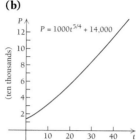

17.
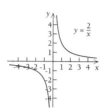

19.
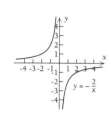

89. ◆   **91.** 2.359   **93.** 1.5, 9.5   **95.** −2.646, 2.646
97. None   **99.** −10.153, −1.871, −0.821, −0.303, 0.098,
0.535, 1.219, 3.297   **101.** (1.585, 5.721)   **103.** (1.19, 270),
(3.25, −660), (5.88, −1840), (−7.30, 4083)

Exercise Set 1.6, p. 73

1. 3:43.2   **3.** **(b)** Linear;   **(c)** $S = 5t + 80$;
(d) $120 billion; $150 billion   **5.** **(b)** Constant, linear;
(c) $S = -20t + 100,330$;   **(d)** Let $S = b$, where b is the
average of sales totals: $100,296.25.   **7.** **(b)** Quadratic;
(c) $N = -25t^2 + 300t$;   **(d)** 144, 394, 594
9. **(a)** $y = -0.67x^3 + 4x^2 - 2.33x + 84$
(b) −$1003.1 billion dollars; $150 billion. The answer found
with the cubic is probably not meaningful.

21.

23.
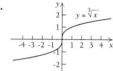

Summary and Review: Chapter 1, p. 75

1. [1.1] (2.01)(2.01), or 4.0401   **2.** [1.1] y^3 **3.** [1.1] y^{-7}
4. [1.1] $243t^{-10}m^{20}$ **5.** [1.1] $9x^2 - 6xt + t^2$
6. [1.1] $9x^2 - t^2$ **7.** [1.1] $12x^2 + 11xt - 5t^2$
8. [1.1] $(x - 8)(x + 6)$ **9.** [1.1] $(5x - 4t)(5x + 4t)$
10. [1.1] $(a - 4b)^2$ **11.** [1.1] $(3x - 1)(7x - 4)$
12. [1.1] $1212.75   **13.** [1.1] $5017.60   **14.** [1.2] $x \geq 1$
15. [1.2] 1   **16.** [1.2] $\frac{5}{4}, -\frac{5}{4}$ **17.** [1.2] 0, 3, $-\frac{5}{2}$
18. [1.2] (−6, 1]   **19.** [1.2] (0, ∞)   **20.** [1.3] 13
21. [1.3] $2h^2 + 3h + 4$ **22.** [1.3] 3   **23.** [1.3] 36
24. [1.3] $h^2 - 2h + 1$ **25.** [1.3] 9
26. [1.3]   **27.** [1.3]

25.
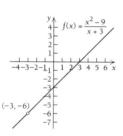

27.
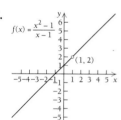

29. $1 \pm \sqrt{3}$ **31.** $-3 \pm \sqrt{10}$ **33.** $\dfrac{1 \pm \sqrt{2}}{2}$

35. $\dfrac{-4 \pm \sqrt{10}}{3}$ **37.** $x^{3/2}$ **39.** $a^{3/5}$ **41.** $t^{1/7}$

43. $t^{-4/3}$ **45.** $t^{-1/2}$ **47.** $(x^2 + 7)^{-1/2}$ **49.** $\sqrt[5]{x}$

51. $\sqrt[3]{y^2}$ **53.** $\dfrac{1}{\sqrt[5]{t^2}}$ **55.** $\dfrac{1}{\sqrt[3]{b}}$ **57.** $\dfrac{1}{\sqrt[6]{e^{17}}}$

59. $\dfrac{1}{\sqrt{x^2 - 3}}$ **61.** 27   **63.** 16   **65.** 8   **67.** All real

numbers except 5   **69.** All real numbers except 2, 3

71. $\left[-\frac{4}{5}, \infty\right)$ **73.** ($4, 2)   **75.** ($50, 500)   **77.** ($5, 1)

79. ($1, 4)   **81.** ($2, 3)   **83.** $20.36   **85.** 84; 220; 364

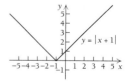

28. [1.5]

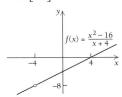

$f(x) = \frac{x^2 - 16}{x + 4}$

29. [1.3] No

30. [1.3] Yes   **31.** [1.3] Yes
32. [1.4]

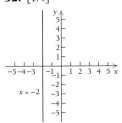

$x = -2$

33. [1.4]

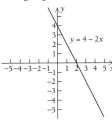

$y = 4 - 2x$

34. [1.4] $y = -\frac{7}{11}x + \frac{6}{11}$, or $7x + 11y = 6$
35. [1.4] $y = 8x + 7$ **36.** [1.4] $m = -\frac{1}{6}$,
y-intercept: $(0, 3)$ **37.** [1.4] 72 lb
38. [1.4] **(a)** $R(x) = 28x$;   **(b)** $C(x) = 16x + 80,000$;
(c) $P(x) = 12x - 80,000$;   **(d)** \$16,000 profit;   **(e)** 6667
39. [1.5] $\sqrt[6]{y}$ **40.** [1.5] $x^{3/20}$ **41.** [1.5] 125
42. [1.5] $(-\infty, 3]$ **43.** [1.5] All real numbers except 1 and
-1 **44.** [1.5] (\$3, 27)   **45.** [1.6] $y = -\frac{11}{6}x + 10$
46. [1.6] $y = x^2 - 2x + 5$ **47.** [1.3]

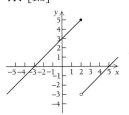

48. [1.5] $\frac{1}{32}$
49. [1.3]

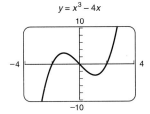

$y = x^3 - 4x$

50. [1.3], [1.5]

$y = \sqrt[3]{|9 - x^2|} - 1$

51. [1.5] $-2, 0, 2$ **52.** [1.5] $-3.162, -2.828, 2.828, 3.162$
53. [1.5] $(-2.73, -1.06)$

Test: Chapter 1, p. 77

1. $\frac{1}{e^k}$ **2.** e^{-13}, or $\frac{1}{e^{13}}$ **3.** $x^2 + 2xh + h^2$
4. $(5x - t)(5x + t)$ **5.** \$920   **6.** $x > -4$ **7. (a)** 5;
(b) $x^2 + 2xh + h^2 - 4$ **8.** $m = -3$; y-intercept: $(0, 2)$
9. $y + 5 = \frac{1}{4}(x - 8)$, or $y = \frac{1}{4}x - 7$ **10.** $m = 6$
11. $F = \frac{2}{3}W$ **12. (a)** $C(x) = 0.5x + 10,000$;
(b) $R(x) = 7.5x$;   **(c)** $P(x) = 7x - 10,000$;   **(d)** 1429
13. (\$3, 16)   **14.**

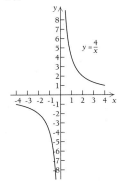

$y = \frac{4}{x}$

15. $t^{-1/2}$ **16.** $\frac{1}{\sqrt[5]{t^3}}$ **17.**

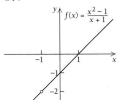

$f(x) = \frac{x^2 - 1}{x + 1}$

18. All real numbers except 2, -7 **19.** $[-2, \infty)$
20. $y = 4x - 1$ **21.** $y = -4.5x^2 + 17.5x - 8$
22. $[c, d)$ **23.**

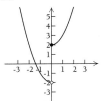

24. (a) 525;   **(b)** $4.31

25.

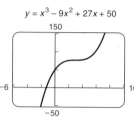

$y = x^3 - 9x^2 + 27x + 50$

26.

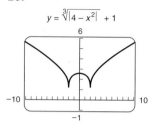

$y = \sqrt[3]{|4 - x^2|} + 1$

27. -1.25 **28.** None   **29.** ($12.50, 250)

CHAPTER 2

Exercise Set 2.1, p. 91

1. No   **3.** Yes   **5. (a)** -1, 2, does not exist;   **(b)** -1;
(c) no;   **(d)** 3;   **(e)** 3;   **(f)** yes   **7. (a)** 2;   **(b)** 2;
(c) yes;   **(d)** 0;   **(e)** 0;   **(f)** yes   **9. (a)** 3;   **(b)** 3;   **(c)** 3;
(d) 2;   **(e)** no;   **(f)** yes   **11. (a)** F;   **(b)** T;   **(c)** F;
(d) F;   **(e)** F;   **(f)** T;   **(g)** T;   **(h)** F;   **(i)** F;   **(j)** T
13. No; yes; no; yes   **15.** 32¢, 55¢, does not exist
17. 78¢   **19.** $20,000, $28,000, does not exist
21. No, yes   **23.** 100, 30, does not exist   **25.** No, yes
27. ◆   **29. (a)** Does not exist;   **(b)** 2   **31.** ◆

Exercise Set 2.2, p. 100

1. -2 **3.** Does not exist   **5.** 11   **7.** -10 **9.** $-\frac{5}{2}$
11. 5   **13.** 2   **15.** 3   **17.** Does not exist   **19.** $\frac{2}{7}$
21. $\frac{5}{4}$ **23.** ∞ **25.** $6x^2$ **27.** $\dfrac{-2}{x^3}$ **29.** $\frac{2}{5}$ **31.** 5
33. $\frac{1}{2}$ **35.** $\frac{2}{3}$ **37.** 0   **39.** ∞ **41.** 0   **43.** ∞
45. (a) 1;   **(b)** 1;   **(c)** 1;   **(d)** 1;   **(e)** yes, no
47. (a) $800, $736, $677.12, $622.95, $573.11;   **(b)** $5656.12;
(c) $10,000   **49.** $13 per unit. This means that as more
and more units are produced, the average cost gets closer and
closer to $13 per unit.   **51. (a)** 4.00, 4.50, 5.14, 6.00, 7.20,
9.00, 12.00, 18.00, 36.00, 54.00, 108.00;   **(b)** ∞;   **(c)** ∞
53. Does not exist   **55.** $-\infty$ **57.** $\frac{3}{2}$ **59.** $-\infty$ **61.** 6
63. -0.2887 **65.** 0.75   **67.** 0.25

Exercise Set 2.3, p. 109

1. (a) $7(2x + h)$;   **(b)** 70, 63, 56.7, 56.07
3. (a) $-7(2x + h)$;   **(b)** $-70, -63, -56.7, -56.07$
5. (a) $7(3x^2 + 3xh + h^2)$;   **(b)** 532, 427, 344.47, 336.8407
7. (a) $\dfrac{-5}{x(x + h)}$;   **(b)** $-0.2083, -0.25, -0.3049, -0.3117$

9. (a) -2;   **(b)** all -2 **11. (a)** $2x + h - 1$;   **(b)** 9, 8,
7.1, 7.01   **13. (a)** 70, 39, 29, 23 pleasure units/unit of
product;   **(b)** The more you get, the less pleasure you get
from each additional unit.   **15. (a)** $300,093.99;
(b) $299,100;   **(c)** $993.99;   **(d)** $993.99
17. (a) $2406 million/year;   **(b)** $3441.53 million/year
19. (a) 1.25 words/minute, 1.25 words/minute, 0.625 words/
minute, 0 words/minute; 0 words/minute;   **(b)** You have
reached a saturation point; you cannot memorize any more.
21. (a) 144 feet;   **(b)** 400 feet;   **(c)** 128 feet/second
23. (a) 125 million people/year for each;   **(b)** no;
(c) A: 290 million people/year, -40 million people/year,
-50 million people/year, 300 million people/year;
B: 125 million people/year in all intervals;   **(d)** A
25. 3182 divorces/year   **27.** $2ax + b + ah$
29. $\dfrac{1}{\sqrt{x + h} + \sqrt{x}}$ **31.** $\dfrac{-2x - h}{x^2(x + h)^2}$
33. $\dfrac{1}{(x + 1)(x + 1 + h)}$ **35.** ◆

Exercise Set 2.4, p. 125

1. (a), (b)

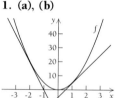

(c) $f'(x) = 10x$;
(d) $-20, 0, 10$

3. (a), (b)

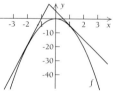

(c) $f'(x) = -10x$;
(d) $20, 0, -10$

5. (a), (b)

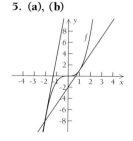

(c) $f'(x) = 3x^2$;
(d) 12, 0, 3

7. (a), (b) All the tangent lines are identical to the original function.

(c) $f'(x) = 2$;
(d) 2, 2, 2

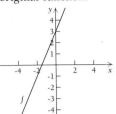

9. (a), (b) All the tangent lines are identical to the original function.

(c) $f'(x) = -4$;
(d) $-4, -4, -4$

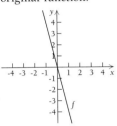

11. (a), (b)

(c) $f'(x) = 2x + 1$;
(d) $-3, 1, 3$

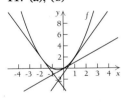

13. (a), (b)

(c) $f'(x) = \dfrac{-1}{x^2}$;

(d) $-\dfrac{1}{4}$, does not exist, -1

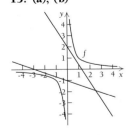

15. $f'(x) = m$ **17.** $y = 6x - 9$; $y = -2x - 1$; $y = 20x - 100$ **19.** $y = -5x + 10$; $y = -5x - 10$; $y = -0.0005x + 0.1$ **21.** $y = 2x + 5$; $y = 4$; $y = -10x + 29$ **23.** $x_0, x_3, x_4, x_6, x_{12}$

25. 1, 2, 3, 4, and so on   **27.** $\dfrac{-2}{x^3}$ **29.** $\dfrac{1}{(1 + x)^2}$

31. $x = -3$ **33.** ◈

Exercise Set 2.5, p. 135

1. $7x^6$ **3.** 0   **5.** $600x^{149}$ **7.** $3x^2 + 6x$ **9.** $\dfrac{4}{\sqrt{x}}$

11. $0.07x^{-0.93}$ **13.** $\dfrac{2}{5\sqrt[5]{x}}$ **15.** $\dfrac{-3}{x^4}$ **17.** $6x - 8$

19. $\dfrac{1}{4\sqrt[4]{x^3}} + \dfrac{1}{x^2}$ **21.** $1.6x^{1.5}$ **23.** $\dfrac{-5}{x^2} - 1$ **25.** 4

27. 4   **29.** x^3 **31.** $-0.02x - 0.5$

33. $-2x^{-5/3} + \frac{3}{4}x^{-1/4} + \frac{6}{5}x^{1/5} - 24x^{-4}$ **35.** $-\dfrac{2}{x^2} - \dfrac{1}{2}$

37. $-16x^{-2} + 24x^{-4} - 4x^{-5}$

39. $\frac{1}{2}x^{-1/2} + \frac{1}{3}x^{-2/3} - \frac{1}{4}x^{-3/4} + \frac{1}{5}x^{-4/5}$ **41.** $y = 10x - 15$; $y = x + 3$; $y = -2x + 1$ **43.** $(0, 0)$ **45.** $(0, 0)$

47. $\left(\frac{5}{6}, \frac{23}{12}\right)$ **49.** $(-25, 76.25)$ **51.** There are none.

53. The tangent is horizontal at all points on the graph.

55. $\left(\frac{5}{3}, \frac{148}{27}\right)$, $(-1, -4)$ **57.** $(\sqrt{3}, 2 - 2\sqrt{3})$, $(-\sqrt{3}, 2 + 2\sqrt{3})$ **59.** $\left(\frac{19}{2}, \frac{399}{4}\right)$ **61.** $(60, 150)$

63. $\left(-2 + \sqrt{3}, \frac{4}{3} - \sqrt{3}\right), \left(-2 - \sqrt{3}, \frac{4}{3} + \sqrt{3}\right)$

65. $(0, -4), \left(\sqrt{\frac{2}{3}}, -\frac{40}{9}\right), \left(-\sqrt{\frac{2}{3}}, -\frac{40}{9}\right)$ **67.** $2x - 1$

69. $2x + 1$ **71.** $3x^2 - \dfrac{1}{x^2}$ **73.** $-192x^2$ **75.** $\dfrac{2}{3\sqrt[3]{x^2}}$

77. $3x^2 + 6x + 3$

79. $\dfrac{F(x + h) - F(x)}{h}$

$$= \dfrac{[f(x + h) - g(x + h)] - [f(x) - g(x)]}{h}$$

$$= \dfrac{f(x + h) - f(x)}{h} - \dfrac{g(x + h) - g(x)}{h}.$$

As $h \to 0$, the two terms on the right approach $f'(x)$ and $g'(x)$, respectively, so their difference approaches $f'(x) - g'(x)$. Thus, $F'(x) = f'(x) - g'(x)$.   **81.** ◈

83. $(-0.692, -2.639)$, $(0.692, -4.761)$

$y = 1.6x^3 - 2.3x - 3.7$

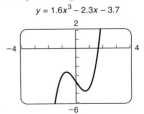

85. $(-0.346, -2.191)$, $(1.929, 2.358)$

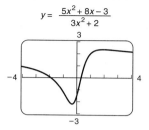

$$y = \frac{5x^2 + 8x - 3}{3x^2 + 2}$$

87.

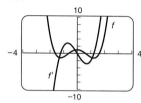

89.

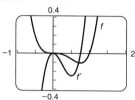

91.

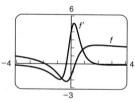

Exercise Set 2.6, p. 143

1. (a) $v(t) = 3t^2 + 1$;   **(b)** $a(t) = 6t$;   **(c)** $v(4) = 49$ ft/sec, $a(4) = 24$ ft/sec^2 **3. (a)** $P(x) = -0.001x^2 + 3.8x - 60$;
(b) $R(100) = \$500$, $C(100) = \$190$, $P(100) = \$310$;
(c) $R'(x) = 5$, $C'(x) = 0.002x + 1.2$,
$P'(x) = -0.002x + 3.8$;   **(d)** $R'(100) = \$5$ per unit,
$C'(100) = \$1.40$ per unit, $P'(100) = \$3.60$ per
unit   **5. (a)** $N'(a) = -2a + 300$;   **(b)** 2906;
(c) 280 units/thousand dollars   **7. (a)** 630 units per
month, 980 units per month, 1430 units per month,
630 units per month;   **(b)** $M'(t) = -4t + 100$;   **(c)** 80, 60,
0, -80 **9. (a)** $\dfrac{dD}{dp} = -\dfrac{1}{2\sqrt{p}}$;   **(b)** 95;   **(c)** $-\dfrac{1}{10}$

11. $\dfrac{dC}{dr} = 2\pi$ **13. (a)** $T'(t) = -0.2t + 1.2$;   **(b)** 100.175°;

(c) 0.9 degrees/day   **15.** $\dfrac{dT}{dW} = 1.31W^{0.31}$

17. $\dfrac{dR}{dQ} = kQ - Q^2$ **19.** $A'(t) = 0.08$ **21.** ◈

23.

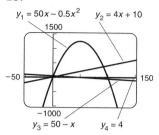

$y_1 = 50x - 0.5x^2$ $y_2 = 4x + 10$

$y_3 = 50 - x$ $y_4 = 4$

25.

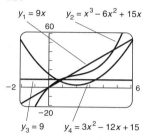

$y_1 = 9x$ $y_2 = x^3 - 6x^2 + 15x$

$y_3 = 9$ $y_4 = 3x^2 - 12x + 15$

Exercise Set 2.7, p. 151

1. $11x^{10}$ **3.** $\dfrac{1}{x^2}$ **5.** $3x^2$

7. $(8x^5 - 3x^2 + 20)\left(32x^3 - \dfrac{3}{2\sqrt{x}}\right) +$
$(40x^4 - 6x)(8x^4 - 3\sqrt{x})$ **9.** $300 - 2x$
11. $(4\sqrt{x} - 6)(3x^2 - 2) + (2x^{-1/2})(x^3 - 2x + 4)$
13. $2x + 6$ **15.** $6x^5 - 32x^3 + 32x$
17. $(5x^{-3})(4x^3 - 15x^2 + 10) - 15x^{-4}(x^4 - 5x^3 + 10x - 2)$,
or $5 - 100x^{-3} + 30x^{-4}$

19. $\left(x + \dfrac{2}{x}\right)(2x) + \left(1 - \dfrac{2}{x^2}\right)(x^2 - 3)$, or $3x^2 + \dfrac{6}{x^2} - 1$

21. $\dfrac{300}{(300 - x)^2}$ **23.** $\dfrac{17}{(2x + 5)^2}$ **25.** $\dfrac{-x^4 - 3x^2 - 2x}{(x^3 - 1)^2}$

27. $\dfrac{1}{(1 - x)^2}$ **29.** $\dfrac{2}{(x + 1)^2}$ **31.** $\dfrac{-1}{(x - 3)^2}$

33. $\dfrac{-2x^2 + 6x + 2}{(x^2 + 1)^2}$ **35.** $\dfrac{-18x + 35}{x^8}$

37. $\dfrac{4\sqrt{x} - \frac{1}{2}x^{-1/2}(4x + 3)}{x}$, or $\dfrac{4x - 3}{2x^{3/2}}$ **39.** $y = 2$;

$y = \frac{1}{2}x + 2$ **41. (a)** $D'(p) = \dfrac{-2978}{(10p + 11)^2}$;   **(b)** $-\dfrac{2978}{2601}$

43. (a) $R(p) = p(400 - p) = 400p - p^2$;
(b) $R'(p) = 400 - 2p$ **45. (a)** $R(p) = 4000 + 3p$;

(b) $R'(p) = 3$ **47.** $A'(x) = \dfrac{xC'(x) - C(x)}{x^2}$

49. (a) $T'(t) = \dfrac{4 - 4t^2}{(t^2 + 1)^2}$;   **(b)** $T(2) = 100.2°$F;

(c) $-0.48°$F/hour   **51.** $\dfrac{5x^3 - 30x^2\sqrt{x}}{2\sqrt{x}(\sqrt{x} - 5)^2}$

53. $\dfrac{-3(1 + 2v)}{(1 + v + v^2)^2}$ **55.** $\dfrac{2t^3 - t^2 + 1}{(1 - t + t^2 - t^3)^2}$

57. $\dfrac{5x^3 + 15x^2 + 2}{2x\sqrt{x}}$

59. $[x(9x^2 + 6) + (3x^3 + 6x - 2)](3x^4 + 7) +$
$12x^4(3x^3 + 6x - 2)$ **61.** $\dfrac{6t^2(t^5 + 3)}{(t^3 + 1)^2} + \dfrac{5t^4(t^3 - 1)}{t^3 + 1}$

63.
$$\dfrac{\{(x^7 - 2x^6 + 9)[(2x^2 + 3)(12x^2 - 7) + 4x(4x^3 - 7x + 2)]}{\qquad - (7x^6 - 12x^5)(2x^2 + 3)(4x^3 - 7x + 2)\}}{(x^7 - 2x^6 + 9)^2}$$

65.

67. No points at which the tangent line is horizontal.

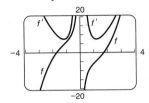

69. $(-0.2, -0.75)$, $(0.2, 0.75)$ **71.** $(-1, -2)$, $(1, 2)$

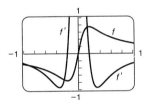   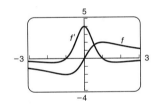

Exercise Set 2.8, p. 159

1. $-55(1 - x)^{54}$ **3.** $\dfrac{4}{\sqrt{1 + 8x}}$ **5.** $\dfrac{3x}{\sqrt{3x^2 - 4}}$

7. $-240x(3x^2 - 6)^{-41}$ **9.** $\sqrt{2x + 3} + \dfrac{x}{\sqrt{2x + 3}}$, or

$\dfrac{3(x + 1)}{\sqrt{2x + 3}}$ **11.** $2x\sqrt{x - 1} + \dfrac{x^2}{2\sqrt{x - 1}}$, or $\dfrac{5x^2 - 4x}{2\sqrt{x - 1}}$

13. $\dfrac{-6}{(3x + 8)^3}$ **15.** $(1 + x^3)^2(-3x^2 - 12x^5)$, or
$-3x^2(1 + x^3)^2(1 + 4x^3)$ **17.** $4x - 400$

19. $2(x + 6)^9(x - 5)^3(7x - 13)$

21. $4(x - 4)^7(3 - x)^3(10 - 3x)$ **23.** $4(2x - 3)^2(3 - 8x)$

25. $\left(\dfrac{1 - x}{1 + x}\right)^{-1/2} \cdot \dfrac{-1}{(x + 1)^2}$ **27.** $\left(\dfrac{3x - 1}{5x + 2}\right)^3 \cdot \dfrac{44}{(5x + 2)^2}$

29. $\frac{1}{3}(x^4 + 3x^2)^{-2/3}(4x^3 + 6x)$

31. $100(2x^3 - 3x^2 + 4x + 1)^{99}(6x^2 - 6x + 4)$

33. $\left(\dfrac{2x + 3}{5x - 1}\right)^{-5} \cdot \dfrac{68}{(5x - 1)^2}$ **35.** $\left(\dfrac{x^2 + 1}{x^2 - 1}\right)^{-1/2} \cdot \dfrac{-2x}{(x^2 - 1)^2}$

37. $\dfrac{(2x + 3)^3(-6x - 61)}{(3x - 2)^6}$

39. $16(2x + 1)^{-1/3}(3x - 4)^{5/4} + 45(2x + 1)^{2/3}(3x - 4)^{1/4}$, or
$(2x + 1)^{-1/3}(3x - 4)^{1/4}(138x - 19)$ **41.** $\frac{1}{2}u^{-1/2}$, $2x$,

$\dfrac{x}{\sqrt{x^2 - 1}}$ **43.** $50u^{49}$, $12x^2 - 4x$,

$50(4x^3 - 2x^2)^{49}(12x^2 - 4x)$ **45.** $2u + 1$, $3x^2 - 2$,
$(2x^3 - 4x + 1)(3x^2 - 2)$ **47.** $y = \frac{5}{4}x + \frac{3}{4}$

49. (a) $\dfrac{2x - 3x^2}{(1 + x)^6}$;   (b) $\dfrac{2x - 3x^2}{(1 + x)^6}$;   (c) same

51. $12x^2 - 12x + 5$, $6x^2 + 3$ **53.** $\dfrac{16}{x^2} - 1$, $\dfrac{2}{4x^2 - 1}$

55. $x^4 - 2x^2 + 2$, $x^4 + 2x^2$ **57.** $f(x) = x^5$,

$g(x) = 3x^2 - 7$ **59.** $f(x) = \dfrac{x + 1}{x - 1}$, $g(x) = x^3$

61. -216 **63.** $-4(-11)^{-2/3} \approx -0.8087$

65. $C'(x) = \dfrac{1500x^2}{\sqrt{x^3 + 2}}$, $C'(10) = 4738.68$

67. $P'(x) = \dfrac{2000x}{\sqrt{x^2 + 3}} - \dfrac{1500x^2}{\sqrt{x^3 + 2}}$ **69.** $\$3000(1 + i)^2$

71. (a) $D(t) = \dfrac{80,000}{1.6t + 9}$;   (b) $D'(t) = \dfrac{-128,000}{(1.6t + 9)^2}$;

(c) $\dfrac{-128,000}{28,561}$, or about -4.48 **73.** $\dfrac{x^2 - 2}{\sqrt[3]{(x^3 - 6x + 1)^2}}$

75. $\dfrac{x - 2}{2(x - 1)^{3/2}}$ **77.** $\dfrac{-4(1 + 2v)^3}{v^5}$

79. $\dfrac{1}{\sqrt{1 - x^2}\,(1 - x)}$, or $\dfrac{\sqrt{1 - x^2}}{(1 - x^2)(1 - x)}$

81. $3\left(\dfrac{x^2 - x - 1}{x^2 + 1}\right)^2 \cdot \dfrac{x^2 + 4x - 1}{(x^2 + 1)^2}$, or

$\dfrac{3(x^2 - x - 1)^2(x^2 + 4x - 1)}{(x^2 + 1)^4}$

83. $\dfrac{1}{\sqrt{t}(1 + \sqrt{t})^2}$ **85.**

87. $(-2.14476, -7.728)$, $(2.14476, 7.728)$

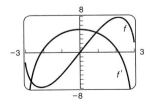

89.

$y_1 = f(x) = x\sqrt{4 - x^2}$ $y_2 = f'(x) = \dfrac{4 - 2x^2}{\sqrt{4 - x^2}}$

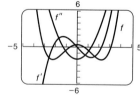

Exercise Set 2.9, p. 164

1. 0   **3.** $-\dfrac{2}{x^3}$ **5.** $-\dfrac{3}{16}x^{-7/4}$ **7.** $12x^2 + \dfrac{8}{x^3}$ **9.** $\dfrac{12}{x^5}$

11. $n(n - 1)x^{n-2}$ **13.** $12x^2 - 2$ **15.** $-\frac{1}{4}(x - 1)^{-3/2}$, or

$\dfrac{-1}{4\sqrt{(x - 1)^3}}$ **17.** $2a$

19. $86(x^2 - 8x)^{41}(85x^2 - 680x + 1344)$

21. $200x^2(x^4 - 4x^2)^{48}(199x^4 - 798x^2 + 792)$

23. $-\frac{2}{9}x^{-4/3}$ **25.** $-\frac{3}{16}(x - 8)^{-5/4}$ **27.** $6x^{-4} + 24x^{-5}$

29. 24   **31.** $720x$ **33.** $20(x^2 - 5)^8[19x^2 - 5]$

35. $a(t) = 6t + 2$ **37.** $P''(t) = 200,000$

39. $y' = -x^{-2} - 2x^{-3}$, $y'' = 2x^{-3} + 6x^{-4}$,

$y''' = -6x^{-4} - 24x^{-5}$ **41.** $y' = \dfrac{1 + 2x^2}{\sqrt{1 + x^2}}$,

$y'' = \dfrac{2x^3 + 3x}{(1 + x^2)^{3/2}}$, $y''' = \dfrac{3}{(1 + x^2)^{5/2}}$ **43.** $y' = \dfrac{11}{(2x + 3)^2}$,

$y'' = \dfrac{-44}{(2x + 3)^3}$, $y''' = \dfrac{264}{(2x + 3)^4}$ **45.** $y' = \dfrac{x - 2}{2(x - 1)^{3/2}}$,

$y'' = \dfrac{4 - x}{4(x - 1)^{5/2}}$, $y''' = \dfrac{3(x - 6)}{8(x - 1)^{7/2}}$ **47.** $\dfrac{2}{(x - 1)^3}$

49. $\dfrac{3}{(x + 2)^2}$, $\dfrac{-6}{(x + 2)^3}$, $\dfrac{18}{(x + 2)^4}$, $\dfrac{-72}{(x + 2)^5}$, $\dfrac{360}{(x + 2)^6}$

51.

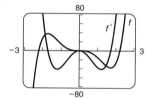

53.

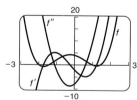

55. ◈

Summary and Review: Chapter 2, p. 166

1. [2.1], [2.2] -4 **2.** [2.1], [2.2] 10   **3.** [2.1], [2.2] -10
4. [2.2] 5   **5.** [2.2] 0   **6.** [2.1] No   **7.** [2.1] Yes
8. [2.1] -4 **9.** [2.1] -4 **10.** [2.1] Yes   **11.** [2.1] Does
not exist   **12.** [2.1] -2 **13.** [2.1] No   **14.** [2.3] 1

15. [2.3] -3 **16.** [2.3] $4x + 2h$ **17.** [2.4] $y = x - 1$
18. [2.5] $(4, 5)$ **19.** [2.5] $(5, -108)$ **20.** [2.5] $20x^4$
21. [2.5] $x^{-2/3}$ **22.** [2.5] $64x^{-9}$ **23.** [2.5] $6x^{-3/5}$
24. [2.5] $0.7x^6 - 12x^3 - 3x^2$ **25.** [2.5] $x^5 + 32x^3 - 5$
26. [2.7] $2x$ **27.** [2.7] $\dfrac{-x^2 + 16x + 8}{(8 - x)^2}$
28. [2.8] $2(5 - x)(2x - 1)^4(26 - 7x)$
29. [2.8] $35x^4(x^5 - 2)^6$
30. [2.8] $3x^2(4x + 3)^{-1/4} + 2x(4x + 3)^{3/4}$, or
$x(4x + 3)^{-1/4}(11x + 6)$ **31.** [2.9] $240x^{-6}$
32. [2.9] $840x^3$ **33.** [2.6] **(a)** $1 + 4t^3$;   **(b)** $12t^2$;
(c) 33, 48   **34.** [2.6] **(a)** $-8x^2 + 47x + 10$;   **(b)** $800,
$3050, -$2250;   **(c)** 40, $16x - 7$, $-16x + 47$;   **(d)** $40 per
unit, $313 per unit, $-$273 per unit   **35.** [2.6] **(a)** $100t$;
(b) 30,000;   **(c)** 2000 per year   **36.** [2.8] $4x^2 - 4x + 6$,
$-2x^2 - 9$ **37.** [2.2] 0   **38.** [2.8] $\dfrac{-9x^4 - 4x^3 + 9x + 2}{2\sqrt{1 + 3x}(1 + x^3)^2}$
39. [2.2] 0.1667   **40.** [2.2] -0.25
41. [2.5] $(-1.7137, 37.445)$, $(0, 0)$, $(1.7137, -37.445)$

Test: Chapter 2, p. 167

1. Yes   **2.** No   **3.** Does not exist   **4.** 1   **5.** No
6. 3   **7.** 3   **8.** Yes   **9.** 6   **10.** $\frac{1}{2}$ **11.** Does not exist
12. 4   **13.** ∞ **14.** $3(2x + h)$ **15.** $y = \frac{3}{4}x + 2$
16. $(0, 0)$, $(2, -4)$ **17.** $84x^{83}$ **18.** $\dfrac{5}{\sqrt{x}}$ **19.** $\dfrac{10}{x^2}$
20. $\frac{5}{4}x^{1/4}$, or $\frac{5}{4}\sqrt[4]{x}$ **21.** $-x + 0.61$ **22.** $x^2 - 2x + 2$
23. $\dfrac{-6x + 20}{x^5}$ **24.** $\dfrac{5}{(5 - x)^2}$
25. $(x + 3)^3(7 - x)^4(-9x + 13)$
26. $-5(x^5 - 4x^3 + x)^{-6}(5x^4 - 12x^2 + 1)$
27. $\sqrt{x^2 + 5} + \dfrac{x^2}{\sqrt{x^2 + 5}}$, or $\dfrac{2x^2 + 5}{\sqrt{x^2 + 5}}$ **28.** $24x$
29. **(a)** $P(x) = -0.001x^2 + 48.8x - 60$;   **(b)** $R(10) = 500$,
$C(10) = 72.10, $P(10) = 427.90;   **(c)** $R'(x) = 50$,
$C'(x) = 0.002x + 1.2$, $P'(x) = -0.002x + 48.8$;
(d) $R'(10) = 50 per unit, $C'(10) = 1.22 per unit,
$P'(10) = 48.78 per unit   **30.** **(a)** $\dfrac{dM}{dt} = -0.003t^2 + 0.2t$;

(b) 9;   **(c)** 1.7 words/minute   **31.** $x^3 + x^6$, $x^3 + 3x^4 + 3x^5 + x^6$ **32.** 12

33. $-2\left(\dfrac{1+3x}{1-3x}\right)^{1/3} + \left(\dfrac{1-3x}{1+3x}\right)^{2/3}$ **34.** $\dfrac{1}{2}$

35. (1.0836, 25.1029), (2.9503, 8.62466)

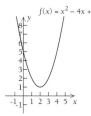

CHAPTER 3

Exercise Set 3.1, p. 184

1. Relative minimum at (2, 1)

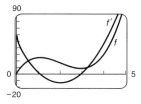

3. Relative maximum at $\left(\frac{1}{2}, \frac{21}{4}\right)$

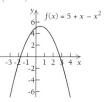

5. Relative minimum at (−1, −2)

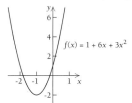

7. Relative minimum at (1, 1); relative maximum at $\left(-\frac{1}{3}, \frac{59}{27}\right)$

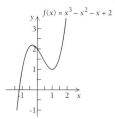

9. Relative maximum at (−1, 8); relative minimum at (1, 4)

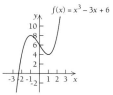

11. Relative minimum at (0, 0); relative maximum at (1, 1)

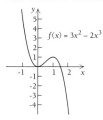

13. None

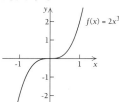

15. Relative maximum at (0, 10); relative minimum at (4, −22)

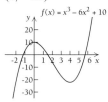

17. Relative maximum at $\left(\frac{3}{4}, \frac{27}{256}\right)$

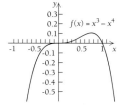

19. Relative minima at $(-2, -13)$ and $(2, -13)$; relative maximum at $(0, 3)$

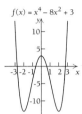

$f(x) = x^4 - 8x^2 + 3$

21. Relative maximum at $(0, 1)$

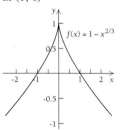

$f(x) = 1 - x^{2/3}$

23. Relative minimum at $(0, -8)$

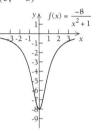

$f(x) = \dfrac{-8}{x^2 + 1}$

25. Relative minimum at $(-1, -2)$; relative maximum at $(1, 2)$

$f(x) = \dfrac{4x}{x^2 + 1}$

27. None

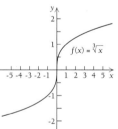

$f(x) = \sqrt[3]{x}$

29. Relative maximum at $(150, 22{,}506)$

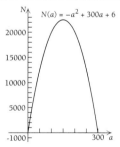

$N(a) = -a^2 + 300a + 6$

31. Relative maximum at $(200, 86.6)$

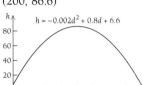

$h = -0.002d^2 + 0.8d + 6.6$

33. ◈

35. Relative minima at $(-5, 425)$ and $(4, -304)$; relative maximum at $(-2, 560)$

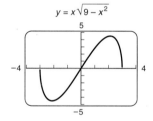

$y = x^4 + 4x^3 - 36x^2 - 160x + 400$

37. Relative maximum at $(2.12, 4.5)$; relative minimum at $(-2.12, -4.5)$

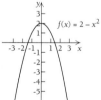

$y = x\sqrt{9 - x^2}$

Exercise Set 3.2, p. 198

1. Relative maximum at $(0, 2)$

$f(x) = 2 - x^2$

3. Relative minimum at $\left(-\frac{1}{2}, -\frac{5}{4}\right)$

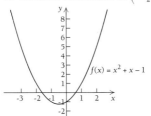

5. Relative maximum at $\left(\frac{3}{8}, -\frac{7}{16}\right)$

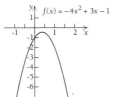

7. Relative maximum at $(-2, 72)$; relative minimum at $(3, -53)$

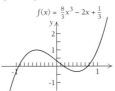

9. Relative maximum at $\left(-\frac{1}{2}, 1\right)$; relative minimum at $\left(\frac{1}{2}, -\frac{1}{3}\right)$

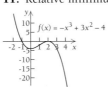

11. Relative minimum at $(0, -4)$; relative maximum at $(2, 0)$

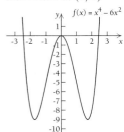

13. Relative minima at $(0, 0)$ and $(3, -27)$; relative maximum at $(1, 5)$

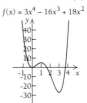

15. Relative minimum at $(-1, 0)$

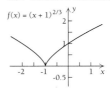

17. Relative minima at $(-\sqrt{3}, -9)$ and $(\sqrt{3}, -9)$; relative maximum at $(0, 0)$

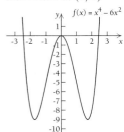

19. Relative maximum at $\left(-\frac{2}{3}, \frac{121}{27}\right)$; relative minimum at $(2, -5)$

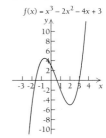

21. Relative minimum at $(-1, -1)$

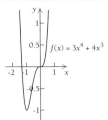

23. Relative maximum at $(-5, 400)$; relative minimum at $(9, -972)$

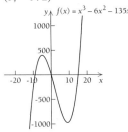

25. Relative minimum at $\left(-1, -\frac{1}{2}\right)$; relative maximum at $\left(1, \frac{1}{2}\right)$

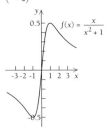

27. Relative maximum at $(0, 3)$ **29.** None

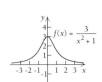

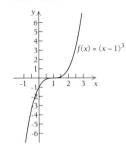

31. Relative minima at $(0, 0)$ and $(1, 0)$; relative maximum at $\left(\frac{1}{2}, \frac{1}{16}\right)$

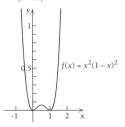

33. Relative minimum at $(-2, -64)$; relative maximum at $(2, 64)$

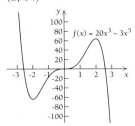

35. Relative minimum at $(-\sqrt{2}, -2)$; relative maximum at $(\sqrt{2}, 2)$

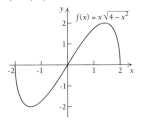

37. None

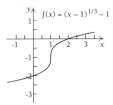

39. $(0, 1)$ **41.** $\left(\frac{1}{2}, \frac{1}{6}\right)$

43.

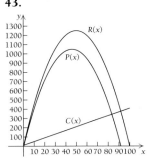

45. An object of radius $13\frac{1}{3}$ mm   **47.** $g = h'$ **49.** ◈
51. Relative maximum at $(1, 1)$; relative minimum at $(0, 0)$

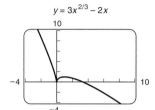

$y = 3x^{2/3} - 2x$

53. Relative maximum at $(0, 0)$; relative minimum at
$(0.8, -1.12)$

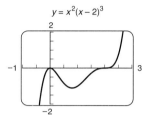

$y = x^2(x - 2)^3$

55. Relative minimum at $\left(\frac{1}{4}, -\frac{1}{4}\right)$

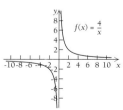

$y = x - \sqrt{x}$

Exercise Set 3.3, p. 212

1.

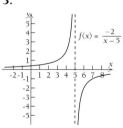

$f(x) = \frac{4}{x}$

3.

$f(x) = \frac{-2}{x - 5}$

5.

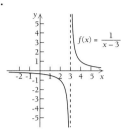

$f(x) = \frac{1}{x - 3}$

7.

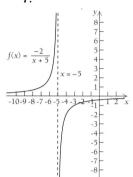

$f(x) = \frac{-2}{x + 5}$

$x = -5$

9.

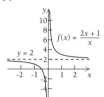

$f(x) = \frac{2x + 1}{x}$

$y = 2$

11.

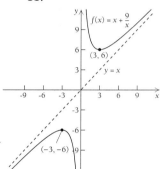

$f(x) = x + \frac{9}{x}$

$(3, 6)$

$y = x$

$(-3, -6)$

13.

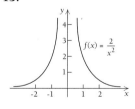

$f(x) = \frac{2}{x^2}$

15.

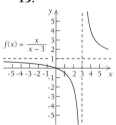

$f(x) = \frac{x}{x - 3}$

17.

$f(x) = \frac{1}{x^2 + 3}$

$\left(0, \frac{1}{3}\right)$

19.

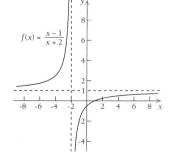

$f(x) = \frac{x - 1}{x + 2}$

21.

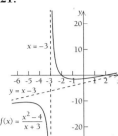

23.

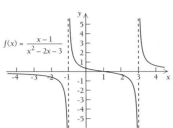

35. ◈

37.

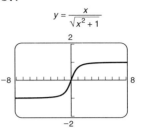

39.

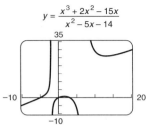

25.

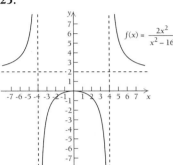

41.

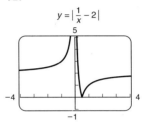

43. (a) $(-1.7, 0)$, $(1.7, 0)$;
(b) $(0, 0.75)$;
(c) $x = 2$, $y = 0.5x + 1$

27.

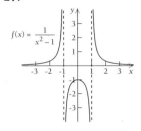

29.

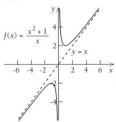

31. (a) $50, $37.24, $32.64, $26.37;
(b) Maximum = 50 at $x = 0$;
(c)

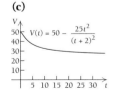

(d) 25;   (e) yes, $25

33. (a) $480, $600, $2400, $4800;
(b) ∞;
(c)

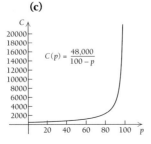

(d) no

Exercise Set 3.4, p. 223

1. (a) 41 mph;   (b) 80 mph;   (c) 13.5 mpg;
(d) 16.5 mpg;   (e) about 22%
3. Maximum = $5\frac{1}{4}$ at $x = \frac{1}{2}$; minimum = 3 at $x = 2$
5. Maximum = 4 at $x = 2$; minimum = 1 at $x = 1$
7. Maximum = $\frac{59}{27}$ at $x = -\frac{1}{3}$; minimum = 1 at $x = -1$
9. Maximum = 1 at $x = 1$; minimum = -5 at $x = -1$
11. Maximum = 15 at $x = -2$; minimum = -13 at $x = 5$
13. Maximum = minimum = -5 for all x in $[-1, 1]$
15. Maximum = 4 at $x = -1$; minimum = -12 at $x = 3$
17. Maximum = $3\frac{1}{5}$ at $x = -\frac{1}{5}$; minimum = -48 at $x = 3$
19. Minimum = -4 at $x = 2$; maximum = 50 at $x = 5$
21. Minimum = -110 at $x = -5$; maximum = 2 at $x = -1$
23. Maximum = 513 at $x = -8$; minimum = -511 at $x = 8$
25. Maximum = 17 at $x = 1$; minimum = -15 at $x = -3$
27. Maximum = 32 at $x = -2$; minimum = $-\frac{27}{16}$ at $x = \frac{3}{2}$
29. Maximum = 13 at $x = -2$ and $x = 2$; minimum = 4 at $x = -1$ and $x = 1$
31. Minimum = -5 at $x = -3$; maximum = -1 at $x = 5$
33. Minimum = 2 at $x = 1$; maximum = $20\frac{1}{20}$ at $x = 20$
35. Maximum = $\frac{4}{5}$ at $x = -2$ and $x = 2$; minimum = 0 at $x = 0$
37. Minimum = -1 at $x = -2$; maximum = 3 at $x = 26$
39. Maximum = 1225 at $x = 35$

41. Minimum $= 200$ at $x = 10$
43. Maximum $= \frac{1}{3}$ at $x = \frac{1}{2}$ **45.** Maximum $= \frac{289}{4}$ at $x = \frac{17}{2}$
47. Maximum $= 2\sqrt{3}$ at $x = -\sqrt{3}$; minimum $= -2\sqrt{3}$ at $x = \sqrt{3}$
49. Maximum $= 5700$ at $x = 2400$
51. Minimum $= -55\frac{1}{3}$ at $x = 1$
53. Maximum $= 2000$ at $x = 20$; minimum $= 0$ at $x = 0$ and $x = 30$
55. Minimum $= 24$ at $x = 6$
57. Minimum $= 108$ at $x = 6$
59. Maximum $= 3$ at $x = -1$; minimum $= -\frac{3}{8}$ at $x = \frac{1}{2}$
61. Maximum $= 2$ at $x = 8$; minimum $= 0$ at $x = 0$
63. None
65. Maximum $= -1$ at $x = 1$; minimum $= -5$ at $x = -1$
67. None
69. Minimum $= 0$ at $x = 0$; maximum $= 1$ at $x = -1$ and $x = 1$ **71.** None
73. Maximum $= -\frac{10}{3} + 2\sqrt{3}$ at $x = 2 - \sqrt{3}$; minimum $= -\frac{10}{3} - 2\sqrt{3}$ at $x = 2 + \sqrt{3}$
75. Minimum $= -1$ at $x = -1$ and $x = 1$
77. Maximum $= 1430$ units per month at $t = 25$ years of service
79. Maximum $= \$1500$ at $x = 3$ **81.** 61.64 mph
83. Maximum $= 3\sqrt{6}$ at $x = 3$; minimum $= -2$ at $x = -2$
85. 7 **87.** ◈
89. Minimum $= -17$ at $x = 3$; maximum $= 10$ at $x = 0$ and $x = 4$
91. Minimum $= 0$ at $x = -1$ and $x = 1$
93. $y = x + 8.857$; 15.857 millimeters

Exercise Set 3.5, p. 240

1. 25 and 25; maximum product $= 625$ **3.** No; $Q = x(50 - x)$ has no minimum   **5.** 2 and -2; minimum product $= -4$ **7.** $x = \frac{1}{2}$, $y = \sqrt{\frac{1}{2}}$; maximum $= \frac{1}{4}$
9. $x = 10$, $y = 10$; minimum $= 200$ **11.** $x = 2$, $y = \frac{32}{3}$; maximum $= \frac{64}{3}$ **13.** 30 yd by 60 yd; maximum area $= 1800$ yd^2 **15.** 13.5 ft by 13.5 ft; 182.25 ft^2
17. 20 in. by 20 in. by 5 in.; maximum $= 2000$ in^3
19. 5 in. by 5 in. by 2.5 in.; minimum $= 75$ in^2
21. 46 units; maximum profit $= \$1048$ **23.** 70 units; maximum profit $= \$19$ **25.** Approximately 1667 units; maximum profit $\approx \$5481$ **27. (a)** $R(x) = 150x - 0.5x^2$;
(b) $P(x) = -0.75x^2 + 150x - 4000$;   **(c)** 100;   **(d)** $\$3500$;
(e) $\$100$ **29.** $\$5.75$, $72{,}500$ (Will the stadium hold that many?)   **31.** 25 **33.** 4 ft by 4 ft by 20 ft   **35.** 9%
37. Reorder 5 times per year; lot size $= 20$ **39.** Reorder

12 times per year; lot size $= 30$ **41.** Reorder about 13 times per year; lot size $= 28$
43. 14 in. by 14 in. by 28 in.   **45.** $x = y = \dfrac{24}{4 + \pi}$ ft
47. $\sqrt[3]{\dfrac{1}{10}}$ **49.** Reorder $\dfrac{Q}{\sqrt{2bQ/a}}$ times per year; lot size $= \sqrt{2bQ/a}$ **51.** Minimum at $x = \dfrac{24\pi}{\pi + 4} \approx 10.56$ in., $24 - x = \dfrac{96}{\pi + 4} \approx 13.45$ in. There is no maximum if the string is to be cut. One would interpret the maximum to be at the endpoint, with the string uncut and used to form a circle.   **53.** S should be about 4.25 miles downshore from A.   **55. (a)** $C'(x) = 8 + \dfrac{3x^2}{100}$;
(b) $A(x) = 8 + \dfrac{20}{x} + \dfrac{x^2}{100}$;   **(c)** $A'(x) = \dfrac{x}{50} - \dfrac{20}{x^2}$;
(d) minimum $= 11$ at $x_0 = 10$, $C'(10) = 11$;
(e) $A(10) = 11$, $C'(10) = 11$ **57.** Minimum $= 6 - 4\sqrt{2}$ at $x = 2 - \sqrt{2}$ and $y = -1 + \sqrt{2}$ **59.** ◈

Exercise Set 3.6, p. 250

1. $\Delta y = 0.0401$; $f'(x)\,\Delta x = 0.04$ **3.** 0.2816, 0.28
5. -0.556, -1 **7.** 6, 6 **9.** $\Delta C = \$2.01$, $C'(70) = \$2$
11. $\Delta R = \$2$, $R'(70) = \$2$
13. (a) $P(x) = -0.01x^2 + 1.4x - 30$;   **(b)** $\Delta P = -\$0.01$, $P'(70) = 0$ **15.** 4.375 **17.** 10.1 **19.** 2.167
21. $dy = 9x^2(2x^3 + 1)^{1/2}\,dx$ **23.** $dy = \frac{1}{5}(x + 27)^{-4/5}\,dx$
25. $dy = (4x^3 - 6x^2 + 10x + 3)\,dx$ **27.** $dy = 3.1$
29. -0.01 **31.** 100 **33.** 2.512 cm^3 **35.** $\dfrac{5}{\pi} \approx 1.6$ ft
37. ◈

Exercise Set 3.7, p. 256

1. $\dfrac{1 - y}{x + 2}$; $-\dfrac{1}{9}$ **3.** $-\dfrac{x}{y}$; $-\dfrac{1}{\sqrt{3}}$ **5.** $\dfrac{6x^2 - 2xy}{x^2 - 3y^2}$; $-\dfrac{36}{23}$
7. $-\dfrac{y}{x}$ **9.** $\dfrac{x}{y}$ **11.** $\dfrac{3x^2}{5y^4}$ **13.** $\dfrac{-3xy^2 - 2y}{3x + 4x^2y}$
15. $\dfrac{dp}{dx} = \dfrac{-2}{2p + 1}$ **17.** $\dfrac{dp}{dx} = -\dfrac{p + 4}{x + 3}$ **19.** $-\dfrac{3}{4}$
21. $\$400$ per day, $\$80$ per day, $\$320$ per day   **23.** $\$16$ per day, $\$8$ per day, $\$8$ per day   **25.** 0.1728π cm^3/day $\approx$ 0.54 cm^3/day   **27. (a)** $1000R \cdot \dfrac{dR}{dt}$;
(b) -0.01125 mm^3/min   **29.** 65 mph   **31.** $-\dfrac{\sqrt{y}}{\sqrt{x}}$

33. $\dfrac{2}{3y^2(x+1)^2}$ **35.** $-\dfrac{9}{4}\sqrt[3]{y}\sqrt{x}$ **37.** $\dfrac{dy}{dx}=\dfrac{1+y}{2-x}$,

$\dfrac{d^2y}{dx^2}=\dfrac{2+2y}{(2-x)^2}$ **39.** $\dfrac{dy}{dx}=\dfrac{x}{y}$, $\dfrac{d^2y}{dx^2}=\dfrac{y^2-x^2}{y^3}$ **41.**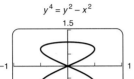

43.

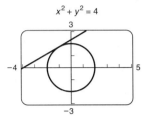

$x^2+y^2=4$

45.

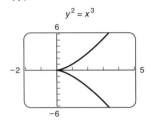

$y^4=y^2-x^2$

47.

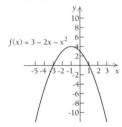

$y^2=x^3$

Summary and Review: Chapter 3, p. 259

1. [3.1], [3.2] Relative maximum at $(-1, 4)$

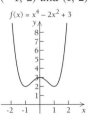

$f(x)=3-2x-x^2$

2. [3.1], [3.2] Relative maximum at $(0, 3)$; relative minima at $(-1, 2)$ and $(1, 2)$

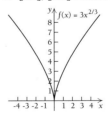

$f(x)=x^4-2x^2+3$

3. [3.1], [3.2] Relative maximum at $(-1, 4)$; relative minimum at $(1, -4)$

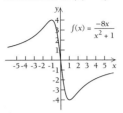

$f(x)=\dfrac{-8x}{x^2+1}$

4. [3.1], [3.2] None

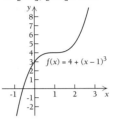

$f(x)=4+(x-1)^3$

5. [3.1], [3.2] Relative maximum at $(-1, 4)$; relative minimum at $\left(\frac{1}{3}, \frac{76}{27}\right)$

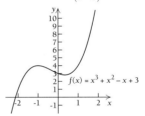

$f(x)=x^3+x^2-x+3$

6. [3.1], [3.2] Relative minimum at $(0, 0)$

$f(x)=3x^{2/3}$

7. [3.1], [3.2] Relative maximum at $(-1, 17)$; relative minimum at $(2, -10)$

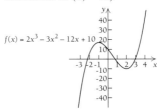

8. [3.1], [3.2] Relative maximum at $(-1, 4)$; relative minimum at $(1, 0)$

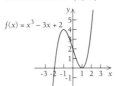

9. [3.3]

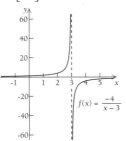

10. [3.3]

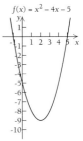

11. [3.3]

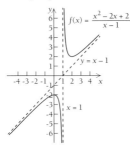

12. [3.3]

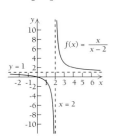

13. [3.4] None   **14.** [3.4] Maximum $= 32$ at $x = -2$; minimum $= -17$ at $x = 5$ **15.** [3.4] Maximum $= 19$ at $x = -1$; minimum $= -35$ at $x = 2$
16. [3.4] Maximum $= -2$ at $x = 1$; minimum $= -12$ at $x = -1$ **17.** [3.4] Minimum $= 3$ at $x = -1$

18. [3.4] None   **19.** [3.4] Maximum $= 163$ at $x = -3$; minimum $= -1$ at $x = -1$ **20.** [3.4] Minimum $= 10$ at $x = 1$ **21.** [3.4] Maximum $= 256$ at $x = 4$; minimum $= 0$ at $x = 0$ and $x = 5$
22. [3.4] Maximum $= \frac{53}{4}$ at $x = \frac{5}{2}$ **23.** [3.5] 30, 30
24. [3.5] $Q = -1$ at $x = -1$, $y = -1$ **25.** [3.5] 30 units; \$451 profit   **26.** [3.5] 10 ft $\times$ 10 ft $\times$ 25 ft; \$1500
27. [3.5] 12 times; 30 bicycles   **28.** [3.6] $\Delta y = -10.875$; $f'(x)\, \Delta x = -13$ **29.** [3.6] **(a)** $dy = (3x^2 - 1)\, dx$;
(b) $dy = 0.11$ **30.** [3.6] 8.3125   **31.** [3.7] $\dfrac{2x^2 + 3y}{-2y^2 - 3x}$;
$\dfrac{4}{5}$ **32.** [3.7] $-1\frac{3}{4}$ ft/sec   **33.** [3.7] \$600 per day, \$450
per day, \$150 per day   **34.** [3.4] Minimum $= 0$ at $x = 3$
35. [3.7] $\dfrac{3x^5 - 4x^3 - 12xy^2}{12x^2y + 4y^3 - 3y^5}$ **36.** [3.1], [3.2] Relative
minimum at $(-9, -9477)$; relative maximum at $(0, 0)$; relative minimum at $(15, -37{,}125)$ **37.** [3.1], [3.2] Relative maximum at $(-1.714, 37.445)$; relative minimum at $(1.714, -37.445)$ **38.** [3.1], [3.2] Relative minima at $(-3, -1)$ and $(3, -1)$; relative maximum at $(0, 1.08)$

Test: Chapter 3, p. 260

1. Relative minimum at $(2, -9)$

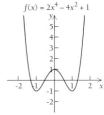

2. Relative minima at $(-1, -1)$ and $(1, -1)$; relative maximum at $(0, 1)$

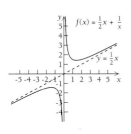

3. Relative minimum at $(2, -4)$

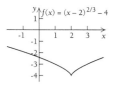

4. Relative maximum at $(0, 4)$

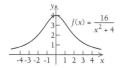

5. Relative maximum at $(-1, 2)$; relative minimum at $\left(\frac{1}{3}, \frac{22}{27}\right)$

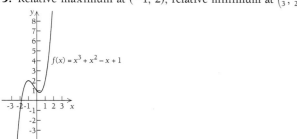

6. Relative minimum at $(-1, 2)$;   **7.** None
relative maximum at $(1, 6)$

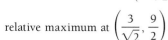

8. Relative minimum at $\left(-\dfrac{3}{\sqrt{2}}, -\dfrac{9}{2}\right)$;

relative maximum at $\left(\dfrac{3}{\sqrt{2}}, \dfrac{9}{2}\right)$

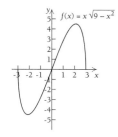

9.

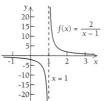

10.

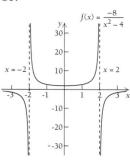

11.

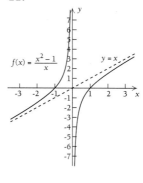

12.

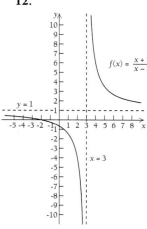

13. Maximum $= 9$ at $x = 3$ **14.** Maximum $= 2$ at $x = -1$; minimum $= -1$ at $x = -2$
15. Maximum $= 28.49$ at $x = 4.3$ **16.** Maximum $= 7$ at $x = -1$; minimum $= 3$ at $x = 1$ **17.** None
18. Minimum $= -\frac{13}{12}$ at $x = \frac{1}{6}$ **19.** Minimum $= 48$ at $x = 4$ **20.** 4 and -4 **21.** $x = 5$, $y = -5$; minimum $= 50$ **22.** Maximum profit $= \$24{,}980$; 500 units   **23.** 40 in. by 40 in. by 10 in.; maximum volume $= 16{,}000$ in^3 **24.** 35 times at lot size 35
25. $\Delta y = 1.01$, $f'(x)\,\Delta x = 1$ **26.** 10.2
27. (a) $\dfrac{x\,dx}{\sqrt{x^2 + 3}}$; (b) 0.0075593   **28.** $-\dfrac{x^2}{y^2}$, $-\dfrac{1}{4}$
29. -0.96 ft/sec   **30.** Maximum $= \frac{1}{3} \cdot 2^{2/3}$ at $x = \sqrt[3]{2}$; minimum $= 0$ at $x = 0$
31. (a) $A(x) = \dfrac{C(x)}{x} = 100 + \dfrac{100}{\sqrt{x}} + \dfrac{\sqrt{x}}{100}$;
(b) minimum $= 102$ at $x = 10{,}000$ **32.** Relative maximum at $(1.09, 25.1)$; relative minimum at $(2.97, 8.6)$

CHAPTER 4

Exercise Set 4.1, p. 276

1.

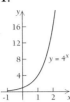

3.

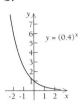

5.

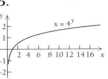

7. $3e^{3x}$ **9.** $-10e^{-2x}$

11. e^{-x} **13.** $-7e^x$ **15.** e^{2x} **17.** $x^3e^x(x+4)$

19. $\dfrac{e^x(x-4)}{x^5}$ **21.** $(-2x+7)e^{-x^2+7x}$ **23.** $-xe^{-x^2/2}$

25. $\dfrac{e^{\sqrt{x-7}}}{2\sqrt{x-7}}$ **27.** $\dfrac{e^x}{2\sqrt{e^x-1}}$

29. $(1-2x)e^{-2x}-e^{-x}+3x^2$ **31.** e^{-x} **33.** ke^{-kx}

35.

37.

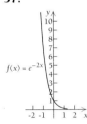

39.

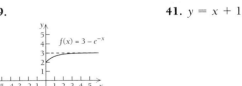

41. $y = x + 1$

43. (a) $C'(t) = 50e^{-t}$;   **(b)** \$50 million;   **(c)** \$0.916 million

45. (a) Approximately 335, 280, 173;
(b)

(c) $D'(p) = -1.44e^{-0.003p}$

47. (a) 0 ppm, 3.7 ppm, 5.4 ppm, 4.48 ppm, 0.05 ppm;
(b)

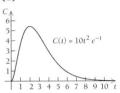

(c) $C'(t) = 10te^{-t}(2-t)$ **(d)** $C \approx 5.4$ ppm at $t = 2$ hr

49. $15e^{3x}(e^{3x}+1)^4$ **51.** $-e^{-t}-3e^{3t}$

53. $\dfrac{e^x(x-1)^2}{(x^2+1)^2}$ **55.** $\dfrac{e^{\sqrt{x}}}{2\sqrt{x}}+\dfrac{1}{2}\sqrt{e^x}$

57. $\dfrac{1}{2}e^{x/2}\left[\dfrac{x}{\sqrt{x-1}}\right]$ **59.** $\dfrac{4}{(e^x+e^{-x})^2}$ **61.** 2, 2.25,

2.48832, 2.59374, 2.71692   **63.** Maximum $= 4e^{-2} \approx 0.54$
at $x = 2$ **65.** ◈

67. Relative minimum at $(0, 0)$; relative maximum at
$(2, 0.54)$

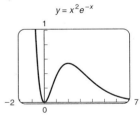

69. See text prior to Example 9. The graphs of $f(x)$ and $f'(x)$
are both the graph of $y = e^x$.

71.

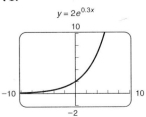

73.

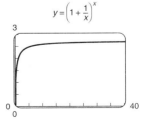

Exercise Set 4.2, p. 293

1. $2^3 = 8$ **3.** $8^{1/3} = 2$ **5.** $a^J = K$ **7.** $b^v = T$

9. $\log_e b = M$ **11.** $\log_{10} 100 = 2$ **13.** $\log_{10} 0.1 = -1$

15. $\log_M V = p$ **17.** 2.708   **19.** −1.609   **21.** 2.609

23. 2.9957   **25.** −1.3863   **27.** 4   **29.** 8.681690

31. −4.006334   **33.** 8.999619   **35.** $t \approx 4.6$

37. $t \approx 4.1$ **39.** $t \approx 2.3$ **41.** $t \approx 141$ **43.** $-\dfrac{6}{x}$

45. $x^3(1 + 4 \ln x) - x$ **47.** $\dfrac{1 - 4 \ln x}{x^5}$ **49.** $\dfrac{1}{x}$

51. $\dfrac{10x}{5x^2 - 7}$ **53.** $\dfrac{1}{x \ln 4x}$ **55.** $\dfrac{x^2 + 7}{x(x^2 - 7)}$

57. $e^x\left(\dfrac{1}{x} + \ln x\right)$ **59.** $\dfrac{e^x}{e^x + 1}$ **61.** $\dfrac{2 \ln x}{x}$

63. (a) 1000;   **(b)** $N'(a) = \dfrac{200}{a}$, $N'(10) = 20$;

(c) minimum $= 1000$ at $a = 1$ **65.** 58 days

67. (a) \$58.69, \$78.00;   **(b)** $\$63.80e^{-1.1t}$;   **(c)** 2.7

69. (a) 18.1%, 69.9%;   **(b)** $P'(t) = 20e^{-0.2t}$;   **(c)** 11.5

71. (a) 68%;   **(b)** 36%;   **(c)** 3.6%;   **(d)** 5%;   **(e)** $-\dfrac{20}{t + 1}$;

(f) maximum $= 68\%$ at $t = 0$ **73. (a)** 2.37 ft/sec;

(b) 3.37 ft/sec;   **(c)** $v'(p) = \dfrac{0.37}{p}$, $v'(p) =$ the acceleration

of the walker   **75.** $\dfrac{-4(\ln x)^{-5}}{x}$ **77.** $\dfrac{15t^2}{t^3 + 1}$

79. $\dfrac{4[\ln (x + 5)]^3}{x + 5}$ **81.** $\dfrac{5t^4 - 3t^2 + 6t}{(t^3 + 3)(t^2 - 1)}$ **83.** $\dfrac{24x + 25}{8x^2 + 5x}$

85. $\dfrac{2(1 - \ln t^2)}{t^3}$ **87.** $x^n \ln x$ **89.** $\dfrac{1}{\sqrt{1 + t^2}}$

91. $\dfrac{3}{x \ln x}$ **93.** 1   **95.** $\sqrt[e]{e} \approx 1.444667861$; $\sqrt[e]{e} > \sqrt[x]{x}$

for any $x > 0$ such that $x \neq e$ **97.** $t = \dfrac{\ln P - \ln P_0}{k}$

99. ∞ **101.** Let $a = \ln x$; then $e^a = x$, so $\log x =$

$\log e^a = a \log e$. Then $a = \ln x = \dfrac{\log x}{\log e} \approx \dfrac{\log x}{0.4343} \approx$

$2.3026 \log x$.   **103.** ◆

105. $y_1 = f(x) = x \ln x$ $y_2 = f'(x) = 1 + \ln x$

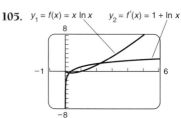

107. $y_1 = f(x) = \dfrac{\ln x}{x^2}$ $y_2 = f'(x) = \dfrac{1 - 2 \ln x}{x^3}$

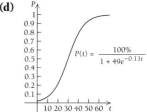

109. $-\dfrac{1}{2e} \approx -0.184$

Exercise Set 4.3, p. 305

1. $Q(t) = Q_0 e^{kt}$ **3. (a)** $P(t) = P_0 e^{0.065t}$;   **(b)** \$1067.16,

\$1138.83;   **(c)** 10.7 yr   **5.** 6.9%   **7. (a)** $N(t) = 50e^{0.1t}$;

(b) 369;   **(c)** 6.9 yr   **9.** 6.9 yr (2002)

11. (a) $k = 0.161602$; $V(t) = \$84,000e^{0.161602t}$;

(b) \$271,283,000;   **(c)** 4.3 yr;   **(d)** 58 yr

13. (a) $k = 0.061248$, $P(t) = \$100e^{0.061248t}$;   **(b)** \$628.04;

(c) 11.3 yr (1978)   **15.** $k = 0.505745$; $C(t) = 0.2036e^{0.505745t}$;

401 million   **17.** Approximately \$202,000,000,000,000

19. $k = 0.15468$, or 15.468%; approximately \$1,908,456;

approximately \$3,035,373   **21. (a)** 2%;   **(b)** 3.8%, 7.0%,

21.6%, 50.2%, 93.1%, 98.03%;   **(c)** $P'(t) = \dfrac{6.37e^{-0.13t}}{(1 + 49e^{-0.13t})^2}$;

(d)

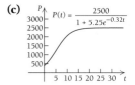

$P(t) = \dfrac{100\%}{1 + 49e^{-0.13t}}$

23. 19.8 yr   **25. (a)** $P(t) = 241e^{0.009t}$;   **(b)** 268 million;

(c) 77 yr   **27.** 0.20%   **29. (a)** 400, 520, 1214, 2059,

2396, 2478;   **(b)** $P'(t) = \dfrac{4200e^{-0.32t}}{(1 + 5.25e^{-0.32t})^2}$;

(c)

$P(t) = \dfrac{2500}{1 + 5.25e^{-0.32t}}$

31. (a) 0%, 33.0%, 55.1%, 69.9%, 86.5%, 99.2%, 99.8%;
(b) $P'(t) = 0.4e^{-0.4t}$;
(c)

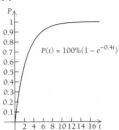

$P(t) = 100\%(1 - e^{-0.4t})$

33. 7.573%   **35.** 9%   **37.** $T_3 = \dfrac{\ln 3}{k}$ **39.** Answers

depend on particular data.   **41.** $\dfrac{\ln 2}{24} \approx 2.9\%$ per hour

43. ◆
45. (a) $N(t) = 25.65e^{0.4012t}$;
(b)

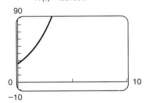

$N(t) = 25.65e^{0.4012t}$

(c) 7,053,000,000;   **(d)** 1994

Exercise Set 4.4, p. 317

1. $1505.97   **3.** $35,957.75   **5. (a)** $40,000;   **(b)** $5413
7. $27,194.82   **9.** 7.2 days   **11.** 23% per minute
13. 10.1 g   **15.** 19,109 yr   **17.** 7604 yr
19. (a) $A = A_0 e^{-kt}$;   **(b)** 9 hr   **21. (a)** 0.8%;
(b) 78.7% W_0 **23. (a)** 25%I_0, 6%I_0, 1.5%I_0;
(b) 0.00008%I_0 **25. (a)** 25°;   **(b)** $k = 0.05$;   **(c)** 84.2°;
(d) 32 min   **27.** 7 P.M.   **29. (a)** $k = 0.0162649$,
$P(t) = 453{,}000e^{-0.0162649t}$, where $t =$ the number of years
since 1970;   **(b)** 278,090;   **(c)** 801 yr   **31.** 233 million,
173 million   **33. (a)** 11 watts;   **(b)** 173 days;
(c) 402 days;   **(d)** 50 watts   **35. (a)** $k = 0.514$;
$S(t) = 205e^{-0.514t}$;   **(b)** 32,877;   **(c)** 2020
37. ($166.16, 292)   **39.** ◆

Exercise Set 4.5, p. 324

1. $e^{6.4378}$ **3.** $e^{12.238}$ **5.** $e^{k \cdot \ln 4}$ **7.** $e^{kT \cdot \ln 8}$ **9.** $(\ln 6)6^x$
11. $(\ln 10)10^x$ **13.** $(6.2)^x[x \ln 6.2 + 1]$

15. $10^x x^2[x \ln 10 + 3]$ **17.** $\dfrac{1}{\ln 4} \cdot \dfrac{1}{x}$ **19.** $\dfrac{2}{\ln 10} \cdot \dfrac{1}{x}$

21. $\dfrac{1}{\ln 10} \cdot \dfrac{1}{x}$ **23.** $x^2\left[\dfrac{1}{\ln 8} + 3 \log_8 x\right]$

25. $-250{,}000(\ln 4)\left(\frac{1}{4}\right)^t$ **27.** 5   **29. (a)** $I = 10^7 \cdot I_0$;

(b) $I = 10^8 \cdot I_0$;   **(c)** the intensity in (b) is 10 times that in

part (a);   **(d)** $\dfrac{dI}{dR} = (I_0 \cdot \ln 10) \cdot 10^R$ **31.** $\dfrac{1}{\ln 10} \cdot \dfrac{1}{I}$

33. $\dfrac{m}{\ln 10} \cdot \dfrac{1}{x}$ **35.** $2(\ln 3)3^{2x}$ **37.** $(\ln x + 1)x^x$

39. $x^{e^x} e^x\left(\ln x + \dfrac{1}{x}\right)$ **41.** $\dfrac{1}{\ln a} \cdot \dfrac{f'(x)}{f(x)}$ **43.** ◆

Exercise Set 4.6, p. 332

1. (a) $E(p) = \dfrac{p}{400 - p}$;   **(b)** $E(125) = \frac{5}{11}$, inelastic;

(c) $p = $200   **3. (a)** $E(p) = \dfrac{p}{50 - p}$;   **(b)** $E(46) = \frac{23}{2}$,

elastic;   **(c)** $p = $25   **5. (a)** $E(p) = 1$, for all $p > 0$;
(b) $E(50) = 1$, unit elasticity;   **(c)** Total revenue $= R(p) =$
400, for all $p > 0$. It has 400 as a maximum for all $p > 0$.

7. (a) $E(p) = \dfrac{p}{1000 - 2p}$;   **(b)** $E(400) = 2$, elastic;

(c) $p = \$\frac{1000}{3} \approx 333.33 **9. (a)** $E(p) = \dfrac{p}{4}$;

(b) $E(10) = \frac{5}{2}$, elastic;   **(c)** $p = $4

11. (a) $E(p) = \dfrac{2p}{p + 3}$;   **(b)** $E(1) = \frac{1}{2}$, inelastic;   **(c)** $p = $3

13. (a) $E(p) = \dfrac{25p}{967 - 25p}$;   **(b)** $p = 19.34$¢;

(c) $p > 19.34$¢;   **(d)** $p < 19.34$¢;   **(e)** $p = 19.34$¢;

(f) decrease   **15. (a)** $E(p) = \dfrac{3p^3}{2(200 - p^3)}$;

(b) $\dfrac{81}{346} \approx 0.234$;   **(c)** increases it   **17. (a)** $E(p) = n$;

(b) no, E is a constant n;   **(c)** only when $n = 1$
19. $E(p) = -pL'(p)$ **21.** ◆

Summary and Review: Chapter 4, p. 334

1. $[4.2]\ \dfrac{1}{x}$ **2.** $[4.1]\ e^x$ **3.** $[4.2]\ \dfrac{4x^3}{x^4 + 5}$ **4.** $[4.1]\ \dfrac{e^{2\sqrt{x}}}{\sqrt{x}}$

5. $[4.2]\ \dfrac{6}{x}$ **6.** $[4.1]\ 4e^{4x} + 4x^3$ **7.** $[4.2]\ \dfrac{1 - 3\ln x}{x^4}$

8. [4.1], [4.2] $e^{x^2}\left(\dfrac{1}{x} + 2x \ln 4x\right)$ **9.** [4.1], [4.2] $4e^{4x} - \dfrac{1}{x}$

10. [4.2] $8x^7 - \dfrac{8}{x}$ **11.** [4.1], [4.2] $\dfrac{1 - x}{e^x}$

12. [4.2] 6.9300   **13.** [4.2] −3.2698   **14.** [4.2] 8.7601
15. [4.2] 3.2698   **16.** [4.2] 2.54995   **17.** [4.2] −3.6602
18. [4.3] $Q(t) = Q_0 e^{kt}$ **19.** [4.3] 4.3%   **20.** [4.3] 10.2 yr
21. [4.3] **(a)** $k = 0.050991$, $C(t) = \$4.65e^{0.050991t}$;
(b) \$27.70, \$46.13   **22.** [4.3] **(a)** $N(t) = 60e^{0.12t}$;   **(b)** 199;
(c) 5.8 yr   **23.** [4.4] 5.3 yr   **24.** [4.4] 18.2% per day
25. [4.4] **(a)** $A(t) = 800e^{-0.07t}$;   **(b)** 197 g;   **(c)** 9.9 days
26. [4.3] **(a)** 0.5034, 0.7534, 0.9698, 0.9991, 0.9999;
(b) $p'(t) = 0.7e^{-0.7t}$;
(c)

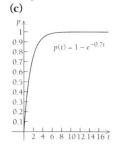

27. [4.4] \$34,735.26   **28.** [4.5] $(\ln 3)3^x$

29. [4.5] $\dfrac{1}{\ln 15} \cdot \dfrac{1}{x}$ **30.** [4.6] **(a)** $E(p) = \dfrac{2p}{p + 4}$;

(b) $E(\$1) = \frac{2}{5}$, inelastic;   **(c)** $E(\$12) = \frac{3}{2}$, elastic;

(d) decrease;   **(e)** $p = \$4$ **31.** [4.1] $\dfrac{-8}{(e^{2x} - e^{-2x})^2}$

32. [4.2] $-\dfrac{1}{1024e}$

33. [4.1]   **34.** [4.1] 0

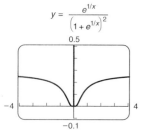

$$y = \dfrac{e^{1/x}}{\left(1 + e^{1/x}\right)^2}$$

Test: Chapter 4, p. 335

1. e^x **2.** $\dfrac{1}{x}$ **3.** $-2xe^{-x^2}$ **4.** $\dfrac{1}{x}$ **5.** $e^x - 15x^2$

6. $3e^x\left(\dfrac{1}{x} + \ln x\right)$ **7.** $\dfrac{e^x - 3x^2}{e^x - x^3}$ **8.** $\dfrac{\dfrac{1}{x} - \ln x}{e^x}$, or

$\dfrac{1 - x \ln x}{xe^x}$ **9.** 1.0674   **10.** 0.5554   **11.** 0.4057

12. $M(t) = M_0 e^{kt}$ **13.** 17.3% per hour   **14.** 10 yr
15. **(a)** 0.0330326; $C(t) = \$0.54e^{0.0330326t}$;   **(b)** \$3.79,
\$5.28   **16.** **(a)** $A(t) = 3e^{-0.1t}$;   **(b)** 1.1 cc;   **(c)** 6.9 hr
17. 63 days   **18.** 0.000069% per year   **19.** **(a)** 4%;
(b) 5.2%, 14.5%, 40.7%, 73.5%, 91.8%, 99.5%, 99.9%;

(c) $P'(t) = \dfrac{6.72e^{-0.28t}}{(1 + 24e^{-0.28t})^2}$;

(d)

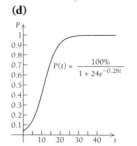

20. \$28,503.49   **21.** $(\ln 20)20^x$ **22.** $\dfrac{1}{\ln 20} \cdot \dfrac{1}{x}$

23. **(a)** $E(p) = \dfrac{p}{5}$;   **(b)** $E(\$3) = \frac{3}{5}$, inelastic;

(c) $E(\$18) = \frac{18}{5}$, elastic;   **(d)** increase;   **(e)** $p = \$5$
24. $(\ln x)^2$ **25.** Maximum = $256/e^4 \approx 4.69$ at $x = 4$;
minimum = 0 at $x = 0$
26.   **27.** 0

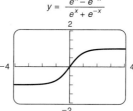

$$y = \dfrac{e^x - e^{-x}}{e^x + e^{-x}}$$

CHAPTER 5

Exercise Set 5.1, p. 347

1. $\dfrac{x^7}{7} + C$ **3.** $2x + C$ **5.** $\dfrac{4}{5}x^{5/4} + C$

7. $\dfrac{x^3}{3} + \dfrac{x^2}{2} - x + C$ **9.** $\dfrac{t^3}{3} - t^2 + 3t + C$

11. $\dfrac{5}{8}e^{8x} + C$ **13.** $\dfrac{x^4}{4} - \dfrac{7}{15}x^{15/7} + C$

15. $1000 \ln x + C$ **17.** $-x^{-1} + C$ **19.** $\dfrac{2}{3}x^{3/2} + C$

21. $-18x^{1/3} + C$ **23.** $-4e^{-2x} + C$

25. $\dfrac{1}{3}x^3 - x^{3/2} - 3x^{-1/3} + C$ **27.** $f(x) = \dfrac{x^2}{2} - 3x + 13$

29. $f(x) = \dfrac{x^3}{3} - 4x + 7$ **31.** $C(x) = \dfrac{x^4}{4} - x^2 + 100$

33. **(a)** $R(x) = \dfrac{x^3}{3} - 3x$;   **(b)** If you sell no products, you

make no money.   **35.** $D(p) = \dfrac{4000}{p} + 3$

37. **(a)** $E(t) = 30t - 5t^2 + 32$;   **(b)** $E(3) = 77\%$,
$E(5) = 57\%$ **39.** $s(t) = t^3 + 4$ **41.** $v(t) = 2t^2 + 20$
43. $s(t) = -\frac{1}{3}t^3 + 3t^2 + 6t + 10$
45. $s(t) = -16t^2 + v_0 t + s_0$ **47.** $\frac{1}{4}$ mi
49. **(a)** $M(t) = -0.001t^3 + 0.1t^2$;   **(b)** 6

51. $f(t) = \dfrac{t^{\sqrt{3}+1}}{\sqrt{3}+1} + 8$ **53.** $\dfrac{x^6}{6} - \dfrac{2}{5}x^5 + \dfrac{1}{4}x^4 + C$

55. $\dfrac{2}{5}t^{5/2} + 4t^{3/2} + 18\sqrt{t} + C$

57. $\dfrac{t^4}{4} + t^3 + \dfrac{3}{2}t^2 + t + C$, or $\dfrac{(t+1)^4}{4} + C$

59. $\dfrac{b}{a}e^{ax} + C$ **61.** $\dfrac{12}{7}x^{7/3} + C$ **63.** $\dfrac{t^3}{3} - t^2 + 4t + C$

65. ◈

Exercise Set 5.2, p. 358

1. 8   **3.** 8   **5.** $41\frac{2}{3}$ **7.** $\frac{1}{4}$ **9.** $10\frac{2}{3}$
11. $e^3 - 1 \approx 19.086$ **13.** $\ln 3 \approx 1.0986$ **15.** 51
17. An antiderivative, velocity   **19.** An antiderivative,
energy used in time t **21.** An antiderivative, total revenue
23. An antiderivative, amount of drug in blood
25. **(a)** $C(x) = 50x$;   **(b)** $A(x) = 50x$, same;   **(c)** As x
increases, the area over $[0, x]$ increases. Also, as x increases,
total cost increases.

27. **(a)** $C(x) = 100x - 0.1x^2$;   **(b)** $R(x) = 100x + 0.1x^2$;
(c) $P(x) = 0.2x^2$;   **(d)** $P(1000) = \$200,000$

29. **(a)** $s(t) = t^3 + t^2$;   **(b)** 150   **31.** $3\frac{1}{2}$ **33.** $359\frac{7}{15}$
35. $6\frac{3}{4}$ **37.** ◈

Exercise Set 5.3, p. 368

1. $\dfrac{1}{6}$ **3.** $\dfrac{4}{15}$ **5.** $e^b - e^a$ **7.** $b^3 - a^3$ **9.** $\dfrac{e^2}{2} + \dfrac{1}{2}$

11. $\dfrac{2}{3}$ **13.** $\dfrac{5}{34}$ **15.** 4   **17.** $9\dfrac{5}{6}$ **19.** 12

21. $e^5 - \dfrac{1}{e}$ **23.** $7\dfrac{1}{3}$ **25.** $17\dfrac{1}{3}$ **27.** **(a)** \$2948.26;

(b) \$2913.90;   **(c)** $k = 6.9$, so it will be on the 7th day
29. \$3600   **31.** \$68,676,000   **33.** 90
35. **(a)** 2,253,169;   **(b)** 21.8 days   **37.** 9   **39.** $\frac{307}{6}$
41. $\frac{15}{4}$ **43.** 8   **45.** 12   **47.** ◈   **49.** 4   **51.** 529.3556
53. 11.9879   **55.** 3.1416   **57.** 520.8333   **59.** 1.50733

Exercise Set 5.4, p. 377

1. $\frac{1}{4}$ **3.** $\frac{9}{2}$ **5.** $\frac{125}{6}$ **7.** $\frac{9}{2}$ **9.** $\frac{1}{6}$ **11.** $\frac{125}{3}$ **13.** $\frac{32}{3}$
15. 3   **17.** 128   **19.** 9   **21.** **(a)** B;   **(b)** 2

23. $\dfrac{(e-1)^2}{e}$ **25.** $\dfrac{2}{3}$ **27.** 96   **29.** $\dfrac{\pi p R^4}{8Lv}$ **31.** ◈

33. 24.961   **35.** 16.7083
37. **(a)**

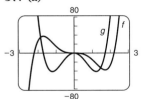

(b) $a = -1.8623$, $b = 0$,
$c = 1.4594$;   **(c)** 64.5239;
(d) 17.683

Exercise Set 5.5, p. 384

1. $\ln(7 + x^3) + C$ **3.** $\frac{1}{4}e^{4x} + C$ **5.** $2e^{x/2} + C$

7. $\frac{1}{4}e^{x^4} + C$ **9.** $-\frac{1}{3}e^{-t^3} + C$ **11.** $\dfrac{(\ln 4x)^2}{2} + C$

13. $\ln(1 + x) + C$ **15.** $-\ln(4 - x) + C$
17. $\frac{1}{24}(t^3 - 1)^8 + C$ **19.** $\frac{1}{8}(x^4 + x^3 + x^2)^8 + C$
21. $\ln(4 + e^x) + C$ **23.** $\frac{1}{4}(\ln x^2)^2 + C$, or $(\ln x)^2 + C$

25. $\ln(\ln x) + C$ **27.** $\dfrac{2}{3a}(ax + b)^{3/2} + C$

29. $\dfrac{b}{a}e^{ax} + C$ **31.** $\dfrac{-1}{4(1 + x^3)^4} + C$

33. $-\frac{21}{8}(4 - x^2)^{4/3} + C$ **35.** $e - 1$ **37.** $\frac{21}{4}$

39. $\ln 4 - \ln 2 = \ln \frac{4}{2} = \ln 2$ **41.** $\ln 19$ **43.** $1 - \dfrac{1}{e^b}$

45. $1 - \dfrac{1}{e^{mb}}$ **47.** $\dfrac{208}{3}$ **49.** $\dfrac{1640}{6561}$ **51.** $\dfrac{315}{8}$

53. $D(p) = 2000\sqrt{25 - p^2} + 5000$

55. (a) $P(T) = 2000T^2 + [-250{,}000(e^{-0.1T} - 1)] -$
$250{,}000$, or $2000T^2 - 250{,}000e^{-0.1T}$; **(b)** $P(10) = \$108{,}030$

57. $\dfrac{128}{3}$ **59.** $-\dfrac{5}{12}(1 - 4x^2)^{3/2} + C$ **61.** $-\dfrac{1}{3}e^{-x^3} + C$

63. $-e^{1/t} + C$ **65.** $-\dfrac{1}{3}(\ln x)^{-3} + C$

67. $\dfrac{2}{9}(x^3 + 1)^{3/2} + C$ **69.** $\dfrac{3}{4}(x^2 - 6x)^{2/3} + C$

71. $\dfrac{1}{8}[\ln (t^4 + 8)]^2 + C$ **73.** $x + \dfrac{9}{x + 3} + C$

75. $t - \ln (t - 4) + C$ **77.** $-\ln (1 + e^{-x}) + C$

79. $\dfrac{1}{n + 1}(\ln x)^{n+1} + C$ **81.** $\ln [\ln (\ln x)] + C$

83. $\dfrac{9(7x^2 + 9)^{n+1}}{14(n + 1)} + C$

Exercise Set 5.6, p. 393

1. $xe^{5x} - \dfrac{1}{5}e^{5x} + C$ **3.** $\dfrac{1}{2}x^6 + C$ **5.** $\dfrac{x}{2}e^{2x} - \dfrac{1}{4}e^{2x} + C$

7. $-\dfrac{x}{2}e^{-2x} - \dfrac{1}{4}e^{-2x} + C$ **9.** $\dfrac{x^3}{3}\ln x - \dfrac{x^3}{9} + C$

11. $\dfrac{x^2}{2}\ln x^2 - \dfrac{x^2}{2} + C$, or $x^2 \ln x - \dfrac{x^2}{2} + C$

13. $(x + 3) \ln (x + 3) - x + C$. Let $u = \ln (x + 3)$,
$dv = dx$, and choose $v = x + 3$ for an antiderivative of v.

15. $\left(\dfrac{x^2}{2} + 2x\right) \ln x - \dfrac{x^2}{4} - 2x + C$

17. $\left(\dfrac{x^2}{2} - x\right) \ln x - \dfrac{x^2}{4} + x + C$

19. $\dfrac{2}{3}x(x + 2)^{3/2} - \dfrac{4}{15}(x + 2)^{5/2} + C$

21. $\dfrac{x^4}{4}\ln 2x - \dfrac{x^4}{16} + C$ **23.** $x^2e^x - 2xe^x + 2e^x + C$

25. $\dfrac{1}{2}x^2e^{2x} + \dfrac{1}{4}e^{2x} - \dfrac{1}{2}xe^{2x} + C$

27. $e^{-2x}\left(-\dfrac{1}{2}x^3 - \dfrac{3}{4}x^2 - \dfrac{3}{4}x - \dfrac{3}{8}\right) + C$

29. $e^{3x}\left[\dfrac{1}{3}(x^4 + 1) - \dfrac{4}{9}x^3 + \dfrac{4}{9}x^2 - \dfrac{8}{27}x + \dfrac{8}{81}\right] + C$

31. $\dfrac{8}{3}\ln 2 - \dfrac{7}{9}$ **33.** $9 \ln 9 - 5 \ln 5 - 4$ **35.** 1

37. $C(x) = \dfrac{8}{3}x(x + 3)^{3/2} - \dfrac{16}{15}(x + 3)^{5/2}$

39. (a) $10[e^{-T}(-T - 1) + 1]$; **(b)** ≈ 9.084

41. $\dfrac{2}{9}x^{3/2}[3 \ln x - 2] + C$ **43.** $\dfrac{e^t}{t + 1} + C$

45. $2\sqrt{x} \ln x - 4\sqrt{x} + C$

47. $(13t^2 - 48)\left[\dfrac{5}{16}(4t + 7)^{4/5}\right] - t\left[\dfrac{325}{288}(4t + 7)^{9/5}\right] +$
$\dfrac{1625}{16{,}128}(4t + 7)^{14/5} + C$ **49.** Let $u = x^n$ and $dv = e^x \, dx$.
Then $du = nx^{n-1} \, dx$ and $v = e^x$. Then use integration by
parts.   **51.** ◈   **53.** $355{,}986$

Exercise Set 5.7, p. 397

1. $\dfrac{1}{9}e^{-3x}(-3x - 1) + C$ **3.** $\dfrac{5^x}{\ln 5} + C$

5. $\dfrac{1}{8}\ln\left(\dfrac{4 + x}{4 - x}\right) + C$ **7.** $5 - x - 5 \ln (5 - x) + C$

9. $\dfrac{1}{5(5 - x)} + \dfrac{1}{25}\ln\left(\dfrac{x}{5 - x}\right) + C$

11. $(\ln 3)x + x \ln x - x + C$

13. $\dfrac{x^4e^{5x}}{5} - \dfrac{4}{25}x^3e^{5x} + \dfrac{12}{125}x^2e^{5x} - \dfrac{24}{3125}e^{5x}(5x - 1) + C$

15. $x^4\left(\dfrac{1}{4}\ln x - \dfrac{1}{16}\right) + C$ **17.** $\ln (x + \sqrt{x^2 + 7}) + C$

19. $\dfrac{2}{5 - 7x} + \dfrac{2}{5}\ln\left(\dfrac{x}{5 - 7x}\right) + C$

21. $-\dfrac{5}{4}\ln\left(\dfrac{x - \frac{1}{2}}{x + \frac{1}{2}}\right) + C$

23. $m\sqrt{m^2 + 4} + 4 \ln (m + \sqrt{m^2 + 4}) + C$

25. $\dfrac{5 \ln x}{2x^2} + \dfrac{5}{4x^2} + C$

27. $x^3e^x - 3x^2e^x + 6xe^x - 6e^x + C$

29. $S(p) = 100\left[\dfrac{20}{20 - p} + \ln (20 - p)\right]$

31. $-4 \ln\left(\dfrac{x}{3x - 2}\right) + C$

33. $\dfrac{1}{-2(x - 2)} + \dfrac{1}{4}\ln\left(\dfrac{x}{x - 2}\right) + C$

35. $\dfrac{-3}{(e^{-x} - 3)} + \ln (e^{-x} - 3) + C$

Exercise Set 5.8, p. 407

1. (a) 1.4914; **(b)** 0.8571 **3.** 0 **5.** $e - 1$ **7.** $\dfrac{4}{3}$

9. 13 **11.** $\dfrac{1}{n + 1}$ **13. (a)** 90 words per minute;

(b) 96 words per minute at $t = 1$;   **(c)** 70 words per minute
15. 254.5 million   **17. (a)** 100 after 10 hr;   **(b)** $33\frac{1}{3}$
19. 1.0016 **21.** 5.333 **23.** 2.159

Summary and Review: Chapter 5, p. 410

1. [5.1] $\frac{8}{5}x^5 + C$ **2.** [5.1] $3e^x + 2x + C$
3. [5.1] $t^3 + \frac{7}{2}t^2 + \ln t + C$ **4.** [5.2], [5.3] 9
5. [5.2], [5.3] $\frac{62}{3}$ **6.** [5.2] Total number of words written in t minutes; an antiderivative   **7.** [5.2] Total amount of sales at the end of t days; an antiderivative   **8.** [5.3] $\frac{1}{6}(b^6 - a^6)$
9. [5.4] $-\frac{2}{5}$ **10.** [5.3] $e - \frac{1}{2}$ **11.** [5.3] $3 \ln 3$
12. [5.4] 0   **13.** [5.4] Negative   **14.** [5.4] Positive
15. [5.4] $\frac{27}{2}$ **16.** [5.5] $\frac{1}{4}e^{x^4} + C$
17. [5.5] $\ln (4t^6 + 3) + C$ **18.** [5.5] $\frac{1}{4}(\ln 4x)^2 + C$
19. [5.5] $-\frac{2}{3}e^{-3x} + C$ **20.** [5.6] $e^{3x}\left(x - \frac{1}{3}\right) + C$
21. [5.6] $x \ln x^7 - 7x + C$ **22.** [5.6] $x^3 \ln x - \frac{x^3}{3} + C$
23. [5.7] $\frac{1}{14} \ln \left(\frac{7 + x}{7 - x}\right) + C$
24. [5.7] $\frac{1}{5}x^2 e^{5x} - \frac{2}{25}x e^{5x} + \frac{2}{125}e^{5x} + C$
25. [5.7] $\frac{1}{49} + \frac{x}{7} - \frac{1}{49}\ln (7x + 1) + C$
26. [5.7] $\ln (x + \sqrt{x^2 - 36}) + C$
27. [5.7] $x^7\left(\frac{\ln x}{7} - \frac{1}{49}\right) + C$ **28.** [5.7] $\frac{1}{64}e^{8x}(8x - 1) + C$
29. [5.8] 3.667   **30.** [5.8] $\frac{1}{2}(11 - e^{-2})$ **31.** [5.3] 80
32. [5.5] $e^{12} - 1$, or about \$162,754
33. [5.6] $e^{0.1x}(10x^3 - 300x^2 + 6000x - 60,000) + C$
34. [5.5] $\ln (4t^3 + 7) + C$
35. [5.7] $\frac{2}{75}(5x - 8)\sqrt{4 + 5x} + C$ **36.** [5.5] $e^{x^5} + C$
37. [5.5] $\ln (x + 9) + C$ **38.** [5.5] $\frac{1}{96}(t^8 + 3)^{12} + C$
39. [5.6] $x \ln 7x - x + C$ **40.** [5.6] $\frac{x^2}{2}\ln 8x - \frac{x^2}{4} + C$
41. [5.5], [5.7] $\frac{1}{10}[\ln (t^5 + 3)]^2 + C$
42. [5.5] $-\frac{1}{2}\ln (1 + 2e^{-x}) + C$
43. [5.5], [5.6], [5.7] $(\ln \sqrt{x})^2 + C$
44. [5.6], [5.7] $\frac{x^{92}}{92}\left(\ln x - \frac{1}{92}\right) + C$
45. [5.6] $(x - 3)\ln (x - 3) - (x - 4)\ln (x - 4) + C$
46. [5.5] $-\frac{1}{3(\ln x)^3} + C$ **47.** [5.4] 1.333

Test: Chapter 5, p. 411

1. $x + C$ **2.** $200x^5 + C$ **3.** $e^x + \ln x + \frac{8}{11}x^{11/8} + C$
4. $\frac{1}{6}$ **5.** $4 \ln 3$ **6.** An antiderivative, total number of

words typed in t minutes   **7.** 12   **8.** $-\frac{1}{2}\left(\frac{1}{e^2} - 1\right)$
9. $\ln b - \ln a$, or $\ln \dfrac{b}{a}$ **10.** Positive
11. $\ln (x + 8) + C$ **12.** $-2e^{-0.5x} + C$
13. $\frac{1}{40}(t^4 + 1)^{10} + C$ **14.** $\frac{x}{5}e^{5x} - \frac{e^{5x}}{25} + C$
15. $\frac{x^4}{4}\ln x^4 - \frac{x^4}{4} + C$ **16.** $\frac{2^x}{\ln 2} + C$
17. $\frac{1}{7}\ln \left(\frac{x}{7 - x}\right) + C$ **18.** 6   **19.** $\frac{1}{3}$ **20.** 95
21. \$49,000   **22.** 94 words   **23.** $\frac{1}{10}\ln \left(\frac{x}{10 - x}\right) + C$
24. $e^x(x^5 - 5x^4 + 20x^3 - 60x^2 + 120x - 120) + C$
25. $\frac{1}{6}e^{x^6} + C$ **26.** $\frac{2}{3}x^{3/2}\ln x - \frac{4}{9}x^{3/2} + C$
27. $\frac{1}{15}(3x^2 - 8)(x^2 + 4)^{3/2} + C$ **28.** $\frac{1}{16}\ln \left(\frac{8 + x}{8 - x}\right) + C$
29. $e^{0.1x}(10x^4 - 400x^3 + 12,000x^2 - 240,000x + 2,400,000) + C$ **30.** $\frac{x^2}{2}\ln 13x - \frac{x^2}{4} + C$
31. $\frac{(\ln x)^4}{4} - \frac{4}{3}(\ln x)^3 + 5 \ln x + C$
32. $(x + 3)\ln (x + 3) - (x + 5)\ln (x + 5) + C$
33. $\frac{3}{10}(8x^3 + 10)(5x - 4)^{2/3} - \frac{108}{125}x^2(5x - 4)^{5/3} + \frac{81}{625}x(5x - 4)^{8/3} - \frac{243}{34,375}(5x - 4)^{11/3} + C$ **34.** 8

CHAPTER 6

Exercise Set 6.1, p. 422

1. (a) (6, \$5); **(b)** \$15; **(c)** \$9   **3. (a)** (1, \$9);
(b) \$3.33; **(c)** \$1.67   **5. (a)** (3, \$9); **(b)** \$36; **(c)** \$18
7. (a) (50, \$500); **(b)** \$12,500; **(c)** \$6250
9. (a) (2, \$3); **(b)** \$2; **(c)** \$0.35   **11. (a)** (5, \$0.61);
(b) \$86.36; **(c)** \$2.45   **13.** ◆
15. (a) (1, \$9);
(b)   **(c)** \$3.33; **(d)** \$1.67

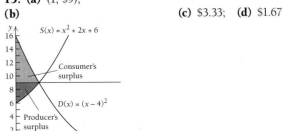

Exercise Set 6.2, p. 429

1. $131   **3.** $5610.72   **5.** $340,754.12   **7.** $949.94
9. $259.37   **11.** $8264.94   **13.** $29,676.18
15. $17,802.91   **17.** $966,112.17   **19.** 3,674,343,000 tons
21. After 38 yr (2028)   **23.** 19.994 lb
25. 112,891.1 million barrels   **27.** $2,383,120
29. $125.30   **31.** ◈   **33.** The area and the amount of
continuous money flow are the same. Both are 5610.72.

Exercise Set 6.3, p. 436

1. Convergent, $\frac{1}{3}$ **3.** Divergent   **5.** Convergent, 1
7. Convergent, $\frac{1}{2}$ **9.** Divergent   **11.** Convergent, 5
13. Divergent   **15.** Divergent   **17.** Divergent
19. Convergent, 1   **21.** $\frac{1}{2}$ **23.** 1   **25.** $45,000
27. $6250   **29.** $4500   **31.** 33,333 lb   **33.** Divergent
35. Convergent, 2   **37.** Convergent, $\frac{1}{2}$

39. $\dfrac{1}{k^2}$; the total dose of the drug   **41.** ◈

43.

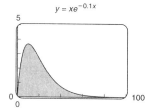

$y = xe^{-0.1x}$

45. 1.25

Exercise Set 6.4, p. 447

1. $\displaystyle\int_0^1 2x\,dx = [x^2]_0^1 = 1^2 - 0^2 = 1$

3. $\displaystyle\int_4^7 \frac{1}{3}\,dx = \left[\frac{1}{3}x\right]_4^7 = \frac{1}{3}(7-4) = 1$

5. $\displaystyle\int_1^3 \frac{3}{26}x^2\,dx = \left[\frac{3}{26}\cdot\frac{x^3}{3}\right]_1^3 = \frac{1}{26}(3^3 - 1^3) = 1$

7. $\displaystyle\int_1^e \frac{1}{x}\,dx = [\ln x]_1^e = \ln e - \ln 1 = 1 - 0 = 1$

9. $\displaystyle\int_{-1}^1 \frac{3}{2}x^2\,dx = \left[\frac{3}{2}\cdot\frac{1}{3}x^3\right]_{-1}^1$

$= \frac{1}{2}(1^3 - (-1)^3)$

$= \frac{1}{2}(1 + 1) = 1$

11. $\displaystyle\int_0^\infty 3e^{-3x}\,dx = \lim_{b\to\infty}\int_0^b 3e^{-3x}\,dx$

$= \lim_{b\to\infty}\left[\frac{3}{-3}e^{-3x}\right]_0^b$

$= \lim_{b\to\infty}\left[-e^{-3x}\right]_0^b$

$= \lim_{b\to\infty}\left[-e^{-3b} - (-e^{-3\cdot 0})\right] = \lim_{b\to\infty}\left(1 - \frac{1}{e^{3b}}\right)$

$= 1$

13. $\frac{1}{4}$; $f(x) = \frac{1}{4}x$ **15.** $\frac{3}{2}$; $f(x) = \frac{3}{2}x^2$ **17.** $\frac{1}{5}$; $f(x) = \frac{1}{5}$
19. $\dfrac{1}{2}$; $f(x) = \dfrac{2-x}{2}$ **21.** $\dfrac{1}{\ln 3}$; $f(x) = \dfrac{1}{x\ln 3}$
23. $\dfrac{1}{e^3 - 1}$; $f(x) = \dfrac{e^x}{e^3 - 1}$ **25.** $\dfrac{8}{25}$ **27.** $\dfrac{1}{2}$ **29.** 0.3297
31. 0.99995   **33.** 0.3935   **35.** 0.9502   **37.** $b = \sqrt[4]{4}$, or
$\sqrt{2}$ **39.** ◈

Exercise Set 6.5, p. 458

1. $\mu = E(x) = \frac{7}{2}$, $E(x^2) = 13$, $\sigma^2 = \frac{3}{4}$, $\sigma = \frac{1}{2}\sqrt{3}$
3. $\mu = E(x) = 2$, $E(x^2) = \frac{9}{2}$, $\sigma^2 = \frac{1}{2}$, $\sigma = \sqrt{\frac{1}{2}}$
5. $\mu = E(x) = \frac{14}{9}$, $E(x^2) = \frac{5}{2}$, $\sigma^2 = \frac{13}{162}$, $\sigma = \sqrt{\frac{13}{162}} = \frac{1}{9}\sqrt{\frac{13}{2}}$
7. $\mu = E(x) = -\frac{5}{4}$, $E(x^2) = \frac{11}{5}$, $\sigma^2 = \frac{51}{80}$, $\sigma = \sqrt{\frac{51}{80}} = \frac{1}{4}\sqrt{\frac{51}{5}}$
9. $\mu = E(x) = \dfrac{2}{\ln 3}$, $E(x^2) = \dfrac{4}{\ln 3}$, $\sigma^2 = \dfrac{4\ln 3 - 4}{(\ln 3)^2}$,
$\sigma = \dfrac{2}{\ln 3}\sqrt{\ln 3 - 1}$ **11.** 0.4964   **13.** 0.3665
15. 0.6442   **17.** 0.0078   **19.** 0.1716   **21.** 0.0013
23. (a) 0.6826;   **(b)** 68.26%   **25.** 0.2898   **27.** 0.4514
29. 0.62%   **31.** 0.1115   **33.** $\mu = E(x) = \dfrac{a+b}{2}$,
$E(x^2) = \dfrac{b^3 - a^3}{3(b-a)}$, or $\dfrac{b^2 + ba + a^2}{3}$, $\sigma^2 = \dfrac{(b-a)^2}{12}$,
$\sigma = \dfrac{b-a}{2\sqrt{3}}$ **35.** $\sqrt{2}$ **37.** $\dfrac{\ln 2}{k}$ **39.** 5.801 oz   **41.** ◈

Exercise Set 6.6, p. 463

1. $\dfrac{\pi}{3}$ **3.** $\dfrac{7\pi}{3}$ **5.** $\dfrac{\pi}{2}(e^{10} - e^{-4})$ **7.** $\dfrac{2\pi}{3}$ **9.** $\pi \ln 3$
11. 32π **13.** $\dfrac{32\pi}{5}$ **15.** 56π **17.** $\dfrac{32\pi}{3}$ **19.** $2\pi e^3$
21. π

Exercise Set 6.7, p. 471

1. $y = x^4 + C$; $y = x^4 + 3$, $y = x^4$, $y = x^4 - 796$; answers

may vary   **3.** $y = \frac{1}{2}e^{2x} + \frac{1}{2}x^2 + C$; $y = \frac{1}{2}e^{2x} + \frac{1}{2}x^2 - 5$,
$y = \frac{1}{2}e^{2x} + \frac{1}{2}x^2 + 7$; $y = \frac{1}{2}e^{2x} + \frac{1}{2}x^2$; answers may vary
5. $y = 3 \ln x - \frac{1}{3}x^3 + \frac{1}{6}x^6 + C$;
$y = 3 \ln x - \frac{1}{3}x^3 + \frac{1}{6}x^6 - 15$, $y = 3 \ln x - \frac{1}{3}x^3 + \frac{1}{6}x^6 - 7$,
$y = 3 \ln x - \frac{1}{3}x^3 + \frac{1}{6}x^6$; answers may vary
7. $y = \frac{1}{3}x^3 + x^2 - 3x + 4$ **9.** $f(x) = \frac{3}{5}x^{5/3} - \frac{1}{2}x^2 - \frac{61}{10}$
11. $y'' = \dfrac{1}{x}$. Then

$$
\begin{array}{c|c}
y'' - \dfrac{1}{x} & = 0 \\
\hline
\dfrac{1}{x} - \dfrac{1}{x} & 0 \\
0 &
\end{array}
$$

13. $y' = 4e^x + 3xe^x$, $y'' = 7e^x + 3xe^x$. Then

$$
\begin{array}{c|c}
y'' - 2y' + y & = 0 \\
\hline
(7e^x + 3xe^x) - 2(4e^x + 3xe^x) + (e^x + 3xe^x) & 0 \\
7e^x + 3xe^x - 8e^x - 6xe^x + e^x + 3xe^x & \\
0 &
\end{array}
$$

15. $y = C_1 e^{x^4}$, where $C_1 = e^C$ **17.** $y = \sqrt[3]{\frac{3}{2}x^2 + C}$
19. $y = \sqrt{2x^2 + C_1}$, $y = -\sqrt{2x^2 + C_1}$, where $C_1 = 2C$
21. $y = \sqrt{6x + C_1}$, $y = -\sqrt{6x + C_1}$, where $C_1 = 2C$
23. $y = -3 + 8e^{x^2/2}$ **25.** $y = \sqrt[3]{15x - 3}$
27. $y = C_1 e^{3x}$, where $C_1 = e^C$ **29.** $P = C_1 e^{2t}$, where
$C_1 = e^C$ **31.** $f(x) = \ln x - 2x^2 + \frac{2}{3}x^{3/2} + 9$
33. $C(x) = 2.6x - 0.01x^2 + 120$,
$A(x) = 2.6 - 0.01x + \dfrac{120}{x}$

35. (a) $P(C) = \dfrac{400}{(C + 3)^{1/2}} - 40$;   **(b)** \$97
37. (a) $R = k \cdot \ln (S + 1) + C$;   **(b)** $R = k \cdot \ln (S + 1)$;
(c) no units, no pleasure from them   **39.** $x = 200 - p$
41. $x = \dfrac{C_1}{p^n}$ **43.** $R = C_1 \cdot S^k$, where $C_1 = e^C$

45.
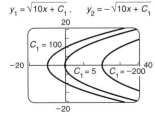
$y_1 = \sqrt{10x + C_1}$, $y_2 = -\sqrt{10x + C_1}$

Summary and Review: Chapter 6, p. 473

1. [6.1] (2, \$16)   **2.** [6.1] \$18.67   **3.** [6.1] \$5.33
4. [6.2] \$83,560.54   **5.** [6.2] \$251.50   **6.** [6.2] 282,732
thousand tons   **7.** [6.2] 183 yr   **8.** [6.2] \$247.88
9. [6.2] \$44,517   **10.** [6.2], [6.3] \$53,333
11. [6.3] Convergent, 1   **12.** [6.3] Divergent
13. [6.3] Convergent, $\dfrac{1}{2}$ **14.** [6.4] $\dfrac{8}{3}$; $f(x) = \dfrac{8}{3x^3}$
15. [6.4] 0.6   **16.** [6.5] $\frac{3}{5}$ **17.** [6.5] $\frac{3}{4}$ **18.** [6.5] $\frac{3}{4}$
19. [6.5] $\frac{3}{80}$ **20.** [6.5] $\dfrac{\sqrt{15}}{20}$ **21.** [6.5] 0.4678
22. [6.5] 0.8827   **23.** [6.5] 0.1002   **24.** [6.5] 0.5000
25. [6.5] 0.0918   **26.** [6.6] $\dfrac{127\pi}{7}$ **27.** [6.6] $\dfrac{\pi}{6}$
28. [6.7] $y = C_1 e^{x^{11}}$, where $C_1 = e^C$
29. [6.7] $y = \pm\sqrt{4x + C_1}$, where $C_1 = 2C$
30. [6.7] $y = 5e^{4x}$ **31.** [6.7] $v = \sqrt[3]{15t + 19}$
32. [6.7] $y = \pm\sqrt{3x^2 + C_1}$, where $C_1 = 2C$
33. [6.7] $y = 8 - C_1 e^{-x^2/2}$, where $C_1 = e^{-C}$
34. [6.7] $x = 100 - p$ **35.** [6.7] $V = \$36.37 - \$6.37e^{-kt}$
36. [6.4] $c = \sqrt[9]{4.5}$ **37.** [6.3] Convergent, $-\frac{1}{5}$
38. [6.3] Convergent, 3   **39.** [6.3] 1

Test: Chapter 6, p. 475

1. (3, \$16)   **2.** \$45   **3.** \$22.50   **4.** \$29,192
5. \$344.66   **6.** 37,121,870 thousand tons   **7.** 39 yr
8. \$2465.97   **9.** \$30,717.71   **10.** \$34,545.45
11. Convergent, $\dfrac{1}{4}$ **12.** Divergent   **13.** $\dfrac{1}{4}$; $f(x) = \dfrac{x^3}{4}$
14. 0.8647   **15.** $E(x) = \dfrac{13}{6}$ **16.** $E(x^2) = 5$
17. $\mu = \dfrac{13}{6}$ **18.** $\sigma^2 = \dfrac{11}{36}$ **19.** $\sigma = \dfrac{\sqrt{11}}{6}$ **20.** 0.4332
21. 0.4420   **22.** 0.9071   **23.** 0.4207   **24.** $\pi \ln 5$
25. $\dfrac{5\pi}{2}$ **26.** $y = C_1 e^{x^8}$, where $C_1 = e^C$
27. $y = \sqrt{18x + C_1}$, $y = -\sqrt{18x + C_1}$, where $C_1 = 2C$
28. $y = 11e^{6t}$ **29.** $y = 5 - C_1 e^{-x^3/3}$, where $C_1 = e^{-C}$
30. $v = \pm\sqrt[4]{8t + C_1}$, where $C_1 = 4C$ **31.** $y = C_1 e^{4x + x^2/2}$,
where $C_1 = e^C$ **32.** $x = \dfrac{C_1}{p^4}$ **33. (a)** $V(t) = 36(1 - e^{-kt})$;
(b) $k = 0.12$;   **(c)** $V(t) = 36(1 - e^{-0.12t})$;
(d) $V(12) \approx \$27.47$;   **(e)** $t \approx 14.9$ months   **34.** $b = \sqrt[6]{6}$
35. Convergent, $-\frac{1}{4}$ **36.** 3.14

Chapter 7

Exercise Set 7.1, p. 487

1. $f(0, -2) = 0, f(2, 3) = -8, f(10, -5) = 200$
3. $f(0, -2) = 1, f(-2, 1) = -13\frac{8}{9}, f(2, 1) = 23$
5. $f(e, 2) = \ln e + 2^3 = 1 + 8 = 9, f(e^2, 4) = 66,$
$f(e^3, 5) = 128$ 7. $f(-1, 2, 3) = 6, f(2, -1, 3) = 12$ ◈
9. \$151,571.66   11. 7.7%   13. 244.7 mph   15. ◈
17. $0°$ 19. $-22°$
21.

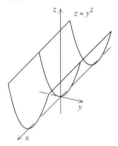

$z = y^2$

23.

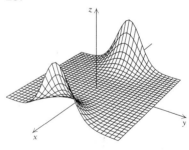

$z = (x^4 - 16x^2)e^{-y^2}$

25.

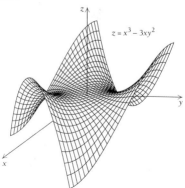

$z = x^3 - 3xy^2$

Exercise Set 7.2, p. 493

1. $\dfrac{\partial z}{\partial x} = 2 - 3y, \dfrac{\partial z}{\partial y} = -3x, \dfrac{\partial z}{\partial x}\Big|_{(-2, -3)} = 11, \dfrac{\partial z}{\partial y}\Big|_{(0, -5)} = 0$

3. $\dfrac{\partial z}{\partial x} = 6x - 2y, \dfrac{\partial z}{\partial y} = -2x + 1, \dfrac{\partial z}{\partial x}\Big|_{(-2, -3)} = -6,$

$\dfrac{\partial z}{\partial y}\Big|_{(0, -5)} = 1$ 5. $f_x = 2, f_y = -3, f_x(-2, 4) = 2,$

$f_y(4, -3) = -3$ 7. $f_x = \dfrac{x}{\sqrt{x^2 + y^2}}, f_y = \dfrac{y}{\sqrt{x^2 + y^2}},$

$f_x(-2, 1) = \dfrac{-2}{\sqrt{5}}, f_y(-3, -2) = \dfrac{-2}{\sqrt{13}}$ 9. $f_x = 2e^{2x+3y},$

$f_y = 3e^{2x+3y}$ 11. $f_x = ye^{xy}, f_y = xe^{xy}$ 13. $f_x = \dfrac{y}{x + y},$

$f_y = \dfrac{y}{x + y} + \ln(x + y)$ 15. $f_x = 1 + \ln xy, f_y = \dfrac{x}{y}$

17. $f_x = \dfrac{1}{y} + \dfrac{y}{x^2}, f_y = -\dfrac{x}{y^2} - \dfrac{1}{x}$
19. $f_x = 12(2x + y - 5), f_y = 6(2x + y - 5)$
21. $\dfrac{\partial f}{\partial b} = 12m + 6b - 30, \dfrac{\partial f}{\partial m} = 28m + 12b - 64$
23. $f_x = 3y - 2\lambda, f_y = 3x - \lambda, f_\lambda = -(2x + y - 8)$
25. $f_x = 2x - 10\lambda, f_y = 2y - 2\lambda, f_\lambda = -(10x + 2y - 4)$

27. (a) 3,888,064 units;   (b) $\dfrac{\partial p}{\partial x} = 1117.8\left(\dfrac{y}{x}\right)^{0.379}$

$\dfrac{\partial p}{\partial y} = 682.2\left(\dfrac{x}{y}\right)^{0.621}$;   (c) 965.8, 866.8   29. $99.6°$
31. $121.3°$ 33. $1.09(T - 58)$ 35. 78.244

37. -0.846 39. $f_x = \dfrac{-4xt^2}{(x^2 - t^2)^2}, f_t = \dfrac{4x^2t}{(x^2 - t^2)^2}$

41. $f_x = \dfrac{1}{\sqrt{x}(1 + 2\sqrt{t})}, f_t = \dfrac{-1 - 2\sqrt{x}}{\sqrt{t}(1 + 2\sqrt{t})^2}$

43. $f_x = 4x^{-1/3} - 2x^{-3/4}t^{1/2} + 6x^{-3/2}t^{3/2},$
$f_t = -4x^{1/4}t^{-1/2} - 18x^{-1/2}t^{1/2}$ 45. ◈

Exercise Set 7.3, p. 497

1. $\dfrac{\partial^2 f}{\partial x^2} = 6, \dfrac{\partial^2 f}{\partial y \, \partial x} = \dfrac{\partial^2 f}{\partial x \, \partial y} = -1, \dfrac{\partial^2 f}{\partial y^2} = 0$ 3. $\dfrac{\partial^2 f}{\partial x^2} = 0,$

$\dfrac{\partial^2 f}{\partial y \, \partial x} = \dfrac{\partial^2 f}{\partial x \, \partial y} = 3, \dfrac{\partial^2 f}{\partial y^2} = 0$ 5. $\dfrac{\partial^2 f}{\partial x^2} = 20x^3y^4 + 6xy^2,$

$\dfrac{\partial^2 f}{\partial y \, \partial x} = \dfrac{\partial^2 f}{\partial x \, \partial y} = 20x^4y^3 + 6x^2y, \dfrac{\partial^2 f}{\partial y^2} = 12x^5y^2 + 2x^3$

7. $f_{xx} = 0, f_{yx} = 0, f_{xy} = 0, f_{yy} = 0$ 9. $f_{xx} = 4y^2e^{2xy},$
$f_{yx} = f_{xy} = 4xye^{2xy} + 2e^{2xy}, f_{yy} = 4x^2e^{2xy}$ 11. $f_{xx} = 0,$

$f_{yx} = f_{xy} = 0, f_{yy} = e^y$ **13.** $f_{xx} = -\dfrac{y}{x^2}, f_{yx} = f_{xy} = \dfrac{1}{x}$,

$f_{yy} = 0$ **15.** $f_{xx} = \dfrac{-6y}{x^4}, f_{yx} = f_{xy} = \dfrac{2(y^3 - x^3)}{x^3 y^3}, f_{yy} = \dfrac{6x}{y^4}$

17. $\dfrac{\partial^2 f}{\partial x^2} = \dfrac{2y^2 - 2x^2}{(x^2 + y^2)^2}, \dfrac{\partial^2 f}{\partial y^2} = \dfrac{2x^2 - 2y^2}{(x^2 + y^2)^2}$, so the sum is 0

19. (a) $-y$; **(b)** x; **(c)** $f_{yx}(0, 0) = 1, f_{xy}(0, 0) = -1$; so $f_{yx}(0, 0) \neq f_{xy}(0, 0)$

Exercise Set 7.4, p. 506

1. Minimum $= -\frac{1}{3}$ at $\left(-\frac{1}{3}, \frac{2}{3}\right)$ **3.** Maximum $= \frac{4}{27}$ at $\left(\frac{2}{3}, \frac{2}{3}\right)$
5. Minimum $= -1$ at $(1, 1)$ **7.** Minimum $= -7$ at $(1, -2)$
9. Minimum $= -5$ at $(-1, 2)$ **11.** None
13. 6 thousand of the \$17 radio and 5 thousand of the \$21 radio   **15.** Maximum of $P = 35$ (million dollars) when $a = 10$ (million dollars) and $p = \$3$
17. The dimensions of the bottom are 8 ft by 8 ft, and the height is 5 ft.
19. (a) $R(p_1, p_2) = 64p_1 - 4p_1^2 - 4p_1 p_2 + 56p_2 - 4p_2^2$;
(b) $p_1 = 6(\$60), p_2 = 4(\$40)$; **(c)** $q_1 = 32$ (hundreds), $q_2 = 28$ (hundreds); **(d)** \$304,000   **21.** None; saddle point at $(0, 0)$ **23.** Minimum $= 0$ at $(0, 0)$; saddle points at $(2, 1)$ and $(-2, 1)$ **25.** ◈   **27.** Minimum $= -5$ at $(0, 0)$ **29.** None

Exercise Set 7.5, p. 516

1. (a) $y = 0.835x - 0.017$; **(b)** \$7.5 million; \$24.2 million
3. (a) $y = 1.97x + 69.09$; **(b)** 80.91; 82.88
5. (a) $y = 1.068421x - 1.236842$; **(b)** 85.3   **7.** ◈
9. (a) $y = -0.005925x + 15.54734$; **(b)** 3:42.54; 3:40.40;
(c) 3:44.32   **11. (a)** $y = 0.498e^{0.551x}$; **(b)** \$70.94 million; \$4,333,321 million; **(c)** This model predicts higher sales.
13. (a) $y = -0.075x^3 + 0.582x^2 + 0.757x + 69.7$;
(b) 78.99; 77.79; **(c)** This model predicts lower life expectancies.

Exercise Set 7.6, p. 525

1. Maximum $= 8$ at $(2, 4)$ **3.** Maximum $= -16$ at $(2, 4)$
5. Minimum $= 20$ at $(4, 2)$ **7.** Minimum $= -96$ at $(8, -12)$ **9.** Minimum $= \frac{3}{2}$ at $\left(1, \frac{1}{2}, -\frac{1}{2}\right)$ **11.** 35 and 35
13. 3 and -3 **15.** $9\frac{3}{4}$ in., $9\frac{3}{4}$ in.; $95\frac{1}{16}$ in^2; no
17. $r = \sqrt[3]{\dfrac{27}{2\pi}} \approx 1.6$ ft; $h = 2 \cdot r \approx 3.2$ ft; minimum surface area ≈ 48.3 ft^2 **19.** Maximum of $S = 800$ at $L = 20, M = 60$ **21. (a)** $C(x, y, z) = 7xy + 6yz + 6xz$;
(b) $x = 60$ ft, $y = 60$ ft, $z = 70$ ft; \$75,600   **23.** 10,000 on A, 100 on B **25.** Minimum $= -\frac{155}{128}$ at $\left(-\frac{7}{16}, -\frac{3}{4}\right)$

27. Maximum $= \dfrac{1}{27}$ at $\left(\pm\dfrac{1}{\sqrt{3}}, \pm\dfrac{1}{\sqrt{3}}, \pm\dfrac{1}{\sqrt{3}}\right)$

29. Maximum $= 2$ at $\left(\dfrac{1}{2}, \dfrac{1}{2}, \dfrac{1}{2}, \dfrac{1}{2}\right)$

31. $\lambda = \dfrac{p_x}{c_1} = \dfrac{p_y}{c_2}$ **33.** ◈

Exercise Set 7.7, p. 531

1. 1   **3.** 0   **5.** 6   **7.** $\frac{3}{20}$ **9.** 4   **11.** $\frac{4}{15}$ **13.** 1
15. 39   **17.** $\frac{13}{240}$ **19.** ◈

Summary and Review: Chapter 7, p. 532

1. [7.1] 1   **2.** [7.2] $3y^3$ **3.** [7.2] $e^y + 9xy^2 + 2$
4. [7.3] $9y^2$ **5.** [7.3] $9y^2$ **6.** [7.3] 0
7. [7.3] $e^y + 18xy$ **8.** [7.2] $2x \ln y$ **9.** [7.2] $\dfrac{x^2}{y} + 4y^3$
10. [7.3] $\dfrac{2x}{y}$ **11.** [7.3] $\dfrac{2x}{y}$ **12.** [7.3] $2 \ln y$
13. [7.3] $-\dfrac{x^2}{y^2} + 12y^2$ **14.** [7.4] Minimum $= -\frac{549}{20}$ at $\left(5, \frac{27}{2}\right)$ **15.** [7.4] Minimum $= -4$ at $(0, -2)$
16. [7.4] Maximum $= \frac{45}{4}$ at $\left(\frac{3}{2}, -3\right)$
17. [7.4] Minimum $= 29$ at $(-1, 2)$
18. [7.5] **(a)** $y = 0.118x + 0.303$; **(b)** \$1.13 million; \$1.48 million   **19.** [7.5] **(a)** $y = 0.6x + 6.7$;
(b) 9.1 million   **20.** [7.6] Minimum $= \frac{125}{4}$ at $\left(-\frac{3}{2}, -3\right)$
21. [7.6] Maximum $= 300$ at $(5, 10)$ **22.** [7.7] $\frac{97}{30}$
23. [7.7] $\frac{1}{60}$ **24.** [7.7] 0
25. [7.6] The cylindrical container
26. [7.1]

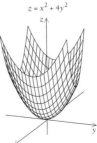

$z = x^2 + 4y^2$

Test: Chapter 7, p. 534

1. $e^{-1} - 2$ **2.** $e^x + 6x^2 y$ **3.** $2x^3 + 1$ **4.** $e^x + 12xy$
5. $6x^2$ **6.** $6x^2$ **7.** 0   **8.** Minimum $= -\frac{7}{16}$ at $\left(\frac{3}{4}, \frac{1}{2}\right)$

9. None   **10. (a)** $y = \frac{9}{2}x + \frac{17}{3}$;   **(b)** \$23.67 million
11. Maximum $= -19$ at $(4, 5)$ **12.** 1   **13.** \$400,000 for
labor, \$200,000 for capital   **14.** $f_x = \dfrac{-x^4 + 4xt + 6x^2t}{(x^3 + 2t)^2}$,
$f_t = \dfrac{-2x^2(x + 1)}{(x^3 + 2t)^2}$
15.

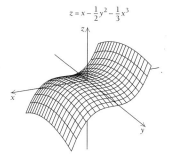

$z = x - \frac{1}{2}y^2 - \frac{1}{3}x^3$

CHAPTER 8

Exercise Set 8.1, p. 550

1. I   **3.** III   **5.** $\dfrac{\pi}{6}$ **7.** $\dfrac{\pi}{3}$ **9.** $\dfrac{5\pi}{12}$ **11.** $270°$

13. $-45°$ **15.** $1440°$ **17.** $\left(\dfrac{180}{\pi}\right)° \approx 57.3°$ **19.** $\dfrac{\sqrt{3}}{2}$

21. 0   **23.** $\dfrac{1}{2}$ **25.** Undefined   **27.** $\dfrac{\sqrt{3}}{3}$ **29.** -1

31. $\tan(u + v) = \dfrac{\sin(u + v)}{\cos(u + v)}$. Then use identities (6) and
(4).   **33.** 0.5210   **35.** -0.8632 **37.** -0.3327
39. Undefined   **41.** \$0, \$0.9 thousand, \$3.5 thousand,
\$7 thousand, \$14 thousand, \$0, \$7 thousand   **43.** ◈
45. Yes   **47.** No   **49.** 0.5   **51.** 2

Exercise Set 8.2, p. 560

1. $\sec x = \dfrac{1}{\cos x}$, so $\dfrac{d}{dx}\sec x = \dfrac{\cos x \cdot 0 - (-\sin x) \cdot 1}{\cos^2 x} =$
$\dfrac{\sin x}{\cos^2 x} = \dfrac{\sin x}{\cos x} \cdot \dfrac{1}{\cos x} = \tan x \sec x$ **3.** $x \cos x + \sin x$

5. $e^x(\cos x + \sin x)$ **7.** $\dfrac{x \cos x - \sin x}{x^2}$

9. $2 \sin x \cos x$ **11.** $\cos^2 x - \sin^2 x$ **13.** $\dfrac{1}{1 + \cos x}$

15. $\dfrac{2 \sin x}{\cos^3 x}$ **17.** $\dfrac{-\sin x}{2\sqrt{1 + \cos x}}$ **19.** $-x^2 \sin x$

21. $\cos x \cdot e^{\sin x}$ **23.** $-\sin x$
25. $(2x + 3x^2) \cos(x^2 + x^3)$
27. $(4x^3 - 5x^4) \sin(x^5 - x^4)$ **29.** $-\frac{1}{2}x^{-1/2} \cdot \sin\sqrt{x}$

31. $-\sin x \cdot \cos(\cos x)$ **33.** $\dfrac{-2 \sin 4x}{\sqrt{\cos 4x}}$

35. $\frac{2}{3}[\csc^2(5 - 2x)^{1/3}](5 - 2x)^{-2/3}$
37. $-12(\tan 3x)(\sec 3x)^2$ **39.** $\cot x$

41. $40{,}000(\cos t - \sin t)$ **43.** $\dfrac{3\pi}{8}\cos\dfrac{\pi}{8}t$

45. $-\dfrac{1000\pi}{9}\left[\sin\dfrac{\pi}{45}(t - 10)\right]$ **47. (a)** Amplitude $= 5$,
period $= \dfrac{\pi}{2}$, phase shift $= -\dfrac{\pi}{4}$;   **(b)** $20 \cos(4t + \pi)$
49. (a) $L = (a^{2/3} + b^{2/3})^{3/2}$;   **(b)** $16\sqrt{2}$

51. $\dfrac{dy}{dx} = 0.03594\pi \cos 1.198\pi x$ **53.** $90°$

55. $f_x = -ye^x$, $f_y = 2\cos 2y - e^x$, $f_{xx} = -ye^x$,
$f_{yx} = f_{xy} = -e^x$, $f_{yy} = -4\sin 2y$
57. $f_x = -2\sin(2x + 3y)$, $f_y = -3 \cdot \sin(2x + 3y)$,
$f_{xx} = -4\cos(2x + 3y)$, $f_{yx} = f_{xy} = -6\cos(2x + 3y)$,
$f_{yy} = -9\cos(2x + 3y)$
59. $f_x = 5x^3y \cdot \sec^2(5xy) + 3x^2 \cdot \tan(5xy)$,
$f_y = 5x^4 \cdot \sec^2(5xy)$, $f_{xx} = 50x^3y^2 \cdot \sec^2(5xy) \cdot \tan(5xy) + 30x^2y\sec^2(5xy) + 6x\tan(5xy)$,
$f_{yx} = f_{xy} = 50x^4y \cdot \sec^2(5xy) \cdot \tan(5xy) + 20x^3 \cdot \sec^2(5xy)$,
$f_{yy} = 50x^5 \cdot \sec^2(5xy) \cdot \tan(5xy)$

61. $\dfrac{\cos y}{1 + x \sin y}$ **63.** ◈

65.

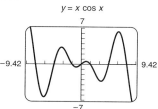

$y = x \cos x$

7

-9.42 9.42

-7

67.

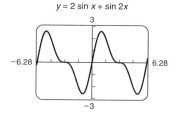

$y = 2 \sin x + \sin 2x$

3

-6.28 6.28

-3

69.

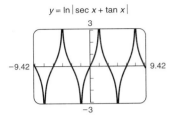

$y = \ln |\sec x + \tan x|$

71.

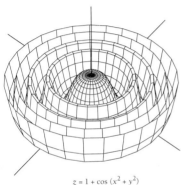

$z = 1 + \cos (x^2 + y^2)$

73. $-5.4978, -2.3562, 0.7854, 3.927$

Exercise Set 8.3, p. 565

1. $\dfrac{1}{2}$ **3.** 0   **5.** 1   **7.** $\dfrac{\sin^5 x}{5} + C$ **9.** $\dfrac{\cos^3 x}{3} + C$

11. $\sin (x + 3) + C$ **13.** $-\frac{1}{2} \cos 2x + C$

15. $\frac{1}{2} \sin x^2 + C$ **17.** $-\cos (e^x) + C$

19. $-\ln |\cos x| + C$, or $\ln |\sec x| + C$

21. $\frac{1}{4} x \sin 4x + \frac{1}{16} \cos 4x + C$

23. $3x \sin x + 3 \cos x + C$

25. $2x \sin x - (x^2 - 2) \cos x + C$ **27.** $\tan x - x + C$

29. \$84 thousand   **31.** $\frac{1}{2} x[\sin (\ln x) - \cos (\ln x)] + C$

33. $\dfrac{e^x}{2} (\cos x + \sin x) + C$

35. $-x^3 \cos x + 3x^2 \sin x + 6x \cos x - 6 \sin x + C$

37. $\frac{1}{7} \tan 7x + C$ **39.** $\tan x + \sec x + C$

41. $-\csc u + C$ **43.** $\cos x - \ln |\csc x + \cot x| + C$

45. $x + 2 \ln |\sec x + \tan x| + \tan x + C$ **47.** $\dfrac{\sqrt{2}}{2}$

49. π **51.** $\blacklozenge$ **53.** 0   **55.** 1   **57.** 2

Exercise Set 8.4, p. 569

1. $\dfrac{\pi}{4}$ **3.** $\dfrac{\pi}{2}$ **5.** $\dfrac{\pi}{4}$ **7.** $-\dfrac{\pi}{6}$ **9.** $\tan^{-1} (e^t) + C$

11. $\dfrac{1}{5} \tan^{-1} (5x) + C$ **13.** Use substitution. Let
$u = \sin^{-1} x$; then $x = \sin u$ and $dx = \cos u \, du$. Then the integral

$$\int \frac{1}{\sqrt{1 - x^2}} \, dx = \int \frac{1}{\sqrt{1 - \sin^2 u}} \cos u \, du$$

$$= \int \frac{1}{\cos u} \cdot \cos u \, du$$

$$= \int du = u + C = \sin^{-1} x + C.$$

15. 1.412 radians   **17.** 0.049 radian
19. $x \tan^{-1} x - \frac{1}{2} \ln (1 + x^2) + C$

Summary and Review: Chapter 8, p. 570

1. [8.1] $\dfrac{4\pi}{3}$ **2.** [8.1] $\dfrac{7\pi}{4}$ **3.** [8.1] $150°$ **4.** [8.1] $420°$

5. [8.1] $\dfrac{\sqrt{3}}{2}$ **6.** [8.1] 0   **7.** [8.1] 1   **8.** [8.1] Undefined

9. [8.2] $(6x - 1) \sec^2 (3x^2 - x)$ **10.** [8.2] $6 \sin^5 x \cos x$

11. [8.2] $\dfrac{1}{\cos x \sin x}$ **12.** [8.2] $3e^{\sin 3x} \cos 3x$

13. [8.2] $\dfrac{-2(1 + \cos 2x)}{\sin^2 2x}$ **14.** [8.2] $-\dfrac{\sin \sqrt{t}}{2\sqrt{t}} - \dfrac{\sin t}{2\sqrt{\cos t}}$

15. [8.2] $\dfrac{\sec^2 x}{2\sqrt{\tan x}}$ **16.** [8.2] $e^x (\cos x + \sin x)$

17. [8.2] $3(\cos^2 3x - \sin^2 3x)$ **18.** [8.2] $x \sin x$

19. [8.2] $-9 \sin (3t - 2\pi)$ **20.** [8.2] $\dfrac{5\pi}{3} \cos \dfrac{\pi}{3} t$

21. [8.3] $\frac{1}{2}$ **22.** [8.3] $\sin e^x + C$

23. [8.3] $-\cos (x - 1) + C$ **24.** [8.3] $\frac{1}{8} \sin 8x + C$

25. [8.3] $\frac{3}{2} x \sin 2x + \frac{3}{4} \cos 2x + C$

26. [8.3] $-\frac{1}{3} x \cos 3x + \frac{1}{9} \sin 3x + C$

27. [8.4] $\frac{1}{8} \sin^{-1} 8t + C$ **28.** [8.3] $\ln |5x - \sin x| + C$

29. [8.2] $e^{\sin x} \cos x (\cos^2 x - 3 \sin x - 1)$

30. [8.2] $f_x = -\sin y \sin x, f_{xy} = -\sin x \cos y$

31. [8.1] No   **32.** [8.1] Yes

Test: Chapter 8, p. 571

1. $\dfrac{2\pi}{3}$ **2.** $300°$ **3.** $\dfrac{1}{2}$ **4.** -1 **5.** $-\sin t$

6. $(6x - 5) \cos (3x^2 - 5x)$ **7.** $\dfrac{\sin x - x \cdot \cos x}{\sin^2 x}$

8. $2 \tan t \cdot \sec^2 t$ **9.** $\dfrac{\cos x - \sin x}{2\sqrt{\sin x + \cos x}}$

10. $\dfrac{-2}{1 - 2 \sin x \cos x}$ **11.** $S'(t) = \dfrac{5\pi}{2} \cos \dfrac{\pi}{8} t$

12. $1 - \dfrac{\sqrt{3}}{2}$ **13.** $\frac{1}{10}(\sin x)^{10} + C$ **14.** $\frac{1}{5} \sin 5t + C$

15. $\frac{1}{5} \sin 5x - x \cos 5x + C$ **16.** $\frac{1}{4} \tan^{-1} 4t + C$

17. $\ln |2x - \sin x| + C$ **18.** 1

19. $\dfrac{dy}{dx} = -2 (\sin x)e^{\cos x}, \dfrac{d^2y}{dx^2} = 2e^{\cos x}(\sin^2 x - \cos x)$

20. $f_x = -\dfrac{\sin x}{\cos y}, f_{xy} = -\dfrac{\sin x \sin y}{\cos^2 y}$

21. Does not exist; $-\infty$

22.

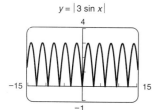

$y = |3 \sin x|$

23.

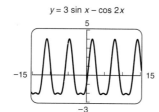

$y = 3 \sin x - \cos 2x$

CUMULATIVE REVIEW, p. 575 _____

1. $y = -4x - 27$ **2.** $x^2 + 2xh + h^2 - 5$
3. (a)   **(b)** 3;   **(c)** -3;   **(d)** no

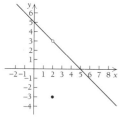

4. -8 **5.** 3   **6.** Does not exist   **7.** 4   **8.** $3x$ **9.** 0
10. -9 **11.** $2x - 7$ **12.** $\frac{1}{4}x^{-3/4}$ **13.** $-6x^{-7}$
14. $(x - 3)5(x + 1)^4 + (x + 1)^5$, or $2(x + 1)^4(3x - 7)$
15. $\dfrac{5 - 2x^3}{x^6}$ **16.** $\dfrac{1 - x}{e^x}$ **17.** $\dfrac{2x}{x^2 + 5}$ **18.** 1
19. $3e^{3x} + 2x$ **20.** $e^{-x}[2x \cos (x^2 + 3) - \sin (x^2 + 3)]$
21. $-\tan x$ **22.** $2(\sec^2 x)(\tan x)$ **23.** $3xy^2 + \dfrac{y}{x}$
24. $y = x - 2$

25. Relative maximum at $(-1, 3)$; relative minimum at $(1, -1)$

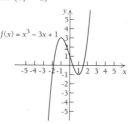

$f(x) = x^3 - 3x + 1$

26. Relative maxima at $(-1, -2)$ and $(1, -2)$; relative minimum at $(0, -3)$

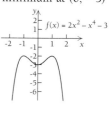

$f(x) = 2x^2 - x^4 - 3$

27. Relative minimum at $(-1, -4)$; relative maximum at $(1, 4)$

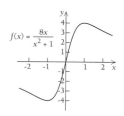

$f(x) = \dfrac{8x}{x^2 + 1}$

28. Relative maximum at $(0, -2)$. For graph, see Example 7 in Section 3.3.   **29.** Minimum $= -7$ at $x = 1$ **30.** None
31. Maximum $= 6\frac{2}{3}$ at $x = -1$; minimum $= 4\frac{1}{3}$ at $x = -2$
32. 15   **33.** 30 times; lot size 15
34. $\Delta y = 3.03, f'(x) \Delta x = 3$ **35. (a)** $N(t) = 8000e^{0.1t}$;
(b) 17,804;   **(c)** 6.9 yr   **36. (a)** $E(p) = \dfrac{p}{12 - p}$;
(b) $E(2) = \frac{1}{5}$, inelastic;   **(c)** $E(9) = 3$; elastic;   **(d)** increase;
(e) $p = \$6$ **37.** $\dfrac{1}{2}x^6 + C$ **38.** $3 - \dfrac{2}{e}$
39. $\dfrac{7}{9(7 - 3x)} + \dfrac{1}{9} \ln (7 - 3x) + C$ **40.** $\dfrac{1}{4}e^{x^4} + C$
41. $\left(\dfrac{x^2}{2} + 3x\right) \ln x - \dfrac{x^2}{4} - 3x + C$
42. $-\cos (x^2 + 3) + C$ **43.** $\sin (e^x) + C$ **44.** $\frac{232}{3}$
45. $\$22,679.49$ **46.** $\$60,792.27$ **47.** Convergent, $\frac{1}{4374}$
48. $\frac{3}{2} \ln 3$ **49.** 0.6826   **50.** (169, \$7); \$751.33
51. $-\dfrac{\pi}{2}\left(\dfrac{1}{e^{10}} - 1\right)$ **52.** $y = C_1e^{x^2/2}$, where $C_1 = e^C$
53. (a) $P(t) = 1 - e^{-kt}$;   **(b)** $k = 0.12$;
(c) $P(t) = 1 - e^{-0.12t}$;   **(d)** $P(20) = 0.9093$, or 90.93%
54. $8xy^3 + 3$ **55.** $e^y + 24x^2y$ **56.** None
57. Maximum $= 4$ at (3, 3)   **58.** $3(e^3 - 1)$ **59.** $\dfrac{\sqrt{2}}{2}$
60. $\frac{1}{2}$ **61.** 0   **62.** $\frac{1}{8} \ln |\sec 8x + \tan 8x| + C$
63. $\frac{1}{3} \tan^{-1} (3x) + C$
64. $[-\csc^2 (\sin (x^3))][\cos (x^3)][3x^2]$ **65.** $\dfrac{1}{\sqrt{1 - x^2}}$

Index

CHAPTER 5

Summary of Important Formulas

18. $\int k \, dx = kx + C$

19. $\int x^r \, dx = \dfrac{x^{r+1}}{r+1} + C, \quad r \neq -1$

$\int (r+1)x^r \, dx = x^{r+1} + C, \quad r \neq -1$

20. $\int \dfrac{dx}{x} = \ln x + C, \quad x > 0;$

$\int \dfrac{dx}{x} = \ln |x| + C, \quad x < 0;$

21. $\int e^x \, dx = e^x + C$

22. $\int kf(x) \, dx = k \int f(x) \, dx$

23. $\int [f(x) \pm g(x)] \, dx = \int f(x) \, dx \pm \int g(x) \, dx$

24. $\int u \, dv = uv - \int v \, du$

Further integration formulas occur in Table 1 at the back of the book.